T0220938

Population and Community Ecology of
Ontogenetic Development

MONOGRAPHS IN POPULATION BIOLOGY
EDITED BY SIMON A. LEVIN AND HENRY S. HORN

A complete series list follows the index.

Population and Community Ecology of Ontogenetic Development

ANDRÉ M. DE ROOS

LENNART PERSSON

PRINCETON UNIVERSITY PRESS

Princeton and Oxford

Copyright © 2013 by Princeton University Press
Published by Princeton University Press, 41 William Street,
Princeton, New Jersey 08540
In the United Kingdom: Princeton University Press, 6 Oxford Street,
Woodstock, Oxfordshire OX20 1TW

press.princeton.edu

All Rights Reserved

Library of Congress Cataloging-in-Publication Data

Roos, André M. de, 1961–
    Population and community ecology of ontogenetic development / Andre M. de Roos,
Lennart Persson.
        p. cm. — (Monographs in population biology)
    Includes bibliographical references and index.
    ISBN-13: 978-0-691-13757-5 (cloth : alk. paper)
    ISBN-10: 0-691-13757-9 (cloth : alk. paper) 1. Animal populations. 2. Niche (Ecology)
3. Ontogeny. I. Persson, Lennart, 1948– II. Title.
    QL752.R66 2013
    591.7'88—dc23   2012017291

British Library Cataloging-in-Publication Data is available

This book has been composed in Times LT Std

Printed on acid-free paper. ∞

Printed in the United States of America

10 9 8 7 6 5 4 3 2 1

Contents

PART IV
EXTENSIONS AND PERSPECTIVES

Technical Appendices

Preface

The richness and diversity of biological systems in general, and ecological systems in particular, are a source for both inspiration and frustration among scientists. It is an inspiration to search for general principles and processes that underlie the complexity of ecological systems. Yet, frustration may develop, because at times that search seems to advance at such a slow pace. Both our research careers have largely focused on an area of ecological complexity that most ecologists have ignored or avoided: the implications of intraspecific size variation owing to ontogenetic development on population and community processes. Moving into this research area has undoubtedly been a major endeavor for us. The changes in size and morphology of individuals over ontogeny translate into changes in the type and intensity of interactions among them and thus increase the complexity of the interaction network that forms the backbone of an ecological system. Even among unicellular organisms like phytoplankton and bacteria, individual organisms undergo significant changes in size (growth over the cell cycle). Actually, we can think of no organism on earth in which the individual, be it a single cell or a multicellular organism, starts to reproduce directly after birth. Because of this total ubiquity of ontogenetic development among organisms, we considered it a fundamental and also unavoidable question for ecology as a discipline to investigate the effects that ontogenetic development may have on ecological processes.

The fact that inclusion of ontogenetic development in the study of population and community processes undoubtedly increases complexity made the search for generality a challenging task. Still, it relatively soon became clear to us that general patterns and principles regarding the dynamics of size-structured populations were present, despite the added complexities in both model and empirical developments and analyses. In particular, we found that the scaling of three basic individual ecological traits—search rate, maximum food intake, and metabolism—was a major determinant of consumer-resource dynamics. The population dynamics of size-structured consumer-resource systems turned out to be possible to understand (and predict) based on how the resource density necessary for maintenance (the critical resource density) scaled with body size.

The finding that a complex, fully size-structured model can be stringently reduced to a two-stage (juvenile-adult) model under equilibrium conditions represented a major theoretical breakthrough in our studies of the community dynamics of size/stage-structured communities. This substantially simpler two-stage model also turned out to preserve the major mechanisms by which ontogenetic development affects community dynamics. At a more fundamental level, this two-stage approach allowed us to transparently show how the balance of two rates—population maturation and population reproduction—determines how ontogenetic variation affects both population and community dynamics. Furthermore, it made clear to us exactly how limited the set of conditions is that unstructured ecological theory applies to. This set of conditions forms only a borderline case, in which mass-specific rates of new biomass production through somatic growth and reproduction and biomass loss through mortality are independent of the individual's body mass. It separates two broad domains of conditions, in which either juveniles or adults are energetically superior. We had already in previous papers advocated that unstructured ecological theory represents a special case of a more general and richer theory. Still, it was only during the course of writing this book that we have started to grasp the full implications of this idea for the discipline of ecology as a whole.

Our research on size-structured interactions has naturally not occurred in isolation. For the development of the mathematical framework of physiologically structured population models, which we use in our analysis, we owe a lot to the work of Odo Diekmann, Mats Gyllenberg, and Hans Metz. Odo Diekmann and Hans Metz are the main architects of this framework and in addition inspired and supervised one of us (A.M.dR.) during the early stages of his career. Without the tools and techniques they developed (albeit only in theory), this book could not have been written. Discussions and collaboration with Mats Gyllenberg on the first discrete-continuous consumer-resource model were quite important in taking our work on consumer-resource dynamics further. A major source of inspiration for us has always been the pioneering conceptual and experimental work by Earl Werner and his students on the dynamics of size-structured communities. For many years, one of us (A.M.dR.) was deeply involved in a research collaboration with Bill Gurney, Ed McCauley, Bill Murdoch, and Roger Nisbet on consumer-resource dynamics, particularly targeted at the dynamics of the water flea *Daphnia pulex*. To understand and quantitatively predict the dynamics of this model organism turned out to be much tougher than anticipated, but the endeavor itself was hugely inspiring and a great deal was learned from it, both theoretically and experimentally. Among many other scientists involved in our work, we want to thank especially Kjell Leonardsson and Gary Mittelbach.

Most important have been the stimulating interactions with PhD students and postdocs in everyday discussions and at the frequent workshops that we have organized over the years from the area of Abisko, in northern Sweden, to the North Sea coast of the Netherlands (for some reason we never considered places south of the Netherlands). The work carried out by PhD students and postdocs also constitutes a major contribution to the development of the research field of size-structured dynamics. Our thanks go to Jens Andersson, David Boukal, Pär Byström, Bent Christensen, David Claessen, Vincent Hin, Joakim Hjelm, Magnus Huss, Sido Mylius, Karin Nilsson, Tim Schellekens, Arne Schröder, Karen van de Wolfshaar, Tobias van Kooten, and Anieke van Leeuwen.

Our research has over the years been generously supported from different sources. The Universities of Amsterdam and Umeå have provided salaries for us and also for a number of PhD students involved in different research projects. The Swedish Research Council has constantly supported the research of L.P. for more than thirty years. Two larger research grants made it possible to substantially extend and develop our research on size-structured dynamics: the PIONIER grant "Towards a Theory of Life-History Dependent, Food-Web Interactions Using Structured-Population Models" from the Netherlands Organisation for Scientific Research (NWO) to A.M.dR. and the excellence project "Lake Ecosystem Response to Environmental Change (LEREC)" from the Swedish Research Council for Environment, Agricultural Sciences, and Spatial Planning to L.P. During the early phase of our research collaboration, the EU-FAIR project "Size-Based Tools for Managing Freshwater Fish Communities" was important in getting us working on size-structured community dynamics. The EU-FAIR, the PIONIER, and the LEREC projects also made it possible to organize a number of workshops, both with the scientists directly involved in the projects and others who were invited for specific meetings. We thank all these researchers for stimulating and productive discussions. At the time when the idea of synthesizing our results and ideas into a book started to emerge, L.P. was on a sabbatical at the University of Calgary on a fellowship supported by the Killam Foundation, which is gratefully acknowledged. During the writing of the very first chapters, A.M.dR. was generously supported by a sabbatical fellowship from the National Center for Ecological Analysis and Synthesis (NCEAS) in Santa Barbara, California.

Many people have allowed us access to their published or unpublished data, for which we are most grateful. They include Anders Angerbjörn, Malcolm Elliott, Anders Forsman, Karl Gottard, Magnus Huss, Frank Johansson, Martin Lind, Thomas Massie, Ed McCauley, Karin Nilsson, Tomas Pärt, and Earl Werner.

A substantial part of the work of finalizing the different chapters of this book was carried out in the countryside at Kussjön in the inlands of Västerbotten, Sweden. Kussjön has for many years served as a retreat for writing papers and for numerous discussions on both scientific as well as nonscientific topics. Many ideas ultimately originated during long walks in the silent setting of this beautiful place. Last but not least, A.M.dR. wants to thank Mirjam for the years of love and family life with their three wonderful sons, Joshua, Daniel, and Kristian, and thanks Liliana for the love during the last years, in which this monograph was written. Similarly, L.P. wants to thank Ingrid for many years of love and for many discussions on the philosophy and theory of science in general.

PART I
SUMMARY AND INTRODUCTION

CHAPTER ONE

Summary

A Bird's-Eye View of Community and Population Effects of Ontogenetic Development

Why start with summarizing the contents of a book? In the present case we see at least two good reasons. First, the amount of information provided in our book is without doubt quite massive. A summary provides an overview of the many topics that we cover and, we hope, reduces the risk that the reader will get lost in details and can no longer see the forest through the trees. Second, and partly related to the first reason, a number of, in our mind, novel and fundamental insights (some of them were not even known to us when we started to write the book!) are advanced. Here a summary serves the purpose of clearly showing how different chapters fit together in a general framework with respect to model approaches as well as results obtained. Reading this summary chapter will show you the different types of community modules that we analyze (summarized in figure 1.3) and give you a clear impression of the results and insights that we present in this book. Most of all, we hope it will serve as an encouragement to delve in more detail into the chapters that follow.

HISTORICAL BACKGROUND

Ecology has a long tradition of building theory about the dynamics of populations and the structure of communities that emerge from it. The early foundations for this theory were established by the pioneering work of Alfred Lotka and Vito Volterra, who formulated the most simple models for the dynamics of populations engaged in competitive and predator-prey interactions. Lotka-Volterra models have been used widely and form the basis of most, if not all, of our current theory on population and community processes. Because of their simplicity Lotka-Volterra models have often been viewed as leading to predictions that are general and representative for many different systems. And yet, these models completely ignore one of the most important processes in an individual's life history—one that is, moreover, unique to biology: development.

Lotka-Volterra models use population abundance as the state variable to characterize a population and describe changes therein using rather abstract demographic and community parameters such as population growth rates and coefficients of interaction between species. The dynamics of a population is described as the balance between reproduction and mortality, which increase and decrease abundance, respectively. In essence, these two processes are not unique to biological systems, as synthesis of particles from substituent elements and degradation of particles also occurs in, for example, chemical systems. Analogously, increases and decreases in population abundance result from the reproduction and mortality, respectively, of individual organisms. However, no individual organism of any species can reproduce right after it has been born, nor is its chance to die the same throughout all stages of its life history. Individual organisms go through an ontogenetic development, which is a major component of their life history. In fact, given that many individuals die before they manage to reproduce, ontogenetic development can be considered the most prominent life history process after mortality.

The intricacies of an individual's life history, including its ontogenetic development, have forever fascinated ecologists and have been studied extensively. Nonetheless, the absence of ontogenetic development as an important life history process from the core models of ecological theory has not been cause for major concern, nor has it been considered an important omission. The progression through different life stages has been accounted for in matrix models that focus on the potential of population growth of single populations, but when considering interactions between species and models of larger communities, the simplification of the population to a number of individuals without distinguishing between these individuals has been the rule. Basic ecological models thus ignore without much qualm the most prominent process in an individual's life history, one that is unique to biological systems and has no counterpart in physical and chemical systems.

The most important aspect of ontogenetic development in virtually all species is an increase in body size. Body size, in turn, determines to a large extent an individual's ecology in terms of its feeding, growth and reproduction rate, the food sources it can exploit, and the predators to which it is exposed. Moreover, in most species growth in body size is plastic, dependent on the environmental conditions the individual experiences, in particular, the availability of food resources that are required for its maintenance, growth, and reproduction. The consequences of such plastic and food-dependent ontogenetic development for the dynamics of populations and the structure of communities are the subject of this book. We investigate these population and community consequences using a collection of models that differ in the amount of detail with which the individual life history is represented (figure 1.1). All models account for growth in body size as the most important aspect of ontogenetic development and rigorously adhere to a mass-conservation principle. Thus, they explicitly account for the processes by which energy is acquired (feeding) and through which energy is spent (maintenance, growth, and reproduction). We also confront the predictions derived from these models as much as possible with experimental and empirical data that we either collected ourselves or that are available in the literature. More specifically, we present experimental and/ or empirical evidence for the majority of phenomena that we deduce from our model analysis.

We use two basic model formulations: a fully size-structured model formulation with an explicit handling of continuous size distributions and a stage-based formulation with two stages (juveniles, adults). Most important, the latter can be shown to be a faithful approximation to the fully size-structured model formulation, as model results of the two formulations are completely identical under equilibrium conditions. In the stage-based formulation, the juvenile stage increases in biomass as a result of adult reproduction and somatic growth of juvenile individuals, whereas adult biomass increases as a result of maturation of juvenile biomass into adult biomass (figure 1.1, top panel). Significantly, all these rates are dependent on the resource biomass level. Biomass is lost from the system through mortality and energetic costs for maintenance. In the fully size-structured model formulation, individuals grow in body size over ontogeny to reach maturity at a specific size (figure 1.1, bottom panel). The rate of increase in size is a function of net intake (intake minus maintenance costs). Net intake and reproduction rates are also in this case dependent on resource biomass density.

A crucial question to ask when considering community effects of ontogenetic development is whether it matters that basic ecological models ignore this process when modeling population dynamics. In our opinion, the answer is obviously yes, as we otherwise would not have written this book. The book underpins this assertion by presenting the major consequences of ontogenetic development for community structure and population dynamics. Furthermore, we will show that the conditions under which development does *not* have a significant community influence represent only a limiting case. We hence

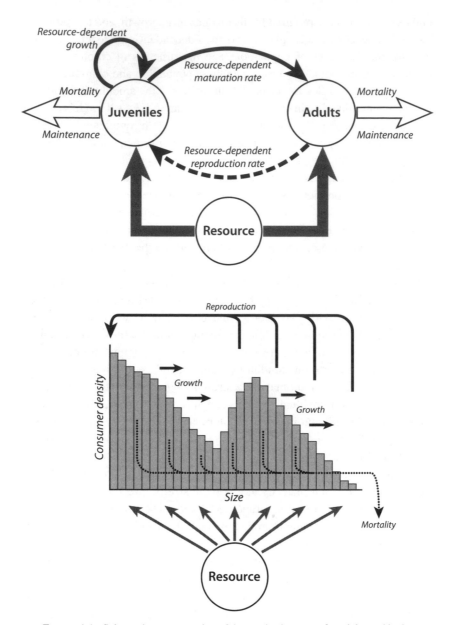

FIGURE 1.1. Schematic representation of the two basic types of models used in the different chapters. *Top panel*: The consumer life cycle as a dynamic system represented by a two-stage biomass model where biomass is generated from ingested resources (*solid gray arrows*), lost through mortality and maintenance (*open arrows*), and channeled between the juvenile and the adult stages through maturation and reproduction, respectively (*solid black arrow, dashed black arrow*). In

postulate that the ecological theory based on Lotka-Volterra models is all but a limiting case of a more general population and community theory.

BIOMASS OVERCOMPENSATION

Arguably the most important finding presented in the book is that an increase in mortality of a population can lead to an increase in its biomass, which runs counter to all our intuitive ideas about the consequences of mortality. Figure 1.2 illustrates this phenomenon using the stage-based formulation schematically presented above for a consumer population, but in contrast to what is assumed in figure 1.1, juvenile and adult individuals forage on two different resources with otherwise exactly identical mass-specific feeding rates (this model is analyzed further in chapter 6). When individual consumers forage to an equal extent on both resources (50 percent foraging effort on both resource 1 and 2, respectively), irrespective of whether they are juvenile or adult, an increase in mortality of all consumers leads to the intuitively obvious result that total consumer biomass decreases (figure 1.2, left). Furthermore, the increase in mortality does not change the ratio between juvenile and adult consumer biomass in equilibrium. The relative composition of the population hence remains the same. Note that we have assumed in this model that individual consumers do vary in their body size as they grow from their size at birth to mature at an adult body size, but that per unit of biomass, the rates of feeding, maintenance, and mortality are the same for all consumers. The fact that both juveniles and adults exploit both resources to an equal extent makes them identical on a mass-specific basis.

When juvenile consumers exclusively feed on resource 1 and at maturation switch to exclusively feeding on resource 2, the response of the population to an increase in mortality of all individuals is quite different: a higher mortality translates into a higher total biomass of the population, mainly because the

addition, juvenile biomass increases through somatic growth (*circular solid black arrow*). *Bottom panel*: The consumer life cycle as a dynamic system represented by a physiologically or fully size-structured population model. Biomass is generated from a size-dependent intake of the resource (*gray arrows*) and lost through mortality (*dotted arrow*) and maintenance costs. Individuals grow as a function of energy intake minus maintenance costs (*black horizontal arrows*). At a fixed size they start to reproduce (*vertical and horizontal black arrow*). In both formulations, all biomass growth and recruitment processes depend on resource availability.

biomass density of juvenile consumers increases significantly, whereas adult consumer biomass only decreases with mortality, as before (figure 1.2, right). This positive relationship between mortality and biomass density in equilibrium we refer to as *biomass overcompensation*. Biomass overcompensation comes about because the relative composition of the population changes with changing conditions, that is, mortality. Note that we refer to an increase in standing stock equilibrium biomass and *not* to an increase in a population dynamic rate process, such as the population reproduction rate, with increasing mortality. The latter is in fact more understandable and has been discussed before in the literature.

ONTOGENETIC (A)SYMMETRY IN ENERGETICS

When does biomass overcompensation occur? This question brings us to another important concept presented in the book, that of *ontogenetic symmetry in energetics*. In classical Lotka-Volterra models all individuals are by definition the same and thus characterized by equal rates of feeding, maintenance, reproduction, and mortality. As we will show, the population dynamics models that account for individual growth in body size throughout life history and hence for the population size structure, presented in figure 1.1 above, can under specific conditions be simplified to a single ordinary differential equation for the changes in total population biomass over time. More specifically, the classical, bioenergetics model presented by Yodzis and Innes (1992) represents an example of such a model, which looks like an unstructured population model that characterizes the population with a single state variable: population biomass. At the same time, the Yodzis and Innes model can also be shown to exactly describe the dynamics of a fully size-structured population, in which all individuals experience the same mortality and per unit of body mass produce the same amount of new biomass through either somatic growth in body size or production of offspring. The condition that allows for simplifying a size-structured model to a simple model for total population biomass we refer to as *ontogenetic symmetry in energetics*, because of the symmetry between individuals of different body sizes (i.e., in different stages in their development) in the efficiency with which they convert ingested food into new biomass. Under conditions of ontogenetic symmetry, size-structured population dynamics do not differ from the dynamics of unstructured populations: increases in mortality lead to the expected decrease in total population biomass, and changing external conditions such as mortality or food supply do not change the relative population composition.

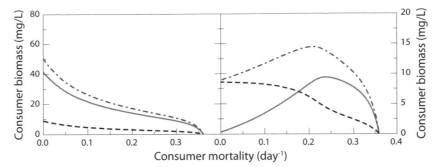

FIGURE 1.2. Changes in juvenile (*light-gray solid lines*), adult (*black dashed lines*), and total consumer biomass (*dark-gray dotted-dashed lines*), with increasing mortality targeting all consumers equally in a consumer-resource model in which juvenile and adult consumers either forage to the same extent on two different resources (*left*; $q_{J,1} = q_{J,2} = q_{A,1} = q_{A,2} = 0.5$, see box 6.2) or switch from exclusively foraging on resource 1 as juvenile to foraging exclusively on resource 2 as adult (*right*; $q_{J,1} = q_{A,2} = 1$, $q_{J,2} = q_{A,1} = 0$, see box 6.2). See boxes 6.1 and 6.2 for model equations and parameters, respectively. Results pertain to a consumer population with characteristic adult body weight equal to $W_A = 0.0001$ gram and maximum density of resources 1 and 2 equal to $R_{max,1} = 100$ mg/L and $R_{max,2} = 18$ mg/L, respectively.

The results shown in the right panel of figure 1.2 hence occur because of *ontogenetic asymmetry* between juvenile and adult consumers. In particular, juveniles and adults differ because we have assumed that they forage on different resources that are produced at different rates. Figure 1.2 represents a situation with a high productivity of resource 1, on which the juveniles forage, whereas resource 2, exclusive to adult consumers, is in short supply. Juvenile and adult consumers thus experience very different feeding regimes. At low mortality this leads to a domination of the population by adults, because adult resource is limited in equilibrium. Adult consumers then use most of their intake to cover their maintenance costs, and reproduction is low. At the same time, juvenile resource density is high, leading to rapid growth of juveniles from their size at birth to their maturation size and early maturation.

Hence, at low mortality a bottleneck occurs in the adult stage of the consumer life history. An increase in mortality relaxes this bottleneck, as it decreases adult biomass and increases the availability of resource for the surviving adults. The latter, positive effect is so substantial that total reproduction by all adults together increases with increasing mortality. In turn, this increase in total reproduction translates into an increase in biomass of juveniles, which continue to experience high food availability and can cash in on the high supply of their resource. The biomass overcompensation hence occurs because individual consumers differ in their energetics, which results in a differential change in maturation and reproduction rate when an increase in mortality relaxes the intraspecific competition and resource densities increase. The differential change in maturation and reproduction rate leads to a change in population composition and ultimately to a more efficient use of the supplied resources.

Biomass overcompensation occurs when the ontogenetic symmetry in energetics is broken. In physics, symmetry breaking is known to bring a system from a disorderly state, which occurs in the symmetric case, into one of two definite states that are robust against small changes as opposed to the symmetric state. Analogously, we can identify two distinct regimes of ontogenetic asymmetry in energetics. First, if juveniles are energetically less efficient than adults, a bottleneck in development during the juvenile stage will mostly limit the population at equilibrium when mortality is low. Then, the population tends to be dominated at equilibrium by juveniles. Increases in mortality will generically decrease juvenile biomass but increase adult biomass. Second, if adults are energetically less efficient, a bottleneck in reproduction will limit the population at equilibrium when mortality is low, leading to adults making up the largest part of population biomass under these circumstances. Increases in mortality will then decrease adult biomass but increase juvenile biomass. These two regimes of ontogenetic asymmetry, either juvenile development or adult reproduction limiting a population at equilibrium, are like two sides of a coin. A state of ontogenetic symmetry represents the dividing line between them. Just as in physics, the system states that are observed in the symmetric case, which in our case refers to the type and nature of community equilibria or the type of population dynamics predicted by a model, are sensitive to small changes in the energetic status of juveniles and adults. This is the reason why we postulate that classic ecological theory based on Lotka-Volterra models only covers a limiting case of the more general theory, which also encompasses cases with asymmetric conditions of individual energetics.

Biomass overcompensation occurs for almost all parameter combinations in different size-structured, consumer population models irrespective of the precise details of the individual life history that are accounted for in these models. Figure 1.2 does represent an extreme example, in that even total population biomass increases with increasing mortality. In general, however, it is more likely to find that total population biomass decreases and only the biomass density of juveniles or that of adults increases in response to mortality. Stage-specific biomass overcompensation, for example, occurs when adults and juveniles share the same resource but differ in their mass specific intake rate (figure 1.3, top row, left module). It is hence a generic phenomenon, although the precise form of biomass overcompensation may depend on model details (see chapter 3 for details). Crucial for its occurrence is the fact that individuals require energy to cover the costs for maintaining themselves, an aspect of individual life history that is often ignored in Lotka-Volterra models.

EMERGENT COMMUNITY EFFECTS OF BIOMASS OVERCOMPENSATION

Under conditions of population regulation at equilibrium by limited juvenile development, the equilibrium density of adult biomass will not only increase in response to increases in mortality that targets all consumers equally but also in response to increases in mortality that target only juveniles or only adults. The same holds for the increase in juvenile equilibrium biomass with increasing mortality: it occurs irrespective of whether mortality targets only juveniles, only adults, or both juveniles and adults equally (chapter 3; figure 1.3, top row, left module). Biomass overcompensation hence occurs irrespective of the precise size-dependency of the mortality imposed. This result immediately reveals the possible community consequences of biomass overcompensation. For example, consider that juvenile equilibrium biomass increases in response to increased juvenile mortality. If this mortality is imposed by a predator species foraging on these juveniles, predators will experience a higher availability of food (i.e., more juvenile consumer biomass) the higher the mortality they impose on these juveniles. Thus, a positive feedback emerges between the stage-specific mortality that stage-specific predators impose and their food availability.

At the population level, this positive feedback gives rise to an *emergent Allee effect* (chapter 4; figure 1.3, top row, middle module): low densities of predators fail to build up a population, as the predation mortality they impose on juvenile consumers is not sufficient to raise juvenile biomass density to levels that are high enough for predator persistence. In contrast, a high predator density will lead to an increase in juvenile consumer biomass that is sufficient for predator persistence. Over significant ranges of resource productivity and predator mortality, this threshold behavior will imply that both an equilibrium without predators as well as a coexistence equilibrium of consumers and predators is stable. The Allee effect, as embodied by the threshold predator density for persistence and the bistability between equilibria with and without predators present, does not result from any specific model ingredient but

Community structure

Chapter 3
Biomass overcompensation

Increasing mortality

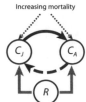

Chapter 4
Emergent Allee effects

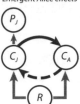

Chapter 5
Emergent facilitation

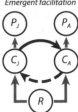

Chapter 6
Ontogenetic niche shifts

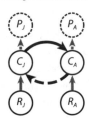

Chapter 6
Ontogenetic niche shifts

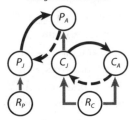

Chapter 7
Mixed interactions

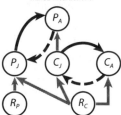

Chapter 8
Interspecific competition and predation

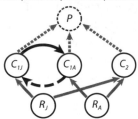

Population dynamics

Chapter 9
Consumer-resource dynamics

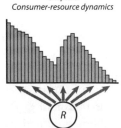

Chapter 10
Consumers with pulsed reproduction

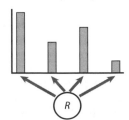

Chapter 11
Cannibalism

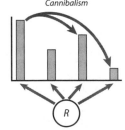

emerges as a consequence of the ontogenetic development of the consumers that the predators prey on. Emergent Allee effects can occur for predators that specialize either on juvenile consumers or on adult consumers, depending on whether the consumer population in absence of predators is regulated by limited reproduction or development, respectively. They also occur in a far more detailed size-structured model that accounts for size-structure in both prey and predator population and in which prey vulnerability is a continuous function of both prey and predator body size (figure 1.1, bottom panel; see chapter 4 for details). Furthermore, there is convincing evidence from an experimental manipulation of the fish community in an entire lake that corroborates the occurrence of emergent Allee effects in natural systems.

A second type of emergent community effect, *emergent facilitation* between different stage-specific predators (chapter 5; figure 1.3, top row, right module), occurs because juvenile biomass in equilibrium also increases in response to an increase in adult mortality. Consider the effect of an adult-specialized predator species imposing predation mortality on adult consumers. This will translate into a higher density of juvenile biomass in equilibrium and thus to a higher food availability for a predator species that specializes on juvenile consumers. Through the changes in the consumer population stage-structure, the adult-specialized predator thus benefits the juvenile-specialized predator. We will show that as a consequence the juvenile-specialized predator may only be able to persist in the presence of the adult-specialized predator that competes with it for the same prey species (chapter 5). Extinction of the adult-specialized predator then necessarily leads to extinction of the juvenile-specialized predator as well. Such facilitation of the juvenile-specialized predator by the

FIGURE 1.3. Principal trophic modules (*R* for resource, *C* for consumer, *P* for predator) considered in the different chapters. Gray arrows represent ingestion, black solid arrows maturation, and black dashed arrows reproduction. Dashed gray arrows and dashed circles represent variations on the basic feeding relations and trophic modules (indicated by solid gray arrows and solid circles) investigated in the different chapters. Dotted black arrows represent different types of additional mortality to which consumers are exposed. In chapters 3, 4, 6, and 7, both types of model, illustrated in figure 1.1, are used to investigate community structure, whereas chapters 5 and 8 use the stage-structured model only. Chapters 9, 10, and 11 use the physiologically structured model (figure 1.1, bottom panel) with either continuous (chapter 9) or pulsed (chapters 10 and 11) reproduction.

adult-specialized predator occurs over significant ranges of predator mortality. The reverse phenomenon, facilitation of the adult-specialized predator by a juvenile-specialized predator, also occurs readily, but for conditions in which the consumer population in absence of predators is mostly regulated in equilibrium by limited juvenile development. Biomass overcompensation in consumer populations in response to mortality thus indirectly gives rise to positive effects at higher trophic levels, either within the same predator population (emergent Allee effect) or among predators that forage on different size ranges of consumers (emergent facilitation).

ONTOGENETIC NICHE SHIFTS IN CONSUMER LIFE HISTORY

In many species the ecology of individuals changes drastically with increases in body sizes. Some of these changes are quantitative in nature, such as increases in the rate of foraging and maintenance. Individuals may, however, also switch diet and forage on completely different resources as juveniles and adults. Such qualitative changes in ecology throughout individual life history are referred to as ontogenetic niche or diet shifts. Figure 1.2 already illustrates that more complex outcomes of population dynamics can be expected in cases where individuals change their diet at maturation, as it even allows for overcompensation in total population biomass.

In cases where juveniles and adults share a single resource, their energetic efficiency determines whether at low mortality the population at equilibrium is regulated by slow juvenile development or by low adult reproduction. In contrast, these two modes of population regulation support two alternative stable states when juveniles and adults have different diets. Under the same conditions of resource productivity and mortality, the population may equilibrate either in a stable state in which juvenile development is limited or in one in which adult reproduction is limited. This bistability between a development-controlled and a reproduction-controlled equilibrium of the consumer-resource system generically occurs over a wide range of resource productivities (chapter 6).

Ontogenetic niche shifts in consumer life history and the ensuing bistability between a development-controlled and a reproduction-controlled equilibrium have implications for the occurrence of emergent effects at higher trophic levels (chapter 6; figure 1.3, second row, left module). In particular, because total consumer biomass can increase with an increase in mortality that targets all consumers equally (figure 1.2) an emergent Allee effect can in this case occur for a generalist predator that feeds on both juvenile and adult consumers, which is impossible when juvenile and adult consumers share a single resource. For stage-specific predators the bistability between a development-controlled and a reproduction-controlled equilibrium of consumers implies that predators may be able to invade and persist only in one of these two stable states, in which the consumer stage they forage on dominates the consumer population, whereas the food density they encounter in the alternative consumer equilibrium is too low for population growth. As a consequence, increasing the productivity of, for example, the resource that juvenile consumers forage on, eventually leads to extinction of a stage-specific predator species that specializes on the juvenile consumers. Similarly, increasing productivity of the resource for adult consumers leads to extinction of an adult-specialized predator species. Together, however, the juvenile-specialized and adult-specialized predator species can persist for all combinations of resource productivity that lead to extinction of one of the predators when alone. Thus, through the ontogenetic niche shift in consumer life history, the emergent facilitation between two stage-specific predators becomes reciprocal, whereas this facilitation is always only one-sided (either juvenile-specialized predators facilitate adult-specialized ones or vice versa) in cases where juvenile and adult consumers share a resource (chapter 6).

ONTOGENETIC NICHE SHIFTS
IN PREDATOR LIFE HISTORY

Many predator species do not prey on consumers throughout their entire life; as a juvenile they forage on smaller-size alternative resources. Predators hence often also exhibit an ontogenetic niche shift in their life history from basic resource–feeding to predation (figure 1.3, second row, right module). The occurrence of ontogenetic niche shifts in the life history of predators of size-structured consumers allows for the possibility of multiple (up to three) stable community equilibria under the same conditions of resource productivity. An ontogenetic niche shift of predators from feeding on their exclusive, alternative resource as a juvenile to preying on juvenile consumers as an adult allows for the occurrence of a stable coexistence equilibrium of consumers and predators that is regulated by limited predator reproduction, next to a stable coexistence equilibrium that is regulated by limited predator development (chapter 6). If the predator preys on a consumer population that in absence of predators is regulated by limited reproduction, it may in addition experience an emergent Allee effect. In combination with the two types of coexistence equilibria, the emergent Allee effect allows for the occurrence of three different stable community states over certain ranges of resource productivity: (1) a stable consumer-only equilibrium without predators; (2) a stable coexistence equilibrium regulated by limited predator development during its juvenile stage; and (3) a stable coexistence equilibrium regulated by limited predator reproduction.

The type of regulation of the predator population at equilibrium has consequences for both the composition of the consumer species it preys on as well as for the life history of the predators themselves (chapter 6). When limited development regulates the predator population at equilibrium, most intraspecific competition for food among predators occurs in the niche during which they feed on their alternative resource, while intraspecific competition in the predatory niche is negligible. As a consequence, the predation mortality imposed on consumers is generally low, such that the consumer population tends to be regulated by limited reproduction and dominated by adults (note that we concentrate here on the more likely scenario of predators foraging on juvenile consumers, which in turn are more efficient in their energetics than adult consumers).

If individual predators can grow in body size even after maturation, this high adult consumer density allows individual predators to reach very large body sizes, albeit only a few individuals manage to do so. In contrast, in the coexistence equilibrium with limited reproduction regulating the predator population, intraspecific competition is negligible in the resource-feeding niche of juvenile predators and mostly occurs in the predatory niche. Even if predators can grow in body size after maturation, they will not manage to do so, because food for adult predators is limited. Juvenile consumers experience high predation mortality in this coexistence equilibrium. This in turn leads to high resource densities for consumers, fast development of surviving juveniles, and high reproduction of adult consumers. If they are capable of somatic growth after maturation, the few surviving adult consumer individuals can in this coexistence state reach body sizes that are close to their maximum. In cases where consumers and predators can both grow in body size after maturation these two stable coexistence equilibria can be aptly distinguished as the "stunted-consumer" and the "stunted-predator" equilibrium, on the grounds that in either equilibrium individuals of one of the two species reach body sizes that are only slightly larger than their maturation size, whereas the other species grows to body sizes close to its maximum.

When juvenile predators forage on a basic resource before switching to predation on consumers, the possibility arises that they compete in this resource-feeding niche with their future prey. This type of mixed interaction between consumers and predators, which can also be classified as intraguild predation or life-history omnivory, is considered to be a major force structuring natural communities (figure 1.3, third row, left module). Diet overlap between consumers and juvenile predators restricts the coexistence possibilities of consumers and predators (chapter 7). However, this outcome crucially depends on the extent to which adult predators need to feed on consumers to successfully reproduce. If adult predators can feed on the basic resource as well and gain sufficient energy from it for reproduction, they may drive consumers to extinction over large ranges of resource productivity (chapters 6 and 7). On the other hand, if preying on consumers is necessary for predator reproduction, consumers are never driven to extinction by predators (chapter 7).

Consumers and predators then coexist in stable equilibrium, in which predators dominate and overrule the competitive advantages of consumers through the predation mortality they impose. Consumers may become dominant owing to their advantage in resource competition, in which case the density of the resource that consumers and juvenile predators share becomes so low that juvenile predators experience a severe bottleneck in their development. Ultimately, this drives the predator to extinction. Rather than leading to a balance between the competitive advantage of the consumer and the top-down suppression by the predator, mixed interactions, in which predators rely on the presence of consumers for persistence, thus lead to the occurrence of bistability between a competition-driven community state without predators present and a predation-driven coexistence state (chapter 7).

COMPETITION BETWEEN CONSUMERS WITH AND WITHOUT ONTOGENETIC NICHE SHIFTS

Figure 1.2 suggests that consumers that experience a niche shift in their life history from feeding on resource 1 as a juvenile to feeding on resource 2 as an adult may be able to exploit the available resources more efficiently at higher levels of mortality than at lower mortality. This change in efficiency with increasing mortality turns out to have major consequences for coexistence of different consumer species that compete for the same resources. Consumers that specialize on different resources in different stages of their life history are referred to as ontogenetic specialists. In contrast, ontogenetic generalists do not change their pattern of resource usage throughout their life history.

In competition for two basic resources an ontogenetic specialist and an ontogenetic generalist can coexist in a stable equilibrium for certain combinations of resource productivity (chapter 8; figure 1.3, third row, right module), while for other combinations of resource productivity either the ontogenetic specialist outcompetes the generalist or vice versa. The diet shift in itself thus leads to niche partitioning between the specialist and the generalist, even in cases where both consumers are energetically equally efficient (chapter 8). In general, the range of resource productivities for which the ontogenetic specialist outcompetes the generalist is larger when specialists do not switch completely from foraging on resource 1 as a juvenile to foraging on resource 2 as an adult, but rather when both juveniles and adults exploit both resources to a varying extent.

Ontogenetic specialists also outcompete ontogenetic generalists over larger ranges of resource productivity when mortality is high, which is not surprising given the increase in energetic efficiency of specialists with increasing mortality illustrated in figure 1.2. This increase in competitive ability with increasing mortality has very counterintuitive consequences for the coexistence of ontogenetic specialists and generalists in the presence of shared predators. The classic idea is that shared predators can promote the coexistence of an inferior and a superior competitor species, only if in the presence of predators the dominance of superior competitors is countered by higher predation mortality. Such predator-mediated coexistence of competitors thus relies on a trade-off for the two competitors between their competitive ability and vulnerability to predation.

When consumers go through an ontogenetic niche shift, however, this trade-off is not necessary for coexistence of the two competitors in the presence of a shared predator, as a consequence of the increase in energetic efficiency with higher mortality of ontogenetic specialists. More specifically, we show that an ontogenetic specialist with a significant competitive disadvantage, such that it is always outcompeted by the ontogenetic generalist in the absence of predators, can outcompete the generalist when predators are present. This reversal of competitive dominance occurs even if the ontogenetic specialist is the preferred prey of the predators and predators attack specialists at a higher rate than generalists (chapter 8). Hence, even with a double handicap, both a competitive and a predatory disadvantage, the ontogenetic specialist outcompetes the generalist over a significant range of resource productivities when predators are present as a consequence of the fact that the adaptability of its population stage structure allows it to exploit resources more efficiently when mortality increases.

ONTOGENETIC (A)SYMMETRY IN ENERGETICS
AND POPULATION DYNAMICS

In simple but generic models that account for the size structure of a consumer population, ontogenetic symmetry in energetics leads to the same type of dynamics as observed in Lotka-Volterra predator-prey models: If resource growth is logistic, classical predator-prey cycles occur at high resource carrying capacity. If resource growth follows semichemostat dynamics and resource productivity is hence independent of resource density, the size-structured consumer-resource system approaches a stable equilibrium, as unstructured Lotka-Volterra models would. In contrast, size-structured consumer-resource models predict qualitatively different types of dynamics to occur in the two different regimes of ontogenetic asymmetry in energetics, which are characterized by either juveniles or adults being more efficient in acquiring food and/or their use of assimilated energy.

With semichemostat resource dynamics (i.e., constant resource productivity independent of resource density) size-structured consumer-resource models (figure 1.3, bottom row, left module) predict regular population cycles to occur in both regimes of ontogenetic asymmetry (chapter 9). The characteristics of these population cycles are, however, quite different in the two regimes. In cases where juvenile consumers are more efficient in their acquisition and use of energy, juveniles tend to outcompete all older and larger conspecifics, because their higher efficiency allows them to survive at lower resource densities. As a consequence, cycles in population density with large amplitude occur, in which the population is dominated by one and the same cohort of individuals throughout the entire cycle. In these so-called *juvenile-driven* cycles adult consumers are not continuously present in the population but only occur when the dominant cohort matures. Following maturation a dominant cohort produces a pulse of offspring that makes up the next dominant cohort. Their more efficient use of energy allows the offspring cohort to drive their parents to extinction, thereby monopolizing the population entirely. The life history of individual consumers in these population cycles resembles the consumer life history in a population equilibrium that is regulated by limited adult reproduction: juvenile growth in body size is fast, leading to short juvenile periods and high juvenile survival, while adult fecundity is low and adult lifespan is short. Moreover, the ratio between juvenile and adult consumer density is low during these cycles, as it is in a reproduction-controlled equilibrium state.

In cases where adult consumers are more efficient in their acquisition and use of energy, population cycles also occur over large ranges of parameters, but the amplitude of these so-called *adult-driven* cycles is much smaller (chapter 9). The composition of the consumer population during these cycles is relatively constant, with juveniles and adults coexisting continuously throughout the cycle. Juvenile growth in body size is in this case slow and is further retarded when a pulse of consumers matures. The retardation of juvenile development and the consequent long juvenile period and low juvenile survival are major factors giving rise to the cycles. Adult fecundity and the ratio between juvenile and adult density are both high. These characteristics closely resemble the characteristics of an equilibrium state regulated by limited juvenile development.

In many aspects juvenile-driven and adult-driven population cycles thus resemble the consumer equilibrium states, regulated by limited reproduction and limited development, respectively. Both types of cycles have a period that is close to the duration of the juvenile stage, which distinguishes them from the classic predator-prey cycles occurring in unstructured Lotka-Volterra type models. When resource growth is logistic as opposed to following semichemostat dynamics, the size-structured consumer population exhibits complex dynamics as a result of an interaction between the cyclic tendency originating from the ontogenetic asymmetry in energetics and the cyclic tendency embodied in classic predator-prey cycles (chapter 9). In cases where adult consumers are more efficient in their energy use, this interplay results in the occurrence of two types of cycles over large ranges of parameters: one large-amplitude cycle that resembles the classic predator-prey cycle and another, with smaller amplitude, resembling the adult-driven ontogenetic cycle. In cases where juvenile consumers are more efficient in their use of energy, however, such bistability does not occur, and the resulting dynamics are more of a mixed type, in-between the juvenile-driven ontogenetic cycles and classic predator-prey cycles.

Juvenile-driven and adult-driven population cycles also occur in size-structured consumer-resource models that in more detail represent specific ecological systems. These cycles therefore seem to be robust population dynamic patterns, which occur independently of rather drastic changes in model assumptions—for example, whether reproduction occurs continuously throughout the year or only as a pulsed event in the beginning of summer (chapter 10; figure 1.3, bottom row, middle module). The type of ontogenetic asymmetry, whether juveniles or adults are energetically more efficient, determines which type of cycle ensues.

Other types of intraspecific interaction may annul the size-dependent, intraspecific competitive relations between juveniles and adults and thus suppress the occurrence of population cycles. As a particularly well-studied example, cannibalism of juveniles by adult conspecifics may stabilize juvenile-driven population cycles that result from ontogenetic asymmetry in energetics (chapter 11; figure 1.3, bottom row, right module). However, the interplay between size-dependent competition for resources and size-dependent cannibalism also gives rise to new population dynamics phenomena. At low intensity, cannibalism of juveniles by adults completely cancels out the competitive advantage of juveniles, leading to rapid juvenile growth and low juvenile survival, while adult cannibals reach body sizes that are only slightly larger than their size at maturation. In a number of aspects this type of cannibal-driven dynamics resembles the "stunted-predator" equilibrium state that we observed in mixed interaction systems, in which juvenile predators compete for resources with consumers they prey on as adults (chapter 7). The cannibal-driven dynamics can be viewed as a single-species version of this "stunted-predator" dynamics.

At higher intensities of cannibalism, we also find a single-species analogue of the "stunted-consumer" equilibrium that occurred in mixed interaction systems (chapters 7 and 11). In this case a few adult cannibals can reach large (giant) body sizes, while the population as a whole is mainly regulated by resource competition among small juveniles. This competition leads to slow juvenile development, long juvenile periods, and relatively high juvenile survival. Because the mixed interaction now occurs within one and the same cannibalistic population, the two states are not simultaneously stable. Rather, the population alternates over time between these two regimes, giving rise to a new type of dynamics that we refer to as "dwarf-and-giant" cycles (chapter 11). In such cycles, periods of cannibal-driven and competition-driven dynamics alternate, in which the population is mainly regulated by top-down suppression of juveniles through cannibalism and competition for resources among juveniles, respectively. An extensive dataset on experimental lake systems supports model predictions for the occurrence of these dynamics in a number of cannibalistic systems (chapter 11).

Formulation and parameterization of detailed size-structured population models, as exemplified by the consumer-resource model (chapter 10) and the cannibal-resource model (chapter 11) that we study, requires the availability of large amounts of experimental data to carefully parameterize the various functions (for example, size-dependent feeding and maintenance rates, and energy allocation rates to growth and reproduction) describing individual life history. This data "hunger" of detailed models can be seen as a disadvantage, but detailed models also provide the advantage that they allow for extensive confrontation of model predictions against empirical data. In particular, they allow for more than just testing empirical observations on population-level statistics, such as the amplitude and periodicity of cycles. A major, but rather unknown advantage of detailed size-structured models is that they make predictions about individual life history under conditions of density dependence and population feedback. Because they do not directly result from and hence are rather independent of the model assumptions, these predictions can be used for tests of model performance against observations on life history in natural situations. For both the size-structured, consumer-resource and the cannibalism model, such tests reveal the close match between model predictions and empirical data (chapters 10 and 11).

GENERALIZATION

The population and community effects of ontogenetic development that are discussed above have all been derived for systems in which the growth rate in body size during the ontogeny of individual organisms strongly depends on the availability of food. Although this is valid for the large majority of species, the growth patterns of mammals and many bird species is much less influenced by food availability and is more under genetic control. Several lines of evidence suggest, however, that the insights we obtained also hold more generally, for example, for systems in which growth rate in body size is largely independent of food availability and genetically more predetermined.

When growth in body size is genetically predetermined, it is not the supply (i.e., availability) of food that determines the energetics of individual organisms but the demands on energy for growth, maintenance, and reproduction. Such demand-driven energetics necessitate that energy storage of individual organisms in the form of reserve (i.e., fat) tissue is explicitly accounted for in population dynamic models, as the predetermined costs for somatic growth and maintenance have to be covered from these reserves in times of food shortage. We show that as a rule biomass overcompensation also occurs in a size-structured consumer-resource model based on demand-driven energetics of individual organisms (chapter 12). This biomass overcompensation again results from differential responses in maturation and reproduction rate, when intraspecific competition is relaxed and resource density increases at higher mortality levels. The changes in maturation rate, however, do not result from a shorting of the juvenile period, which is constant given the predetermined growth rate in body size. Rather, the changes in maturation and reproduction rate now result from differential changes in the amount of energy storage by juvenile and adult consumers when resource density increases.

This size-structured consumer-resource model based on demand-driven energetics also displays the same two types of population cycles that we identified for size-structured models for consumers with supply-driven energetics (compare chapters 9 and 12). Juvenile-driven and adult-driven cycles occur when juveniles and adults are more efficient in their acquisition and use of energy, respectively. Moreover, the cycle characteristics in the demand-driven size-structured model in most major aspects resemble the characteristics of the two types of cycles in supply-driven, size-structured consumer-resource models. Lastly, theoretical considerations and experimental evidence also suggest that adult-driven cycles may be an important dynamic feature of populations of unicellular organisms as well.

Altogether, these lines of evidence hint at the generality of the distinct phenomena that we have revealed for the two different regimes of ontogenetic asymmetry, governed by a higher energetic efficiency of either juveniles or adults. Above all, we hope it will convince readers also to consider ontogenetic development in their own research as a major and important life history process shaping population and community dynamics.

Enjoy reading the details in the subsequent chapters!

CHAPTER TWO

Life History Processes, Ontogenetic Development, and Density Dependence

BACK TO DARWIN

Standard textbooks in ecology (e.g., Begon, Harper, and Townsend 1996) define population ecology as "the study of the variations in time and space in the size and densities of populations, and of the factors causing those variations." In this context a population is defined as "a group of individuals of one species in an area, though the size and nature of the area is defined, often arbitrarily, for the purposes of the study undertaken." But how have we, as ecologists, conceptualized populations? In other words, how have we in our research looked at these "group[s] of individuals of one species"? One way to shed light on this question is to reflect on the classical models that form the theoretical basis of population ecology: the Lotka-Volterra competition model, the Lotka-Volterra predator-prey model, and the Fretwell-Oksanen food chain model. These core models represent three fundamental types of species interactions: the antagonistic interaction between two competitors, the consumer-resource interaction between a predator and its prey, and the indirect mutualism in chains of species. These models indeed account for changes in population abundance over time and hence seem completely in line with the definition of population ecology. At the same time these models do not explicitly account for aspects of these "group[s] of individuals" other than how many of them there are. In particular, they completely ignore any possible variation among the individuals in the population.

From a historical point of view, the focus on population abundance and the neglect of variation among individuals should not surprise us, given that these models were in fact developed by nonbiologists (Vito Volterra was a brilliant mathematician, and Alfred Lotka was a chemist). Much of our thinking in ecology, especially when it comes to developing models and theory, has been inspired by physics and chemistry, with their long tradition of studying interactions within and between groups of particles. However, in these fields one particle is generally the same as any other. In contrast, every individual organism is unique in its own right and therefore different from almost any other. Of course, this uniqueness of the individual organism has been catapulted into our thinking by Charles Darwin. Although he mainly focused on an evolutionary time scale, Darwin disposed of the idea that species are unchanging, uniform entities. He moreover pointed out the far-reaching consequences that variation among individuals might have on the dynamics of ecological communities. In fact, Lewontin (1964) has suggested that the most important advance of the Darwinian revolution relates to its recognition of the importance of the "unique individual." It replaced the Aristotelian view of "ideal types" to which observed objects were "imperfect approximations." Neglecting the variation among individuals implies that we neglect the uniqueness of individual characteristics, as well as the fact that individuals change in these characteristics during life with concomitant changes in ecological performance. In doing so, population ecology overlooks the absolutely unique feature that makes an ecological community different from say a chemical reaction network! It is somewhat surprising that we as biologists, who usually have a keen interest in, and fascination with, the rich detail and complexity of our study objects, have accepted this neglect of individual variation for so long and have not been seriously worried about the apparent lack of biological realism in the core of ecological theory.

Considering the scientific developments in our field over the past century reveals that the special features of ecological systems are well appreciated and that populations are not at all seen as collections of simple particles. The interactions among individual organisms are highly complex and often characterized by subtle details. For example, that predators do not simply eat on encounter but may be limited in their foraging by satiation was accounted for in the predator-prey model proposed by Rosenzweig and MacArthur (1963), which led to the discovery of the paradox-of-enrichment (Rosenzweig 1971). Similarly, the Lotka-Volterra competition model fails to account for the fact that competition is always about something, be it resources, space, or sexual partners. MacArthur and Levins (1967; MacArthur 1972) and later Tilman (1982) explicitly accounted for competition for resources in population models and thereby established the R^* theory as a cornerstone of population and community ecology. Polis and Holt (Holt and Polis 1997; Polis, Myers, and Holt 1989) emphasized the notion that interactions among individuals may not be simply competitive or predatory but are often a mixture of both, such as in case of intraguild predation.

The recognition that distributions of populations are never homogeneous over space and that ecological interactions are modified by spatial heterogeneity can be seen as another example: it has given rise to the broad field of metapopu-lation theory (Hanski 1999) and more recently to spatially explicit (simulation) models (e.g., de Roos, McCauley, and Wilson 1991; Hassell, Comins, and May 1991; Rand and Wilson 1995). More recently a lot of attention has been paid to the fact that individual behavior has a significant influence on interactions between individuals (Abrams 1996; Abrams, Menge, et al. 1996; see Bolker, Holyoak, et al. 2003 for a review; Werner 1992). A meta-analysis (Preisser, Bolnick, and Benard 2005) has shown that in interactions between predators and their prey these indirect effects attributable to behavioral responses may be just as important as the direct effects of killing and eating the prey. Not-withstanding their importance, these intricacies of ecological interactions have been invariably investigated in settings in which variation between individuals was almost entirely ignored. For example, population models that examine the effect of optimal foraging on population dynamics (e.g., Fryxell and Lundberg 1993; Gleeson and Wilson 1986; Krivan 1996, 1997) in essence assume that all foragers are behaving exactly identically, which intuitively does not seem to resemble much of an optimal foraging strategy at all. Despite our progress in understanding the subtleties of ecological interactions, we have thus neverthe-less been unable to move forward from the Aristotelian view of "ideal types"!

One can rightfully question whether it is necessary to account for the uniqueness of individual organisms if we want to capture the essence of the two fundamental processes that change population abundances, namely, re-production and mortality. Indeed the earliest population dynamics models of Malthus (1798) and Verhulst (1838) make little reference to individual organ-isms at all and describe population dynamics solely as a function of popula-tion abundance itself. The rates at which reproduction and mortality proceed, however, are the cumulative result of all birth and death events in the life his-tory of the individual organisms making up the population. The population and individual level represent two scales of biological organization that are linked because events at the lower scale of organization (the individual) underlie pro-cesses at the higher scale (the population). It is not until we realize that there is a lot of variation among individual organisms, which directly impinges on their chances to give birth or die, that we face the necessity to account for individu-als and variation among them explicitly.

Individuals differ from each other in many ways, ranging from genetic dif-ferences to morphological or behavioral differences. From the point of view of population dynamics, however, the most important source of variation surely relates to the individual's state of development or maturation. Life progresses from birth until death, with reproduction, which is crucial for population per-sistence, possibly happening somewhere in between. We cannot think of any species on earth, though, for which the timing of reproduction between its birth and death is completely random (although the extent and consequences of this nonrandomness is naturally expected to differ between a bacteria and an elephant!). This fact underlies the distinction between juveniles and adults and immediately establishes the individual state of development as an aspect of individual variation with direct population dynamics consequences. It is this obvious connection between ontogenetic development and population dynamics that has set us on the research track of which this book is a summary. In many places we identify the state of ontogenetic development of an individual with its body size. The reason for taking size rather than age as a representative measure of ontogenetic development will be discussed in more detail later in this chapter, but one important corollary of this choice is that the rate of ontogenetic development may be variable, in contrast to ageing which always progresses at unit rate. However obvious this may seem, this distinction is often lost on superficial readers of our work, and we have been regularly accused of working on age-structured models (e.g., Allendorf, England, et al. 2008, p. 334). Whether ontogenetic development is plastic or not makes a crucial and fundamental difference, a fact that we hope will be clear to the reader after reading this book.

We are hence not only interested in the abundance of a population but also in its composition in terms of the relative frequency of individuals with different states of ontogenetic development or body size. We will consistently use the term "population structure" to refer to this (relative) composition of the population. The state of any biological population can then be completely characterized in terms of its structure and its abundance, whereas the state or structure of a community is completely determined by the abundance and structure of its constituting populations. Models of population dynamics that account for both abundance and population structure we will hence refer to as "structured population models" (Metz and Diekmann 1986). Analogously, models that account only for population abundance and ignore its structure are generally referred to as "unstructured." Unstructured models thus conceptualize the population as a collection of identical individuals, although an alternative interpretation might be that they are based on the assumption that all individuals in the population can be represented by some "average" type (however that average may be defined). In this book we are concerned with the question of what factors shape the structure and abundance of populations and the communities of which they are a part. How will population structure and abundance change when environmental conditions deteriorate or improve? And what are the intrinsic dynamics of the community itself, even in the absence of any changes in the environmental factors? Of course we are not the first to write a book with this aim, but we do have a specific focus that we hope sets this book apart from previous monographs: we focus on how individual life history, and specifically the plastic ontogenetic development of individuals during their life history, affect population structure and dynamics.

Compared with other biological disciplines, such as organismal or cell biology, ecology has a relatively strong tradition of employing theory, either verbal or embodied in mathematical models, to further the understanding about the systems we study. One can bring forward some obvious reasons for this importance of theory in ecology: first, because of the complexity of ecological systems, their study really benefits from the use of mathematical approaches. However, we do not want to go as far as claiming that ecological systems are more complex than other biological systems. Another reason may offer a better explanation for the prominence of theory in ecology: because experiments at the level of populations and communities over many generations are costly in terms of time and money, a theoretical analysis of a question before an experiment is designed may avoid wasting these valuable resources. Whatever the reason, the fact is that much of our understanding about populations and communities and their dynamics is couched in terms of mathematical models. We clearly build on this tradition and phrase large parts of our presentation in terms of models or theoretical considerations. At the same time, we switch to discussing data or observations whenever that is possible. To readers who are not especially mathematically inclined, we apologize for regularly reverting to mathematical language. Be assured that we do prefer and provide more intuitive, nonmathematical explanations wherever possible. Also, the aim of our work is not to study and analyze some interesting mathematical models but to come to a better understanding of the dynamics and structure of populations and communities. The mathematics we have hence tried to put in its appropriate place: as a tool, which we sometimes need for its precision of expression and its rigor of reasoning.

INDIVIDUAL- VERSUS POPULATION-LEVEL ASSUMPTIONS

Next to the characteristics of an individual itself—in terms of, for example, its stage of ontogenetic development, its body size, or the amount of reserves it has at its disposal—the chance to give birth or to die is often dependent on the condition of the environment the individual is living in. Moreover, a population of individuals may be able to affect this environment and hence indirectly change birth and death rates. The availability of food or resources is one aspect of that environmental condition with a significant influence on birth and death rates, which is influenced by population foraging. As soon as we want to construct a model for the dynamics of a population, the dependence of reproduction and mortality on an environmental variable like food density confronts us with a choice: we can either model this influence mechanistically or phenomenologically. With a mechanistic representation of an environmental influence we mean an explicit description how the life history process in question depends on the environment the individual lives in. This might, for example, require specifying individual fecundity as an explicit function of food density. Well-known examples of mechanistic models in the sense as used here are the predator-prey model analyzed by Rosenzweig and MacArthur (1963), Tilman's resource competition models (Tilman 1982), and the bioenergetics model proposed by Yodzis and Innes (1992). In contrast, a phenomenological model of population dynamics would not explicitly specify the mechanism by which the environment influences reproduction and mortality but would model the rates of these processes in a density-dependent manner. That is, it would specify the per-capita fecundity and death rate of individual organisms as an explicit function of population abundance. Examples of such phenomenological models are the logistic growth model, the Ricker model, and the Lotka-Volterra competition model.

The distinction between mechanistic and phenomenological models is sometimes subtle and seems in the context of unstructured population dynamics models also rather trivial or even a matter of taste. For example, consider two populations of consumers, whose densities are given by N_1 and N_2, respectively, competing for a resource, the density of which is R. If we assume that the resource grows logistically and the consumers are characterized by linear functional responses and constant death rates, the dynamics of consumers and resource can be described by the following system of ordinary differential equations (ODEs):

$$\frac{dR}{dt} = \delta R \left(1 - \frac{R}{R_{max}} \right) - a_1 RN_1 - a_2 RN_2 \tag{2.1a}$$

$$\frac{dN_1}{dt} = \varepsilon_1 a_1 RN_1 - \mu_1 N_1 \tag{2.1b}$$

$$\frac{dN_2}{dt} = \varepsilon_2 a_2 RN_2 - \mu_2 N_2 \tag{2.1c}$$

In these equations we have used δ and R_{max} to represent the population growth rate and carrying capacity of the logistically growing resource, while a_1 and a_2, ε_1 and ε_2, and μ_1 and μ_2 are the attack rates, conversion efficiencies, and mortality rates of the two consumer species, respectively. Now, if the dynamics of the resource is fast in comparison with the time scale on which abundances of the two consumers change, we can assume that dR/dt approximately equals 0 at all times. This, so-called quasi–steady state assumption allows us to express the resource density as a function of the densities of the two consumers: $R = R_{max}(1 - a_1 N_1/\delta - a_2 N_2/\delta)$.

Substituting this expression into the ODEs (2.1b) and (2.1c) transforms the system into the following two ODEs:

$$\frac{dN_1}{dt} = r_1 N_1 \left(1 - \frac{N_1 + \alpha_{12} N_2}{K_1}\right) \qquad (2.2a)$$

$$\frac{dN_2}{dt} = r_2 N_2 \left(1 - \frac{\alpha_{21} N_1 + N_2}{K_2}\right) \qquad (2.2b)$$

Here we have introduced new, composite parameters r_1 and r_2, representing the population growth rates, and K_1 and K_2, representing the carrying capacities of the two consumer species, respectively, whereas α_{21} and α_{12} are the competitive effect of consumer species 1 on species 2, and vice versa, respectively. In terms of the original parameters, the population growth rate is defined as $r_1 = \varepsilon_1 a_1 R_{max} - \mu_1$; the carrying capacity is $K_1 = \delta(1 - \mu_1/(\varepsilon_1 a_1 R_{max}))/a_1$; while the competitive coefficient is defined as $\alpha_{21} = a_1/a_2$. Analogous expressions relate the composite parameters r_2, K_2, and α_{12} to the original ones. The bottom line of this derivation is that by assuming fast time-scale dynamics of the resource we have transformed a mechanistic model for resource competition between two consumer species into the classical but phenomenological Lotka-Volterra model for interspecific competition.

Why then do we prefer mechanistic model assumptions over phenomenological ones? One reason is that by its very definition the phenomenological dependence of, for example, fecundity on population abundance can only be assessed in a population context. Phenomenological models hence use as ingredients population-level observations that they aim to predict or analyze, which in some sense introduces circularity into the model formulation and analysis. In contrast, the dependence of fecundity on food density in the environment relates to an individual-level process that can in principle be measured in experiments with individual organisms in isolation.

Second, predictions between mechanistic and phenomenological models of the same system may differ. As became clear while transforming the mechanistic resource competition model between two consumers into the phenomenological Lotka-Volterra model, the latter is a kind of short-circuited version of the former: the feedback of the population on individual fecundity, which operates via population foraging and the consequent depression in resource density, is assumed to operate instantaneously. In the mechanistic model this feedback operates with a delay, the length of which depends on the rate at which the resource density actually declines through the population foraging. In a size-structured consumer-resource model this difference between a delayed and instantaneous population feedback has been shown to make the difference between either finding population cycles over a large range of model parameters or a stable equilibrium (de Roos and Persson 2003; see also chapter 9).

A third reason to favor mechanistic model assumptions relates directly to our focus on ontogenetic development and population dynamics, as mechanistic assumptions make it easier to formulate realistic population models if more than a single process in individual life history is influenced by the environmental condition and is hence indirectly affected by density dependence. For example, if both growth in body size as well as individual fecundity are affected by population feedback on resources, a phenomenological model of population dynamics would require explicit relationships of these two processes as a function of population abundance. Determining the rate of both growth and reproduction at a range of different population abundances involves a set of laborious and probably difficult population-level experiments. In contrast, mechanistic model ingredients represent these processes as a function of environmental food condition, which dependence can be measured using individual-level, laboratory experiments at different food densities. The downside, however, of specifying the model ingredients mechanistically is that we not only have to model the individual response to the environment in terms of, for example, growth and fecundity, but in addition we also have to specify how an individual impacts its environment and how the environment itself changes over time in the absence of individuals impacting it.

A last reason to favor mechanistic models has only become apparent to us while analyzing structured population models: mechanistic model assumptions provide more scope for confronting model predictions with empirical observations. When multiple life history processes depend on an environmental variable like food density, it is a priori impossible to predict the dynamics of these processes in a population context, when food density is controlled by population feedback. For example, whether growth in body size will be strongly retarded while potential fecundity is high or vice versa (somatic growth is rapid, but fecundity strongly suppressed) is an outcome of the model in a population context that is only weakly related to the assumptions on which the model is based. These model predictions at the individual level under conditions of population feedback can hence be used to contrast the model with empirical observations, in addition to comparing the magnitude and fluctuations in population abundance. In chapters 10 and 11 we provide several examples of such confrontations between predictions derived from structured models and empirical data on individual-level processes under the influence of population feedback. This larger scope for model testing is almost completely neglected in the discussion about model complexity versus generality (Bolker, Holyoak, et al. 2003; de Roos and Persson 2005; Murdoch, McCauley, et al. 1992). It partly offsets for the undeniable fact that complex models are based on a larger number of assumptions and involve a larger number of parameters, a fact that has often been put forward to argue that complex models are less general and testable than simple ones. In the final chapter we present a more in-depth discussion of this relationship between model assumptions, model predictions, and empirical observations.

THE POPULATION DYNAMICAL TRIAD

Modeling population dynamics on the basis of mechanistic assumptions about individual life history and its dependence on environmental conditions leads to an interdependence between population, individual, and environment that we will refer to as the population dynamical triad. Figure 2.1 illustrates the population dynamical triad for the case of a population of size-structured consumers that feed on a shared food resource. The population of consumers is fully characterized by its size distribution, which is hence usually referred to as the population or p-state (de Roos 1997; Metz and Diekmann 1986). The population state changes only as a result of processes that are part of the life history of the individual. In this particular example these processes include growth in body size, mortality, and reproduction. The dynamics of the population is therefore obtained by keeping track of all events happening to individual consumers. The rates at which life history processes of individual consumers progress depend both on the state of the individual itself and on the current condition of the environment it is living in. As we have adopted body size to fully characterize the individual consumer, it is referred to as the individual or i-state (de Roos 1997; Metz and Diekmann 1986). Finally, the condition of the environment changes as a result of the impact that the population of consumers exerts on it. Here, this condition is the density of food in the environment, which changes owing to the foraging impact of the entire population. Notice that the feedback of the population on its own dynamics is rather indirect, as it operates via the changes in the environment and their subsequent effects on individual life history processes.

To the best of our knowledge, Metz and Diekmann (1986) were the first to introduce the clear distinction between individual or i-state, population or

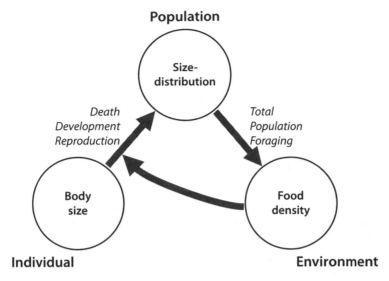

FIGURE 2.1. The population dynamical triad exemplified for a population of size-structured consumers feeding on a shared food resource. Body size hence characterizes the state of an individual consumer, which in conjunction with the ambient food density also determines its ecological performance in terms of resource feeding, growth in body size, reproduction, and mortality.

p-state, and the condition of the environment (which for some subtle reasons is not really a "state" in the systems theoretical interpretation of the word). This distinction may seem academic at first, but it has certainly proven its worth. For example, when considering figure 2.1, one should notice that if the condition of the environment were to remain constant and unaffected by the population, all individuals would be independent of one another. This means that their life history would progress at a rate that is unaffected by their conspecifics. This independence of individuals in a constant environment allows one to calculate even for rather complex individual life histories at which exponential rate a population of such individuals would eventually be growing in a constant environment (de Roos 2008). It furthermore allows for the calculation of the steady state of structured population models using a generic procedure (Diekmann, Gyllenberg, and Metz 2003). And last but not least, it is an important element in evolutionary analyses that focus on the invasion of a mutant in an environment occupied by a resident population (Diekmann, Gyllenberg, and Metz 2003). Because on invasion mutant individuals invariably occur at a low density, interactions among mutants themselves occur so infrequently that they can be ignored. Hence, whether or not a mutant invades can then be assessed by calculating its population growth rate in an environment that is completely determined by the resident population. We have exploited this feature to study the evolution of size at maturation induced by harvesting in a size-structured consumer population feeding on a shared resource (de Roos, Boukal, and Persson 2006).

GROWTH PATTERNS AND THE ECOLOGY
OF ONTOGENETIC DEVELOPMENT

In the previous sections we identified the body size of an individual organism with its state of development. At the population level this assumption implies that between-individual variation in developmental state is synonymous with within-population variation in body size. At the individual level it implies that the process of development itself is related to growth in individual body size. Different patterns of growth in body size are observed in different organisms, usually related to basic energetic processes such as rates of ingestion and metabolism. A generic feature of any organism is that its energetic status depends on the rate of intake relative to the rate of metabolic demands (Kleiber 1961). If energy intake exceeds energy demands, surplus energy can be used for either somatic or gonad growth or both. Even for unicellular organisms like bacteria and phytoplankton, it holds that single organisms use energy surplus to grow in size from one cell division to the next. Still, the growth of each individual unicellular unit is limited (less than an order of magnitude in weight) in comparison with metazoans, such as, for example, fish individuals that can easily increase seven orders of magnitude in body weight over their life period from hatching.

An overview of the animal kingdom shows that the overwhelming majority of all species undergoes substantial development from the time that the individual organism searches independently for food until it matures. Werner (1988) pointed out that in twenty-five out of thirty-three phyla, major ontogenetic development is a prominent feature of the life cycle, in particular because around 80 percent of all animal organisms on earth undergo metamorphosis. This figure is evidently a result of the taxonomical dominance of insects. However, substantial growth and development after the individual becomes independent of its parents also occurs in 75 percent of all vertebrate taxa. This predominance of growth and development among vertebrates results from the taxonomical dominance of fish (30,000 species), reptiles (8,200 species), and amphibians (5,700 species).[*] In the remaining vertebrates, birds (9,800 species) and mammals (5,400 species), the offspring are supported by their

---

[*] Number of species in different animal phyla is based on information from the International Union for the Conservation of Nature.

parents until they reach a weight in the range of 1–100 percent of their parents for birds and 3–50 percent for mammals (Werner 1988). In both groups the size of the juveniles at the moment they become independent of their parents is often closer to the upper limit of the ranges indicated, and their increase in size afterward is at least less than two orders of magnitude. Already at this point it can thus be concluded that substantial growth and development as an independent forager is the rule among most animal organisms with some birds and mammals as the only exceptions. The plant kingdom is totally dominated by multicellular forms, with seed plants (275,000 species), mosses (24,000 species), and ferns (10,000 species), constituting about 309,000 out of a total of 330,000 species. Among these forms, individual growth from the size of a seed to adult size is many orders of magnitude in weight. Similar to the animal kingdom, substantial individual growth and development thus characterizes most plants as well.

Growth and development can take a variety of different forms. In its simplest form, growth and development involve an increase in size that merely preserves the same basal body architecture as occurs in many plants, reptiles, and fish (the latter with the exception of the transition from the larval to the juvenile stage). In other groups, such as crustaceans and spiders, the individual organism progresses through a series of discrete stages by molting. Finally, the body architecture may be totally reorganized as is the case for organisms undergoing metamorphosis. Metamorphosis generally involves discrete and drastic shifts in resource use as exemplified by many amphibian species. Such drastic ontogenetic niche shifts can, however, also be present in continuously growing organisms like fish, through discrete habitat shifts (Werner 1988). The implications of the substantial increase in body size during ontogeny exhibited by most organisms have been the topic of extensive discussions about the limitations of isometric growth over large size ranges (Werner 1988). These limitations result from a hierarchical series of constraints. First, physical and mechanical constraints set feasibility limitations on size and morphology, on top of which ecological factors impose more restrictions (Werner 1988). A thorough treatment of this issue is provided in Werner (1988) and is beyond the scope of this book. We hence refer the reader to that work and the references therein for in-depth coverage of this interesting topic.

Another aspect of growth patterns that has been the focus of substantial discussion is the extent to which growth is determinate/nonplastic or nondeterminate/plastic (Sebens 1987; Stearns 1992). This issue has, for example, been related to life history characteristics, such as whether or not an individual will approach an asymptotic size (Sebens 1987). Stearns (1992) defined organisms with indeterminate growth as those that continue to grow after maturation, as occurs in most perennial plants, fish, amphibians, reptiles, crustaceans, worms, and mollusks. Conversely, he defined organisms with determinate growth as those that stop growing in somatic mass at or before maturation, as is the case for many insects, mammals, and annual plants. With this definition, the demarcation line between determinate and indeterminate growth thus lies in the much simpler rule that in the case of determinate growth after maturation all energy is allocated to reproduction. A more detailed categorization of different growth types was made by Sebens (1987), but he also used the occurrence of individual growth in somatic mass after maturation as an important criterion for distinguishing different growth types.

As we will find in the following chapters, Stearns's definition of determinate and indeterminate growth is not very relevant in the population and community ecology context we are interested in, because (1) it does not differentiate between whether individual growth is assumed to be food dependent or not and (2) we will show that most of the effects of intraspecific size variation on community structure and population dynamics are totally independent of whether or not individual organisms stop growing in somatic mass after maturation. To grow in size an individual by necessity needs energy; hence individual growth is in an absolute sense *always* dependent on food. A classification of growth patterns based on the presence or absence of its dependence on food thus seems rather meaningless. With the purpose of making definitions and concepts both clear and meaningful we will therefore take as a point of departure the extent to which the growth of an individual is a simple reflection of food availability in the environment. More specifically, we will differentiate between consumers that feed according to the amount of food available (supply feeders) and consumers that require a certain energy quota over a given time period in order to sustain themselves (demand feeders) (Kooijman 2000).

Before giving a more precise definition of supply versus demand feeders, we will first provide an overview of energetic patterns with respect to growth rates, maximum ingestion rates, and metabolic demands in poikilotherms and homeotherms, respectively. As it turns out, the extent to which consumers can be categorized as either supply feeders or demand feeders is largely determined by whether they are poikilotherms or homeotherms. Reviews of growth patterns have shown that homeotherms like birds and mammals have much higher observed growth rates than poikilotherms such as fish and reptiles, with marsupials having a growth rate intermediate between these groups (Case 1978; Peters 1983; Ricklefs 1976; Starck and Ricklefs 1998) (figure 2.2). The overall higher growth rate of homeotherms has been related not only to the provision of energy from the parents to the egg and offspring in the form of yolk, maternal fluids, and bodies of prey (Werner 1988) but also to the

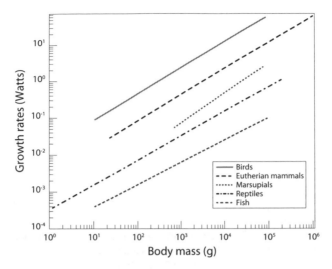

FIGURE 2.2. The relationship between adult body size and somatic growth rate of an individual per day—calculated as the average rate of increase as the animal grows from 10 percent to 90 percent of its adult size—for birds, eutherian mammals, marsupials, reptiles, and fish. Data from Case (1978) and Peters (1983).

higher body temperature of homeotherms, which increases their capacity to ingest and assimilate food (Stearns 1992) (figure 2.3, left panel). Support for the importance of a high body temperature per se for the occurrence of high growth rates comes from insect colonies, where social thermoregulation may lead to growth rates substantially higher than expected for ectotherms (Danks 2002). The fact that at high temperatures butterfly larvae can grow at higher rates than birds also reflects the importance of high temperature (Gotthard 2008; Reynolds 1990). Furthermore, among homeotherms the overall higher growth rates in birds compared with mammals can be related to the former's higher body temperature. Among birds themselves differences occur between altricial and precocial birds, as the former have growth rates up to an order of magnitude higher than the latter. One reason for this pattern is that the size of the digestive tract is larger in altricial birds and hence allows for faster uptake of energy (Lilja 1983; Starck and Ricklefs 1998).

The observed higher growth rates of homeotherms compared with poikilotherms can thus be related to their higher consumption capacity. Overall, the maximum ingestion rate is about fourteen times higher for carnivorous and herbivorous homeotherms than for poikilotherms (Peters 1983) (figure 2.3,

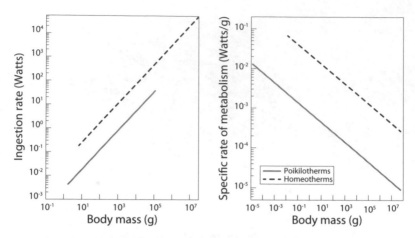

FIGURE 2.3. *Left*: Ingestion rate of carnivorous homeotherms and poikilotherms (tetrapods). *Right*: Specific metabolic rate/standard metabolism of homeotherms and poikilotherms. Data from Farlow (1976) and Peters (1983).

left panel). Besides being a result of a higher body temperature per se, Pough (1980) related the higher activity of birds and mammals to the cardiovascular and respiratory system of endotherms, which enables them to increase the supply of oxygen to active tissues and thereby maintain a high and sustained level of aerobic energy production. In contrast, in ectotherms like amphibians and reptiles aerobic metabolism only accounts for a small portion of the total energy used to sustain high activity levels, and periods of high activity are therefore constrained.

Homeotherms and poikilotherms also differ substantially in metabolic costs, as the metabolic rate (standard metabolism) is thirty times higher in homeotherms than in poikilotherms (Peters 1983; Stearns 1992) (figure 2.3, right panel). At high food availability the net energy balance of homeotherms is nevertheless higher than that of poikilotherms, leading to the high growth rates shown in figure 2.2. For example, based on figure 2.3, the net energy balance of a homeotherm with a weight of 100 kg is 337 watts, whereas that of a similarly sized poikilotherm is only 30 watts. Although at ample food conditions a homeotherm has a higher growth capacity, it is in contrast more sensitive to poor food conditions, as it operates with high metabolic costs. The low energy approach to life that is characteristic of, for example, amphibians and reptiles and their reliance on anaerobic metabolism for sustained high activity may here be an advantage. It provides ectotherms the capacity to rapidly generate energy for brief periods of time when needed without paying the costs of maintaining the high resting metabolism of homeotherms (Pough 1980). Without saying that ectothermy represents an adaption for life under poor food conditions, it can be concluded that ectothermy increases an organism's ability to sustain itself under a large range of food conditions. Correspondingly, it has been theoretically shown that poikilotherms may sustain themselves under a much larger range of environmental productivities than homeotherms (Shurin and Seabloom 2005; Yodzis and Innes 1992).

After this short exposé of generic patterns of energetics in different organism groups, we return to the distinction between supply versus demand feeders. Kooijman (2000) distinguished these two extreme types as, on the one hand, systems where the output is the result of the state of the system and its input (supply-driven) and, on the other hand, systems where the input results from the state and the output (demand-driven). In the former case (supply-driven systems) energy intake is purely driven by energy available in the environment, and allocation to somatic growth is not fixed but instead depends on food conditions. In the latter case (demand-driven systems) the lead is taken by some energy-consuming process like somatic growth, which requires a target level of energy intake to be met, such that the organism "cats what it needs." In reality, every organism has elements of both supply-driven and demand-driven systems. Still, given the more restricted scope of homeotherms to sustain themselves under different environmental conditions, they can be viewed as more demand-driven organisms, whereas poikilotherms are more supply-driven organisms. Birds are consequently typical examples of demand-driven organisms that can only survive under relatively high food densities (Kooijman 2000). As a result, the range of different growth curves that are possible over ontogeny is much more restricted in this group than among more supply-driven organisms such as amphibians and reptiles.

The focus of our book is on supply feeders, which form the overwhelming majority of animal taxa on earth. Still, the theoretical framework that we will use is also highly appropriate for demand feeders. In chapter 12 we will specifically discuss how many of the results we obtain for supply-driven systems generalize to demand-driven systems, where the amount of energy used for somatic growth is fixed. In supply feeders we include both organisms in which individuals continue to allocate energy to somatic growth after maturation, as in many fish species, and organisms in which individuals allocate all their surplus energy to reproduction after maturation, such as is the case in many holometabolous insects (see chapter 3). As we hope will become apparent to the reader of this book, the different ways in which organisms use ingested energy can better be discussed in terms of variations between supply- and demand-driven systems than as a useful approach to distinguish between the growth patterns observed in different species.

BODY-SIZE SCALING AND MAGNITUDE
OF BODY-SIZE CHANGES

An important aspect affecting interactions between organisms is how different individual-level, ecological processes scale with body size. It is long since recognized that ecological processes such as food intake and metabolism are heavily affected by body size (Peters 1983; Werner 1988). But when will *intraspecific* scalings in these ecological processes actually matter for population and community dynamics? This question will in one way or another be addressed in all chapters of this book. We here simply postulate that intraspecific size variation will matter whenever the body-size scaling of different individual-level processes is *asymmetric*. What we mean by *ontogenetic symmetry* and *ontogenetic asymmetry* we now present here because of the central role of these concepts in our book. In chapter 3, however, we will delve more deeply into the issue of asymmetric, as opposed to symmetric, body-size scaling.

To do so, we consider an organism that ingests food following a Holling type II functional response and assume that both attack rate and handling time are size dependent and scale with body mass following a power function. Additionally, we assume that metabolic demands also scale with body mass according to a power function (Peters 1983). With these premises, we now express the change in body mass, dB/dt, as the difference between mass intake and metabolic costs:

$$\frac{dB}{dt} = \frac{ca_1 B^{a_2} R}{1 + a_1 B^{a_2} h_1 B^{h_2} R} - m_1 B^{m_2}, \tag{2.3}$$

where a_1 is the scalar and a_2 the exponent of the body mass–dependent attack rate; h_1 and h_2 the scalar and exponent, respectively, of the body mass–dependent handling time; and m_1 and m_2 the scalar and exponent, respectively, of the body mass–dependent metabolic rate. R is resource biomass density, and c is a conversion factor. The resource density R^*, for which energy intake exactly equals metabolic demands, can be derived by setting $dB/dt = 0$ and rearranging equation (2.3):

$$R^* = \frac{m_1 B^{m_2}}{a_1 B^{a_2}(c - h_1 m_1 B^{(h_2 + m_2)})} \tag{2.4}$$

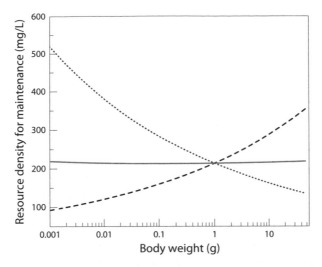

FIGURE 2.4. Resource density needed to cover maintenance requirements for differently sized consumers for three different size-scaling exponents of the attack rate (*solid line*: $a_2 = 0.725$; *dotted line*: $a_2 = 0.85$; *dashed line*: $a_2 = 0.60$). Other parameters: $c = 0.5$, $a_1 = 0.4 \text{ L} \cdot \text{day}^{-1}$, $h_1 = 0.005 \text{ day} \cdot \text{g}^{-h_2}/\text{mg}$, $h_2 = -0.8$, $m_1 = 30 \text{ mg} \cdot \text{g}^{-m_2} \cdot \text{day}^{-1}$, and $m_2 = 0.75$.

We call R^* the critical resource density and define *ontogenetic symmetry* as the case in which different combinations of the size scaling of the attack rate, handling time, and metabolic demands result in R^* being largely *independent* of body mass. *Ontogenetic asymmetry* then implies that R^* changes significantly with body size. Figure 2.4 illustrates the relationship between R^* and body mass for three different values of the size scaling of attack rate and the same size scaling of handling time and metabolic rate. In this case ontogenetic symmetry occurs for $a_2 = 0.725$. With a lower size scaling of the attack rate (0.60), R^* will increase with body size, and smaller individuals will be competitively superior; whereas with a higher size scaling of the attack rate (0.85, figure 2.4), R^* is a decreasing function of body mass, and larger individuals are competitively superior.

The above definition of ontogenetic symmetry and asymmetry in terms of critical resource demands emphasizes the importance of resource requirements at different body sizes. Ontogenetic symmetry and asymmetry can, however, also be considered from an alternative perspective, which emphasizes the net-energy production and the synthesis of new biomass at different body sizes. From this latter perspective ontogenetic symmetry is defined as the condition in which the *mass-specific* rate, at which new biomass is produced through somatic growth and/or reproduction, is the same for all individuals, irrespective of their body size. In the case of ontogenetic asymmetry this mass-specific rate of new biomass production is hence different for differently sized individuals.

As we will show, the concept of ontogenetic asymmetry is extremely useful, because its occurrence has major ramifications for both community structure and population dynamics. The concept of ontogenetic asymmetry is actually so central to our lines of argumentation that an alternative title of our book could have been "The Ecology of Ontogenetic Asymmetry." It needs to be pointed out, though, that in our discussion of ontogenetic (a)symmetry the search rate, handling time, and metabolic rate of an individual are always considered to be size-dependent, even in the case of ontogenetic symmetry. Ontogenetic symmetry only implies that the combined effect of these size-dependent rates is such that individuals, independent of their body size, require similar food levels to sustain themselves. In contrast, most of the existing theory about population dynamics and community structure is based on unstructured models that by assumption stipulate that individuals are identical and, consequently, that ontogenetic symmetry prevails. These models will therefore faithfully capture ecological dynamics only when the minimum resource requirements to meet maintenance costs are independent of body size.

From this we can draw two conclusions: First, unstructured theory represents a limiting case of a general ecological theory for the specific size-scaling conditions that lead to ontogenetic symmetry. Second, intraspecific size variation per se does not immediately necessitate the use of size-structured models, as we can still have ontogenetic symmetry for specific conditions (e.g., $a_2 = 0.725$ in figure 2.4; see also chapter 3). Also, when ontogenetic (a)symmetry is considered from a perspective of net-energy production, it is important to note that the rates *per individual organism* are different for differently sized individuals even under conditions of ontogenetic symmetry, the latter stipulating only that the *mass-specific* rates of net biomass production do not differ. We will in fact use this latter interpretation of ontogenetic symmetry in most chapters, although in chapter 10 we also use an approach based on equations (2.3) and (2.4) and show that it leads to the same results.

In addition to the size scaling of individual ecological processes, one intuitively expects that the extent to which size variation is present in a population should also be important for whether or not intraspecific size variation matters for population and community dynamics. To address this question more specifically, we investigate the influence on population and community dynamics of the ratio between the individual's mass at birth and its mass at maturation. We do so because this ratio will influence the time delay between the moment that an individual is born and the onset of its reproduction. Such time delays per se are known to induce cycles related to stage structure (Gurney and Nisbet 1985). The influence of the birth mass/maturation mass ratio, which we will refer to with the parameter z, on population dynamics will turn out to depend on whether juveniles or adults are competitively superior. If juveniles are competitively superior, a large z value will reduce the time delay between birth and maturation and thus lead to a decreased influence of ontogenetic asymmetry on both community structure and population dynamics. In chapters 3 and 9 we carry out a more thorough analysis of this question, but as a rule of thumb, a z value of 0.18 or less will lead to dynamics that is heavily influenced by ontogenetic asymmetry. On the other hand, if adults are competitively superior, the influence of z is smaller. In this case, a high z value may even reinforce the effects of ontogenetic asymmetry on community dynamics, as we will see in chapter 3. The different effect of z on population processes for different stage-dependent competitive abilities can be related to the fact that it is the magnitude of the time delay per se that is important when juveniles are competitively superior, whereas it is the variability in this magnitude that is important when adults are competitively superior. We will come back to the question about the influence of time delays and whether they are fixed or variable in time in chapters 3, 9, and 12.

A compilation of birth mass/maturation mass ratios for insects, crustaceans, fish, amphibians, reptiles, birds, and mammals is shown in figure 2.5. For organisms undergoing metamorphosis (mainly insects and amphibians) we used the mass just before metamorphosis as "maturation" mass. For birds and mammals we did not use actual birth mass, but rather adopted as "birth" mass the mass at which the offspring becomes independent of its parents in terms of ingesting food through its own feeding activities. The latter measure will be somewhat imprecise, as the development to independence is gradual, but we are here only interested in broad patterns. It is striking that the birth mass/maturation mass ratio for all species from all groups except mammals and birds is substantially smaller than 0.14, in fact in most cases smaller than 0.01. The "birth" mass/maturation mass ratios among birds and mammals are substantially larger than in the other groups and also show a large variation between species. In birds the range in ratios is 0.23–1.3 and in mammals 0.1–0.88 (figure 2.5). The fact that the "birth" mass/maturation mass ratio is higher than 1 for some bird species is because the mass of a fledging individual may be larger when it leaves the nest than when it becomes an adult. The high variation in "birth" mass/maturation mass ratio can in both birds and mammals be related to the time during and degree to which the parents provide their young with food (e.g., precocial versus altricial birds). Overall, the patterns in z for

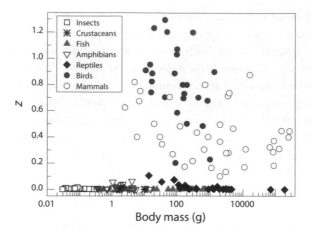

FIGURE 2.5. Ratio between the birth mass and maturation mass (z) for insects, crustaceans, fish, amphibians, reptiles, birds, and mammals. For insects and amphibians, maturation mass was identified with the mass of the largest larval stage. For mammals and birds, birth mass was taken as the size at which the offspring becomes independent of its parents in terms of relying on itself for finding and ingesting food. Data were obtained from an extensive literature search, covering many papers.

all groups except birds and mammals point to a high likelihood for population and community consequences of size variation when either juvenile or adults are superior competitors, whereas for most birds and mammals this should be less likely when juveniles are competitively superior.

CHANGES IN ECOLOGICAL ROLES OVER ONTOGENY

In addition to the size scaling of individual processes and the extent of the body-size increase per se, a change in body size over ontogeny inevitably involves a change in the ecological role of the individual. Ontogenetic asymmetry can have major consequences for community structure and population dynamics even when the individual feeds on the same resource, albeit at different intensities, over its whole life span. Still, an increase in size over ontogeny is generally also associated with changes in the resources used, generally referred to as ontogenetic niche shifts (Werner and Gilliam 1984). These changes can take place as an expansion of the niche when larger-size resources become accessible, while the smallest resource items ingested remain more or less of the same size (niche expansion), or these changes may entail that smaller resources are dropped from the diet, leading to distinct ontogenetic niche shifts (Bern 1994; Schellekens, de Roos, and Persson 2010). In many cases, a shift in resource use also involves a shift in habitat use, as is observed in organisms such as fish (Dahlgren and Eggleston 2000; Mittelbach and Osenberg 1993; Persson, Byström, et al. 2000), crustaceans (Childress and Herrnkind 2001), and reptiles (Keren-Rotem, Bouskila, and Geffen 2006). The change in habitat use may also be associated with a complete morphological reconstruction, as in dragonflies and amphibians that shift from an aquatic to a terrestrial life style in connection with metamorphosis. Shifts in ontogenetic niche mean that the intensity of interactions will change. For example, among many fish species juveniles share a common resource, whereas the adults may not overlap with each other in resource use at all (cf. Mittelbach and Osenberg 1993). Conversely, aquatic amphibian and insect larvae may shift from zero interactions to strong interactions with members of the terrestrial community when they emerge from aquatic systems (Sabo and Power 2002).

Shifts in ontogenetic niche with increasing size will in many cases also involve role reversals such that a competitive interaction early in life changes to a predator-prey interaction later on (Wilbur 1988). A typical example of this is *life history omnivory*, where small stages of two species compete for a shared resource, whereas one species preys on the small stages of the other when large (Persson, de Roos, and Byström 2007; Polis, Myers, and Holt 1989). Although omnivory in most cases has been discussed in connection with interspecific interactions, omnivory is in size-structured systems also common as a purely intraspecific interaction, that is, cannibalism (Claessen, de Roos, and Persson 2004; Persson, de Roos, et al. 2003; Rudolf 2008).

STEPPING BACK—SOME PERSPECTIVES

Given the additional complications in interactions that result from intraspecific size variation, as reflected in the changes in both the nature and intensity of interactions as well as in role reversals, one may at first sight be perplexed by such complexity per se. This might easily lead to the conclusion that one can gain only very limited general insights about the dynamics of size-structured systems. We argue, however, and hope to be able to convey, the opposite. Moreover, we will advocate that a general theory of ecology, which is couched in terms of biomass dynamics but is rigorously based on individual life history as well as on bioenergetics principles, can be developed. This perspective contrasts with current ecological theory, which is dominated by unstructured models and hence only represents a special case of a more general theory (see also de Roos, Persson, and McCauley 2003). In the final chapter of this book we will present in more depth our ideas about how a theory that stringently links individual-based models of different complexities in a logical hierarchy can be developed. Although the models we analyze in the following chapters differ in their degree of richness in details (particularly the models discussed in chapters 10 and 11), they nevertheless form a set of models within the same framework and consistently describe population dynamics in terms of the underlying processes in individual life history. This fact also leads us to discuss how model simplifications from complex to more simple can be achieved within an individual-based framework by making more simple assumptions about individual energetics, in contrast with a simplification through a chain of complex-to-simple models that are obtained by population-level aggregation without preserving the individual basis (Murdoch and Nisbet 1996; Murdoch, Briggs, and Nisbet 2003).

Finally, we have already pointed out that physiologically structured population models, by their recognition of and distinction between the individual and the population state, lend themselves to a more complete and critical testing of model predictions. But this is only one side of the coin, as the ability to study in a modeling context individual-level processes that are shaped by population feedback will in itself be a major asset in unraveling the mechanisms driving the dynamics under different ecological conditions.

PART II
ONTOGENETIC DEVELOPMENT
AND COMMUNITY STRUCTURE

Biomass Overcompensation

Underlying many, if not all, of the effects of ontogenetic growth on the structure of ecological communities is a phenomenon that we refer to as "biomass overcompensation" (de Roos, Schellekens, et al. 2007). Whenever the mortality that a population is exposed to changes, not only its total abundance is likely to change, but also its composition in terms of densities of individuals of different body sizes. This change in population structure may lead to an increase in the biomass density in a particular size-class of the population when mortality increases. We use the term "biomass overcompensation" for such an increase in biomass in response to an increase in mortality. As we will show, the total population biomass will generically not increase when individuals that share a single resource are exposed to higher mortality, but the biomass density of juveniles or adults may. Hence, the biomass overcompensation is often stage specific.

In a recent review Abrams (2009) discusses three mechanisms giving rise to an increase in population size in response to higher mortality, a phenomenon he refers to as the "hydra effect" (following Abrams and Matsuda 2005). First, when populations fluctuate, their time-averaged density may be lower than the (unstable) equilibrium density. Increasing mortality may then stabilize fluctuations and bring the two densities closer together. Second, if mortality and density dependence occur segregated in time, as in, for example, the Ricker model (Ricker 1954), an increase in density-independent mortality preceding the density-dependent process may lead to an increase in population density, when the latter is measured *before* density dependence operates (but to a decrease when measured *afterward*!). Finally, in consumer-resource interactions, an increase in consumer mortality will increase equilibrium resource density, which may lead to an increase in resource productivity when resource growth is either exponential or logistic. Unfortunately, the models discussed by Abrams (2009) do not consistently account for mass- or energy-flows between resource and consumer populations and incorporate only phenomenological descriptions of density dependence without an explicit reference to the mechanisms that give rise to it. Abrams's analysis therefore does not show whether and how the larger population density can be sustained energetically. In addition, because he does not distinguish between numbers and total biomass of individuals in a population, it remains unclear whether the larger numerical population abundances also represent larger biomass densities.

In contrast, we follow a bioenergetics approach and formulate population models based on meticulous representations of the energy budget of individual consumers to account explicitly and consistently for all input and output of biomass and its use and turnover in the consumer-resource systems we study. We investigate what the necessary conditions are for the biomass in particular size ranges of a population to increase in response to an increase in mortality and how the overcompensation comes about through a change in the population energetics. Ultimately, this overcompensation in stage-specific biomass solely results from the interplay between mortality and intraspecific, exploitative competition for resources among consumers. As a direct effect, mortality decreases overall density, but indirectly it benefits the survivors through the relaxation of competition for resources. This change in resource use by the consumer population may lead to biomass overcompensation, even in the absence of any increase in resource productivity.

A STAGE-STRUCTURED, BIOENERGETICS MODEL

Yodzis and Innes (1992) formulated a bioenergetics model for the change in biomass of a consumer population feeding on a shared resource. Their model is based on the assumption that both resource intake and energetic costs to cover maintenance requirements are directly proportional to the body size of individual consumers. Note that this assumption pertains to individuals *within* the same species. Readers of Yodzis and Innes's original paper quite often overlook this assumption, as it also discusses the scaling of ingestion and maintenance costs per unit biomass *across* different species. With respect to the latter scaling, Yodzis and Innes assume that for a consumer species with an average adult body size W_A the rates of energy assimilation and maintenance per unit of biomass are proportional to $W_A^{-0.25}$, as in the more recently developed metabolic theory of ecology (Brown, Gillooly, et al. 2004). Yodzis and Innes do acknowledge that intake and maintenance tend to scale with a power of body weight less than 1 within individual species as well. However, they argue that "the mass-specific rates for the population as a whole should not differ greatly from the specific rates for a typical individual" and hence state, "[t]herefore, we are going to make the following approximation: as an estimate of the rates T and J per unit biomass in the population, we will use the corresponding rates per unit biomass of a typical adult individual" (Yodzis and Innes 1992, p. 1154).

We use the Yodzis and Innes model as a baseline but reformulate it using our own notation. As in the original model, we assume that the energetic costs to cover maintenance requirements equal an amount T_C per unit biomass (the subscript "C" in this and all following equations is used to label all functions and parameters that relate to consumers to distinguish them from analogous, predator-related functions and parameters that occur in following chapters). The maximum rate per unit biomass, with which food is ingested, we refer to as M_C, as opposed to the parameter J_{max} used by Yodzis and Innes. The rate at which consumers ingest resource furthermore follows a type II functional response as a function of resource density R. The mass-specific ingestion rate of resource is hence given by:

$$\omega_C(R) = M_C \frac{R}{H_C + R} \tag{3.1}$$

In this equation the parameter H_C is the resource density at which the ingestion rate reaches half of its maximum value. Like Yodzis and Innes, we will assume that there is some energetic loss when ingested food is converted into new biomass, and we hence introduce a conversion efficiency σ_C. The balance between assimilation through resource feeding and maintenance costs determines the net production of biomass, which per unit of biomass equals:

$$\nu_C(R) = \sigma_C \omega_C(R) - T_C = \sigma_C M_C \frac{R}{H_C + R} - T_C. \tag{3.2}$$

Yodzis and Innes neglect mortality of individual consumers, arguing that their model does not account for other populations that might be responsible for deaths through predation or disease. We will, however, explicitly account for consumer mortality and assume that consumers experience a background mortality rate that we refer to as μ_C.

To complete the model we only have to specify what the dynamics of the resource R are in the absence of consumers. We will represent these autonomous dynamics of resource by so-called semichemostat dynamics:

$$G(R) = \rho(R_{max} - R).$$

Here, the parameter R_{max} is the density that the resource attains in the absence of consumers, while the parameter ρ reflects the turnover rate of the resource. We prefer to use semichemostat dynamics over the more familiar logistic resource growth, because the latter models an exponentially expanding resource mass at low resource density and hence represents a spontaneous source of biomass to the system that is not based on any explicit energy or mass input. The logistic growth formula therefore does not conform to basic principles of energy or mass conservation. In contrast, semichemostat resource dynamics does conform to such principles, as it predicts that resource productivity is constant and independent of resource density, equaling ρR_{max}. As a consequence, possible increases in consumer biomass will result not from a change in resource productivity but from a change in population energetics.

The Yodzis and Innes model (1992) for the dynamics of consumer biomass C and resource biomass R can now be expressed in terms of the following ODEs:

$$\frac{dR}{dt} = G(R) - \omega_C(R)C \tag{3.3a}$$

$$\frac{dC}{dt} = \nu_C(R)C - \mu_C C \tag{3.3b}$$

This model looks like an unstructured model, which does not seem to take into account that individuals within the consumer species may differ in body size. Structurally, it is also very similar to the predator-prey models first formulated by Volterra (1926) and later extended with logistic prey growth and a predator functional response by Rosenzweig and MacArthur (1963; Rosenzweig 1971), as well as to the consumer-resource models formulated and analyzed by Tilman (1982). There is a difference, though, in that the model above is an *exact* representation of the biomass dynamics of a size-structured population model, which accounts for a full and continuous distribution of consumers over an arbitrary range of body sizes, if the following assumptions about consumer life history hold: (1) consumer ingestion and net production of new consumer biomass scale linearly with consumer body size, (2) the production of new biomass through somatic growth and reproduction is equally efficient, and (3) consumer mortality is independent of consumer body size.

In other words, even though the model is only formulated in terms of two ODEs, its dynamics are completely identical to the dynamics of a size-structured, individual-based model (an *i*-state distribution model, sensu Caswell and John 1992), which is formulated in terms of a partial differential equation (PDE) and explicitly accounts for consumer individuals of all different body sizes. This equivalence of the ODE model (3.3) and the corresponding size-structured model is discussed in more detail in section 2 of the technical appendices. It should be noted that the equivalence does not require any specific assumptions about when consumers start producing offspring; a consumer individual of a particular body size is just assumed to allocate an arbitrary, possibly size-dependent fraction of its net biomass production to somatic growth and the remainder to reproduction.

The Yodzis and Innes model (3.3) can hence be interpreted as an individual-based population model that accounts for a simple consumer life history, in which in particular juvenile and adult consumers differ neither in their mass-specific resource intake rate nor in the per-capita mortality they experience. (Note that consumers *do* differ in their per-capita resource intake, as larger individuals have larger body sizes and hence consume more). In the following chapters we will nonetheless refer to the Yodzis and Innes model as an *unstructured* model, even though it differs essentially from the truly unstructured models, such as those formulated by Lotka, Volterra, and Tilman, which assume individuals to be identical and represent populations by their number. Apart from the fact that introducing an additional, more appropriate type classification for the Yodzis and Innes model (for example, single-stage structured model) would unnecessarily complicate our presentation, we do this for two reasons. First, the Yodzis and Innes model has all the dynamical properties of unstructured models with respect to both population and community dynamics, given that it characterizes the population by a single state variable, total population biomass. Second, using the Yodzis and Innes model as a proxy for truly unstructured models that only account for population numbers yields a stringent link between unstructured models and size-structured models, where the Yodzis and Innes model can be considered a special case of both the former and the latter (for the reasons given in the previous paragraph).

What if juvenile and adult consumers do differ in their feeding or their mortality rate, while still competing with each other for the same shared resource? There are many different types of population models of varying complexity that we could formulate to account for and investigate the consequences of such differences between juvenile and adult consumers, but an appealing approach would be to introduce population stage structure into the Yodzis and Innes model. Assume therefore that we distinguish between juvenile biomass J and adult biomass A, which together sum up to the total consumer biomass C in the Yodzis and Innes model. As we want to account for juveniles and adults that differ in their ingestion of resources, we introduce the notation $\omega_J(R)$ and $\omega_A(R)$ to discriminate between mass-specific foraging rates of juveniles and adults, respectively. Consequently, net biomass production of juveniles and adults will also be stage-specific, which we will refer to as $\nu_J(R)$ and $\nu_A(R)$. Furthermore, differences in mortality will be represented by introducing separate parameters for the juvenile mortality rate μ_J and the adult mortality rate μ_A.

If we now make the rather natural assumption that juvenile individuals do not reproduce but use their net biomass production for growth and development, while adult individuals use their net biomass production solely for reproduction, we can intuitively write down the following set of ODEs for the biomass dynamics of resource, juvenile consumers, and adult consumers:

$$\frac{dR}{dt} = G(R) - \omega_J(R)J - \omega_A(R)A$$

$$\frac{dJ}{dt} = \nu_A(R)A + \nu_J(R)J - \Phi - \mu_J J$$

$$\frac{dA}{dt} = \Phi - \mu_A A$$

All of the terms occurring in these ODEs are rather straightforward to interpret: As opposed to a single term representing resource feeding by consumers, the ODE for the resource dynamics now includes separate terms $\omega_J(R)J$ and $\omega_A(R)A$ for the feeding by juveniles and adults, respectively. The net-biomass-production rates by adults and juveniles occur as the first two terms in the ODE describing the dynamics of juvenile consumer biomass, because reproduction by adults as well as somatic growth by juveniles both contribute to an increase in juvenile biomass and do not change adult biomass. Finally, the biomass of juveniles and adults decreases owing to stage-specific mortality, represented by the last term in each of the two ODEs describing consumer stage dynamics.

The new and unknown element in the model equations above is the term Φ, describing the biomass flow from the juvenile to the adult consumer stage as a result of the maturation of juveniles. In principle there are many different ways in which we could model the maturation term, but most of these will be merely phenomenological. In order to come up with a formulation that is tightly linked to the life history of the consumer individuals, we assumed that juveniles are born with a size at birth referred to as s_b and mature on reaching an adult size threshold s_m, in addition to the basic assumptions that consumer ingestion and net production of new consumer biomass scale linearly with consumer body size and that juveniles use all their net biomass production for somatic growth (see above). Notice that these assumptions imply that the model will account for a size distribution of juvenile individuals between their size at birth s_b and their size at maturation s_m, but that adult consumers will all be of the same size and not grow larger, as they are assumed to use all their net biomass production for reproduction only. Straightforward biomass conservation considerations now allow us to derive the following expression for the biomass maturation rate out of the juvenile stage (see box 3.1):

$$\Phi = \frac{\nu_J(R) - \mu_J}{1 - z^{1 - \mu_J/\nu_J(R)}} J$$

BOX 3.1

DERIVATION OF THE MATURATION FUNCTION
IN THE STAGE-STRUCTURED BIOMASS MODEL

In the following derivation we consider only equilibrium conditions. Hence, when referring to variables and/or functions, we in fact refer to their equilibrium values. To keep the notation as simple as possible, however, we do not introduce additional notation to label them as such.

A juvenile consumer is assumed to be born at size s_b. As it uses all its net biomass production for somatic growth, it will grow at a mass-specific rate equal to $v_J(R)$. This will translate into an exponential growth in body size after birth, and thus at age a it will have reached a body size $s(a) = s_b e^{v_J(R)a}$. As it matures on reaching a size threshold s_m, this expression for the size-age relation allows us to compute the duration of the juvenile period as $a_m = \ln(s_m/s_b)/v_J(R) = -\ln z/v_J(R)$, in which we have introduced the parameter z as the ratio between the size at birth and the size at maturation s_b/s_m. Because juvenile consumers experience a background mortality rate equal to μ_J, their probability to survive until maturation equals $e^{-\mu_J a_m} = z^{\mu_J/v_J(R)}$.

Assume now that in terms of biomass the reproduction rate of the total consumer population is denoted by Θ and the maturation rate by Φ. The maturation rate is related to the reproduction rate following $\Phi = \Theta(s_m/s_b)e^{-\mu_J a_m}$. In the right-hand side of this expression the second term reflects the fact that individuals increase in body size by a factor s_m/s_b during their juvenile period, while the last term represents their probability to survive the juvenile period. The relation can also be expressed as $\Theta = \Phi z^{1 - \mu_J/v_J(R)}$. In addition, simple conservation considerations tell us that in equilibrium the biomass outflow from the juvenile stage equals the inflow into that stage plus the difference between the biomass production through somatic growth and the loss through mortality. In other words, $\Phi = \Theta + v_J(R)J - \mu_J J$. These two relations involving Θ and Φ allow us to eliminate the unknown parameter Θ and derive the following expression for the population maturation rate:

$$\Phi = \frac{v_J(R) - \mu_J}{1 - z^{1 - \mu_J/v_J(R)}} J$$

In this expression z represents the ratio between the size at birth s_b and the size at maturation s_m. We will rewrite the above equation as $\Phi = \gamma(v_J(R), \mu_J)J$ by introducing the function $\gamma(v_J(R), \mu_J)$ to represent the mass-specific rate at which juveniles mature. It is defined as:

$$\gamma(\nu_J(R),\mu_J) = \frac{\nu_J(R) - \mu_J}{1 - z^{1 - \mu_J/\nu_J(R)}}. \tag{3.4}$$

The dependence of $\gamma(\nu_J(R), \mu_J)$ on the juvenile net-biomass-production rate $\nu_J(R)$, as well as the juvenile mortality rate μ_J makes intuitive sense, because a high juvenile net-biomass-production rate implies that juveniles grow fast, which in turn translates into a high recruitment rate to adult size-classes. Similarly, a high juvenile mortality rate implies that only few juveniles survive the juvenile period and hence decreases the maturation rate to adult size-classes. Using this expression for the maturation rate, we can now represent the stage-structured extension of the Yodzis and Innes model with the following system of ODEs:

$$\frac{dR}{dt} = G(R) - \omega_J(R)J - \omega_A(R)A \tag{3.5a}$$

$$\frac{dJ}{dt} = \nu_A(R)A + \nu_J(R)J - \gamma(\nu_J(R),\mu_J)J - \mu_J J \tag{3.5b}$$

$$\frac{dA}{dt} = \gamma(\nu_J(R),\mu_J)J - \mu_A A \tag{3.5c}$$

In the derivation of the expression for the mass-specific maturation rate $\gamma(\nu_J(R), \mu_J)$ (see box 3.1) we did, however, require the population to be in equilibrium, and hence the model formulation above does not automatically allow for nonequilibrium conditions. A closer consideration of the model ODEs (3.5) also shows that they do not make much sense under starvation conditions. For example, if resource ingestion by adults is insufficient to cover energy requirements for maintenance, adult net biomass production $\nu_A(R)$ is negative, but according to the equations above this loss rate would decrease juvenile and not adult biomass present! The model can be extended to cover starvation conditions by assuming that juvenile growth stops if juvenile net biomass production $\nu_J(R)$ turns negative and similarly that reproduction takes place only as long as adult net biomass production $\nu_A(R)$ is positive. Furthermore, juveniles and adults experience increased mortality in cases where their net biomass reproduction is negative. The equations specifying dynamics of this more general variant of the stage-structured biomass model under both equilibrium as well as nonequilibrium conditions are presented in box 3.2. Their mathematical derivation we present in section 3 of the technical appendices. The default values for the model parameters are listed in box 3.3. Box 3.4 provides an explanation how these default values were derived from published data on individual energetics.

The basic formulation of the stage-structured biomass model in principle allows us to choose any functional form for $\omega_J(R)$ and $\omega_A(R)$, the functions representing the mass-specific ingestion rate of juveniles and adults, respectively,

BOX 3.2

EQUATIONS AND FUNCTIONS OF THE STAGE-STRUCTURED
BIOMASS MODEL

Dynamic equations	Description
$\dfrac{dR}{dt} = G(R) - \omega_J(R)J - \omega_A(R)A$	Resource biomass dynamics
$\dfrac{dJ}{dt} = v_A^+(R)A - \gamma(v_J^+, \mu_J)J + v_J(R)J - \mu_J J$	Biomass dynamics of juveniles
$\dfrac{dA}{dt} = \gamma(v_J^+, \mu_J)J + (v_A(R) - v_A^+(R))A - \mu_A A$	Biomass dynamics of adults

Function	Expression	Description
$G(R)$	$\rho(R_{max} - R)$	Intrinsic resource turnover
$\omega_J(R)$	$M_C R/(H_C + R)$	Resource intake rate by juveniles
$\omega_A(R)$	$qM_C R/(H_C + R)$	Resource intake rate by adults
$v_J(R)$	$\sigma_C \omega_J(R) - T_C$	Net energy production of juveniles
$v_A(R)$	$\sigma_C \omega_A(R) - T_C$	Net energy production of adults
$\gamma(v_J^+, \mu_J)$	$(v_J^+(R) - \mu_J)/(1 - z^{(1 - \mu_J/v_J^+(R))})$	Maturation rate of juveniles

$v_J^+(R)$ and $v_A^+(R)$ represent the value of $v_J(R)$ and $v_A(R)$, respectively, but restricted to non-negative values. For brevity, the dependence of v_J^+ on resource density is suppressed in the maturation function $\gamma(v_J^+, \mu_J)$. Notice that $\gamma(v_J^+, \mu_J)$ is positive as long as $v_J^+(R)$ is positive. Also, $v_A^+(R) - v_A(R)$ cancels as long as $v_A(R)$ is positive and otherwise equals $-v_A(R)$. Hence, the use of $v_J^+(R)$ and $v_A^+(R)$ in the model equations implies that growth, maturation, and fecundity are positive as long as juvenile and adult net production, respectively, are positive and otherwise equal 0.

as long as they depend only on resource density and not on consumer body size. The same holds for the mass-specific rate of net biomass production by juveniles and adults, that is, the functions $v_J(R)$ and $v_A(R)$, respectively. However, to stay close to the original formulation of the Yodzis and Innes model, we will assume that juveniles and adults differ in their mass-specific ingestion rate by only a factor of q and hence choose $\omega_J(R) = M_C R/(H_C + R)$ and $\omega_A(R) = qM_C R/(H_C + R)$. Furthermore, we will adopt the same relation between ingestion and net biomass production as in the Yodzis and Innes model, that is, $v_J(R) = \sigma_C \omega_J(R) - T_C$ and $v_A(R) = \sigma_C \omega_A(R) - T_C$. These choices ensure that the stage-structured biomass model reduces to the Yodzis and Innes model for $q = 1$ and equal juvenile and adult mortality: $\mu_J = \mu_A = \mu_C$. For $q = 1$, $\omega_J(R)$ equals $\omega_A(R)$ and $v_J(R)$ equals $v_A(R)$, in which case we can replace these functions

BOX 3.3

PARAMETERS OF THE STAGE-STRUCTURED BIOMASS MODEL

Parameter	Default value	Unit	Description
Resource			
ρ	0.1	day^{-1}	Resource turnover rate
R_{max}	30	mg/L	Resource maximum biomass density
Consumer			
M_C	$0.1W_A^{-0.25}$	day^{-1}	Mass-specific maximum ingestion rate
H_C	3.0	mg/L	Ingestion half-saturation resource density
q	0.5 or 2.0	—	Adult-juvenile consumer ingestion ratio
T_C	$0.01W_A^{-0.25}$	day^{-1}	Mass-specific maintenance rate
σ_C	0.5	—	Conversion efficiency
z	0.01, 0.1, or 0.5	—	Newborn-adult consumer size ratio
μ_J	$0.0015W_A^{-0.25}$	day^{-1}	Juvenile background mortality rate
μ_A	$0.0015W_A^{-0.25}$	day^{-1}	Adult background mortality rate

W_A represents the average adult consumer body weight (in grams). Refer to box 3.4 for a discussion of the derivation of these default parameters from published data on individual energetics.

by stage-independent functions $\omega_C(R)$ and $\nu_C(R)$ for resource ingestion and net biomass production, respectively. Defining total consumer biomass as the sum of juvenile and adult biomass, $C = J + A$, and adding equations (3.5b) and (3.5c) of the stage-structured model to obtain an equation for dC/dt then brings us back to the Yodzis and Innes model (3.3).

Summarizing, like the Yodzis and Innes model, the stage-structured bio-mass model captures the dynamics of an individual-based population model, which explicitly accounts for consumers of different body sizes and is hence formulated in terms of a partial differential equation (PDE) for the (continuous) size-distribution of consumers. Under equilibrium conditions this representation is exact, and all equilibria predicted by the stage-structured biomass model correspond one-to-one to equilibria of the fully size-structured model. Under nonequilibrium conditions, the stage-structured and the corresponding PDE model make only slightly different predictions when adults have a higher mass-specific ingestion rate than juveniles ($q > 1$) (de Roos, Schellekens, et al. 2008b). When juveniles have a higher ingestion rate ($q < 1$), however, the PDE model results in so-called single-generation oscillations (Gurney and

BOX 3.4

Derivation of Parameters in the Stage-Structured
Biomass Model

CONSUMER PARAMETERS

Across differently sized consumer species the mass-specific metabolic rate, mass-specific maximum ingestion rate, and mortality rate all tend to be inversely proportional to the quarter power of adult body size of the species (Brown, Gillooly, et al. 2004; Peters 1983; Yodzis and Innes 1992). Hence, all tend to scale as $cW_A^{-0.25}$ with different scaling constants c. W_A in this scaling relation represents the average adult body weight of a particular consumer species. The scaling relations usually differ between invertebrates, ectothermic vertebrates, and endotherms. To derive a default set of parameters, we restrict our parameter estimates to invertebrates. We adopt the general convention here to express all rates per day and characteristic body weights of individual species in grams. Biomass densities of resources, consumers, and predators would then most naturally be expressed in units of grams per liter, but this choice results in many decimals in numerical expressions of these densities. To increase readability, we therefore choose to express biomass densities throughout the book in milligrams per liter instead.

The best-studied relationship is that between mass-specific metabolic rate and adult body weight. Yodzis and Innes (1992, table 1) provide an estimate of $c = 0.5$ when expressing metabolic rate in $kg \cdot kg^{-1} \cdot year^{-1}$ and weight in kg. Similarly, Peters (1983, pp. 43–44) provides an estimate of 0.63 for planktonic crustaceans. Brown, Gillooly, et al. (2004, fig. 1) show regressions between the temperature-corrected, standard metabolic rate, measured in watts, and body weight, measured in grams. For an imaginary invertebrate species with a 1 kg adult body weight at a temperature of $T = 291\,K\,(1/(kT) \approx 40)$, these relations predict a metabolic rate between 0.044 and 0.069 watts (from figs. 1b and 1a, respectively, in Brown, Gillooly, et al. 2004). Standard metabolic rates can be converted to the field metabolic rates used in Yodzis and Innes (1992) by multiplying with a factor of 2.5 (Brown, Gillooly, et al. 2004; Yodzis and Innes 1992). Yodzis and Innes furthermore assume, following Peters (1983), that 1 kg wet mass equals $7 \cdot 10^6$ joules, leading to an equivalence of 1 watt = 1 J/s = 4.5 kg wet mass/year. Using these conversions, the relations presented by Brown, Gillooly, et al. (2004) yield an estimate for c between 0.5 and 0.78. Taken together, these studies all predict a scaling constant for the relationship between mass-specific metabolic rate and adult body weight ranging between 0.5 and 0.8. We adopt 0.65 as a representative value, corresponding to 0.01 when mass-specific metabolic rate is expressed per day and adult body weight in grams.

(Box 3.4 continued)

Estimates for the scaling of mass-specific maximum ingestion rate and adult body weight are much more variable. On the basis of only eight data points for ectothermic invertebrates, Yodzis and Innes (1992) argue that the maximum rate of energy ingestion is roughly eleven to twenty-two times larger than their mass-specific metabolic rate (depending on whether the species is a herbivore losing 55 percent of ingested food to feces or a carnivore losing only 15 percent to feces). Hansen, Bjørnsen, and Hansen (1997) present an extensive data set of maximum ingestion rates and invertebrate predator body volumes for a diverse range of flagellate, dinoflagellate, ciliate, rotifer, copepod, cladoceran, and larvae of meroplankton species. They find support for a common scaling exponent across all these groups equal to -0.23 for the relation between specific ingestion rate per hour and body volume in μm^3 in these species. The scaling constant, however, varies among the different species groups, from 0.53 for meroplankton larvae to 6.06 for copepods (Hansen, Bjørnsen, and Hansen 1997, table 9), with a median value around 2.2. Hansen, Bjørnsen, and Hansen (1997, p. 689) assume that $1 \mu m^3$ predator body volume corresponds to a dry weight of $0.28 \cdot 10^{-12}$ gram, which equals $1.4 \cdot 10^{-12}$ gram wet weight, assuming a 20 percent dry-wet weight ratio. For a species with an adult body size of 1 gram wet weight, the scaling relations and conversion factors hence yield estimates of the maximum specific ingestion rate that range between 0.024 and 0.27 per day, with a median value of 0.1. We adopt this median value for our default parameter estimate, leading to a scaling constant equal to 0.1 for the daily, mass-specific maximum ingestion rate as a function of $W_A^{-0.25}$ (W_A in grams).

Gilooly, Brown, et al. (2001, supporting information) provide data on the observed lifespan for a range of invertebrate species with different body weights. We used the regression relationship for the dependence of lifespan on temperature presented by Gilooly, Brown, et al. (2001, fig. 2) to rescale all lifespan estimates to a reference temperature of $T = 291\,K$. Subsequently, a linear regression of the log-transformed lifespan estimates against the log-transformed body weights showed that daily mortality rate (computed as the inverse of the lifespan) scaled as $0.0015W_A^{-0.25}$. The relation between natural mortality and body weight was recently analyzed in more detail by McCoy and Gilooly (2008). These authors also find substantial support for a scaling of the rate of mortality with $W_A^{-0.25}$. However, the regression relationship for invertebrates provided in figure 2a of McCoy and Gilooly (2008) yields an estimate of the mortality rate per year equal to 3.0, or 0.008 per day, for a species with an adult body size of 1 gram wet weight (corresponding to a dry weight of 0.25 gram; McCoy and Gilooly 2008, supporting information) at a reference temperature of $T = 291\,K$. This more than fivefold higher mortality rate is especially remarkable when it is

(Box 3.4 continued)

compared with the earlier study by Gillooly, Brown, et al. (2001), because the data points in the latter study make up almost 30 percent of all data points used by McCoy and Gillooly (2008). In fact, a comparison of the supporting information of both publications, which presents all the data points used in the analysis, reveals that the mortality estimates for the various invertebrate species presented by Gillooly, Brown, et al. (2001) are also listed by McCoy and Gillooly (2008), but then always exactly multiplied by a factor of six. Because visual inspection of the lifespan data presented by Gillooly, Brown, et al. (2001) suggest that the presented values are much more in line with data presented in other publications, we disregard the estimate presented by McCoy and Gillooly (2008) and assume that the scaling constant relating daily mortality rate to $W_A^{-0.25}$ equals 0.0015.

In their analysis of zooplankton grazing rates, Hansen, Bjørnsen, and Hansen (1997) showed that the half-saturation resource density for ingestion, H_C, was independent of species body weight and equal to $240\,\mu g\,C/L$ when averaged over all groups together. Theoretically, the half-saturation density can be expected to be independent of species body weight, as it equals the ratio of two rates, the maximum ingestion rate and the clearance or attack rate, that both can be argued to scale with $W_A^{-0.25}$ (Brown, Gillooly, et al. 2004; Hansen, Bjørnsen, and Hansen 1997). Based on a 45 percent carbon-dry weight ratio (Hansen, Bjørnsen, and Hansen 1997) and assuming a 20 percent dry-wet weight ratio, the half-saturation constant can be estimated to equal 0.003 gram wet weight per liter, i.e., 3 milligrams per liter.

Regarding the assimilation efficiency of ingested food Yodzis and Innes (1992) claimed that 55 percent and 15 percent of the ingestion by herbivorous and carnivorous species, respectively, is lost to feces and urine. However, these percentages do not qualify as estimates for the conversion efficiency of ingested food to assimilate, as they do not account for energy losses attributable to specific dynamic action, which are higher in carnivores then in herbivores (Peters 1983). Peters (1983, pp. 105–106) therefore argues that the conversion of ingested food by both herbivores and carnivores proceeds with roughly the same efficiency of 50 percent, which we adopt as our default value for this parameter.

Hendriks and Mulder (2008) forwarded a theoretical prediction that offspring size should scale with the square root of the adult body size. An extensive meta-analysis showed that for invertebrates this prediction was supported by available data. Hendriks and Mulder (2008, fig. 3 and table 3) present regression relationships between offspring and adult body size for a variety of invertebrate groups that mostly cover a range of adult body sizes from $1 \cdot 10^{-6}$ to 1 gram. At the median weight of $1 \cdot 10^{-3}$ gram, these regression relationships yield estimates of the offspring body size ranging from $1.9 \cdot 10^{-6}$ to $6.1 \cdot 10^{-4}$ gram, with a midpoint

(Box 3.4 continued)

around $3.4 \cdot 10^{-5}$. Taking the latter as a representative value and adopting the scaling of offspring size with the square root of the adult body size, we derive a relationship $z = 0.001 W_A^{-0.5}$ for the offspring-adult body size ratio as a function of adult body weight in gram wet weight. However, the relation between offspring size and adult body size is very variable owing to the many different strategies and modes of reproduction that different species employ. Over the range of adult body sizes from $1 \cdot 10^{-6}$ to 1 gram, the regression relations presented by Hendriks and Mulder (2008) lead to estimates for the ratio between offspring and adult body size that range from $1 \cdot 10^{-5}$ to 0.6, reflecting the fact that some species produce many tiny eggs, whereas others invest more per offspring produced. Because the offspring-adult body size ratio is also indicative for the importance of the developmental phase, and hence ontogeny, in individual life history relative to the reproductive part of it, we will not adopt a single value for z but will investigate in detail how different values for this parameter influence model predictions to assess how differences in the relative importance of development over reproduction affect community equilibria.

Finally, we will investigate model dynamics while varying the value of q representing the ratio between the adult and juvenile maximum ingestion rate. An estimate for this parameter can hardly be derived from experimental data, as it is only a phenomenological representation of stage-specific differences in resource availability and resource use between juveniles and adults.

RESOURCE PARAMETERS AND SCALING CONSIDERATIONS

Given that our main interest is in the dynamics of the consumer population, we will not devote much attention to the parameterization of the resource dynamics. It should be noted, however, that the half-saturation resource density H_C that occurs in the consumer functional response carries a dimension of resource biomass per unit volume and is the only volume-related consumer parameter. Furthermore, the half-saturation density has the same dimension as the parameter $R_{\max}$ that represents the maximum resource density in the absence of consumers. As a consequence, as long as we keep the ratio of these two parameters constant, their actual values are only relevant if we are interested in the absolute densities of resource and consumers. More precisely, if we were to express both H_C and $R_{\max}$ taking 1/3 liter as our unit of volume, the half-saturation constant would equal 1 milligram per unit volume and $R_{\max}$ would be three times smaller, but all model predictions would remain qualitatively identical. Quantitatively, however, all resource and consumer densities would be scaled down by a factor

---

(Box 3.4 continued)

of three. We will hence directly relate our default value for R_{max} to the default value of H_C and assume that R_{max} is an order of magnitude larger. Furthermore, for the resource turnover rate we will assume a default value of $\rho = 0.1$ per day, which equals the metabolic rate of our default consumer species with an adult body size of 10^{-4} gram. The influence of these choices for the resource parameters will be discussed and investigated in several places throughout the book, but overall their choice guarantees that maximum resource densities are large enough to allow consumer persistence and to rule out the dynamical influence of the resource turnover rate on model predictions.

Nisbet 1985), whereas the stage-structured model shows equilibrium dynamics (see also chapter 9).

The stage-structured model hence mechanistically accounts for a simplified consumer life history in which (1) consumer ingestion and net production of new consumer biomass scale linearly with consumer body size; (2) juveniles are born at size s_b and mature on reaching the maturation size s_m; (3) juveniles and adults invest all their net biomass production into somatic growth and reproduction, respectively; and (4) juvenile and adult consumers possibly differ in their mass-specific ingestion and net-biomass-production rate as well as in their size-independent mortality rate. The model consistently translates these individual-level assumptions on size-dependent ingestion, growth, reproduction, and maintenance to the population level. It differs from the Yodzis and Innes model in that it allows for differences in feeding and mortality between juveniles and adults.

The stage-structured model shares some common features—in particular size-dependent foraging and metabolism, food-dependent individual growth, and energetic differences between juveniles and adults—with alternative models that describe individual acquisition and use of food for growth and reproduction (Gurney, McCauley, et al. 1990; Kooijman 2000; Lika and Nisbet 2000; Persson, Leonardsson, et al. 1998). We hence consider it the simplest, representative model for a consumer-resource interaction that is individual based and accounts for food-dependent reproduction, mortality, *and* development, setting it apart from the classical, unstructured models in terms of population numbers as well as age-structured models. Lastly, even though we will mostly use it only to distinguish between juvenile and adult consumers, the stage-structured model framework is sufficiently flexible to account for more life history stages as well (de Roos, Schellekens, et al. 2008b; see, for an example, van Leeuwen, de Roos, and Persson 2008).

EQUAL INGESTION RATES

Population dynamics is first and foremost concerned with changes in numbers of individuals owing to death and reproduction. Density dependence in either of these two processes can keep populations from growing indefinitely and hence can regulate population densities to remain at or close to some constant equilibrium level. Most standard models in ecology—for example, the consumer-resource models analyzed by Tilman (1982), the predator-prey model first described by Volterra (1926), and its extensions by Rosenzweig and MacArthur (1963; Rosenzweig 1971)—assume mortality of consumers to be density independent and consumer reproduction to depend on available resources. As a consequence, density dependence among consumers arises indirectly, when increases in consumer density translate into a decrease in resource density and hence to reductions in fecundity. These features of classical consumer-resource or predator-prey models lead to two general predictions that can be illustrated by considering the equilibrium conditions of the Yodzis and Innes model (3.3). First, by setting the right-hand side of equation (3.3b) equal to 0, we obtain the condition for the resource density $\tilde{R}$ in an equilibrium, in which consumers are present:

$$v_C(\tilde{R}) = \mu_C \quad \Leftrightarrow \quad \sigma_C\omega_C(\tilde{R}) - T_C = \mu_C \qquad (3.6)$$

The equilibrium resource density is therefore only and completely determined by life history parameters of the consumer and not by any of the parameters that influence resource growth. In particular, increases in consumer mortality μ_C lead to an increase in equilibrium resource density and consequently in net biomass production to compensate for the increased mortality losses. Second, setting both the right-hand side of equation (3.3a) and (3.3b) equal to 0 and adding up the resulting expressions lead—after some manipulation and after substitution of the relation $v_C(R) = \sigma_C\omega_C(R) - T_C$—to an expression for the consumer density in a resource-consumer equilibrium:

$$\tilde{C} = \frac{\sigma_C G(\tilde{R})}{\mu_C + T_C} \qquad (3.7)$$

This condition makes clear that changes in resource productivity will give rise to changes in consumer equilibrium biomass. Furthermore, changes in consumer mortality μ_C have a direct effect, leading to a decrease in consumer equilibrium biomass, but they also indirectly affect equilibrium biomass of consumers through the change in resource growth rate $G(\tilde{R})$ that results from the increase in equilibrium resource biomass $\tilde{R}$. This indirect effect could in principle tend to increase consumer biomass—for example, if the dynamics of the resource in absence of consumers follows a logistic growth equation (Abrams and Matsuda 2005; Rosenzweig and MacArthur 1963); but as we have assumed semichemostat dynamics for the resource, this indirect effect of changes in consumer mortality also leads to decreases in consumer equilibrium biomass.

Increases in consumer mortality are hence expected to decrease (total) consumer biomass when all consumers compete for a single resource (below we will in fact see that total population biomass may increase with mortality when individuals exploit different resources in different stages of their life history). We will refer to this response as the "direct effect" of (increased) mortality. The increase in mortality also increases the resource density and hence indirectly relaxes the competition among consumers for food. This latter effect we will refer to as the "indirect, positive effect" of (increased) mortality. Equation (3.6) makes clear that the relaxation of competition among consumers exactly compensates for the increase in mortality as long as all consumers are equally affected by the increased mortality and consumers do not differ in resource ingestion rate. But how does the increase in mortality change the distribution of biomass of stages? We can analyze this question by studying the equilibrium of the stage-structured biomass model (3.5), assuming that juveniles and adults have equal ingestion rates ($q = 1$) and equal juvenile and adult consumer mortality, $\mu_J = \mu_A = \mu_C$. In this case both ingestion and net-biomass-production rates are the same for juvenile and adult consumers, that is, $\omega_J(R) = \omega_A(R)$ and $\nu_J(R) = \nu_A(R)$, and the system of equations (3.5) is identical to the Yodzis and Innes model (3.3), as shown above.

Setting $dJ/dt = 0$ and $dA/dt = 0$ allows us to equate the right-hand sides of equations (3.5b) and (3.5c) to each other, resulting, after some rearranging of terms, in:

$$(\nu_J(\tilde{R}) - \mu_J)\tilde{J} = -(\nu_A(\tilde{R}) - \mu_A)\tilde{A} \tag{3.8}$$

This condition relates the net turnover in juvenile biomass at equilibrium—that is, the balance between biomass production and biomass loss through mortality—to the net turnover in adult biomass. With $\nu_J(R)$ equal to $\nu_A(R)$ and $\mu_J = \mu_A = \mu_C$, this condition can only be satisfied for non-zero values of $\tilde{J}$ and $\tilde{A}$ if biomass production exactly cancels the losses through mortality in both the juvenile and adult stage—in other words, if $\nu_C(\tilde{R}) = \mu_C$. This derivation reveals an important characteristic of the Yodzis and Innes model, which we will later show to hold more generally: in the absence of any differences between juvenile and adult consumers in terms of ingestion and mortality, the net turnover of biomass in both the juvenile and adult stage exactly vanishes in equilibrium. In other words, neither the juvenile stage nor the adult stage is a net source or a net sink of biomass production.

Because the net-turnover rates in the two stages, $(\nu_J(\tilde{R}) - \mu_J)\tilde{J}$ and $(\nu_A(\tilde{R}) - \mu_A)\tilde{A}$, must have opposite signs if non-zero, equation (3.8) makes clear that any difference between juveniles and adults in ingestion or mortality will break up this neutral or symmetric situation and turn one of the two stages into a net biomass source and the other into a net biomass sink. We will therefore also refer to condition (3.8) as the "source-sink" condition and use the term "source-sink status" to indicate whether a particular stage is a net producer of biomass in equilibrium, that is, whether or not the production rate of new biomass within the stage exceeds the loss rate through mortality. When considered over the entire consumer life history, though, the sum over all biomass-production rates in equilibrium exactly equals the sum over all biomass-loss rates through mortality.

From the equilibrium condition for juvenile consumer biomass $dJ/dt = 0$ we can derive that the source-sink status of the juvenile consumer stage is directly tied to the balance between total population reproduction and maturation rate. Thus, $dJ/dt = 0$ leads to

$$\nu_A(\tilde{R})\tilde{A} - \gamma(\nu_J(\tilde{R}),\mu_J)\tilde{J} = -(\nu_J(\tilde{R}) - \mu_J)\tilde{J}, \qquad (3.9)$$

which implies that the difference between the total population reproduction and maturation rate has the sign opposite to that of the net-turnover rate in the juvenile consumer stage. In other words, if the juvenile stage is a source of biomass production, total maturation rate is necessarily larger than reproduction rate, while the reverse holds in cases where the juvenile stage is a net sink of biomass. If $\nu_J(R)$ equals $\nu_A(R)$ and $\mu_J = \mu_A = \mu_C$, however, these two population rates are always necessarily equal to each other.

To derive an expression for the ratio between juvenile and adult consumer biomass at equilibrium we can equate dA/dt to 0 and calculate:

$$\frac{\tilde{J}}{\tilde{A}} = \frac{\mu_A}{\gamma(\nu_J(\tilde{R}),\mu_J)} \qquad (3.10)$$

If $\nu_J(R) = \nu_A(R) = \nu_C(R)$ and $\mu_J = \mu_A = \mu_C$, $\nu_C(\tilde{R})$ becomes equal to μ_C in equilibrium. The right-hand side of this equation then simplifies to $-\ln(z)$, as the maturation function $\gamma(\nu_C,\mu_C)$ can be shown to have a regular limit equal to $-\nu_C(R)/\ln(z) = -\mu_C/\ln(z)$ for $\nu_C(R) \to \mu_C$. With equal ingestion and mortality of juvenile and adult consumers, the distribution of biomass over these two stages in equilibrium is hence always constant and fully determined by the ratio between consumer size at birth and maturation z.

Therefore, in the Yodzis and Innes model, which results when assuming $q = 1$ and $\mu_J = \mu_A = \mu_C$ in the stage-structured biomass model, biomass gains equal losses through mortality in both the juvenile stage and the adult stage,

total population reproduction in terms of biomass always equals total population maturation rate, and the ratio between juvenile and adult biomass in
equilibrium is constant. Because these assertions are independent of consumer
mortality rate, we can moreover conclude that increases in the consumer mortality rate μ_C will lead to identical changes in population reproduction and
maturation and will not affect the juvenile-adult biomass ratio. Given that according to equation (3.7) total consumer biomass decreases, so does both juvenile and adult biomass with increases in mortality. Without differences in
net biomass production or mortality between juvenile and adult consumers, a
symmetrical situation arises, in which changes in consumer mortality lead to
identical changes in juvenile and adult biomass (figure 3.1, left panel), as well
as in total maturation and reproduction rate.

The middle and right panels of figure 3.1 show that qualitatively different
responses in stage-specific biomass occur if only mortality for one of the two
stages is increased (while still assuming equal juvenile and adult ingestion
rates): increasing juvenile mortality leads to an increase in adult biomass at low
mortality levels, whereas increasing adult mortality leads to an increase in juvenile biomass at low mortality levels. We have verified that the graphs as shown
in figure 3.1 for $z = 0.1$ do not qualitatively change for larger values of z up to
0.5. Hence, the stage targeted by mortality decreases owing to the direct effect
of mortality, whereas the biomass in the nontarget stage increases as a result of
the indirect, positive effect of mortality. At this point we will ignore the fact that

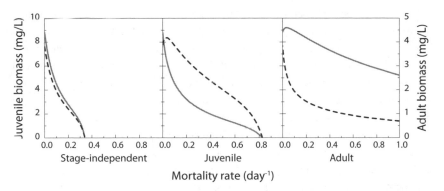

FIGURE 3.1. Changes in juvenile (*solid lines*) and adult (*dashed lines*) biomass
with increases in stage-independent, juvenile, and adult mortality on top of the
background mortality for $q = 1$ and $z = 0.1$. Otherwise, default parameters (see
box 3.3) apply for a consumer with body size $W_A = 0.0001$ gram. Notice that the
maximum mortality rates shown correspond to a loss rate that is ten times the
maintenance rate and sixty-five times the background mortality rate.

these increases in biomass are rather small in absolute value; in following sections we will see that the biomass overcompensation response can be substantial as soon as there are differences in ingestion between juveniles and adults. Our main focus is hence on the qualitative feature that stage-specific biomass can increase. From equation (3.10) we can infer that the adult biomass in equilibrium equals the ratio of the maturation rate and the adult mortality rate:

$$\tilde{A} = \frac{\gamma(\nu_J(\tilde{R}), \mu_J)\tilde{J}}{\mu_A} \tag{3.11}$$

This shows that the increase in adult biomass with increasing juvenile mortality is due to an increase in population maturation rate. The decrease in equilibrium juvenile biomass hence relaxes resource competition to such an extent that it translates into an increase in per-capita juvenile maturation rate, which more than compensates for the lower juvenile biomass density. Similarly, from equation (3.9) we can derive that the equilibrium juvenile biomass follows:

$$\tilde{J} = \frac{\nu_A(\tilde{R})\tilde{A}}{\gamma(\nu_J(\tilde{R}), \mu_J) + \mu_J - \nu_J(\tilde{R})} \tag{3.12}$$

In expression (3.11) for the equilibrium adult biomass the denominator represents the rate at which biomass leaves the adult stage through mortality (the inverse of this denominator hence represents the mean residence time of biomass in the adult stage). In contrast, the denominator in expression (3.12) is a more complicated sum of the rates at which biomass leaves the juvenile stage through maturation ($\gamma(\nu_J(\tilde{R}), \mu_J)$) and mortality ($\mu_J$) and the rate at which biomass in the juvenile stage produces new juvenile biomass ($\nu_J(\tilde{R})$). The mean residence time of biomass in the juvenile stage hence depends on maturation, mortality, and biomass production. With increasing adult mortality and the consequent higher equilibrium resource densities, the adult biomass production rate $\nu_A(\tilde{R})$ as well as the juvenile maturation and biomass production rate, ($\gamma(\nu_J(\tilde{R}), \mu_J)$) and $\nu_J(\tilde{R})$, respectively, increase, which makes it hard to assess from what the increase in juvenile biomass as shown in the right panel of figure 3.1 results.

Using Maple (Waterloo Maple Inc., Waterloo, Ontario) to calculate the derivative of the denominator in the right-hand side of equation (3.12) with respect to $\nu_J(\tilde{R})$, it can be shown that the sign of the denominator is completely determined by the ratio of the juvenile biomass production and the juvenile mortality rate, $\nu_J(\tilde{R})/\mu_J$, and the ratio z between the body size at birth and at maturation. As can be inferred from figure 3.2, the denominator is likely to increase with increasing μ_A for high values of either $\nu_J(\tilde{R})/\mu_J$ as well as z and will decrease otherwise. The bottom line of this analysis is that the increase in equilibrium juvenile biomass with increasing adult mortality cannot directly be tied to an increase in a single

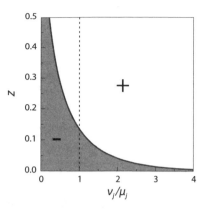

FIGURE 3.2. Combinations of values for v_j/μ_j and z for which the expression $\gamma(v_j,\mu_j) - v_j + \mu_j$ increases (*white area above solid line*) or decreases (*gray area below solid line*) with increasing values of v_j. Notice that to the right of the vertical dashed line, the juvenile stage is a source of biomass production, whereas to the left of it the juvenile stage is a sink of biomass production.

life history process. Instead, depending on parameter values, it results from an increase in population reproduction rate $v_A(\tilde{R})\tilde{A}$ and/or a faster increase in juvenile biomass production rate $v_J(\tilde{R})$ than in the rate at which biomass leaves the juvenile stage through maturation ($\gamma(v_J(\tilde{R}),\mu_J)$) and mortality ($\mu_J$).

Figure 3.1 represents a type of graph known as bifurcation graph (Kuznetsov 1995). In their simplest form bifurcation graphs show equilibrium values of a model, or more generally a dynamical system, as a function of a model parameter. This parameter is referred to as the bifurcation parameter. More complicated versions of bifurcation graphs may also show the occurrence of limit cycles or even complex, chaotic dynamics by plotting all extreme values (i.e., all local maxima and minima) that are observed during the limit cycle or the more complex dynamics for a particular model variable. Figure 4.1 in the next chapter provides an example of such a more complicated bifurcation graph. Bifurcation graphs will figure extensively throughout the entire book and we hence provide in box 3.5 an explanation of the different methods that can be used to construct such graphs.

UNEQUAL INGESTION RATES

As shown by the source-sink condition (3.8) and equation (3.9), in cases of equal juvenile and adult mortality any difference in resource ingestion rate

BOX 3.5

METHODS FOR CONSTRUCTING BIFURCATION GRAPHS

The most easy-to-imagine way to create a bifurcation graph is to carry out a large number of numerical integrations of the model equations, each with a slightly different value for the bifurcation parameter. This approach, however, has its problems, as the computed outcome will depend on the choice of the initial values for the model variables. One way to circumvent the problems associated with the initial value of model variables is to carry out only a single numerical integration of the model equations but over a very long time period, in which the entire integration period is subdivided into intervals during which the value of the bifurcation parameter is constant, while from one interval to the next the value of the bifurcation parameter is increased or decreased by a small amount. In this way, the range of values of the bifurcation parameter is scanned either from low to high or vice versa. This stepwise increase or decrease of the bifurcation parameter implies that the final values of the model variables obtained for a particular value of the bifurcation parameter are used as initial values of the model variables for the subsequent parameter value.

The advantage of this approach can best be explained in the context of stable model equilibria. Consider that after a numerical integration over a sufficiently long time interval for a particular value of the bifurcation parameter all model variables have ended up close their equilibrium value. These final values of the model variables will also be close to their equilibrium value for a setting of the bifurcation parameter that is slightly larger or smaller, provided that the particular equilibrium still exists for this new parameter value. Hence, adopting these final values of the model variables as initial values for a subsequent integration with a slightly different bifurcation parameter value ensures that we continue to follow the curve of a particular model equilibrium as a function of the bifurcation parameter as long as the equilibrium exists and is stable. Only when the equilibrium becomes unstable or does not occur at all any more for the new value of the bifurcation parameter, will the model variables approach an entirely different equilibrium or a different type of dynamics, such as a limit cycle. By scanning a particular interval of the bifurcation parameter with increasing as well as decreasing parameter values in most cases also reveals the co-occurrence of alternative stable equilibria or alternative types of stable dynamics, such as the co-occurrence of different types of limit cycles, for the same value of the bifurcation parameter. Chapter 9 and those that follow it provide various bifurcation graphs that were constructed in this particular manner by long numerical integrations of size-structured models.

A more sophisticated approach to construct a bifurcation graph for a particular model is to use numerical techniques that allow for tracing the curve of

(Box 3.5 continued)

a particular model equilibrium as a function of the bifurcation parameter, even when this equilibrium is unstable. The advantage of these so-called continuation methods over the integration approach described above is that both stable and unstable equilibria can be computed. For models formulated in terms of ODEs, a number of different software packages are available that implement these continuation methods for the analysis of dynamical systems. We have used MAT-CONT, which is a graphical MATLAB software package (Dhooge, Govaerts, and Yu 2003). This package not only allows one to compute curves of equilibria and limit cycles as a function of a single free model parameter but also determines whether a particular equilibrium or limit cycle is stable or not. In addition, the software can also be used to construct boundaries in terms of two model parameters that separate parameter regions with qualitatively different types of model dynamics. In ecological terms, these boundaries distinguish between parameter values that lead to either persistence or extinction of a species or that lead to stable equilibrium or oscillatory population dynamics. Unless stated otherwise, we have carried out all computations for models in terms of ODEs, such as the stage-structured biomass model, with MATCONT. Accordingly, all our statements and conclusions about the stability of equilibria and limit cycles in such models are based on the information provided by this software package.

In the last part of this chapter and in those that follow we will also present bifurcation graphs for models that describe the dynamics of a continuous population size-distribution. These so-called physiologically structured population models (Metz and Diekmann 1986) are traditionally formulated in terms of partial differential equations (PDEs) or integral equations instead of ODEs. Numerical techniques for equilibrium continuation in physiologically structured population models have been developed only in the past decade (Diekmann, Gyllenberg, and Metz 2003; Kirkilionis, Diekmann, et al. 1997, 2001). These techniques also allow for the continuation of stability boundaries as a function of two model parameters, which separate the parameter combinations for which a particular equilibrium is stable from those combinations for which the equilibrium is unstable and population cycles occur (de Roos, Diekmann, et al. 2010; Kirkilionis, Diekmann, et al. 1997, 2001). These continuation techniques for physiologically structured population models are complicated and have up to now rarely been applied to ecological questions (but see Claessen and de Roos 2003). In section 4 of the technical appendices we therefore present a more detailed discussion about the methods we have used to compute equilibria in the physiologically structured models that we present and the continuation of these equilibria as a function of a bifurcation parameter.

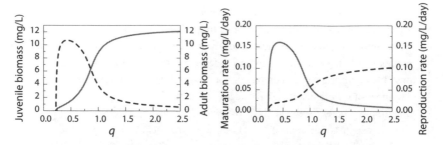

FIGURE 3.3. *Left*: Changes in juvenile (*solid lines*) and adult (*dashed lines*) biomass with increases in the adult-juvenile ingestion ratio q at background mortality
levels for $z = 0.1$. Otherwise, default parameters (see box 3.3) apply for a consumer with body size $W_A = 0.0001$ gram. *Right*: Corresponding changes in total
population maturation (*solid lines*) and reproduction rate (*dashed lines*).

between juveniles and adults implies that one of the two stages is a source
of biomass production, while the other is a sink, and furthermore that population reproduction and maturation rate are no longer equal. Figure 3.3 illustrates these changes by showing the changes in equilibrium juvenile and
adult biomass, as well as in population reproduction and maturation rate, as
a function of the adult-juvenile ingestion ratio q. When adults have a lower
mass-specific ingestion rate than juveniles ($q < 1$), population maturation rate
is higher than population reproduction rate, while the reverse is true in cases
where adults have a higher mass-specific ingestion rate ($q > 1$). Furthermore, if
the adult mass-specific ingestion rate is substantially lower than that of juveniles, adults make up the largest fraction of total population biomass, whereas
juveniles constitute the largest part of total population biomass when adults
have a higher mass-specific ingestion rate than juveniles. The switch from an
adult-dominated to a juvenile-dominated population does, however, in general
not occur when juveniles and adults have equal ingestion rates and experience equal mortality. As discussed before, in the latter case the ratio between
juvenile and adult biomass in the population equals $-\ln(z)$, which means that
juveniles will dominate the population under these conditions for $z < e^{-1}$, and
adults will dominate otherwise.

Intuitively, there is a straightforward explanation for why the life history
stage, which is a net sink of biomass production, will make up the largest
fraction of total population biomass. Consider the consumer population as a
dynamic system, in which biomass is generated from ingested resources, destroyed through mortality and maintenance, and channeled around the individual life cycle through reproduction and maturation (figure 3.4). In cases

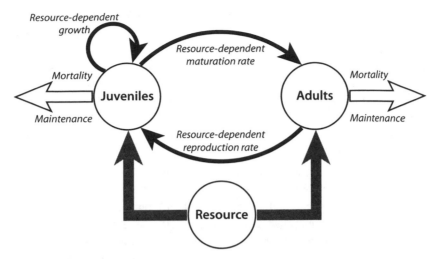

FIGURE 3.4. Schematic representation of the consumer life cycle as a dynamic system, in which biomass is generated from ingested resources (*solid gray arrows*), destroyed through mortality and maintenance (*open arrows*), and channeled from the juvenile to the adult stage through maturation and from the adult back to the juvenile stage through reproduction (*solid black arrows*). In addition, juvenile biomass increases through somatic growth (*circular, solid black arrow*). It is important to note that all these biomass recruitment processes depend on resource availability.

where juveniles and adults do not differ in ingestion rate and mortality and the flow of biomass entering the adult stage through the maturation size threshold exactly equals the flow of biomass entering the juvenile stage through the size at birth threshold, the distribution of biomass over the two stages will obviously depend on their duration relative to the average lifespan of an individual. However, if the population maturation rate becomes more limited than the reproduction rate, a bottleneck develops at the maturation size, which will make biomass to pile up in the juvenile stage preceding the bottleneck, whereas the relative density of adult biomass in the population will be lower.

We will refer to these conditions of a life history bottleneck at the maturation size as an equilibrium state that is mostly governed by "development control." Because development control implies that the adult stage is a net source of biomass production, the lower adult biomass densities nonetheless lead to a population reproduction rate that increases monotonically with an increase in the extent of development control (figure 3.3, right panel, $q > 1$). The biomass that is piled up in the juvenile stage as a consequence of the large recruitment through reproduction and limited maturation out of the stage is mostly lost through mortality and maintenance costs, making the juvenile stage a net sink of biomass production. Vice versa, under conditions that we will refer to as "reproduction control," in which population reproduction becomes more limited than maturation, a bottleneck in the life cycle develops at the size at birth, which will make biomass pile up in the adult stage just preceding this bottleneck. Despite the fact that juvenile biomass is low, the maturation rate into the adult stage is high (figure 3.3, right panel, $q < 1$) because the mass-specific biomass production of juveniles is high and they hence quickly reach mature body sizes. The mass-specific productivity of the biomass in the adult stage is, however, low, and most of this biomass is lost to mortality and maintenance costs before it produces sufficient biomass to compensate for its own demise, making the adult stage a net sink of biomass production. Notice that conditions of development control imply that competition in the juvenile stage is more intense than in the adult stage, whereas conditions of reproduction control are associated with more intense competition for resources among adults than among juveniles.

In cases where a consumer population at equilibrium is under reproduction control and adult biomass hence dominates the population at background mortality levels, any increase in mortality, be it stage-independent or targeted specifically at juveniles or adults, will change the size distribution in the population to such an extent that juvenile biomass will increase with increasing mortality (figure 3.5). Surprisingly, even an increase in mortality that specifically targets juveniles can lead to an increase in juvenile biomass (figure 3.5, middle panels). Vice versa, a population at equilibrium under development control and hence dominated by juvenile biomass, will respond to increases in any type of mortality with a change in the size distribution that may lead to increases in adult biomass with increasing mortality (figure 3.6). Even increases in adult mortality may cause adult biomass to increase (figure 3.6, right panels).

Qualitatively, these responses in stage-specific biomass are similar when juveniles and adults compete for a shared resource and the competitive asymmetry between the stages arises through their innate foraging capacity (i.e., because $q \neq 1$) as well as when juveniles and adults forage on separate, exclusive resources and the competitive asymmetry arises because one of the resources is in larger supply than the other (figures 3.5 and 3.6, top versus bottom panels). As extreme examples, the bottom panels of figure 3.5 and 3.6 show results for the cases in which either juveniles or adults, respectively, do not feed on the dynamic resource ($\omega_J(R) = 0$ and $\omega_A(R) = 0$, respectively) but have their own exclusive resource that allows them to feed at maximum rate ($v_J(R) = \sigma_c M_c - T_c$ and $v_A(R) = \sigma_c M_c - T_c$, respectively). In other words, the increases in stage-specific biomass with increasing mortality will occur

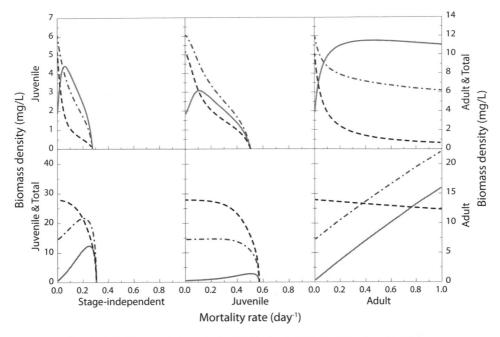

FIGURE 3.5. Changes in juvenile (*solid lines*), adult (*dashed lines*), and total consumer biomass (*dashed-dotted lines*) with increases in stage-independent, juvenile, and adult mortality on top of the background mortality for $q = 0.5$ and $z = 0.01$. Otherwise, default parameters (see box 3.3) apply for a consumer with body size $W_A = 0.0001$ gram. Top panels show results in cases where juveniles compete with adults for the limiting resource, and bottom panels show results for the cases in which juveniles have their own exclusive resource and feed at maximum rate, i.e., for a model in which it is assumed that $\omega_J(R) = 0$ and $\nu_J(R) = \sigma_C M_C - T_C$, while $q = 1$. Notice that the maximum mortality rates shown correspond to a loss rate that is ten times the maintenance rate and sixty-five times the background mortality rate and that total consumer biomass is indicated on the right and left axis in top and bottom panels, respectively.

whenever one of the stages has a foraging advantage over the other, independent of whether this foraging advantage results from environmental circumstances (larger resource availability) or a difference in intrinsic foraging capacity.

Figures 3.5 and 3.6 also show the changes in total consumer biomass with increasing mortality. When juveniles and adults compete for a shared resource, total population biomass invariably decreases with increasing mortality, irrespective of whether the increase in mortality targets all individuals equally or only juveniles or adults (figures 3.5 and 3.6, top panels). The same pattern is found when adult consumers do not feed on the dynamic resource but have their

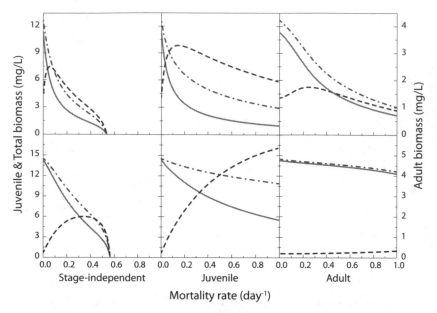

FIGURE 3.6. Changes in juvenile (*solid lines*), adult (*dashed lines*), and total consumer biomass (*dashed-dotted lines*) with increases in stage-independent, juvenile, and adult mortality on top of the background mortality for $q = 2.0$ and $z = 0.5$. Otherwise, default parameters (see box 3.3) apply for a consumer with body size $W_A = 0.0001$ gram. Top panels show results in cases where juveniles compete with adults for the limiting resource, and bottom panels show results for the cases in which adults have their own exclusive resource and feed at maximum rate, i.e., for a model in which it is assumed that $\omega_A(R) = 0$ and $\nu_A(R) = \sigma_C M_C - T_C$. Notice that the maximum mortality rates shown correspond to a loss rate that is ten times the maintenance rate and sixty-five times the background mortality rate.

own exclusive resource, which allows them to feed at maximum rate (figure 3.6, bottom panels). In contrast, when juveniles do not feed on the dynamic resource but have their own exclusive resource allowing them to feed at maximum rate, an increase in mortality also increases total consumer biomass (figure 3.5, bottom panels). This increase in total consumer biomass is most pronounced with an increase in mortality targeting only adult consumers and almost negligible when the increased mortality targets only juveniles. Increases in stage-independent mortality lead to an increase in total consumer biomass that is intermediate between the response to increases in juvenile and adult mortality.

Thus, increasing mortality for all individuals in a consumer population may have the counterintuitive effect that the biomass abundance of that population increases if the food availability of juvenile individuals is large. This increase in total population biomass results from the increase in juvenile biomass with increasing mortality. At background mortality the additional resource for juvenile consumers is only exploited to a limited extent, because juvenile biomass is low and the resource is inaccessible to adults. The resource therefore contributes little to support the total population biomass. With increasing mortality, however, juvenile consumer biomass increases and with it the extent to which their exclusive resource is exploited. This leads to an increased assimilation of juvenile-exclusive resource, supporting a larger total population biomass. Simply put, with increasing mortality there are more juvenile consumers to cash in on the large food supply that is at their disposal. This increase in total population biomass is not observed when adult consumers have their own exclusive resource allowing them to feed at maximum rate (figure 3.6, bottom panels), because quantitatively the increase in adult biomass is more limited, particularly compared with the juvenile biomass density in the population. We have found that overcompensation in total consumer biomass also occurs when juveniles do not feed at maximum rate but exploit a dynamic resource on which they feed exclusively (see chapter 6). The consumers in this case go through a diet shift during their ontogeny, foraging on one resource during the juvenile stage and switching to forage on another resource as adult. For the occurrence of overcompensation in total population biomass in response to an increase in mortality it is hence crucial that juvenile and adult consumers do not compete but rather exploit different resources and that the availability of the juvenile resource is significantly larger than the availability of the adult resource.

The increases in stage-specific biomass with increasing mortality occur mainly because of the skewed biomass distribution over the stages and the bottlenecks in biomass recruitment rates that occur when juveniles and adults differ in ingestion rate. In cases where the population is governed primarily by reproduction control (figure 3.5), both the recruitment to the juvenile stage as well as the biomass in the juvenile stage are low (Figure 3.3, $q < 1$). Even a small release of competition as a result of an increase in mortality will under these conditions lead to a substantial, relative increase in reproduction rate and consequently to an increase in juvenile biomass. Figure 3.7 illustrates these changes by showing the relative change in juvenile, adult, and total biomass, as well as in population maturation rate, in response to a relative increase in stage-independent mortality.

The population reproduction rate reaches levels that are up to almost thirty times higher than the rate at background mortality. In relative terms, this increase in population reproduction rate exceeds the relative increase in mortality until mortality is almost twenty times higher than background levels (figure 3.7, right panel). At this point biomass losses through mortality are roughly

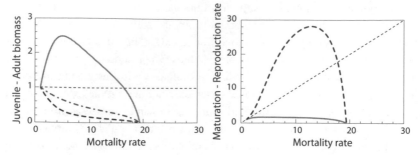

FIGURE 3.7. Changes in juvenile (*left panel, solid line*), adult (*left panel, dashed line*), and total biomass (*left panel, dashed-dotted line*) as well as population maturation (*right panel, solid line*) and reproduction rate (*right panel, dashed line*) with increases in stage-independent mortality for $q = 0.5$ and $z = 0.01$ (cf. the left-top panel in figure 3.5). Otherwise, default parameters (see box 3.3) apply for a consumer with body size $W_A = 0.0001$ gram. All biomass densities (juvenile, total, and adult) and rates (maturation, reproduction, and mortality) are relative measures, as they are scaled by their respective values at the background mortality rate of 0.015 day^{-1}. Here, this reference state corresponds to the starting point of all curves at an x- and y-coordinate equal to 1.

three times higher than losses through maintenance. The increase in equilibrium juvenile biomass is less pronounced, reaching densities only 2.5 times the reference density at background mortality, because the increase in juvenile biomass induced by the increase in population reproduction rate is tempered by the increased loss rate attributable to mortality. Nonetheless, juvenile biomass densities are higher than the reference levels at background mortality until mortality exceeds background levels by a factor of sixteen to seventeen (figure 3.7, left panel). Population maturation rate also increases with increasing mortality owing to the release of resource competition, but in relative terms this increase never exceeds the increase in losses through mortality (figure 3.7, right panel). As a consequence, adult biomass densities only decrease with increasing mortality, a pattern that is closely followed by the total population biomass, because adults constitute its largest part (figure 3.7, left panel).

In cases where the population is governed by development control (figure 3.6) both the maturation rate as well as adult biomass density are low (Figure 3.3, $q > 1$). Even a small release of competition resulting from an increase in mortality will under these conditions lead to a substantial, relative increase in maturation rate and consequently to an increase in adult biomass. As illustrated in figure 3.8, increasing stage-independent mortality increases the population maturation rate up to twenty times background levels. In relative terms, the

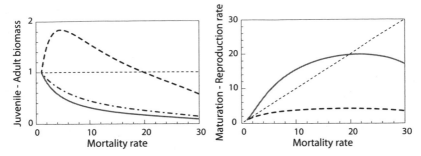

FIGURE 3.8. Changes in juvenile (*left panel, solid line*), adult (*left panel, dashed line*), and total biomass (*left panel, dashed-dotted line*) as well as population maturation (*right panel, solid line*) and reproduction rate (*right panel, dashed line*) with increases in stage-independent mortality for $q = 2.0$ and $z = 0.5$ (cf. the left-top panel in figure 3.6). Otherwise, default parameters (see box 3.3) apply for a consumer with body size $W_A = 0.0001$ gram. All biomass densities (juvenile, total, and adult) and rates (maturation, reproduction, and mortality) are relative measures, as they are scaled by their respective values at the background mortality rate of 0.015 day^{-1}. Here, this reference state corresponds to the starting point of all curves at an x- and y-coordinate equal to 1.

increase in maturation exceeds the relative increase in mortality until mortality is almost twenty times higher than background levels (figure 3.8, right panel), corresponding to an absolute mortality rate that is three times larger than the mass-specific maintenance rate. Adult biomass almost doubles when mortality is increased to roughly five times background levels, while the relative increase in adult biomass exceeds the relative increase in mortality up to the point that mortality is twenty times higher than background levels (figure 3.8, left panel). Even though the release of resource competition also leads to an increase in population reproduction rate, in relative terms this increase never exceeds the increase in losses through mortality (figure 3.8, right panel). Hence, both juvenile and total biomass, of which juveniles constitute the main part, both decrease with increasing mortality (figure 3.8, left panel).

The disproportionally large increases in population maturation and reproduction rate with increasing mortality result from the contrasting facts that between juveniles and adults the most abundant life stage is at the same time the net sink of biomass production and generates new biomass at a rate that is most severely limited by resource availability. In cases of reproduction control, the abundant adults only manage to produce a small reproductive output. This implies that most of the energy that adults derive from foraging is lost on maintenance costs to sustain the large adult biomass. More specifically, at background mortality levels for $q = 0.5$ and $z = 0.1$ (see figure 3.3), adults spend 98 percent of their gross biomass production, $\sigma_C \omega_A(\tilde{R})$, on maintenance costs and only 2 percent on reproduction. In cases where juveniles grow over a more substantial size range between birth and maturation ($q = 0.5$ and $z = 0.01$; figures 3.5 and 3.7) maintenance costs amount to as much as 99.7 percent of adult gross biomass production.

On the other hand, because juveniles are the more efficient foragers, they waste less of their gross biomass production, $\sigma_C \omega_J(\tilde{R})$, on maintenance and invest roughly 50 percent of it into growth and maturation. As a consequence, the population maturation rate is large even though juvenile biomass densities are low (see figure 3.3, $q = 0.5$). Under these conditions a small increase in mortality will translate into a small *decrease* in adult biomass, a small *increase* in resource density, and thus to a small *increase* in *per-capita* gross biomass production by adults. Nonetheless, the small increase in *per-capita* gross biomass production translates into a substantial increase in adult fecundity. More specifically, if in absolute terms the increase in adult ingestion equals $\Delta\omega_A(\tilde{R})$, the *relative* increase in gross adult biomass production is equal to $\Delta\omega_A(\tilde{R})/\omega_A(\tilde{R})$, but the *relative* increase in adult fecundity amounts to

$$\frac{\sigma_C \Delta\omega_A(\tilde{R})}{\sigma_C \omega_A(\tilde{R}) - T_C},$$

which is large if the maintenance costs T_C constitute most of the gross adult biomass production, $\sigma_C \omega_A(\tilde{R})$. This increase in adult fecundity leads to the disproportionately large increase in population reproduction rate and the consequent increase in juvenile biomass with increasing mortality.

In a similar way, under conditions of development control juveniles spend most of their gross biomass production only to maintain the high juvenile biomass densities. More specifically, at background mortality levels for $q = 2.0$ and $z = 0.1$ (see figure 3.3), 93 percent of gross biomass production by juveniles is lost on maintenance costs, while this fraction is even larger (96 percent) if the adult stage makes up a larger part of the entire life cycle ($q = 2.0$ and $z = 0.5$; figures 3.6 and 3.8). Adults, however, invest under these conditions roughly 50 percent of their gross biomass production into reproduction, which forms the basis of the large population reproduction rate despite low adult biomass densities (see figure 3.3, $q = 2.0$). A small increase in mortality under these conditions will translate into a small decrease in juvenile biomass and a small increase in resource density, which will increase gross juvenile biomass production by a small amount, $\sigma_C \Delta\omega_J(\tilde{R})$. Nonetheless, this small increase in gross biomass production translates into a relative increase in net biomass production by juveniles equal to

$$\frac{\sigma_C \Delta \omega_J(\tilde{R})}{\sigma_C \omega_J(\tilde{R}) - T_C},$$

which is disproportionally large if the maintenance costs T_C constitute most of the gross juvenile biomass production, $\sigma_C \omega_J(\tilde{R})$. This substantial increase in the energy that juveniles spend effectively on growth and maturation translates into the disproportionately large increase in population maturation rate and the consequent increase in adult biomass densities.

The foregoing argumentation makes clear that maintenance costs, required to sustain standing-stock biomass densities, play a pivotal role in the occurrence of stage-specific biomass overcompensation. In a later section we will indeed see that maintenance constitutes one of the two processes in individual energetics that may give rise to overcompensation, the other one being indeterminate, food-dependent growth in body size throughout both the juvenile and the adult life stages. It should also be noted that necessarily the life stage that represents a net sink of biomass production and that is hence likely to dominate the population at background mortality levels is the one that will exhibit the disproportionately large increase in net biomass production that can be used effectively for either reproduction or maturation. As a consequence, the life stage that constitutes a net source of biomass production is the one that will exhibit biomass overcompensation in response to increasing mortality.

STAGE-SPECIFIC MORTALITY

The changes in juvenile, adult, and total biomass in response to increases in stage-specific mortality are qualitatively similar to the changes in response to increases in stage-independent mortality. The different types of mortality lead to only quantitative differences in the strength and extent of the biomass overcompensation. We will not discuss overcompensation in total biomass any further, as we have concluded before that it is closely linked to the overcompensation in juvenile biomass in cases where juvenile consumers have an unlimited supply of exclusive resource to feed on. In the following we hence adopt the juvenile and adult biomass density at background mortality levels as the reference biomass densities. The strength of the overcompensation we then define as the ratio between the maximum in the hump-shaped biomass-mortality relationship and the reference biomass density and the extent of the overcompensation as the range of mortality rates over which stage-specific biomass is higher than this reference value. In general, the strength and extent of the biomass overcompensation is largest when the stage-specific mortality targets the stage that is a net sink of biomass production. Mortality that targets only the stage that is a net source of biomass production reduces the strength of the biomass overcompensation, but it tends to increase the extent of its occurrence compared with stage-independent mortality (table 3.1).

More specifically, if the population at equilibrium is governed by reproduction control, increases in adult mortality result in a stronger overcompensatory response in juvenile biomass than with stage-independent mortality (figure 3.7 and table 3.1); juvenile biomass reaches its maximum at a higher mortality rate and is elevated above reference densities over a wider range of mortality. This difference between stage-independent mortality and adult mortality results because in the latter case the increase in juvenile biomass attributable to the disproportionally large increase in population reproduction rate is not countered by increased mortality targeting juveniles. In contrast, increasing juvenile mortality does counter directly the increase in population reproduction rate and consequently leads to a weaker overcompensatory response in juvenile biomass than in cases of stage-independent mortality. Nonetheless, compared to stage-independent mortality, juvenile biomass reaches a maximum at a higher mortality rate and is elevated above reference levels over a wider range of mortality rates when juvenile mortality is increased (table 3.1). Absolute values of stage-independent and stage-specific mortality are, however, hard to compare, as the latter targets only a fraction of the total population. The increase in the extent of the overcompensation with stage-specific mortality is therefore not unambiguous.

If we focus on a population governed by reproduction control, we can derive from equation (3.8) the following expression for the fraction of juvenile biomass in the consumer population at equilibrium:

$$\frac{\tilde{J}}{\tilde{J} + \tilde{A}} = \frac{(\mu_A - \nu_A(\tilde{R}))}{(\mu_A - \nu_A(\tilde{R})) + (\nu_J(\tilde{R}) - \mu_J)}.$$

Because reproduction control implies that the juvenile and adult stage are a net source and sink of biomass production, respectively (i.e., $\nu_J(\tilde{R}) > \mu_J$ and $\nu_A(\tilde{R}) < \mu_A$; table 3.2), all terms in parentheses in the above expression are positive. We can therefore infer that increases in μ_J as well as μ_A increase the right-hand side of the above expression and hence lead to an increase in the fraction of juvenile biomass. These changes in the fraction of juvenile biomass with increasing mortality translate into the observed overcompensation in juvenile biomass as long as the increased mortality does not lead to a strong decline in total consumer biomass (figures 3.5 and 3.7). An increase in juvenile mortality reduces the juvenile net biomass production, $\nu_J(\tilde{R}) - \mu_J$, and eventually turns the juvenile stage into a net-biomass-loss stage. Increases in juvenile mortality

TABLE 3.1. Strength and Extent of Biomass Overcompensation with Different Types of Population Control and Default Parameter Values

Type of control	Reproduction control $(q = 0.5, z = 0.01)$ Increases in juvenile biomass			Development control $(q = 2.0, z = 0.5)$ Increases in adult biomass		
Type of mortality	Stage-independent	Juvenile	Adult	Stage-independent	Juvenile	Adult
Strength of biomass overcompensation						
Maximum biomass increase [†]	2.5	1.7	3.2	1.8	2.4	1.3
Mortality rate at maximum increase [‡]	3.9	7.2	31.2	3.6	9.6	16.5
Extent of biomass overcompensation						
Maximum mortality rate [‡]	17	22.7	580	18.8	103	39.5

[†] Maximum biomass increase is expressed as a multiple of biomass density at background mortality levels.
[‡] Mortality rates are expressed as multiples of the background mortality rate.

TABLE 3.2. Characteristics and Conditions of the Two Modes of Population Control, Including the Changes in Biomass with Increasing Mortality

Type of control	Relationship of population reproduction/maturation rate	Net source of biomass	Net sink of biomass	Most abundant, strongly competing life stage	Changes with increasing mortality
Reproduction control	Reproduction < Maturation $(v_A(\tilde{R})\tilde{A} < \gamma(v_J(\tilde{R}), \mu_J)\tilde{J})$	Juvenile stage $(v_J(\tilde{R}) > \mu_J)$	Adult stage $(v_A(\tilde{R}) < \mu_A)$	Adults	Juvenile biomass increases Adult biomass decreases (figure 3.5)
Development control	Reproduction > Maturation $(v_A(\tilde{R})\tilde{A} > \gamma(v_J(\tilde{R}), \mu_J)\tilde{J})$	Adult stage $(v_A(\tilde{R}) > \mu_A)$	Juvenile stage $(v_J(\tilde{R}) < \mu_J)$	Juveniles	Adult biomass increases Juvenile biomass decreases (figure 3.6)

for $q < 1$ therefore have a comparable effect on the equilibrium as increases in q with equal juvenile and adult mortality (see figure 3.3). In cases of reproduction control, increases in μ_A increase the net biomass loss, $\mu_A - v_A(\tilde{R})$, in the adult stage and hence do not change the type of control in equilibrium. With increases in stage-independent mortality the difference in net biomass production of juveniles and adults, $v_J(\tilde{R}) - \mu_J$ and $v_A(\tilde{R}) - \mu_A$, respectively, depends more and more on these mortality rates. Increasing stage-independent mortality therefore tends to make the net-biomass-production rates of juveniles and adults more similar and equalizes the juvenile-adult biomass ratio in the population. An increase in stage-independent mortality can, however, never change that the equilibrium is governed by reproduction control.

In much the same way, increases in juvenile mortality in cases of development control result in more pronounced overcompensation in adult biomass than stage-independent mortality (figure 3.8 and table 3.1) with biomass densities elevated above reference levels over a significantly larger range of mortality rates, while the maximum in adult biomass is also reached at higher mortality (table 3.1). Again, the stronger response arises because adults do not suffer from additional mortality, and adult biomass hence only responds to the disproportionately large increase in population maturation rate. Weaker overcompensation in adult biomass in this case occurs with increases in adult mortality (table 3.1), as the latter directly counteracts the increasing effect of the higher population maturation rate. Also in cases of development control, the extent of the biomass overcompensation is larger with increasing stage-specific mortality, which might be largely illusionary, however, given that absolute levels of stage-independent and stage-specific mortality are virtually incomparable.

From equation (3.8) we can derive that in cases of development control, the fraction of adult biomass in the population can be expressed as:

$$\frac{\tilde{A}}{\tilde{J} + \tilde{A}} = \frac{(\mu_J - v_J(\tilde{R}))}{(\mu_J - v_J(\tilde{R})) + (v_A(\tilde{R}) - \mu_A)}.$$

All terms in parentheses in the above expression are positive, given that development control implies that the juvenile and adult stage are a net sink and net source of biomass production ($v_J(\tilde{R}) < \mu_J$ and $v_A(\tilde{R}) > \mu_A$; table 3.2), respectively. The expression makes clear that increases in both μ_J and μ_A increase the adult fraction in the population. An increase in adult mortality reduces the adult net biomass production, $v_A(\tilde{R}) - \mu_A$, and eventually turns the adult stage into a net sink of biomass production. Hence, it induces a shift from development to reproduction control, as also occurs when lowering q from $q > 1$ to $q < 1$ (see figure 3.3). Increases in μ_J do not change the type of population control in equilibrium, as they only increase the net biomass loss, $\mu_J - v_J(\tilde{R})$, during the juvenile stage. Increases in stage-independent mortality also increase the adult fraction in the population and in addition tend to equalize the net biomass production of juveniles and adults.

EMPIRICAL EVIDENCE

Our modeling results show that stage-specific overcompensation will occur irrespective of which stage forms the bottleneck as well as which stage is the target of increased mortality. An experimental study testing this prediction was carried out by Schröder, Persson, and de Roos (2009). They exposed small juveniles or adults in populations of least killifish (*Heterandria formosa*) to seven different harvesting levels. They first of all demonstrated the expected presence of a hump-shaped relationship between stage-specific biomass and harvesting mortality for both negative and positive size-selective harvesting, targeting juveniles and adults, respectively (figure 3.9, lower panels). Second, this overcompensation occurred in juvenile biomass, irrespective of whether adults or juveniles were subjected to the harvesting mortality. In contrast, adult biomass either decreased or did not show any significant relationship with harvesting intensity (figure 3.9, upper panels). These results show that the population was reproduction-controlled. Finally, in accordance with model predictions, the hump was more pronounced (maximum overcompensation 22.7 percent) and occurred at a higher harvesting rate for positive size selection (0.015 day^{-1}) than for negative size selection (maximum overcompensation 11.5 percent, harvesting rate 0.008 day^{-1}) (figure 3.9, lower panels).

The work of Schröder, Persson, and de Roos (2009) is the most complete study demonstrating stage-specific overcompensation, as both mortality rate as well as the size selectivity of mortality were varied. Other experimental studies showing evidence for overcompensation have been restricted to either consideration of only one particular mortality regime (juvenile mortality or random mortality), or, if both juveniles and adults were subjected to mortality, to consideration of only one mortality rate, leading to the risk that the imposed mortality rate may have exceeded the level at which overcompensation took place. Slobodkin and Richman (1956) carried out size-selective harvesting experiments with *Daphnia*, in which they removed varying fractions of newborn individuals. The interpretation of their data is complicated owing to the fact that the number of individuals they harvested was based on an estimate of the number of *Daphnia* born over a two- or four-day time period. As a result, the number of individuals to be removed sometimes exceeded the number of small *Daphnia* present at the time of harvesting. Nevertheless, their results show

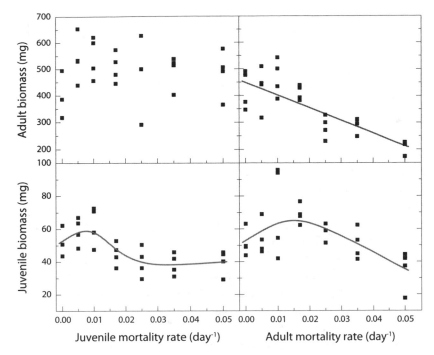

FIGURE 3.9. Response in adult biomass (*upper panels*) and juvenile biomass (*lower panels*) to increased negative (*left*) and positive size-selective (*right*) harvesting in the least killifish *Heterandria formosa* (data from Schröder, Persson, and de Roos 2009). Each symbol represents the estimated long-term average biomass of a replicate population experiment. Solid lines represent the regression relationship of the best-fitting generalized additive model (GAM) (for details, see Schröder, Persson, and de Roos 2009).

that an increase in harvesting of small *Daphnia* resulted in larger densities of this size class. Their results thus suggest that the *Daphnia* population was reproduction-controlled. Recently, Nilsson, Persson, and van Kooten (2010) provided further evidence for reproduction control in *Daphnia*. In an experiment using random mortality, they demonstrated a decrease in adult biomass but a complete compensation in juvenile biomass. In addition, they showed that the complete compensation in juvenile biomass came about through both increased per-capita fecundity of egg-carrying females as well as through an increased proportion of egg-carrying adult females. They attributed the occurrence of complete compensation instead of overcompensation to a harvesting intensity that exceeded the rates at which overcompensation would have occurred.

An experimental demonstration of overcompensation in adults was presented by Cameron and Benton (2004), who carried out stage-specific harvesting of either eggs, juveniles, or adults in populations of soil mites. They found a positive effect on adult density of harvesting eggs, no effect of harvesting juveniles, and a negative effect of harvesting adults. Cameron and Benton argued that adult soil mites are better competitors than juvenile soil mites because of their larger sizes. Under these conditions we would predict that the population is controlled by development and that an increase in juvenile mortality would indeed increase adult density. The negative effect of harvesting adults on adult density can, once again, be related to the fact that the experimental design was restricted to one harvesting rate per harvesting type. In this case the risk of imposing a harvesting intensity exceeding the levels at which overcompensation may have occurred was even higher because the stage showing overcompensation was targeted (cf. figure 3.6).

In the previous section we showed that overcompensation is expected to occur whenever one of the stages has a foraging advantage over the other, independent of whether this foraging advantage results from differences in intrinsic foraging capacity or from environmental circumstances (larger resource availability) (figures 3.5 and 3.6). Nicholson's (1957) classical experiments with the Australian sheep blowfly (*Lucilia cuprina*) represent an impressive set of experiments using the latter approach. By manipulating food density for juveniles and adults, he could shift the blowfly system between development control and reproduction control, respectively, and could also show a number of the predicted compensatory effects of stage-specific mortality on abundances of different life history stages.

First of all, Nicholson showed that switching a blowfly population from being controlled by larval competition to being controlled by adult competition has the theoretically expected effect of approximately quadrupling the adult density (Nicholson 1957, fig. 3). Relaxing adult competition by a change from a limited to an unlimited sugar supply was sufficient to decrease average adult density (Nicholson 1957, fig. 2). These observations correspond with the model prediction that a switch from development control to reproduction control will increase adult biomass and decrease juvenile biomass (figure 3.3, left panel). In an experiment where adults and larvae received the same daily quota of meat, leading to a predominance of larval competition, he further showed that a destruction of 50 percent of immature adults reduced the larval density by 50 percent but had no effect on mean adult density (figure 3.10, left panel). An increase in adult resource supply, by giving adults an exclusive ample supply of ground liver, resulted in a doubling of larval density but hardly any change in adult density (figure 3.10, middle panel). Both

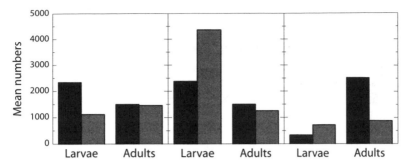

FIGURE 3.10. Experimental results from Nicholson's (1957) classical experiments with the Australian sheep blowfly *Lucilia cuprina*. *Left panel*: Density of larvae and adults in controls (*black bars*) and with 50 percent destruction of adults (*gray bars*) in a treatment with larval competition. *Middle panel*: Density of larvae and adults in controls (*black bars*) and with separate and ample supply of food for adults (*gray bars*) in a treatment with larval competition. *Right panel*: Density of larvae and adults in controls (*black bars*) and with 90 percent destruction of emerging adults (*gray bars*) in a treatment with adult competition.

these results are completely in accordance with the expected responses of a population under development control. In the first case, the reduction in total population reproduction owing to the induced mortality led to a reduction in larval competition, increased larval survival, and thereby a complete compensation in adult numbers. In the second case, increased adult performance led to more eggs and larvae produced, but as a consequence of increased larval competition, few larvae survived to pupation and adult density remained almost unaltered.

In experiments with blowfly populations controlled by larval competition Nicholson showed that a decrease in the water supply shortened adult lifespan but simultaneously increased adult density. This experimental outcome corresponds to our predictions illustrated in figure 3.6—that an increase in adult mortality will increase adult density in populations under development control. In another experiment with larval competition, Nicholson also found that a 50 percent destruction of young larvae each day resulted in a doubling of adult density owing to the reduction in larval competition. He also designed experiments with strong adult competition. In these experiments a substantial destruction (90 percent) of maturing adults resulted in a doubling in larval density in correspondence with the theoretical expectation that juvenile biomass will increase when adult mortality is increased in a population under reproduction control (figure 3.10, right panel).

ASYMMETRY AND LIFE HISTORY EFFECTS

In the foregoing discussion of stage-specific biomass overcompensation we have used different values for the adult-juvenile ingestion ratio q to vary between situations in which the equilibrium state is mostly governed by reproduction control (cf. figure 3.5) and situations with mostly development control (cf. figure 3.6). However, we have at the same time varied the ratio between newborn and adult body size z, using a rather small value for z in cases of reproduction control and a large value of z in cases of development control. This raises questions about what are the separate influences of the foraging asymmetry between juvenile and adult consumers and the size range through which juveniles have to grow from birth until maturation. To disentangle the influence of these two variables for otherwise default parameter values, we have computed which combinations of q and z lead to an increase in stage-specific biomass in response to an increase in mortality. The results of these computations are shown in figure 3.11 as regions in the (q, z)-parameter space, for which either juvenile or adult biomass increases with increases in stage-independent, juvenile and adult mortality, respectively. The computational approach to construct these regions will not be presented here; it is discussed in detail in section 5 of the technical appendices.

Figure 3.11 makes once again clear that overcompensation in juvenile biomass in response to increases in stage-independent mortality occurs only when $q < 1$, corresponding to situations in which the equilibrium is governed through reproduction control. Overcompensation in adult biomass owing to increases in stage-independent mortality, on the other hand, occurs only for $q > 1$, that is, when the equilibrium is mostly under development control. The occurrence of these two different types of overcompensation is mainly determined by the value of q and is only to a limited extent dependent on the newborn-adult size ratio z (figure 3.11). For almost all values of z overcompensation in stage-specific biomass owing to increases in stage-independent mortality will occur as soon as juvenile and adult ingestion rates differ by more than 10 percent. More specifically, whenever $q > 1.1$, adult biomass will increase with stage-independent mortality, whereas juvenile biomass will increase as soon as $q < 0.9$, provided that q is not too low for consumers to persist at all at background mortality rates (i.e., provided that q is larger than approximately 0.25).

As was already shown in figure 3.5 and discussed previously, juvenile biomass overcompensation in response to increases in juvenile mortality is always less pronounced compared with increases in stage-independent mortality because the mortality exclusively targets the stage exhibiting overcompensation. Accordingly, the parameter region in terms of q and z, for which juvenile

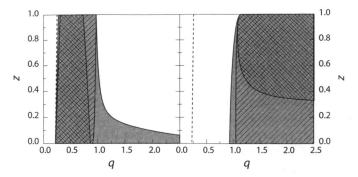

FIGURE 3.11. *Left*: Occurrence of overcompensation in juvenile biomass with increases in adult (*gray parameter region*), stage-independent (*gray, hatched parameter region*), and juvenile mortality (*gray, cross-hatched parameter region*) as a function of the adult-juvenile ingestion ratio q and the newborn-adult consumer body size ratio z. *Right*: Occurrence of overcompensation in adult biomass with increases in juvenile (*gray parameter region*), stage-independent (*gray, hatched parameter region*), and adult mortality (*gray, cross-hatched parameter region*) as a function of the adult-juvenile ingestion ratio q and the newborn-adult consumer body size ratio z. Notice that in both panels the three indicated parameter domains are enclosed within each other in the order mentioned. Otherwise, default parameter values (see box 3.3) were used for a consumer with body size $W_A = 0.0001$ gram. The almost vertical, thin dashed line in both panels near the value $q \approx 0.25$ represents the existence boundary of the equilibrium. For values of q and z to the left of this line, the consumer population cannot persist.

biomass increases in response to increases in juvenile mortality, is smaller than and completely enclosed in the region for which this overcompensation occurs in response to stage-independent mortality (figure 3.11). Under conditions that allow for juvenile biomass overcompensation in response to increased juvenile mortality, we hence also always expect this overcompensation in response to stage-independent mortality.

In contrast, juvenile biomass overcompensation in response to increases in adult mortality was shown to be more pronounced than overcompensation attributable to stage-independent mortality (figure 3.5) and, in particular, to occur over a much wider range of mortality rates. This wider range of conditions that allow for the latter type of overcompensation is also apparent in figure 3.11. Notably, it can even be found if adults have higher ingestion rates than juveniles ($q > 1$; see also figure 3.1, right panel) and occurs whenever the newborn-adult size ratio z is small. As discussed previously in relation to figure 3.5, the larger opportunity for the occurrence of juvenile biomass overcompensation in response to adult mortality is due to the fact that the stage exhibiting overcompensation is not exposed to the decreasing effect of the higher mortality and only benefits from the increased population reproduction rate.

The conditions that allow for adult biomass overcompensation in response to stage-specific mortality mirror the results for juvenile biomass overcompensation (figure 3.11). When the stage-specific mortality targets exclusively the stage exhibiting overcompensation, overcompensation occurs under more restricted conditions. Figure 3.11 shows accordingly that all combinations of q and z allowing for adult biomass overcompensation in response to adult mortality are enclosed in the parameter region for which adult biomass overcompensation with increases in stage-independent mortality is possible. In contrast, adult biomass overcompensation in response to increases in juvenile mortality occurs under a wider range of conditions compared with stage-independent mortality and even occurs when juveniles have a slightly higher ingestion rate than adults ($q < 1$). However, the conditions that allow for this latter type of overcompensation are not as extensive as those that allow for juvenile biomass overcompensation in response to increases in adult mortality (compare left and right panels in figure 3.11).

Figure 3.11 provides information about the conditions that allow for stage-specific biomass overcompensation but does not give any indication about the strength of the response (i.e., the maximum stage-specific biomass density that can be reached by increasing mortality compared with the biomass density at background mortality). Figure 3.12 shows this strength of overcompensation for the cases in which stage-independent mortality is increased. This graph was constructed by computing for every combination of the adult-juvenile ingestion ratio q and the newborn-adult consumer body size ratio z the hump-shaped relationship between either juvenile or adult biomass as a function of stage-independent mortality over a range of mortality values that encompassed the top of this relationship.

The stage-specific biomass density at the top of the hump-shaped relationship was subsequently expressed as a multiple of the biomass density at background mortality for the given values of q and z. For most values of q and z an increase in stage-independent mortality increases stage-specific biomass, with more than 50 percent above densities at background mortality rates. Strong overcompensation in juvenile biomass with up to threefold increases above background densities is observed in cases of pronounced asymmetry in juvenile and adult ingestion and is promoted by lower values of z, that is, when the equilibrium is strongly governed by reproduction control and juveniles grow over a substantial size range before maturation (figure 3.12, left panel). Analogously, strong overcompensation in adult biomass with up to threefold

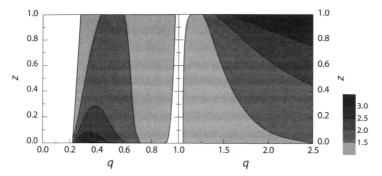

FIGURE 3.12. Strength of overcompensation in juvenile (*left panel*) and adult (*right panel*) biomass in response to increases in stage-independent mortality for different values of the adult-juvenile ingestion ratio q and the newborn-adult consumer body size ratio z. The overcompensation strength is defined as the ratio between the maximum stage-specific biomass that can be observed for stage-independent mortality higher than background mortality levels and the biomass density at this background mortality rate. Notice the different scaling of the x-axis in the left and right panels. Otherwise, default parameter values (see box 3.3) were used for a consumer with body size $W_A = 0.0001$ gram.

increases occurs for large values of z and is more pronounced with higher values of the adult-juvenile ingestion ratio q (figure 3.12, right panel). In other words, more pronounced adult biomass overcompensation occurs when the adult stage makes up a larger part of the entire life cycle of consumers and the equilibrium is strongly governed by maturation control.

MORE COMPLICATED LIFE HISTORIES

In the preceding analysis we only distinguished between two stages of consumers, juveniles and adults. This arguably represents the simplest life history that accounts for individual development next to reproduction and mortality. However, it does raise the question whether biomass overcompensation will also occur and if it will be as prominent in cases of more complex life histories. In itself this question is too general to answer, as there is no limit to the complexity that life histories can exhibit. We will hence restrict ourselves to investigate biomass overcompensation in a limited number of more complicated models. In particular, we will analyze the response to increasing mortality of a model in which the consumer population is subdivided into three distinct life stages, either two juvenile stages and a single adult stage or a single juvenile stage plus a growing and a nongrowing adult stage. Finally, we will investigate the response to increases in mortality of a consumer-resource model based on the Kooijman-Metz energy budget model (Kooijman and Metz 1984). In the latter population model the assumption that resource ingestion is proportional to body size is relaxed, which implies that individuals of different body sizes always differ in the extent to which they are limited by resource. Moreover, in this model the consumer population is represented by a continuous size-distribution not only of juveniles, as in the stage-structured models, but also of adults, as individuals are assumed to grow throughout their entire life.

MULTIPLE JUVENILE STAGES

The stage-structured biomass model that figured extensively in the previous sections can easily be extended to account for more than two consumer life stages (de Roos, Schellekens, et al. 2008b). In the case of multiple consumer stages, the first stage will always represent a stage with juvenile consumers that invest all their net energy production into growth in body size and development. The last consumer stage will always represent a stage of nongrowing adults that have reached their maximum body size and invest all their net energy production into reproduction. The intervening stages, however, may invest part of their net energy production into growth and part of it into reproduction. In cases where we distinguish three consumer life stages, we will represent the biomass in each with the variables C_1, C_2, and C_3. In addition, we label each of the functions or parameters that are stage-specific with the corresponding index of the consumer stage. In particular, we assume that ingestion of consumer stage i is modeled with the function $\omega_i(R) = q_i M_C R/(H_C + R)$ and that μ_i represents the stage-specific mortality rate. Notice that we have now introduced a parameter q_i with $0 < q_i \leq 1$ for each of the three consumer stages, which represents the mass-specific maximum ingestion rate in stage i as a fraction of the default maximum ingestion rate, in contrast to the parameter q that figured prominently in the two-stage biomass model and represented the ratio of adult to juvenile maximum ingestion rate. Analogous to the two-stage biomass model, we assume that net energy production amounts to the difference between the rate of resource assimilation and maintenance, that is:

$$\nu_i(R) = \sigma_C \omega_i(R) - T_C = \sigma_C q_i \frac{M_C R}{H_C + R} - T_C.$$

The mass-specific rate of maturation out of each of the two smallest consumer stages is described by two functions that are analogous to the maturation function given in equation (3.4) for the two-stage biomass model. These functions are defined as:

$$\gamma_1(v_1(R),\mu_1) = \frac{v_1(R) - \mu_1}{1 - z_1^{1 - \mu_1/v_1(R)}}$$

and

$$\gamma_2(\kappa_2 v_2(R),\mu_2) = \frac{\kappa_2 v_2(R) - \mu_2}{1 - z_2^{1 - \mu_2/\kappa_2 v_2(R)}}.$$

Both functions contain stage-specific parameters z_1 and z_2, respectively, which represent the ratio between the smallest and largest size of individuals within the stage, or, in other words, the ratio of body sizes at which individuals enter and leave the particular stage.

Furthermore, the first argument of the function $\gamma_2(\kappa_2 v_2(R),\mu_2)$ contains an additional parameter κ_2, representing the fraction of the net energy production that consumers in the second life stage allocate to growth in body size, with the remaining fraction, $1 - \kappa_2$, of their net energy production being spent on reproduction. Both functions $\gamma_1(v_1(R),\mu_1)$ and $\gamma_2(\kappa_2 v_2(R),\mu_2)$ are hence structurally similar in that the first function argument represents the mass-specific rate of growth in body size, while the second argument represents the mortality rate of the consumers in the particular stage. The product of the two parameters z_1 and z_2 equals the ratio between the smallest possible consumer body size (i.e., its size at birth) and the largest possible body size that a consumer can reach. This product also equals the ratio between the size at birth and at maturation in cases where the first two of the three life history stages are both juvenile stages and do not contribute to reproduction. The ratio between size at birth and at maturation equals the parameter z_1 in cases where the first consumer stage represents a juvenile stage and the other two stages are both adult stages and do contribute to reproduction.

The stage-structured biomass model for a consumer with three distinct life history stages is now described by the following system of four ODEs:

$$\frac{dR}{dt} = G(R) - (\omega_1(R)C_1 + \omega_2(R)C_2 + \omega_3(R)C_3) \tag{3.13a}$$

$$\frac{dC_1}{dt} = (1 - \kappa_2)v_2^+(R)C_2 + v_3^+(R)C_3 - \gamma_1(v_1^+(R),\mu_1)C_1 + v_1(R)C_1 - \mu_1 C_1 \tag{3.13b}$$

$$\frac{dC_2}{dt} = \gamma_1(v_1^+(R),\mu_1)C_1 - \gamma_2(\kappa_2 v_2^+(R),\mu_2)C_2 + v_2(R)C_2$$
$$- (1 - \kappa_2)v_2^+(R)C_2 - \mu_2 C_2 \tag{3.13c}$$

$$\frac{dC_3}{dt} = \gamma_2(\kappa_2 v_2^+(R), \mu_2)C_2 + v_3(R)C_3 - v_3^+(R)C_3 - \mu_3 C_3 \tag{3.13d}$$

These equations are analogous to equations for the two-stage model presented in box 3.2. As before, we will assume that the growth of the resource in the absence of consumers follows semichemostat dynamics: $G(R) = \rho(R_{max} - R)$. Resource density declines through the consorted foraging of consumers in all three stages, as represented by the term $\omega_1(R)C_1 + \omega_2(R)C_2 + \omega_3(R)C_3$ in equation (3.13a). Net energy production of the three consumer stages is described by the terms $v_1(R)C_1$, $v_2(R)C_2$, and $v_3(R)C_3$. If net energy production is positive, the second- and third-stage consumers invest part or all of their production, represented by the terms $(1 - \kappa_2)v_2^+(R)C_2$ and $v_3^+(R)C_3$, respectively, into reproduction. Consumer reproduction is hence the sum of these two contributions, $(1 - \kappa_2)v_2^+(R)C_2 + v_3^+(R)C_3$. All net energy production not invested into reproduction increases stage-specific biomass through somatic growth when positive and decreases stage-specific biomass through starvation mortality when negative (de Roos, Schellekens, et al. 2008b). Maturation of biomass out of the two smallest consumer life stages and recruitment to the following stage is described by the two terms $\gamma_1(v_1^+(R), \mu_1)C_1$ and $\gamma_2(\kappa_2 v_2^+(R), \mu_2)C_2$ occurring in equations (3.13b), (3.13c), and (3.13d). Finally, biomass in all three consumer stages declines through mortality as described by the terms $\mu_1 C_1, \mu_2 C_2$, and $\mu_3 C_3$.

In the current section we focus on a model, in which the second consumer stage represents a juvenile life history stage and does not contribute to reproduction. In the above equations, this model scenario is obtained by choosing κ_2 equal to 1. In the next section we will then focus on a model scenario with an additional life history stage of adult consumers that invest both in somatic growth as well as in reproduction, as represented by a value of κ_2 smaller than 1. We furthermore restrict our analysis to asymmetric situations, in which one of the three consumer stages is more limited in its resource gathering than the other two, in other words, one of the consumer stages has a value of q_i lower than the q_i values of the other two. Out of the three consumer stages the one with the lowest q_i value will always have the smallest net energy production at background mortality levels. Given that it is hence more resource limited than the others, we will refer to this stage as the "bottleneck" stage.

Figure 3.13 illustrates the changes in stage-specific biomass for the model with two juvenile and a single adult consumer stage in response to increases in stage-independent mortality. Invariably, the bottleneck life stage with the lowest q_i value dominates the population in terms of biomass at background mortality, with the other two stages occurring only at low biomass abundance. Increasing stage-independent mortality decreases the biomass in the bottleneck

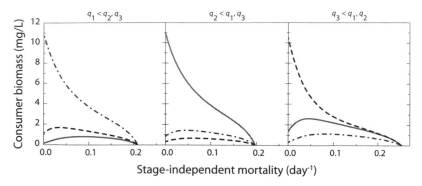

FIGURE 3.13. Changes in biomass of small consumers (small juveniles, *dashed-dotted lines*), medium consumers (large juveniles, *solid lines*), and large consumers (adults, *dashed lines*) with increases in stage-independent mortality in excess of the background mortality for $q_1 = 0.5$, $q_2 = q_3 = 1.0$ (*left*); $q_2 = 0.5$, $q_1 = q_3 = 1.0$ (*middle*); and $q_3 = 0.5$, $q_1 = q_2 = 1.0$ (*right*) in a stage-structured biomass model with two juvenile and one adult consumer stage. Here $z_1 = 0.25$ and $z_2 = 0.2$, while default values are used for all other parameters, as in the stage-structured biomass model with two stages (see box 3.3) for a consumer with body size $W_A = 0.0001$ gram.

stage. In contrast, biomasses in the other two life history stages follow hump-shaped relationships with the increasing stage-independent mortality. We have found that the exact values for q_i of these latter two stages do not qualitatively alter the pattern, as long as these two q_i values are both larger than the q_i value of the bottleneck stage. In other words, as long as there is a single consumer life history stage with a more precarious energy balance than all other stages, we can expect to find the generic pattern of biomass-mortality relations as shown in figure 3.13.

In cases where the additional mortality does not target all consumers equally, but specifically targets one or two of the consumer stages, the changes in stage-specific biomass with increasing mortality are by and large similar to those with stage-independent mortality for most parameter combinations that we investigated: Biomass in the bottleneck stage decreases with increasing mortality, while the two other stages exhibit biomass overcompensation. Only when the adult stage constitutes the energetic bottleneck in consumer life history (reproduction control), do increases in mortality that specifically target the small juveniles or all juveniles together translate into an increase in small-juvenile, but not large-juvenile, biomass (the intermediate consumer life stage). Surprisingly though, if the mortality targets only the large juvenile stage (medium-size consumers), the biomass in both juvenile stages consistently increases with mortality.

When consumers grow over a more extensive range of body sizes during life and the population is characterized by a wider size-distribution, we have already found (in the stage-structured biomass model with a single juvenile and adult stage) that increases in adult mortality under conditions of development control did not lead to overcompensation in adult biomass (cf. the parameter region at low values of z in figure 3.11). In the three-stage consumer-resource model we find a similar result in that increases in large-juvenile or adult mortality do not lead to overcompensation in adult biomass when one of the two juvenile stages constitutes the energetic bottleneck in the consumer life history and consumers can grow over a substantial range of body sizes. Increases in adult mortality under these conditions may, for certain parameter combinations, even produce results such that only the biomass of the medium-size consumer stage follows a slightly hump-shaped relation with increasing mortality, despite the fact that it constitutes the energetic bottleneck in its life history, whereas the biomass in both the small- and large-size (adult) consumer stage decreases.

Stage-specific biomass overcompensation in response to increasing mortality can therefore also readily be observed in a stage-structured biomass model with an additional juvenile stage. As in the two-stage model discussed before, the biomass in the bottleneck stage that is most limited in its energetics decreases while the other stages exhibit hump-shaped biomass relationships with increasing mortality. Distinguishing additional juvenile life history stages therefore increases the number of stages that exhibit biomass overcompensation. This generic pattern of biomass overcompensation holds exactly in cases where the increased mortality targets all consumers equally and with slight modifications when the additional mortality targets only part of the consumer population.

A GROWING ADULT STAGE

The dynamics of a consumer population in which we distinguish between a single juvenile and two adult stages can also be described using the system of ODEs (3.13) by choosing κ_2 smaller than 1. Model predictions for $\kappa_2 = 0.5$ resemble the generic pattern that emerges from both the two-stage model with a single juvenile and adult stage and the three-stage model with two juvenile stages: increases in mortality tend to decrease the biomass in the stage that constitutes the energetic bottleneck in the life cycle and increase the biomass in both nonbottleneck stages. In particular, the biomass in both adult stages increases with increasing stage-independent mortality if the juvenile life stage is most limited in its energetics (q_1 being the smallest of all q_i values). Increases in stage-specific mortality under these conditions have similar effects.

The biomass in the small-adult stage always exhibits a hump-shaped relationship with mortality when the juvenile stage is the bottleneck stage, irrespective of whether mortality is stage-specific or not. Large-adult biomass, on the other hand, may not increase with adult mortality, irrespective of whether it targets small or large adults or both, if consumers exhibit substantial growth in body size during life, that is, in cases where the size at birth is small compared with the maximum body size.

In cases of reproduction control, when one of the adult stages constitutes the energetic bottleneck in the life cycle, the generic pattern of biomass overcompensation also holds as long as consumers exhibit limited growth in body size during life. For example, the biomass in the juvenile and nonbottleneck adult stage always increases with increasing stage-independent mortality when the size at birth is in the order of 50 percent of the maximum body size. Increasing stage-specific mortality leads to roughly similar results except that increases in juvenile mortality may not lead to an increase in small-adult biomass if the large-adult consumer stage is the energetic bottleneck in the life cycle, whereas increases in large-adult mortality do not give rise to an increase in large-adult biomass if the small-adult consumer stage is the energetic bottleneck in the life cycle.

In cases of reproduction control in which consumers exhibit extensive growth in body size during life, that is, when size at birth is small compared with the maximum body size, large adult consumers are absent from the population at background mortality, as long as juveniles are less limited by energetics (i.e., have a larger q_i) than small adults. The latter then make up the largest part of population biomass. Increases in mortality that specifically targets large adult consumers will in this case have no effect whatsoever. Increases in stage-independent mortality result in the generic pattern such that biomass in the most abundant life stage (small adults) decreases, while biomass in the other life stages exhibit hump-shaped responses. This includes the large-adult consumers that are not present at background mortality.

Figure 3.14 (left panel) illustrates this for the case in which limitation by energetics tightens with increasing consumer body size (i.e., $q_1 > q_2 > q_3$). Stage-independent mortality increases juvenile biomass as well as the biomass of large consumers, despite the fact that the latter stage is limited most by energetics. At background mortality small-adult consumers are too much limited by energetics to grow to the large-adult consumer stage, but increases in stage-independent mortality relax competition and make this growth possible. Increases in juvenile mortality will decrease both juvenile and small-adult biomass but will significantly increase the biomass of large-adult consumers (figure 3.14, middle panel). Increases in mortality that targets both adult stages,

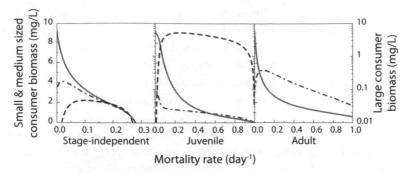

FIGURE 3.14. Changes in biomass of small consumers (juveniles, *dashed-dotted lines*), medium consumers (small adults, *solid lines*), and large consumers (large adults, *dashed lines*) with increases in stage-independent, juvenile, and adult mortality on top of the background mortality for $q_1 = 1.0$, $q_2 = 0.75$, and $q_3 = 0.5$ in a stage-structured biomass model with one juvenile and two adult consumer stages. Here $z_1 = 0.1$, $z_2 = 0.1$, and $\kappa_2 = 0.5$, while default values are used for all other parameters, as in the stage-structured biomass model with two stages (see box 3.3) for a consumer with body size $W_A = 0.0001$ gram. Notice that the right-hand axis uses a logarithmic scale to capture the large differences in biomass of large consumers that occur with changes in the different types of mortality.

however, will only increase juvenile biomass and reduce the biomass of small-adult consumers (figure 3.14, right panel). Moreover, the increased mortality targeting these small adults also implies that the recruitment to the large-adult consumer stage does not increase either. On the other hand, if large-adult consumers are most limited by energetics but juvenile consumers are intermediate between small- and large-adult consumers (i.e., $q_3 > q_1 > q_2$), juvenile biomass always increases and biomass in both adult stages always decrease in response to increases in any type of mortality, be it stage-independent or stage-specific.

A SIZE-STRUCTURED CONSUMER-RESOURCE MODEL: KOOIJMAN-METZ ENERGETICS

The foregoing discussion about a consumer population with more than two life history stages makes clear that biomass overcompensation in response to increases in stage-independent or stage-specific mortality is likely to follow a generic pattern: The biomass in the life history stage, which at background mortality is most limited by its energetics and accordingly makes up the largest part of total population biomass, decreases with increasing mortality whereas other stages are likely to exhibit hump-shaped biomass-mortality relationships.

In the two-stage biomass model this generic pattern boils down to the simple rule that juvenile biomass will increase with increases in any type of mortality when adults are more limited by energetics, while adult biomass will increase when juveniles are more energy limited. The analysis of the three-stage biomass models shows that this is a useful idealization, which carries over with some modifications to situations with more than two distinct life history stages. Deviations from the general pattern may occur especially when the increasing mortality targets only specific stages and when we account for a stage of adult consumers that also grow in body size. In addition, as in the two-stage model, the range of body sizes that the consumers grow through during life influences the occurrence of stage-specific biomass overcompensation.

To elaborate this issue further, we analyze in this section the response to increasing mortality in a model that relaxes some of the assumptions underlying the stage-structured biomass model. In this model, referred to as the Kooijman-Metz consumer-resource model (Kooijman and Metz 1984), resource ingestion is not assumed to scale proportionally with body size, but increases with body size less rapidly than maintenance costs. Moreover, all adult consumers allocate energy to both growth in body size and to reproduction. As a consequence, individual consumers exhibit attenuating but indeterminate growth throughout their entire life. The model hence incorporates a more detailed representation of consumer life history than the stage-structured biomass model. At the heart of the population dynamics model is one of the simplest, dynamic models for the individual energy budget, introduced by Kooijman and Metz (1984). In the following chapters we frequently use this energy budget model as a core part of multispecies models to represent the life history of individual organisms that grow in body size during their entire life. We hence introduce its basic formulation and the resulting consumer-resource model now. We have found that the Kooijman-Metz energy budget model leads to relatively simple, size-structured population models that nonetheless yield population- and community-dynamics results that turn out to be quite generic and by and large independent of more detailed assumptions.

The Kooijman-Metz energy budget model assumes that individual organisms of different body size are isomorphic, such that body weight w is proportional to the cubed length of the individual, $w_C(\ell) = \beta_C \ell^3$ (again the subscript C in this and all following equations is used to label all functions and parameters that relate to consumers to distinguish them from analogous, predator-related functions and parameters that occur in following chapters). Individual consumers are born at length ℓ_b and mature on reaching a length ℓ_j. Resource ingestion by consumers is assumed to be proportional to body surface area and hence to the squared length of an organism. In addition, resource ingestion rate follows a type II functional response to resource biomass with half-saturation constant H_C.

Hence, ingestion rate is modeled with the function $I_C(R, \ell) = M_C R / (H_C + R)\ell^2$, in which R represents resource density and M_C is the scaling coefficient relating maximum ingestion rate to the squared body length of the individual. It is assumed that a fixed fraction κ_C of the assimilated food is used for growth and maintenance requirements, while the remaining fraction $(1 - \kappa_C)$ is channeled to reproduction. Juveniles are assumed to invest the energy channeled to reproduction for the development of reproductive tissue and organs, while adult individuals use it to produce new offspring with a body weight $\beta_C \ell_b^3$. Adult fecundity in terms of the number of new offspring produced per day is therefore described by a function $b_C(R, \ell) = \alpha_C R / (H_C + R)\ell^2$, which is similar to the ingestion rate function except for the scaling coefficient α_C that relates the maximum adult fecundity under very high food conditions to the squared, adult body length. If we use the parameter σ_C to represent the efficiency with which ingested food is assimilated, the total amount of assimilated energy that an individual of body length ℓ channels to growth and maintenance equals $\sigma_C \kappa_C M_C R / (H_C + R)\ell^2$.

Energy requirements for maintenance are assumed to be proportional to body weight and thus scale with the cubed length of an individual following $\xi_C \ell^3$. Individuals change in body weight at a rate that is proportional to the difference between the energy investment into growth plus maintenance and the energy requirements for maintenance:

$$\frac{dw}{dt} = \psi_C^{-1}\left(\sigma_C \kappa_C M_C R / (H_C + R)\ell^2 - \xi_C \ell^3\right)$$

In this equation the parameter ψ_C represents the energetic requirements to grow one unit in body weight. Substituting the weight-length relationship $w_C(\ell) = \beta_C \ell^3$ in the above equation leads, after some manipulation, to the following ODE for the growth in length:

$$\frac{d\ell}{dt} = \gamma_C\left(\ell_m R / (H_C + R) - \ell\right)$$

Here the parameter ℓ_m is defined as $\sigma_C \kappa_C M_C / \xi_C$ and represents the maximum length that individuals can reach under very high food conditions. The parameter γ_C characterizes the growth rate in individual length and is defined as $\gamma_C = \xi_C / (3\beta_C \psi_C)$. The growth rate in length is therefore defined as $g_C(R, \ell) = \gamma_C(\ell_m R / (H_C + R) - \ell)$, which is equivalent with a von Bertalanffy growth law. For a more detailed discussion of the energy budget model and the derivation of the growth and reproduction function, we refer to the original publication by Kooijman and Metz (1984; see also de Roos, Metz, et al. 1990). Finally, for our purpose in this chapter, we assume that consumer mortality is size-independent and equal to a background mortality rate μ_C. The full specification of the model ingredients and equations is presented in box 3.6, and the parameter values are given in box 3.7. The model is parameterized for a specific consumer-resource interaction; in particular, the default parameters mimic the life history characteristics of the planktivorous fish species roach (*Rutilus rutilus*) (de Roos and Persson 2001, 2002), that is common in most Eurasian freshwater systems, feeding on a cladoceran zooplankton, *Daphnia spp.*

Figure 3.15 shows the changes in biomass observed in the Kooijman-Metz model in response to increases in mortality. The model accounts for a continuous size-distribution of consumers between the size at birth (7 mm) and the ultimate size that is determined by the equilibrium resource density. This size distribution is represented by the function $c(t, \ell)$ (see box 3.6). The biomass in a particular size range between a minimum length ℓ_1 and a maximum length

---

BOX 3.6

EQUATIONS AND FUNCTIONS OF THE BASIC KOOIJMAN-METZ MODEL

Dynamic equations	*Description*
$\dfrac{\partial c(t,\ell)}{\partial t} + \dfrac{\partial g_C(R,\ell)c(t,\ell)}{\partial \ell} = -d_C(R,\ell)c(t,\ell)$	Consumer length-distribution dynamics
$g_C(R,\ell_b)c(t,\ell_b) = \displaystyle\int_{\ell_j}^{\ell_m} b_C(R,\ell)c(t,\ell)d\ell$	Consumer total population reproduction
$\dfrac{dR}{dt} = G(R) - \displaystyle\int_{\ell_b}^{\ell_m} I_C(R,\ell)c(t,\ell)d\ell$	Resource biomass dynamics

Function	*Expression*	*Description*
$G(R)$	$\rho(R_{max} - R)$	Resource dynamics in consumer absence
$I_C(R,\ell)$	$M_C R/(H_C + R)\ell^2$	Resource ingestion rate by consumers
$g_C(R,\ell)$	$\gamma_C(\ell_m R/(H_C + R) - \ell)$	Consumer growth rate in length
$b_C(R,\ell)$	$\alpha_C R/(H_C + R)\ell^2$	Consumer fecundity
$d_C(R,\ell)$	μ_C	Consumer death rate
$w_C(\ell)$	$\beta_C \ell^3$	Consumer weight-length scaling relation

BOX 3.7
PARAMETERS OF THE BASIC KOOIJMAN-METZ MODEL

Parameter	Default value	Unit	Description
Resource			
ρ	0.1	day^{-1}	Resource turnover rate
R_{max}	10	mg/L	Resource maximum biomass density
Consumer			
ℓ_b	7	mm	Length at birth
ℓ_j	110	mm	Length at maturation
ℓ_m	300	mm	Maximum length under unlimited food
β_C	0.009	$mg \cdot mm^{-3}$	Weight-length scaling constant
M_C	0.1	$mg \cdot day^{-1} \cdot mm^{-2}$	Maximum ingestion-length scaling constant
H_C	0.015	mg/L	Ingestion half-saturation resource density
γ_C	0.006	day^{-1}	Growth rate in length scaling constant
α_C	0.003	$day^{-1} \cdot mm^{-2}$	Maximum fecundity-length scaling constant
μ_C	0.01	day^{-1}	Consumer background mortality rate

Parameter values mimic the life history of roach (*Rutilus rutilus*), feeding on *Daphnia spp.*

ℓ_2 then equals an integral over this density function $c(t, \ell)$, weighted by the weight-length relationship $w_C(\ell) = \beta_C \ell^3$ (de Roos 1997):

$$\int_{\ell_1}^{\ell_2} \beta_C \ell^3 c(t, \ell) \, d\ell.$$

The ultimate size reached by consumers under equilibrium conditions is given by $\tilde{\ell} = \ell_m \tilde{R}/(H_C + \tilde{R})$. At background mortality, the ultimate size, which is reached at old age, is only slightly larger than the size at which individuals mature (110 mm). With increasing mortality the resource density at equilibrium increases until it reaches its maximum value ($\tilde{R} = R_{max} = 10$ mg/L), close to the mortality threshold above which the consumer population is extinct. Because the equilibrium resource density at these high mortality rates is more than two orders of magnitude larger than the half-saturation resource density

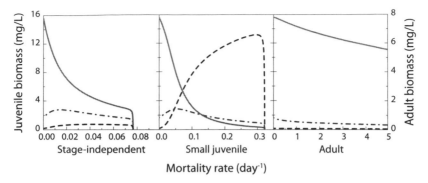

FIGURE 3.15. Changes in biomass of juvenile consumers smaller than 50 mm (*dashed-dotted lines*), juvenile consumers larger than 50 mm (*solid lines*), and adult consumers (*dashed lines*) with increases in stage-independent mortality, mortality targeted at juveniles smaller than 50 mm, and adult mortality on top of the background mortality in the size-structured Kooijman-Metz model for default parameter values (see box 3.7). Notice the very different scaling of the mortality axis in the right-hand panel.

H_C, the ultimate size that individual consumers can reach here is close to their maximum size ℓ_m. The changes in consumer size-distribution are hence rather intricate, especially when mortality targets only a certain size range of consumers. To simplify the representation of these changes, we represent only the changes in biomass density of small-juvenile consumers, defined as all individuals with a body length smaller than 50 mm, large-juvenile consumers, defined as all individuals with a body length between 50 mm and the maturation length $\ell_j = 110$ mm, and all adult consumers that are larger than 110 mm. We have, however, confirmed that the exact choice of the limits to these size ranges does not influence the results presented.

At background mortality rate the population is dominated by large-juvenile individuals that are roughly ten times more abundant (in terms of biomass) than small-juvenile consumers (figure 3.15, left panel). Adults make up only a very small fraction of the total population biomass and, as pointed out before, reach sizes only just above the maturation threshold. The size distribution of the population is hence bell-shaped and can be classified as stunted. Increases in stage-independent mortality decrease the biomass density of large juveniles and increase the biomass density of both small-juvenile as well as adult consumers. These increases in biomass density are partly due to the disproportionately large increase in both the total population reproduction as well as maturation rate (results not shown). More specifically, the relative increase in both these population rates is significantly larger than the relative increase in mortality. The increase in total population reproduction rate gives rise to the increase in biomass of small juveniles. The increase in adult biomass is only partly due to the increased recruitment to the adult stage. Another factor contributing to the increase in adult consumer biomass is their larger scope for somatic growth with increasing mortality, as the higher equilibrium resource densities allow them to grow to much larger body sizes.

The changes in biomass in response to increasing mortality that specifically targets the small-juvenile consumers are qualitatively identical to the changes with stage-independent mortality (figure 3.15, middle panel). Quantitatively, the increases in biomass are more pronounced, especially in adult consumer biomass. The same qualitative pattern of biomass changes is also observed in cases where the mortality targets only juvenile consumers with a length between 50 mm and the maturation length (110 mm, results not shown). With mortality targeting these large-juvenile consumers, the overcompensation in small-juvenile biomass is more pronounced than the overcompensation observed with stage-independent or small-juvenile mortality. These quantitative differences come about because small juveniles are not exposed to the increasing mortality and their biomass density is hence influenced only positively by the increased population reproduction. The overcompensation in adult biomass with increasing mortality targeting large-juvenile consumers is, however, less pronounced than observed with stage-independent or small-juvenile mortality, because the increasing mortality targets individual consumers just before they mature and dampens the increase in total population maturation rate that is one of the factors contributing to the adult biomass overcompensation.

The changes with increases in mortality that targets only adult consumers differ, as biomass densities in all consumer size classes only decrease (figure 3.15, right panel). Moreover, the population can persist over a much wider range of adult mortality rates than with increasing stage-independent or other types of stage-specific mortality. The threshold of adult mortality, at which the population goes extinct, is more than two orders of magnitude larger than the corresponding threshold of stage-independent mortality. Even though the equilibrium resource biomass density increases with increases in adult mortality and individual consumers can thus potentially reach lengths close to their maximum body length ℓ_m, survival in the adult stage is so low that only a negligible fraction of the consumers recruited to the adult stage manage to grow that big. Technically, there hence is an increase in the biomass of larger adults (larger than 120 mm), but quantitatively this increase is insignificant.

These changes in stage-specific biomass density in the Kooijman-Metz model with increases in either stage-independent or stage-specific mortality closely resemble the changes in biomass in the stage-structured biomass model with two juvenile stages and a single adult stage, in which large-juvenile consumers are more limited in their energetics than small juveniles and adults. In fact, the pattern of biomass changes with increasing stage-independent mortality is qualitatively identical to the changes in biomass in the stage-structured model when $q_2 < q_1, q_3$ (figure 3.13, middle panel) and consumers grow over a substantial range of body sizes during life (that is, for small values of the product $z_1 z_2$ in the stage-structured model). Like the Kooijman-Metz model, the stage-structured biomass model predicts overcompensation to occur under these conditions in the biomass of small juveniles and adults with stage-independent as well as stage-specific mortality, except for increases in adult mortality.

With the default parameterization of the Kooijman-Metz model that we have used, individual consumers can indeed grow substantially during life, possibly reaching body weights that are more than four orders of magnitude larger than their weight at birth. Large juveniles are energetically worse off than small juveniles, because energy gains through ingestion scale with weight to the power 2/3, while energy requirements for maintenance are proportional to weight. Overall the net energy production of consumers thus declines with body size. Even though this would suggest that adults would be more energetically limited than large juveniles, the opposite is true, because juveniles are assumed to spend a fraction $1 - \kappa_C$ of their net energy assimilation on the development of reproductive organs. Although the Kooijman-Metz energy budget model assumes that this increase in the state of maturation of juvenile consumers costs energy, it does not contribute to consumer body size and hence does not effectively increase biomass. Adult consumers use this fraction of their assimilated energy to produce newborns, which does effectively contribute to an increase in population biomass. In terms of production of new biomass, large juvenile consumers are therefore more limited in their energetics than adult consumers, despite the fact that relatively speaking net energy production by adult consumers is smaller. Even though in the Kooijman-Metz model adult consumers grow indeterminately and continuously throughout life, the pattern of biomass overcompensation with increasing mortality is therefore adequately captured by the stage-structured biomass model with nongrowing adult consumers.

ONTOGENETIC SYMMETRY AND
BIOMASS OVERCOMPENSATION

The finding that stage-specific biomass can increase in response to an increase in mortality, even when the mortality targets the stage exhibiting overcompensation, is counterintuitive. Furthermore, it has potentially far-reaching consequences for the structure of ecological communities, which will be discussed in more detail in the following chapters. Together, this raises several questions, such as, under what conditions we can expect biomass overcompensation to occur. When should we bother about it? Will it occur in any population model that recognizes differences between individuals in age, stage, or body size? Or what are the necessary ingredients that a model must possess to exhibit stage-specific biomass overcompensation? We can shed more light on these issues by analyzing some properties of the general size-structured population model that is introduced in detail in section 1 of the technical appendices. Such a model accounts for a continuous size-distribution of consumers that are characterized by growth rates in body size, birth rates, and mortality rates, which can be arbitrary functions of individual body size and the resource density in the environment in which the population is living. Because the analysis is rather mathematical, however, and requires a certain level of comfort with size-structured population models formulated in terms of partial differential equations (PDEs), we present the details in sections 6 and 7 of the technical appendices. In this section we will only summarize and discuss the main outcomes of that analysis.

One of the conclusions emerging from the two-stage biomass model is that, generically, in equilibrium the juvenile stage is a source of biomass production and the adult stage is a sink, or vice versa (see table 3.2). It is rather straightforward to show that this conclusion also holds for the general size-structured population model described by the system of equations (A.2) in section 1 of the technical appendices. In particular, irrespective of whether adult individuals grow after maturation and irrespective of the particular form of the growth rate in body size and the scaling of adult fecundity with body size, the general size-structured population model predicts that at equilibrium either the juvenile or the adult life history stage is a source of biomass production with the other stage constituting a biomass production sink. Moreover, the sign of the difference between total population reproduction and maturation rate is an indicator of the source-sink status in terms of biomass production in the juvenile and adult life stage: Whenever the juvenile life stage is a source of biomass production, the total population reproduction rate (i.e., the recruitment rate to the juvenile stage) in terms of biomass is smaller than the total population maturation rate (i.e., the recruitment rate to the adult stage). And, vice versa, whenever the adult life stage is a source of biomass production, the total population reproduction rate is larger than the maturation rate.

The two-stage biomass model does not, however, account for differences within the juvenile or adult stage. The source-sink status in terms of biomass production in the juvenile stage as a whole is thus representative for juveniles of all body sizes. Similarly, ontogenetic symmetry in biomass production between juveniles and adults, characterized by equal mass-specific biomass production rates through either somatic growth or reproduction (i.e., $v_J(R) = v_A(R)$), implies that mass-specific biomass production is the same for all individuals. In the more general size-structured population model presented in section 1 of the technical appendices, however, such ontogenetic symmetry in biomass production for every body size s occurs only if the mass-specific biomass production through reproduction and somatic growth is size-independent, that is, if

$$\frac{b(R,s)s_b + g(R,s)}{s} = h(R) \qquad \text{for all } s \geq s_b \qquad (3.14)$$

Here $h(R)$ represents some arbitrary function that is possibly resource- but not size-dependent. If all individuals experience the same background mortality rate $d(R)$ independent of their body size, the general size-structured model can be shown to simplify under these conditions to the following population model for the dynamics of the total population biomass C:

$$\frac{dC}{dt} = \big(h(R) - d(R)\big)C.$$

This model closely resembles the Yodzis and Innes model with dynamics that are not influenced by the population size distribution but that depend only on total population biomass. On the one hand, this derivation reveals that size-structured models and the Yodzis and Innes model are identical to each other in cases where there is ontogenetic symmetry in biomass production and mortality is independent of body size. In size-structured population models the ontogenetic symmetry requires the special relationship between the growth and reproduction rate expressed by equation (3.14), whereas in the Yodzis and Innes model (Yodzis and Innes 1992) this ontogenetic symmetry is assumed implicitly, as differences between individuals are not accounted for (note also that this is equally the case for truly unstructured models that account only for population abundance in terms of numbers of individuals).

On the other hand, the derivation shows that in equilibrium $h(\tilde{R})$ equals $d(\tilde{R})$, that is, the population will equilibrate in a state that is characterized by ontogenetic symmetry in biomass *turnover*, in which the rate of biomass production through somatic growth and reproduction equals the biomass loss rate through mortality *for every body size s*. The combination of ontogenetic symmetry in biomass production and size-independent mortality is, however, only one set of conditions that will lead to ontogenetic symmetry in biomass turnover at equilibrium. More generally, such ontogenetic symmetry in biomass turnover will occur whenever

$$b(\tilde{R},s)s_b + g(\tilde{R},s) = d(\tilde{R},s)s \qquad \text{for all } s \geq s_b. \qquad (3.15)$$

As before, this condition stipulates that in equilibrium the biomass loss rate through mortality, $d(\tilde{R},s)s$, equals the sum of the biomass production through somatic growth, $g(\tilde{R},s)$, and reproduction, $b(\tilde{R},s)s_b$, irrespective of an individual's body size s.

As in the two-stage biomass model, a state with ontogenetic symmetry in biomass turnover represents a very special condition in the general size-structured population model presented in section 1 of the technical appendices. As shown in section 6 of the technical appendices, such a state necessarily corresponds to an equilibrium state. In this equilibrium the population size distribution in terms of numbers of individuals is inversely proportional to the squared body size s^{-2}. Consequently, the biomass size distribution is inversely proportional to body size itself, and the total biomass within logarithmically spaced size ranges is constant. For no apparent reason, this corresponds to the expected scaling of the body size spectra in marine communities, as conjectured by Sheldon, Prakash, and Sutcliffe (1972; see also Andersen and Beyer 2006).

It can also be shown that the equilibrium biomass density in an arbitrary juvenile size range between s_b and some upper bound $s_0 < s_j$ will *decrease* compared with the symmetric state, if this size range turns into a net source of biomass production, and *increase* when it becomes a net sink. Body size ranges in which net biomass turnover is positive are thus underrepresented, while ranges with negative net biomass turnover are overrepresented in the population biomass at equilibrium. Furthermore, the *decrease* in biomass density in a size range between s_b and s_0 with a positive net biomass production will nonetheless lead to an *increase* in total juvenile biomass as compared with the symmetric state, if the size range between s_0 and s_j is a net-sink of biomass production, such that the biomass losses in this latter range exactly compensate for the gains in the size range between s_b and s_0. In this latter case there is hence ontogenetic symmetry in biomass turnover between the juvenile and adult stage as whole, but not for every body size within those stages.

We have argued that biomass overcompensation results from the breakup of ontogenetic symmetry in net biomass turnover and from distinct asymmetries between differently sized individuals. In the two-stage biomass model the type of biomass overcompensation that is likely to occur is linked to the source-sink status in terms of biomass production of the juvenile and adult life stage. Generically, when the juvenile stage is a net source of biomass production, adults make up the largest part of population biomass at background mortality, and juvenile biomass is likely to exhibit overcompensation in response to increases in any type of mortality. Vice versa, when adults are the net source of biomass production, juveniles dominate population biomass at background mortality, and adult biomass is likely to exhibit overcompensation. We have referred to equilibria characterized by these two different conditions as being under reproduction control and development control.

In contrast, in the Kooijman-Metz consumer-resource model, at background mortality juveniles are most abundant, as their biomass constitutes more than 99 percent of total population biomass. Nevertheless, the juvenile stage is a net source of biomass production with a significantly *larger* total population maturation rate in terms of biomass than total population reproduction rate (results not shown). Despite this larger maturation than reproduction rate, the population is under development control, given the very low abundance of adults and the increase and decrease in adult and total juvenile biomass, respectively, in response to increases in stage-independent mortality. This combination of development control and dominance by juveniles in terms of biomass, together with the juvenile stage as a net source of biomass production, contrasts with the conceptualization that emerged from the stage-structured biomass model with only a juvenile and adult stage.

As discussed above, if small and large juveniles represent net sources and sinks of biomass production, respectively, a population at equilibrium can indeed be under development control and dominated by juveniles in terms of biomass, despite that the juvenile stage is a net source of biomass production. The contrast between the Kooijman-Metz model and the two-stage biomass model arises because of such differences in source-sink status between differently sized juvenile individuals. In the case of the results for the Kooijman-Metz model shown in figure 3.15, the size range of small juveniles, with lengths between 7 and 50 mm, is a net source of biomass production, given that at background mortality the total population reproduction rate in terms of biomass is roughly 150 times smaller than the rate at which biomass leaves this size range and is recruited to the large-juvenile size range.

In contrast, the size range of large juveniles, with lengths between 50 mm and the maturation length (110 mm), is a net sink of biomass production with a recruitment rate that is some fifty times larger than the rate at which biomass is recruited to the next, adult stage. As a whole, this nonetheless makes the juvenile size range a net source of biomass production with development control operating in the large-juvenile size range. Differences in energetics among juveniles or among adults may therefore disrupt the tight link between the source-sink status of the juvenile and adult stage on the hand and the mode of control (development or reproduction), which emerged from the stage-structured biomass model with only one juvenile and one adult stage.

We illustrated the occurrence of stage-specific biomass overcompensation in this chapter using the stage-structured biomass model with a variable number of distinct life history stages and the size-structured Kooijman-Metz model, which accounts for a continuous size-distribution of consumers that grow indeterminately through their entire life. The stage-structured biomass model also accounts for a continuous size-distribution of consumers within stages, except for the last life history stage, which is assumed to represent a class of nongrowing adult individuals. Both model frameworks share the property that growth in body size of individual consumers depends on resource density and thereby indirectly on population feedback. However, the stage-structured biomass model is based on a type of individual energy budget in which the difference between assimilated energy acquired through foraging and the energetic costs to cover maintenance requirements is partitioned over somatic growth and reproduction. In such net-production models (e.g., Lika and Nisbet 2000) maintenance requirements hence take precedence over any other energy-requiring life history process. In contrast, the Kooijman-Metz energy budget model is based on an assumption that net energy assimilation acquired through foraging is partitioned over, on the one hand, growth plus maintenance and, on the other hand, reproduction (Kooijman and Metz 1984; see also Kooijman 2000). In this so-called net-assimilation model maintenance requirements hence do not take complete precedence over other life history processes, and in particular adult fecundity is proportional to resource ingestion.

The analysis in section 7 of the technical appendices now shows that two mechanisms can give rise to biomass overcompensation: (1) resource-dependent growth in body size and (2) adult energetics that follows a net-production energy channeling scheme with non-zero maintenance costs. The analysis focuses on the occurrence of overcompensation in juvenile biomass, as this can be expected to occur most readily. This type of overcompensation is always the consequence of an increase in total population birth rate with increasing mortality. When adult energetics follows a net-production energy channeling scheme, the increase in expected adult lifetime reproduction in response to an increase in resource density can be disproportionately larger than the increase in cumulative food ingestion as an adult. As a consequence, with increasing mortality the expected lifetime reproduction as an adult can remain equal to 1, which is the condition for equilibrium to occur, while expected lifetime food ingestion as an adult *decreases*. This more efficient use of ingested resource forms the basis of the positive response in total population birth rate in equilibrium.

In the discussion of biomass overcompensation in previous sections we have already pointed out the crucial importance of non-zero maintenance requirements and disproportionally large responses in population reproduction or maturation rates in response to decreased competition and increased resource densities. The analysis in section 7 of the technical appendices shows that this type of overcompensation can occur even when growth in body size is independent of resource density. It can therefore also occur in population models, in which individuals are classified by their age or developmental stage, with fixed stage durations. Furthermore, it can occur with increasing juvenile, adult, or stage-independent mortality, that is, in age or size ranges that precede or overlap with the range exposed to additional mortality. It cannot, however, occur in age or size ranges beyond the range exposed to the additional mortality. Hence, adult biomass cannot increase in response to increased juvenile mortality.

The second mechanism that can lead to overcompensation in juvenile biomass is food-dependent growth in body size, irrespective of whether maintenance costs are negligible or individual energetics follow a net-production scheme or not. For this type of overcompensation to occur, juveniles and adults have to share the same resource, and adults should have disproportionally larger increases in body size with an increase in resource than juveniles, as is the case in the size-structured Kooijman-Metz model. As a consequence, with increasing mortality the expected lifetime reproduction as an adult can remain equal to 1, ensuring equilibrium, and the expected lifetime food ingestion as an adult can remain constant, while the expected food ingestion as a juvenile decreases. This latter decrease implies that with increasing mortality juvenile consumers use their ingested energy more efficiently for biomass production, which forms the basis of the positive response in total population birth rate in equilibrium.

Finally, unless growth in individual body size is food dependent, the overcompensation in biomass that arises from an increase in total population birth rate will occur in all age or size intervals that are below a certain threshold. This threshold occurs at that point in the life history where the product of the population birth rate and the individual survival does not change with an increase in additional mortality. Beyond this age or size threshold the decrease in individual survival will outweigh the increase in total population birth rate. Overcompensation in age or size ranges beyond this threshold—in particular, overcompensation in adult biomass, as well as the co-occurrence of a decrease in juvenile biomass and an increase in adult biomass—can occur only if growth in body size is food dependent.

In summary, biomass overcompensation in juvenile size ranges arises through a disproportionately larger increase in lifetime adult offspring production with increases in resource density than the increase in cumulative lifetime resource intake. This leads to an increase in total population birth rate that is the basis of the overcompensation. Such conditions occur when adult energetics follow a net-production energy budget and maintenance costs are non-zero, or when food-dependent growth in body size leads to disproportionately larger increases in adult body size (and thus adult intake) than in juvenile body size. For overcompensation in adult size ranges, food-dependent growth in body size is a necessary condition.

Emergent Allee Effects through Biomass Overcompensation

Intuitively, the community consequences of biomass overcompensation in a particular life stage of a consumer population are relatively straightforward to deduce. In the previous chapter we found, for example, that stage-specific biomass may increase in response to an increase in mortality in the same stage. Now consider that the mortality is imposed by a predator that preys exclusively on the consumer stage exhibiting biomass overcompensation. A larger predator density will imply higher prey mortality from predation. Through biomass overcompensation, however, this will lead to an increase in biomass of the prey stage exposed to predation and consequently to an increase in food availability for the predator. In other words, a positive feedback emerges between the density of predators and the availability of its food, mediated through biomass overcompensation in the prey life history stage that it forages on.

As we will see below, this positive feedback between predation, prey availability, and thus predator population growth rate will manifest itself at the population-level as an Allee effect for the predator: a predator population at low density will decline to extinction, whereas at high densities predators will manage to establish themselves in a community with prey. This positive relation between predator density and its population growth rate does, however, not result from any positively density-dependent interactions among the predators themselves, which generally form the basis of an Allee effect (Allee 1938; Courchamp, Berec, and Gascoigne 2008). Instead, predators only interact with each other through exploitative competition for prey. The Allee effect emerges solely as a consequence of the demographic changes in the prey population, which are induced by the mortality that the predator imposes. For this reason we refer to this phenomenon as an "emergent Allee effect."

In the previous chapter it was also shown that an increase in stage-independent mortality may lead to an increase in either juvenile or adult biomass. By foraging indiscriminately on all prey individuals, a generalist predator may then increase the biomass of either juvenile or adult prey and thereby increase the food availability for another predator species, which specializes on the prey stage exhibiting the biomass overcompensation. Again a positive feedback arises between predation and the food availability for predators mediated through biomass overcompensation in the prey, only now between different predator populations. The generalist predator thus increases the population growth rate of the stage-specific predator and facilitates its persistence. However, as in the emergent Allee effect, the predators only interact through exploitative competition for the same prey population, and the positive influence of one predator species on the other emerges solely from the changes in prey demography induced by predation. Analogously, we refer to this phenomenon as "emergent facilitation."

Emergent Allee effects and emergent facilitation are two of the main consequences of biomass overcompensation in prey populations for community structure. These emergent community effects of biomass overcompensation occurring in prey populations will be discussed in detail in this and the following chapter, starting here with the positive effect of a predator on its own growth rate and focusing in chapter 5 on the positive effect that different predator species may exert on each other.

EMERGENT ALLEE EFFECTS IN STAGE-STRUCTURED BIOMASS MODELS

We will first explore the characteristics and consequences of emergent Allee effects using stage-structured biomass models to describe the dynamics of juvenile and adult prey individuals that are foraged on by a single predator population. We hence use the model as presented in box 3.2 of the previous chapter for both prey and predators. For the predator, however, we will assume that juveniles and adults experience similar mortality rates and prey on juvenile and adult prey following the same functional response. We showed in the previous chapter that under such conditions of equal ingestion and mortality rates for both juvenile and adult individuals, the stage-structured biomass model can be reduced to a single ordinary differential equation (ODE) for the total predator biomass P. This equation for total biomass is analogous to the Yodzis and Innes bioenergetics model (Yodzis and Innes 1992), as presented in equation (3.3b) in the previous chapter:

$$\frac{dP}{dt} = (\nu_P(J,A) - \mu_P)P$$

In this equation the function $\nu_P(J,A)$ represents the net biomass production per unit biomass of the predator, which depends on the density of juvenile and adult prey that it preys on, while μ_P represents the per-capita predator mortality (here and below we label functions and parameters pertaining to predators with the subscript P to distinguish them from analogous, consumer-related functions and parameters, which we label with the subscript C).

We will assume that all predators spend a fraction ϕ and $1 - \phi$ of their time feeding on juvenile and adult prey, respectively. Predator ingestion is hence described by the following type II functional response:

$$M_P \frac{\phi J + (1 - \phi)A}{H_P + \phi J + (1 - \phi)A}$$

The parameters M_P and H_P represent the predator's mass-specific maximum ingestion rate and half-saturation constant, respectively. These assumptions imply that the death rates of juvenile and adult consumers, which we indicate with $d_J(P)$ and $d_A(P)$, respectively, depend on predator biomass P, following:

$$d_J(P) = M_P \frac{\phi P}{H_P + \phi J + (1 - \phi)A}$$

and

$$d_A(P) = M_P \frac{(1 - \phi)P}{H_P + \phi J + (1 - \phi)A},$$

respectively. Notice that juvenile maturation will indirectly also depend on predator biomass through its dependence on the juvenile mortality rate.

As before, we assume net biomass production to equal the difference between assimilation from foraging and energetic costs to cover maintenance requirements (cf. equation (3.2)). Hence, the mass-specific rate of net biomass production equals:

$$v_P(J,A) = \sigma_P M_P \frac{\phi J + (1 - \phi)A}{H_P + \phi J + (1 - \phi)A} - T_P$$

In this equation the parameters σ_P and T_P represent the efficiency with which ingested food is assimilated and the mass-specific maintenance rate of the predator, respectively. The full model specification of this stage-structured, predator-prey biomass model is summarized in box 4.1.

Model parameters for resource, juvenile, and adult consumer dynamics are taken to be as much as possible equal to the default values used in the previous chapter (see boxes 3.3 and 3.4). For predators we assume that the default values of M_P, H_P, σ_P, T_P, and μ_P follow the same scaling relationships with body size as derived in box 3.4 in the previous chapter. That is, we assume that predators are as efficient as their prey when assimilating food so that $\sigma_P = 0.5$, while H_P is taken to be independent of body size and equal to 3.0. M_P, T_P, and μ_P are

BOX 4.1
Equations and Functions of the Stage-Structured Predator-Prey Biomass Model

Dynamic equations	Description
$\dfrac{dR}{dt} = G(R) - \omega_J(R)J - \omega_A(R)A$	Resource biomass dynamics
$\dfrac{dJ}{dt} = v_A^+(R)A - \gamma(v_J^+, d_J)J + v_J(R)J - d_J(P)J$	Biomass dynamics of juveniles
$\dfrac{dA}{dt} = \gamma(v_J^+, d_J)J + (v_A(R) - v_A^+(R))A - d_A(P)A$	Biomass dynamics of adults
$\dfrac{dP}{dt} = (v_P(J,A) - \mu_P)P$	Biomass dynamics of predators

Function	Expression	Description
$G(R)$	$\rho(R_{\max} - R)$	Intrinsic resource turnover
$\omega_J(R)$	$M_C R / (H_C + R)$	Resource intake rate by juveniles
$\omega_A(R)$	$q M_C R / (H_C + R)$	Resource intake rate by adults
$v_J(R)$	$\sigma_C \omega_J(R) - T_C$	Net energy production of juveniles
$v_A(R)$	$\sigma_C \omega_A(R) - T_C$	Net energy production of adults
$d_J(P)$	$\mu_J + M_P \dfrac{\phi P}{H_P + \phi J + (1-\phi)A}$	Mortality rate of juveniles
$d_A(P)$	$\mu_A + M_P \dfrac{(1-\phi)P}{H_P + \phi J + (1-\phi)A}$	Mortality rate of adults
$\gamma(v_J^+, d_J)$	$(v_J^+(R) - d_J(P))/\left(1 - z^{(1 - d_J(P)/v_J^+(R))}\right)$	Maturation rate of juveniles
$v_P(J,A)$	$\sigma_P M_P \dfrac{\phi J + (1-\phi)A}{H_P + \phi J + (1-\phi)A} - T_P$	Net energy production of predators

$v_J^+(R)$ and $v_A^+(R)$ represent the values of $v_J(R)$ and $v_A(R)$, respectively, but restricted to non-negative values. For brevity, the dependence of v_J^+ and d_J on resource and predator density, respectively, is suppressed in the maturation function $\gamma(v_J^+, d_J)$. Notice that $\gamma(v_J^+, d_J)$ is positive as long as $v_J^+(R)$ is positive. Also, $v_A^+(R) - v_A(R)$ cancels as long as $v_A(R)$ is positive and equals $-v_A(R)$ otherwise. Hence, the use of $v_J^+(R)$ and $v_A^+(R)$ in the model equations implies that growth, maturation, and fecundity are positive as long as juvenile and adult net production, respectively, are positive and equal 0 otherwise.

assumed inversely proportional to the quarter power of predator body size with proportionality constants equal to 0.1, 0.01, and 0.0015, respectively (see box 3.4). Predator body size we indicate with W_P, which we take to be equal to 0.01 gram, given that we have used $W_A = 0.0001$ gram as our characteristic, (adult) prey body size and that predators are typically two orders of magnitude larger than their prey (Peters 1983). As a consequence, M_P, T_P, and μ_P have the default values of 0.32, 0.032, and 0.0047, respectively. Notice, however, that the default for the predator per-capita mortality rate μ_P represents a minimum value, as we will investigate model predictions as a function of predator mortality rate *in excess of* this background level. For completeness, box 4.2 summarizes the default model parameter values for resource, consumer, and predator.

PREDATORS SPECIALIZING ON JUVENILE PREY

Varying Resource Productivity

Figure 4.1 shows the changes in equilibrium biomass of basic resource, juvenile and adult prey, and a top predator that forages exclusively on juvenile prey ($\phi = 1$) for increasing values of the maximum resource density R_{max}. For comparison, the figure also shows these changes for a predator that forages with equal intensity on both juvenile and adult prey ($\phi = 0.5$). For these graphs the parameter values for the prey population are chosen such that adult prey dominate the population biomass in the absence of predators ($q = 0.5$, $z = 0.01$; see also figure 3.5 in chapter 3). Hence, with increasing mortality, biomass overcompensation in juvenile prey biomass occurs.

For small R_{max} values, resource densities are too low for the prey population to persist, and hence equilibrium resource density increases linearly with productivity, while equilibrium prey and predator densities are both zero. An expression for the equilibrium resource biomass in the presence of prey can be derived by combining the conditions $dJ/dt = 0$ and $dA/dt = 0$ (see equations (3.5b) and (3.5c) in chapter 3), realizing that juvenile and adult prey mortality rates both equal $\mu_C = 0.0015W_A^{-0.25}$:

$$0 = \frac{1}{J}\frac{dJ}{dt}$$

$$= v_A(\tilde{R})\frac{\tilde{A}}{\tilde{J}} + v_J(\tilde{R}) - \gamma(v_J(\tilde{R}),\mu_C) - \mu_C$$

$$= v_A(\tilde{R})\frac{\gamma(v_J(\tilde{R}),\mu_C)}{\mu_C} + v_J(\tilde{R}) - \gamma(v_J(\tilde{R}),\mu_C) - \mu_C$$

BOX 4.2

PARAMETERS OF THE STAGE-STRUCTURED PREDATOR-PREY BIOMASS MODEL

Parameter	Default value	Unit	Description
Resource			
ρ	0.1	day^{-1}	Resource turnover rate
R_{max}	30	mg/L	Resource maximum biomass density
Prey			
M_C	$0.1W_A^{-0.25}$	day^{-1}	Mass-specific maximum ingestion rate
H_C	3.0	mg/L	Ingestion half-saturation resource density
q	0.5 or 2.0	—	Adult-juvenile consumer ingestion ratio
T_C	$0.01W_A^{-0.25}$	day^{-1}	Mass-specific maintenance rate
σ_C	0.5	—	Conversion efficiency
z	0.01 or 0.5	—	Newborn-adult consumer size ratio
μ_J	$0.0015W_A^{-0.25}$	day^{-1}	Juvenile background mortality rate
μ_A	$0.0015W_A^{-0.25}$	day^{-1}	Adult background mortality rate
Predator			
M_P	$0.1W_P^{-0.25}$	day^{-1}	Mass-specific maximum ingestion rate
H_P	3.0	mg/L	Ingestion half-saturation prey density
T_P	$0.01W_P^{-0.25}$	day^{-1}	Mass-specific maintenance rate
σ_P	0.5	—	Conversion efficiency
μ_P	$0.0015W_P^{-0.25}$	day^{-1}	Predator background mortality rate
ϕ	0.0–1.0	—	Predator foraging preference for juveniles

W_A and W_P represent the average adult body weight (in grams) of consumer and predator, respectively.

In the second step of this derivation the condition $1/J \, dJ/dt = 0$ is simplified by substituting the expression for the adult-juvenile prey ratio that can be derived from the equilibrium condition $dA/dt = 0$. By substitution of the expression for the maturation function $\gamma(v_J(\tilde{R}), \mu_C)$ this equation can be rewritten as:

$$\frac{v_A(\tilde{R})}{\mu_C} z^{\mu_C/v_J(\tilde{R})-1} = 1 \tag{4.1}$$

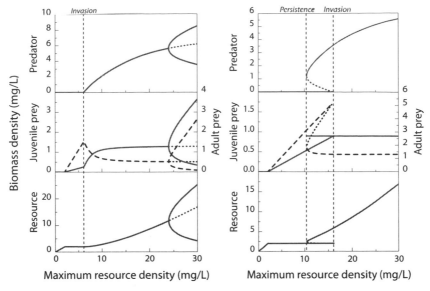

FIGURE 4.1. Changes in equilibrium biomass of basic resource (*bottom*), juvenile (*middle, solid lines*) and adult prey (*middle, dashed lines*), and top predator (*top panels*) with increasing maximum resource density R_{max}. Left and right panels show changes in equilibrium biomass for the cases where predators forage equally on both juvenile and adult prey ($\phi = 0.5$) or exclusively on juvenile prey ($\phi = 1$), respectively. Stable equilibria are indicated with solid or dashed lines, unstable equilibria with dotted lines. For $\phi = 0.5$ and maximum resource densities larger than 24 mg/L, population biomasses exhibit stable oscillations around an unstable equilibrium. Minimum and maximum biomass densities during these cycles are indicated with solid and dashed lines (*left panels*). For $\phi = 1$ and maximum resource densities between 10 and 16 mg/L, one unstable and two stable equilibria co-occur. Throughout, $q = 0.5$ and $z = 0.01$. Otherwise, default parameters apply for a prey and predator with body size $W_A = 0.0001$ and $W_P = 0.01$ gram, respectively.

The left-hand side of the equation that results is composed of three terms with a clear biological interpretation, as it is the product of the probability that a newborn individual reaches maturation, $z^{\mu_C/\nu_J(\tilde{R})}$, the adult fecundity in terms of number of newborn individuals produced per day, $\nu_A(\tilde{R})/z$, and the expected lifetime as an adult after maturation, $1/\mu_C$. The left-hand side of the equilibrium condition (4.1) can hence be interpreted as the average number of offspring that a newborn prey individual will produce throughout its expected lifetime. Obviously, the expected lifetime offspring production should equal 1 in equilibrium, in which case every newborn individual exactly replaces itself during life. Because both adult fecundity and juvenile survival are increasing functions of the resource density R, condition (4.1) determines a unique equilibrium resource density $\tilde{R}_C$. As long as R_{max} is smaller than $\tilde{R}_C$, prey population growth rate is always negative, and prey can hence not survive, whereas prey can establish themselves in an environment without any prey and persist as soon as $R_{max} > \tilde{R}_C$. Hence, when R_{max} exceeds $\tilde{R}_C$, a stable equilibrium of resource and prey is found, in which equilibrium resource biomass is equal to $\tilde{R}_C$ and remains constant with further increases in R_{max}. In this range of R_{max} values equilibrium prey density increases virtually linearly with further increases in R_{max} (see figure 4.1).

The increases in equilibrium prey density with increasing R_{max} allow for the invasion and establishment of a predator population as soon as prey densities are sufficiently high. Predator population growth rate is positive when:

$$\sigma_P M_P \frac{\phi \tilde{J} + (1-\phi)\tilde{A}}{H_P + \phi \tilde{J} + (1-\phi)\tilde{A}} - T_P - \mu_P > 0 \quad \Leftrightarrow$$

$$\phi \tilde{J} + (1-\phi)\tilde{A} > \frac{H_P(T_P + \mu_P)}{\sigma_P M_P - T_P - \mu_P}$$

For a predator that forages with equal intensity on both juvenile and adult prey ($\phi = 0.5$), this happens when total prey biomass $\tilde{J} + \tilde{A}$ fulfills the condition:

$$\tilde{J} + \tilde{A} > 2\frac{H_P(T_P + \mu_P)}{\sigma_P M_P - T_P - \mu_P}$$

With the parameter values used in figure 4.1, this occurs for maximum resource densities larger than $R_{max} \approx 6\,mg/L$ (figure 4.1, left panel, vertical dashed line). We will generally refer to such a threshold value of any particular parameter as the "invasion threshold," irrespective of this parameter being the maximum resource density or the predator mortality rate. The invasion threshold separates two ranges of parameters that do and do not, respectively, allow for invasion and establishment of the predator in a resource-prey equilibrium. Analogously, we will refer to a threshold value of any particular parameter that separates two ranges of parameters for which a resource-prey-predator equilibrium can and cannot occur, respectively, as the "persistence threshold."

For a predator that forages with equal intensity on both juvenile and adult prey ($\phi = 0.5$), the changes in equilibrium biomass conform to the classical pattern of changes with increasing productivity described by Oksanen, Fretwell, et al. (1981). According to their unstructured food chain model, the level of resource productivity uniquely determines the equilibrium state, and the number of trophic levels in the food chain at equilibrium increases in a stepwise and discrete manner with increasing productivity. Furthermore, the threshold for predator invasion and predator persistence coincide, and for productivity levels exceeding the invasion threshold equilibrium predator biomass increases monotonically with productivity. Quantitative characteristics of this pattern are that the equilibrium biomass density of the next-to-highest trophic level does not change with changes in productivity and that biomass densities of nonadjacent trophic levels (i.e., resource and predator) co-vary.

Accordingly, figure 4.1 reveals that the invasion and persistence thresholds coincide: For all R_{max} values above the invasion/persistence threshold, predators can invade a resource-prey equilibrium, leading to a coexistence equilibrium of predator, prey, and resource, whereas predators are unable to invade a resource-prey equilibrium and always go extinct for R_{max} values below the threshold. Furthermore, when prey and predator coexist, total prey biomass remains constant, while both resource and predator biomass increase with further increases in R_{max}. Notice, however, that despite the constant total biomass, the biomasses of juvenile and adult prey do change, with juvenile prey biomass increasing and adult prey biomass decreasing with R_{max}.

At maximum resource densities above $R_{max} \approx 24$ mg/L, the coexistence equilibrium of predator, prey, and resource is unstable, and population biomass exhibits regular population cycles. In addition to the unstable equilibrium, figure 4.1 hence also indicates for this range of parameters the minimum and maximum biomass densities observed during the cycles. These population oscillations resemble the classical predator-prey, paradox-of-enrichment (Rosenzweig 1971), or prey-escape (de Roos, Metz, et al. 1990) cycles that are also observed in unstructured predator-prey models of the Lotka-Volterra type.

Figure 4.1 shows that the pattern of changes in equilibrium biomass when predators forage exclusively on juvenile prey ($\phi = 1$) differs specifically in these two aspects: the threshold levels of productivity that allow for predator invasion in a resource-prey equilibrium and that allow for predator persistence do not coincide, while the changes in equilibrium predator biomass with increasing values of R_{max} are not monotonic. The invasion threshold for predators that forage exclusively on juveniles is determined by the condition that juvenile biomass in the resource-prey equilibrium reaches the density $\tilde{J}_P$ that predators require for persistence. This critical juvenile biomass density is given by:

$$\tilde{J}_P = \frac{H_P(T_P + \mu_P)}{\sigma_P M_P - T_P - \mu_P} \tag{4.2}$$

With the parameter values used in figure 4.1, juvenile prey are less abundant in the absence of predators than adult prey and hence the invasion threshold for a juvenile-specialized predator occurs at a higher value of R_{max} than for a predator that forages on juveniles and adults equally. Furthermore, the curve relating equilibrium predator biomass to the maximum resource density is folded, starting from a zero predator density at the invasion threshold and extending to positive predator densities at *lower* values of R_{max}. The curve reaches a limit point around $R_{max} \approx 10\,mg/L$, which represents the lowest value of the maximum resource density that allows for predator persistence. At this persistence threshold the curve folds forward toward larger values of R_{max}. From the persistence threshold equilibrium predator biomass increases monotonically with R_{max}, as does resource biomass, while adult prey biomass monotonically decreases and juvenile prey biomass stays constant at $\tilde{J}_P$ with increasing R_{max}.

The curve relating equilibrium predator biomass to R_{max} thus consists of an upper section with high predator biomass densities and a lower section between the invasion and persistence threshold with predator densities down to zero. All points that make up this lower curve section correspond to unstable equilibrium points (so-called saddle points), while all points making up the upper section of the equilibrium curve represent stable equilibrium points. Unlike for predators that forage on all prey equally, the equilibrium of juvenile-specialized predators, prey, and resource does not destabilize for higher values of the maximum resource density, and population cycles are hence not observed. Most likely, this absence of classical predator-prey cycles is due to the refuge from predation that prey experience in the adult stage. Such an invulnerable prey or host stage has also been shown to stabilize the equilibrium in other predator-prey models (Abrams and Quince 2005), as well as in host-parasitoid models with stage-specific parasitism (Murdoch, Briggs, and Swarbrick 2005).

The folded shape of the curve relating equilibrium predator biomass to maximum resource density is the direct result of the biomass overcompensation that occurs in juvenile prey biomass with increasing juvenile mortality. Figure 4.2 shows for three different values of R_{max} the relationship between juvenile and adult prey biomass and additional juvenile mortality. The figure moreover shows which predator biomass density is required to impose such levels of additional juvenile mortality. For the R_{max} value that corresponds to the predator persistence threshold ($R_{max} = 10.5\,mg/L$), juvenile prey biomass at background mortality levels is clearly below the density $\tilde{J}_P$ (equation (4.2); horizontal dashed lines in figure 4.2) that is needed to support a predator population in equilibrium.

However, with increasing juvenile mortality, juvenile prey biomass reaches a maximum exactly equal to $\tilde{J}_P$. For all lower values of R_{max}, juvenile prey biomass reaches a maximum that is below $\tilde{J}_P$, while for all higher R_{max} values, the maximum exceeds it. For the R_{max} value that corresponds to the predator invasion threshold ($R_{max} = 16\,mg/L$), juvenile prey biomass exactly equals $\tilde{J}_P$ at background mortality levels. For all R_{max} values that are larger, juvenile prey

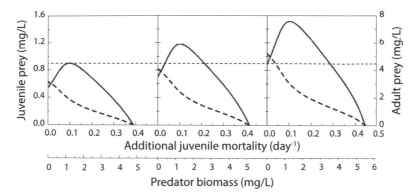

FIGURE 4.2. Changes in juvenile (*solid lines*) and adult (*dashed lines*) biomass with increases in juvenile mortality on top of the background mortality for $q = 0.5$ and $z = 0.01$ at the predator persistence threshold ($R_{max} = 10.5$ mg/L, *left*), its invasion threshold ($R_{max} = 16$ mg/L, *right*), as well as in between these two thresholds ($R_{max} = 13$ mg/L, *middle*). Otherwise, default parameters apply for a prey and predator with body size $W_A = 0.0001$ and $W_P = 0.01$ gram, respectively. The horizontal, thin dashed line in each graph indicates the juvenile biomass density that predators require for persistence. The additional axis at the bottom indicates the predator biomass density that is required to impose the additional juvenile mortality rate as displayed on the primary x-axes of the graphs. To compute the predator mass-specific foraging rate that relates these two different scales, it is assumed that juvenile biomass density equals the density that predators require for persistence.

biomass at background mortality is already sufficient for predators to persist, and hence in a resource-prey equilibrium the predator always finds enough food to grow and establish itself. For R_{max} values in between the predator persistence and invasion threshold, juvenile prey biomass at background mortality in a resource-prey equilibrium is hence insufficient for the predator to grow and invade. Once exposed to increasing mortality, however, juvenile prey biomass does exceed $\tilde{J}_P$. A high enough predator biomass density will therefore be able to raise the juvenile prey biomass and keep it at a level that is needed to support a population of predators.

In the range of R_{max} values between the persistence and the invasion threshold, a predator population at a low biomass density will hence go extinct, whereas it can grow and persist when starting at a sufficiently high biomass density. Such a positive relation between the density of a population and its fitness or capacity to grow is the hallmark of the Allee effect (Allee 1931, 1938; Courchamp, Berec, and Gascoigne 2008; also referred to, in Stephens, Sutherland, and Freckleton 1999, as the demographic Allee effect, as it refers to total fitness as opposed to only a fitness component). Associated with an Allee effect is the occurrence of a lower, unstable population equilibrium and a stable population equilibrium at higher density (see, for example, Courchamp, Berec, and Gascoigne 2008, fig. 1.9). Equilibrium points on the lower and upper curve sections between the persistence and invasion threshold in figure 4.1 (right panel) correspond to this unstable and stable equilibrium, respectively. Demographic Allee effects can be the result of a plethora of different mechanisms, such as predator dilution or saturation; anti-predator vigilance or aggression; cooperation on predation or resource defense; social thermoregulation; collective modification of the environment; increased availability of mates; increased pollination or fertilization success; conspecific enhancement of reproduction; and reduction of inbreeding, genetic drift, and loss of integrity as a result of hybridization (see Courchamp, Berec, and Gascoigne 2008, for an extensive discussion).

However, the positive relationship between predator density and population growth rate discussed here is not based on any such mechanisms. It has no relation whatsoever with the individual predators' behavior or life history, as they are simply exploitative foragers of the prey. Rather, the Allee effect in the predator results from the biomass overcompensation in the prey, which ultimately is the result of the interplay between competition for basic resource among prey individuals and their food-dependent growth in body size as well as reproduction. Therefore, because the demographic Allee effect that the predator experiences emerges as a result of density dependence in prey, we refer to it as an "emergent Allee effect" (de Roos and Persson 2002; de Roos, Persson, and Thieme 2003).

In single-species models for single population dynamics that incorporate an Allee effect, the unstable equilibrium at low population density acts as a threshold density. For population densities above this threshold, the population has a positive growth rate and starts to expand, whereas for densities below it population growth is negative, and the population declines to zero. This would suggest that the curve section between the persistence and invasion threshold in figure 4.1 (right panel) with unstable equilibrium points at low predator densities also represents threshold values, such that for a particular value of R_{max} any predator biomass density that is larger than the predator density in the unstable equilibrium would lead to predator invasion and establishment. This is, however, not true. The key difference is that our model of resource, stage-structured prey and a predator population is formulated in terms of four ODEs and is hence of a dimension larger than 1. The fact that an unstable equilibrium density at the same time separates the conditions of population growth from those of population decline holds only for one-dimensional models, that is,

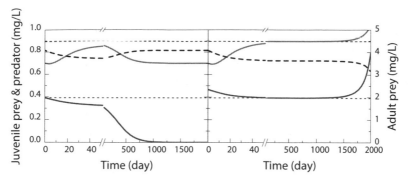

FIGURE 4.3. Dynamics of juvenile prey (*upper solid lines*), adult prey (*dashed lines*), and predator biomass (*lower solid lines*) on invasion of predators into the resource-prey equilibrium at $R_{max} = 13$ mg/L. The initial densities of resource (not shown) and juvenile and adult prey are equal to their values in the equilibrium without the predator on juveniles ($\phi = 1$). Left panel shows dynamics when the initial predator biomass equals the density it occurs at in the unstable resource-prey-predator equilibrium for $R_{max} = 13$ mg/L ($P = 0.392$ mg/L). This density is also indicated by the lower horizontal, thin dashed line in each graph. The upper horizontal, thin dashed line in each graph indicates the juvenile biomass density that predators require for persistence. Right panel shows the dynamics with the lowest predator biomass density that leads to predator establishment ($P = 0.467$ mg/L). Throughout, $q = 0.5$ and $z = 0.01$. Otherwise, default parameters apply for a prey and predator with body size $W_A = 0.0001$ and $W_P = 0.01$ gram, respectively.

models for a single, unstructured species, and is generally not true for models of a higher dimension.[*]

Figure 4.3 illustrates the dynamics of juvenile prey, adult prey, and predator biomass upon invasion of predators into the stable resource-prey equilibrium for $R_{max} = 13$ mg/L. The left panel shows the dynamics when predators start off at exactly the density that they attain in the unstable resource-prey-predator equilibrium, which in the top panel of figure 4.1 corresponds to the predator biomass value for $R_{max} = 13$ mg/L on the lower, unstable curve section. The right panel of figure 4.3 shows the dynamics with the lowest density of initial predator biomass that leads to predator establishment. Predator invasion first

---

[*] In fact, the basic model for Lotka-Volterra competition, in which a coexistence equilibrium exists but is unstable and hence either species might exclude the other depending on initial conditions, already reveals that population densities in an unstable equilibrium do not correspond with threshold densities demarcating population growth and decline. Thus, initial densities that are for both populations above their density in the unstable coexistence equilibrium may lead to exclusion of either one of the two competitors.

decreases the juvenile prey biomass. Even though this decrease is small in absolute value, it is highly significant, as it reduces maturation and hence leads to a *decrease* in adult prey biomass.

The relaxation of competition among adult prey subsequently leads to overcompensation in total population reproduction and hence a *secondary increase* in juvenile prey biomass. Only if this increase leads to a juvenile prey biomass above the density that is needed for predator persistence will predator invasion be successful (as in the right panel of figure 4.3). On invasion, predator biomass initially decreases owing to background mortality, a decrease that slows down when juvenile prey biomass increases. Whether the initial decrease through mortality will lower predator biomass below the density that predators attain in the unstable resource-prey-predator equilibrium determines whether or not juvenile prey biomass will rise above the density that is needed for predator persistence. It is for this reason that successful invasion requires an initial predator density that is 10–20 percent higher (figure 4.3, right) than the density predators attain in the unstable resource-prey-predator equilibrium (figure 4.3, left). We here would like to stress already that predator invasion is only successful if it manages to lower adult prey biomass sufficiently to result in the necessary overcompensation in total population reproduction. Because predators feed on juvenile prey, however, it is only indirectly that they affect adult prey biomass, through a reduction in maturation rate.

Varying Predator Mortality

The emergent Allee effect shows up not only when plotting the possible equilibria as a function of the maximum resource density R_{max}; it generally occurs no matter which parameter is varied, as long as the predator invasion threshold can be crossed by changing the parameter—that is, as long as a change in the parameter can lead to a change in the resource-prey equilibrium from being stable to being unstable against predator invasion. For example, the emergent Allee effect shows up when the conversion efficiency σ_p is varied or when one of the scaling factors relating maximum ingestion, maintenance, and background mortality rate to body size is changed. It does not show up, however, when predator size W_p is varied, as this quantity scales all the rates in the ODE for the dynamics of predator biomass in the same way and hence does not influence the juvenile prey biomass that is required for predator invasion and persistence.

Figure 4.4 shows the changes in equilibrium biomass of basic resource, juvenile and adult prey, and predator as a function of additional predator mortality in excess of background levels. The figure shows these changes both for a predator that forages exclusively on juvenile prey ($\phi = 1$) and for a predator

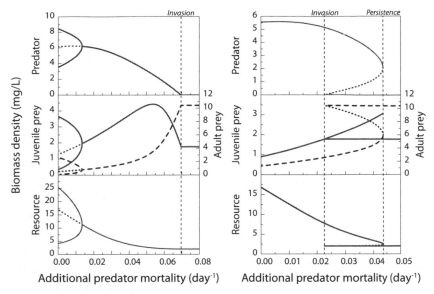

FIGURE 4.4. Changes in equilibrium biomass of basic resource (*bottom*), juvenile (*middle, solid lines*) and adult prey (*middle, dashed lines*), and top predator (*top panels*) with increasing predator mortality in excess of background levels. Left and right panels show changes in equilibrium biomass in cases where predators forage equally on both juvenile and adult prey ($\phi = 0.5$) or exclusively on juvenile prey ($\phi = 1$), respectively. Stable equilibria are indicated with solid or dashed lines, unstable equilibria with dotted lines. For $\phi = 0.5$ and mortality rates below 0.013, population biomasses exhibit stable oscillations around an unstable equilibrium. Minimum and maximum biomass densities during these cycles are indicated with solid and dashed lines (*left panels*). For $\phi = 1$ and mortality rates between 0.023 and 0.044, one unstable and two stable equilibria co-occur. Throughout, $R_{max} = 30$ mg/L, $q = 0.5$, and $z = 0.01$. Otherwise, default parameters apply for a prey and predator with body size $W_A = 0.0001$ and $W_p = 0.01$ gram, respectively.

that forages with equal intensity on both juvenile and adult prey ($\phi = 0.5$). At background mortality (i.e., additional predator mortality is zero), the system with a predator on both juvenile and adult prey exhibits oscillatory dynamics. With increasing predator mortality these cycles shrink in amplitude, and the predator-prey-resource equilibrium becomes stable around an additional mortality of 0.013 day^{-1}. Further increases in predator mortality lead to a monotonic decrease in equilibrium predator and resource biomass and a monotonic increase in total prey biomass in equilibrium. This monotonic increase in total prey biomass mainly results from the monotonic increase in adult prey biomass, whereas juvenile prey biomass shows a hump-shaped relationship with predator mortality. Predators go extinct whenever the additional mortality exceeds 0.07 day^{-1}, which value hence corresponds to the invasion threshold of the predator. For larger mortality rates, only a stable resource-prey equilibrium can occur.

At background mortality the equilibrium of juvenile-specialized predator, prey, and resource is stable. Increasing mortality initially leads to a slight increase in predator biomass and subsequently to a slow decline. The curve relating equilibrium predator biomass to mortality is, however, not monotone, but folded. As before, two sections of this curve can be distinguished, one corresponding to stable equilibrium points of predator, prey, and resource at high predator density (upper solid curve section in top-right panel of figure 4.4), the other corresponding to unstable equilibrium points at low predator density (dotted curve section in top-right panel of figure 4.4). This latter curve section extends from the invasion threshold at zero predator density, which occurs at additional predator mortality around 0.023 day^{-1}, to the limit point in the predator equilibrium curve, which occurs at additional predator mortality around 0.044 day^{-1}. This latter mortality level hence corresponds to the predator persistence threshold. The persistence threshold for predators that forage on juvenile prey occurs at a lower mortality level than for predators that forage on juvenile and adult prey equally (cf. left and right panels in figure 4.4). This results from the fact that with the parameter values of figure 4.4, adults dominate the prey population, and predators that exploit both juveniles and adults hence experience a higher availability of food. However, the persistence threshold of juvenile-specialized predators occurs at mortality rates that are almost twice as high as the maximum predator mortality rate that allows for predator invasion into a resource-prey equilibrium.

The bistability of a resource-prey as well as a resource-prey-predator equilibrium for mortality rates between the invasion and persistence threshold has important implications for the exploitation of predators that forage exclusively on juvenile prey. Figure 4.5 once again shows the changes in predator biomass with increasing mortality rate in excess of background mortality for predators that either forage exclusively on juveniles or forage on both juvenile and adult prey equally. The additional predator mortality rate is, however, now interpreted as the rate at which the predator population is exploited or harvested. In addition, the figure therefore also shows the yield of this exploitation mortality, that is, the amount of predator biomass harvested per day. For predators that forage equally on juvenile and adult prey, the yield is a reasonably symmetric, hump-shaped function of the exploitation rate. At low exploitation rates the population cycles lead to some variability in yield over time, but this variability is small. The maximum yield is reached at an exploitation rate around 0.04 day^{-1}, which corresponds to roughly 60 percent of the mortality rate that

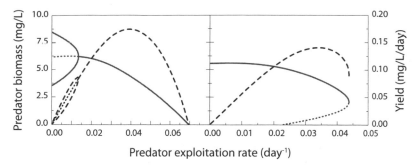

FIGURE 4.5. Changes in equilibrium predator biomass (*solid lines* for stable equilibria and minimum and maximum density during cycles; *dotted lines* for unstable equilibria; same as top panels in figure 4.4) and predator biomass yield (*dashed lines*) with increasing predator exploitation rate. Left and right panels show changes in equilibrium biomass in cases where predators forage equally on both juvenile and adult prey ($\phi = 0.5$) or exclusively on juvenile prey ($\phi = 1$), respectively. Throughout, $R_{max} = 30$ mg/L, $q = 0.5$ and $z = 0.01$. Otherwise, default parameters apply for a prey and predator with body size $W_A = 0.0001$ and $W_P = 0.01$ gram, respectively.

leads to predator extinction. For predators that exclusively forage on juvenile prey, however, the relationship between yield and exploitation rate is far from symmetric. The maximum yield in this case occurs at an exploitation rate of 0.034 day^{-1}, corresponding to almost 80 percent of the exploitation rate that leads to predator extinction.

With increasing exploitation of a juvenile-specialized predator, the harvesting yield hence continues to grow until the population is not too far from extinction. At the same time, the changes in equilibrium predator biomass with increasing exploitation are also not very pronounced. Up to the level of exploitation that leads to the maximum sustainable yield, the equilibrium biomass has only dropped by a mere 25 percent compared with the biomass of an unexploited predator population. Furthermore, at the highest exploitation rate that still allows for predators to persist, the equilibrium biomass is still close to 40 percent of the biomass of an unexploited population. Both the changes in yield as well as the changes in standing-stock biomass of predators with increasing exploitation are poor indicators of the viability of the predator population in terms of the increase in exploitation that the predator can endure before going extinct. For predators that forage on both juvenile and adult prey this is different, because in that case the changes in yield are more gradual and equilibrium predator biomass shows a more pronounced decline with exploitation.

The stage-structured biomass model is of course a rather simplified representation of both prey and predator life history. It is hence fair to question the generality of these results concerning the changes in yield and equilibrium biomass of predators in the face of increasing exploitation. We have found, however, that the qualitative pattern shown in figure 4.5 also occurs in more complex models that represent life histories of prey and predator in more detail. For example, van Leeuwen, de Roos, and Persson (2008) analyze a model for the interaction of cod (*Gadus morhua*) and its main prey, sprat (*Sprattus sprattus*), in the Baltic Sea, which is based on the same framework as the stage-structured biomass model analyzed here but uses for both cod and sprat three stages to represent their life histories. In this model the maximum sustainable yield of cod occurs at an exploitation rate that is only slightly smaller than the maximum exploitation rate that the predator can sustain before going extinct. Furthermore, in the last part of this chapter we present results for a model that accounts for a size-structured predator population foraging on a size-structured consumer population. Also in that model the maximum sustainable yield occurs at an exploitation rate that is close to the extinction threshold. In models that represent life histories in more detail, the phenomenon might hence even be more pronounced than in the two-stage biomass model studied here.

Figure 4.6 shows how different combinations of additional predator mortality rate and maximum resource density affect the occurrence of bistability. In absolute terms, the extent of the mortality interval, in which both a stable resource-prey and a stable resource-prey-predator equilibrium occur, does not change very much with increasing values of the maximum resource density R_{max}: for $R_{max} = 100\,\text{mg/L}$, the extent of the mortality interval is about 0.017 day^{-1}, while it measures about 0.018 day^{-1} for $R_{max} = 20\,\text{mg/L}$. With increasing values of R_{max} the interval just shifts to higher mortality rates. Hence, relative to the entire range of mortality rates that the predator can endure before going extinct, the mortality interval with bistability becomes smaller. Figure 4.6 also shows how different combinations of additional predator mortality rate and predator specialization ϕ affect the occurrence of bistability. For $\phi = 0.5$, which means that predators forage equally on juveniles and adults, we have already seen that bistability does not occur. For values of $\phi > 0.7$, however, mortality rates can be found for which a resource-prey and a resource-prey-predator equilibrium are both stable. Even intermediate levels of predator specialization on juvenile prey thus result in bistability of equilibria and the associated potential of a rapid predator collapse when the persistence threshold is crossed. The extent of the bistability interval clearly grows with increasing specialization of the predator on juveniles, reaching a maximum for $\phi = 1$.

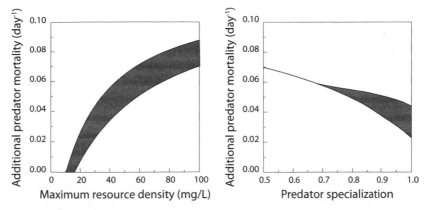

FIGURE 4.6. *Left panel*: Range of additional predator mortality rates in excess of background levels and maximum resource densities R_{max} for which both a stable resource-prey and a stable resource-prey-predator equilibrium occurs, when predators forage on juvenile prey exclusively ($\phi = 1$). For parameter combinations within the gray parameter region, this bistability occurs. *Right panel*: Range of additional predator mortality rates in excess of background levels and the extent to which predators specialize on juvenile prey ϕ, for which a resource-prey and resource-prey-predator equilibrium are both stable. Throughout, $R_{max} = 30$ mg/L (right graph only), $q = 0.5$, and $z = 0.01$. Otherwise, default parameters apply for a prey and predator with body size $W_A = 0.0001$ and $W_P = 0.01$ gram, respectively.

PREDATORS SPECIALIZING ON ADULT PREY

In this section we focus on parameter values for the prey population that result in juvenile prey dominating the population biomass at background mortality levels ($q = 2.0$, $z = 0.5$; see also figure 3.6 in chapter 3). Under these conditions we showed in the previous chapter that overcompensation in adult prey biomass could occur in response to increases in stage-independent, juvenile as well as adult mortality. The latter type of overcompensation can lead to an emergent Allee effect for predators that specialize on adult prey, in the same way as this effect occurred for juvenile-specialized predators under conditions where overcompensation occurs in juvenile prey. In fact, the results for adult-specialized predators under conditions that adult prey biomass exhibits biomass overcompensation closely resemble the results presented in the previous section for juvenile-specialized predators under conditions that juvenile prey exhibit biomass overcompensation. Our presentation of results for adult-specialized predators will therefore be brief.

In particular, for the prey parameter values used in this section we do not present any comparative results for the case that predators forage equally on juvenile and adult prey, as these are qualitatively the same as the results presented for such predators in the previous section (see left panels in figures 4.1 and 4.4): the changes in equilibrium biomass follow the classical pattern of changes with increasing productivity described by Oksanen, Fretwell, et al. (1981), in which the level of resource productivity uniquely determines the equilibrium state and the number of trophic levels in the food chain at equilibrium increases in a stepwise and discrete manner with increasing productivity. The threshold for predator invasion and predator persistence coincide, and for productivity levels exceeding the invasion threshold, equilibrium predator biomass increases monotonically with productivity. Moreover, population cycles occur for high values of the maximum resource density and low values of predator mortality.

Figure 4.7 shows the changes in equilibrium resource, juvenile and adult prey, and predator biomass with changes in maximum resource density R_{max} as well as with increasing additional predator mortality in excess of background levels. The figure makes clear that the predator persistence threshold occurs at a lower value of R_{max} and at a higher additional mortality rate, respectively, than the invasion threshold, analogous to the results for the juvenile-specialized predator presented in figures 4.1 and 4.4. The only qualitative difference between the results presented in figure 4.7 for adult-specialized predators and those presented in figures 4.1 and 4.4 for juvenile-specialized predators is that the roles of juvenile and adult prey are reversed: in figure 4.7 juvenile, as opposed to adult, prey dominate the population when predators are absent and predators keep adult prey biomass constant with increasing maximum resource density, whereas juvenile prey biomass is reduced. Similarly, with increasing predator mortality (figure 4.7, right panel) both juvenile and adult prey biomass increase, but juvenile prey biomass is always higher. More quantitatively, adult-specialized predators seem to go extinct at lower mortality levels and need higher resource productivities for persistence than juvenile-specialized predators. The extent of the parameter ranges in figure 4.7 for which bistability between a resource-prey and a resource-prey-predator equilibrium occurs is similar to the extent of these ranges in figures 4.1 and 4.4. These quantitative comparisons depend, however, on the exact parameter combinations that we have used in this and the previous section and hence may not hold generally.

Figure 4.8 shows how different combinations of additional predator mortality rate and maximum resource density affect the occurrence of bistability. Compared with the corresponding graph in figure 4.6 for juvenile-specialized predators, bistability occurs over a more limited range of parameter

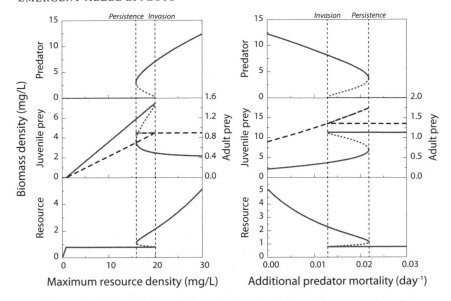

FIGURE 4.7. Changes in equilibrium biomass of basic resource (*bottom*), juvenile (*middle, solid lines*) and adult prey (*middle, dashed lines*), and top predator (*top panels*) with increasing maximum resource density R_{max} (*left panels*) and predator mortality in excess of background levels (*right panels*) when predators forage exclusively on adult prey ($\phi = 0$). Stable equilibria are indicated with solid or dashed lines, unstable equilibria with dotted lines. Population cycles do not occur. For maximum resource densities between 16 and 20 mg/L (*left panel*), as well as for additional predator mortality rates between 0.013 and 0.022 per day (*right panel*), one unstable and two stable equilibria co-occur. Throughout, $R_{max} = 30$ mg/L (right panel only), $q = 2.0$, and $z = 0.5$. Otherwise, default parameters apply for a prey and predator with body size $W_A = 0.0001$ and $W_P = 0.01$ gram, respectively.

combinations for adult-specialized predators. As before, the extent of the mortality interval, in which both a stable resource-prey and a stable resource-prey-predator equilibrium occur, does not change very much with increasing values of the maximum resource density R_{max}; the interval just shifts to higher mortality rates with increasing values of R_{max}. The more limited conditions for the occurrence of bistability are also apparent when all parameter combinations of additional predator mortality rate and predator specialization ϕ are considered, for which a resource-prey and a resource-prey-predator equilibrium are both stable (figure 4.8). Bistability occurs only as long as $\phi < 0.2$. Intermediate levels of predator specialization on adult prey thus do not result in bistability of equilibria and the associated potential of a rapid predator collapse when the persistence threshold is crossed.

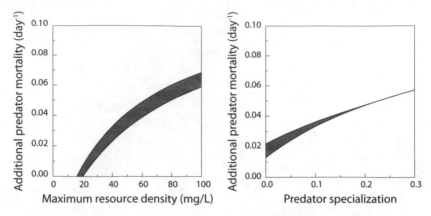

FIGURE 4.8. *Left panel*: Range of additional predator mortality rates in excess of background levels and maximum resource densities $R_{\max}$ for which both a stable resource-prey and a stable resource-prey-predator equilibrium occur, when predators forage on adult prey only ($\phi = 0$). For parameter combinations within the gray parameter region, this bistability occurs. *Right panel*: Range of additional predator mortality rates in excess of background levels and the extent to which predators specialize on juvenile prey ϕ, for which a resource-prey and resource-prey-predator equilibrium are both stable. Throughout, $R_{\max} = 30$ mg/L (right panel only), $q = 2.0$, and $z = 0.5$. Otherwise, default parameters apply for a prey and predator with body size $W_A = 0.0001$ and $W_P = 0.01$ gram, respectively.

In summary, even though the results for juvenile-specialized and adult-specialized predators are qualitatively similar, the comparison suggests that emergent Allee effects are less likely to occur for adult-specialized predators than for juvenile-specialized predators. This is most likely related to the fact, discussed in chapter 3, that biomass overcompensation in adult biomass tends to be less pronounced than biomass overcompensation in juvenile biomass. Such quantitative comparisons are, however, always dependent on the exact parameter combinations that are used in the comparison and hence have only limited generality.

EMERGENT ALLEE EFFECTS
IN THE KOOIJMAN-METZ MODEL

The previous section presented results on emergent Allee effects in the relatively simple setting of a two-stage biomass model. As argued in the introduction of this chapter, however, there is an intimate connection between these Allee effects and the occurrence of biomass overcompensation. Emergent Allee effects will hence occur in any type of model in which biomass overcompensation occurs in a life stage or size class in response to an increase in mortality of that particular stage or class. In particular, in the previous chapter it was shown that overcompensation in the biomass of small-size consumers with increasing mortality of these consumers occurs in the continuously size-structured population model that is based on the Kooijman-Metz energy channeling scheme (figure 3.15). Emergent Allee effects are hence expected to occur as well in this more complex model. In fact, the occurrence of emergent Allee effects was first reported in relation to the Kooijman-Metz consumer-resource model (de Roos and Persson 2002).

The Kooijman-Metz consumer-resource model as presented in the previous chapter (box 3.6) can be extended with another trophic level in a simple manner using the Yodzis and Innes bioenergetics model also used in the first part of this chapter to account for a size-selective predator population. Hence, the dynamics of the total predator biomass P will be described by

$$\frac{dP}{dt} = (\nu_P(C_v) - \mu_P)P \qquad (4.3)$$

Analogous to the model in the first part of this chapter, the function $\nu_P(C_v)$ represents the net biomass production per unit biomass of the predator, which depends on the biomass density C_v of consumers that are vulnerable to predation by the predator, while μ_P represents the per-capita predator mortality. Predator ingestion follows again a type II functional response, $M_P C_v/(H_P + C_v)$, in which M_P and H_P represent the predator's mass-specific maximum ingestion rate and half-saturation constant, respectively. The mass-specific rate of net biomass production equals the difference between the prey biomass that is ingested and assimilated by the predator and the energetic costs to cover maintenance requirements (cf. equation (3.2)), $\nu_P(C_v) = \sigma_P M_P C_v/(H_P + C_v) - T_P$, with σ_P and T_P representing the efficiency with which ingested food is assimilated and the mass-specific maintenance rate of the predator, respectively.

Prey of different body sizes generally differ in their vulnerability to predation mortality, with larger prey individuals less vulnerable than smaller ones. For simplicity, however, we will assume that predators forage only and indiscriminately on consumer individuals that are smaller than a particular threshold length ℓ_v, while larger consumers are assumed to be completely invulnerable to predation. We therefore mimic the size-dependence of prey vulnerability to predation mortality as an all-or-nothing process with characteristic threshold size ℓ_v. The biomass of consumers that are vulnerable to predation now equals:

$$C_v = \int_{\ell_b}^{\ell_v} \beta\ell^3 c(t,\ell)\,d\ell \qquad (4.4)$$

The total amount of consumer biomass consumed by predators per unit of time then equals:

$$M_P \frac{C_v}{H_P + C_v} P = M_P \frac{\displaystyle\int_{\ell_b}^{\ell_v} \beta\ell^3 c(t,\ell)\,d\ell}{H_P + \displaystyle\int_{\ell_b}^{\ell_v} \beta\ell^3 c(t,\ell)\,d\ell} P$$

From this expression it can be seen that every individual in the size range between ℓ_b and ℓ_v experiences an additional predation mortality equal to $M_P P/$ $(H_P + C_v)$. To complete the part of the model describing the dynamics of the size-structured consumer population and its resource, the only function that changes, compared with the formulation presented in box 3.6, is the consumer mortality rate, which is now given by a function $d(P,\ell)$, defined as:

$$d(P,\ell) = \begin{cases} \mu + \dfrac{M_P P}{H_P + C_v} & \text{for } \ell_b \le \ell < \ell_v \\ \mu & \text{otherwise} \end{cases} \qquad (4.5)$$

Together with the size-structured consumer-resource model presented in box 3.6, equations (4.3)–(4.5) complete the model formulation of a tritrophic, predator-prey-resource model in which the unstructured predators forage only on prey individuals smaller than a particular threshold body size ℓ_v.

As the default parameter setting, we make the rather arbitrary assumptions that all prey individuals smaller than $\ell_v = 50$ mm are vulnerable to predation, while the predators are of a representative body length of 300 mm. Assuming the same weight-length relationship for the predators as for the prey, this latter length value translates into a characteristic predator body weight of 250 grams. Using the same scaling relationships with body size as presented in boxes 3.3 and 3.4 yields default values for M_P, T_P, and μ_P equal to 0.025, 0.0025, and 0.00038, respectively. For the parameters H_P and σ_P we adopt the standard values of 3.0 and 0.5, respectively (see box 3.4). Notice, however, that the default value for the predator per-capita mortality rate μ_P represents a minimum value, as we will investigate model predictions as a function of predator mortality rate *in excess of* this background level.

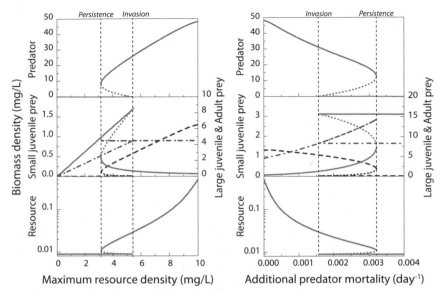

FIGURE 4.9. Changes in equilibrium biomass of basic resource (*bottom*), juvenile consumers smaller (*middle, dashed-dotted line*) and larger (*middle, solid line*) than 50 mm, adult consumers (*middle, dashed line*), and top predator (*top panels*) with increasing maximum resource density R_{max} (*left panels*) and predator mortality on top of background levels (*right panels*) in the size-structured Kooijman-Metz model with an unstructured predator population foraging only on consumers smaller than 50 mm (see box 3.6 and equations (4.3)–(4.5) for model formulation). Stable equilibria are indicated with solid, dashed, or dashed-dotted lines, unstable equilibria with dotted lines. For maximum resource densities between 3.2 and 5.4 mg/L (*left panel*), as well as for additional predator mortality rates between 0.0016 and 0.0032 per day (*right panel*), two stable equilibria co-occur. Throughout, $R_{max} = 10$ mg/L (right panel only). Otherwise, default parameters apply for consumers as given in box 3.7 and for a predator species with body size $W_p = 250$ grams (see text for details).

Figure 4.9 (left) shows the bifurcation diagram of this resource–size-structured consumer–unstructured predator model as a function of maximum resource density. Qualitatively, the diagram shows the same pattern as the corresponding bifurcation diagram for the stage-structured biomass model with predators specializing on juvenile prey (figure 4.1). For R_{max} below 5.4 mg/L, the consumer-resource equilibrium is stable against invasion by the predator. Large juveniles make up the largest part of consumer population biomass, followed by small juveniles, while adult biomass is very low in the absence of predators. The biomass in all three size classes increases approximately linearly with increasing values of R_{max}. As a consequence, for values above $R_{max} = 5.4$ mg/L, the biomass in the small juvenile consumer size class becomes large enough for the predator on these juveniles to have a positive mass-specific population growth rate, $\nu_P(C_\nu) - \mu_P$, in the consumer-resource equilibrium. For $R_{max} > 5.4$ mg/L, the consumer-resource equilibrium is hence unstable, and only a stable predator-consumer-resource equilibrium is found. The bifurcation occurring at the transition point around $R_{max} = 5.4$ mg/L is, however, subcritical such that the curve relating the equilibrium predator density to R_{max} first curves toward lower R_{max} values before turning back toward larger values. This latter turning or critical point occurs at $R_{max} \approx 3.2$ mg/L. A stable equilibrium of predator, consumer, and resource is hence found for all values $R_{max} > 3.2$ mg/L. In this equilibrium the biomasses of predators and adult consumers strongly increase with increasing R_{max}, while the biomass of large juvenile consumers decreases. The biomass of small juvenile consumers remains constant at the density that results in a zero net-growth rate of the predator, $\nu_P(C_\nu) - \mu_P = 0$.

For $R_{max} = 10$ mg/L, figure 4.9 (right) also shows the bifurcation diagram for increasing values of predator mortality in excess of its background mortality rate of 0.00038/day. For additional mortality rates up to 0.0032/day, corresponding to an almost tenfold increase in predator mortality, predators can persist, and a stable predator-consumer-resource equilibrium occurs. As long as predators and consumers coexist, increasing values of the predator mortality rate result in lower biomass densities of predators and adult consumers and larger biomass densities of both small and large juvenile consumers. The increase in small consumer biomass compensates for the increased loss through mortality of the predator, keeping its mass-specific population growth rate, $\nu_P(C_\nu) - \mu_P$, equal to 0. At the highest additional mortality rate of 0.0032/ day that the predator can sustain, predator biomass is roughly 25 percent of its density at background mortality. Any further increase in predator mortality would hence lead to a considerable drop in predator biomass and would result in an approach to an equilibrium with only resource and consumers. This consumer-resource equilibrium is dominated in terms of biomass by large juvenile consumers, while biomass density of small juvenile consumers is too low for predators to invade, as long as predator mortality is larger than 0.0016/ day, that is, five times the predator background mortality. As pointed out before, adult consumers represent only a tiny fraction of equilibrium consumer biomass in the absence of predators.

Like the stage-structured biomass models in the previous sections, the resource-size structured consumer-unstructured predator model that is based on the Kooijman-Metz energy channeling scheme predicts the occurrence of alternative

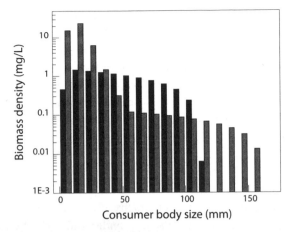

FIGURE 4.10. Biomass-size distribution of the consumer population in the stable consumer-resource equilibrium (*black bars*) and in the coexistence equilibrium with predators present (*gray bars*) for otherwise the same parameter values. $R_{max} = 10$ mg/L. Predators experience an additional mortality rate of 0.0025/day; otherwise, default parameters apply for consumers as given in box 3.7 and for a predator species with body size $W_p = 250$ grams (see text for details).

stable states for substantial ranges of parameters. At background mortality rates, both stable resource-consumer as well as resource-consumer-predator equilibria occur for R_{max} values between 3.2 and 5.4 mg/L. Similarly, for $R_{max} = 10$ mg/L, the coexistence equilibrium and the equilibrium without predators are stable for additional predator mortality rates between 0.0016 and 0.0032/day.

Figure 4.10 compares for $R_{max} = 10$ mg/L and an additional predator mortality rate of 0.0025/day the consumer size-distribution in the equilibrium without predators and in the coexistence equilibrium. In the absence of predators, the consumer size-distribution is compact, also referred to as stunted, as it only covers a size range between the size at birth ($\ell_b = 7$ mm) and a maximum size just above the maturation size of $\ell_j = 110$ mm. In addition, biomass density declines gradually between consecutive size classes at a rate that is roughly similar over the entire size range covered by the distribution. In contrast, when predators are present, the consumer size distribution covers a wider range of body sizes up to a maximum size around 170 mm, with a steeper decline in biomass density between consecutive size classes at body sizes smaller than 50 mm and more gradual declines at larger sizes. Consumer biomass density in size ranges smaller than 40 mm is higher in the coexistence equilibrium compared with the equilibrium without predators. The same holds for the biomass density of adult consumers larger than the maturation size of $\ell_j = 110$ mm (figure 4.10), whereas biomass densities in size ranges between 50 and 110 mm are smaller in the presence of predators.

These higher densities of small juvenile as well as adult biomass result from the biomass overcompensation in the consumer size distribution and resemble the changes that occur in response to increases in small-juvenile mortality shown in figure 3.15. The changes in consumer size distribution are thus due to the differential response in consumer maturation and reproduction when density dependence is relaxed. The predator is simply the agent that relaxes density dependence through the additional mortality it imposes. The steep decline in biomass density over the size range up to 50 mm expresses this mortality imposed by the predator on these consumer sizes.

While constructing the bifurcation graphs for the stage-structured biomass model presented earlier in this chapter, equilibrium stability was assessed using the numerical bifurcation software package MATCONT (Dhooge, Govaerts, and Yu 2003; see also box 3.5). The procedures implemented in this software package for the analysis of (systems of) ordinary differential equations allow us to assess explicitly whether a particular equilibrium is a stable node or focus, an unstable saddle point, or an unstable focus surrounded by a limit cycle (see Kuznetsov 1995, for an introduction to bifurcation theory). No analogous procedures to assess equilibrium stability have yet been developed for the type of size-structured models discussed in this and the following section, which are formulated in terms of partial differential equations. The stability properties of the equilibria shown in the bifurcation graphs of figure 4.9 are therefore not known in full detail, but only to the extent that general principles from bifurcation theory allow us to assess them.

Specifically, to distinguish stable and unstable equilibria, we have exploited the fact that at a bifurcation point where the curve representing consumer-resource equilibria intersects the curve representing predator-consumer-resource equilibria (a so-called transcritical bifurcation or branching point) the stability of the consumer-resource equilibrium changes. The predator-consumer-resource equilibrium is labeled as unstable if the corresponding curve that originates in this point starts off in the direction of parameter values for which the consumer-resource equilibrium is also stable. Otherwise, the equilibrium is indicated as stable. Lastly, at a so-called limit point of an equilibrium curve, where it doubles back on itself, the stability of the equilibrium changes as well. These general principles hence allow us to distinguish with certainty saddle points from stable nodes and foci. However, they do not allow us to assess whether an equilibrium has turned unstable through a so-called Hopf bifurcation with the concomitant development of a limit cycle. These general restrictions should be kept in mind when interpreting bifurcation graphs for structured models that involve a continuous size distribution. For the specific model that is the topic of this section we did carry out numerical simulations over the range of parameter values shown in figure 4.9 (results not shown). These simulation results confirm that the predator-consumer-resource equilibrium is indeed stable as indicated and that population cycles do not occur when predators are present. However, we have carried out such additional simulations for only a few models with continuous size distributions.

The resource-size-structured consumer-unstructured predator model analyzed here is simple enough that it is possible to compute the invasion and persistence threshold of the predator (see figure 4.9) as a function of two parameters of choice. The invasion threshold marks the parameter combination for which the resource-consumer equilibrium changes from being stable to predator invasion into being unstable against such invasion. The persistence threshold represents the parameter combination that marks the limit of predator existence in coexistence with the consumer. The technique to compute these two thresholds as a function of two parameters is rather complicated and will not be discussed in detail. Basically, the invasion threshold can be computed by solving for the resource-consumer equilibrium with the additional condition that the mass-specific population growth rate of the predator, $v_P(C_v) - \mu_P$, equals 0, while predator density is set to 0 as well. The computation of the persistence threshold requires the computation of the coexistence equilibrium of resource, consumer, and predator with the additional condition that the tangent vector along the equilibrium curve is pointing exactly vertical. We carried out both types of computations by numerically solving for the root of the determining system of nonlinear equations, which included up to five conditions (see also section 4 of the technical appendices). The tangent vector to the equilibrium curve, which is required for locating the persistence threshold, was also computed numerically.

Figure 4.11 shows the results of computing the invasion and persistence thresholds in this way as a function of the maximum resource density and the additional predator mortality rate (left) and as a function of this mortality and the maximum consumer length ℓ_v that is vulnerable to predation (right). With increasing values of the maximum resource density, the range of mortality rates, for which both the resource-consumer equilibrium as well as the coexistence equilibrium with resource, consumers, and predators are stable, does not change much in absolute extent. For higher maximum resource densities, however, the bistability range occurs at higher predator mortality rates. At $R_{max} = 10$ mg/L the range of mortality rates for which the alternative stable states can occur is largest when the maximum consumer length ℓ_v that is vulnerable to predation is around 40 mm. For these parameter combinations, the resource-consumer

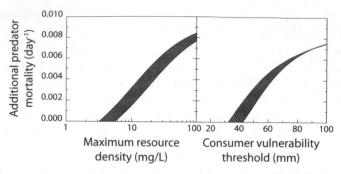

FIGURE 4.11. *Left*: Range (*gray region*) of additional predator mortality rates in excess of background levels and maximum resource densities R_{max} for which both a stable resource-consumer and a stable resource-consumer-predator equilibrium occurs in the resource-size-structured consumer-unstructured predator model based on Kooijman-Metz energy channeling. *Right*: Bistability range (*gray region*) of additional predator mortality rates in excess of background levels and the maximum consumer length ℓ_v that the predator forages on. Throughout, $R_{max} = 10$ mg/L (right panel only). Otherwise, default parameters apply for consumers as given in box 3.7 and for a predator species with body size $W_P = 250$ grams (see text for details).

equilibrium is stable against predator invasion even when the latter only experiences background mortality. On the other hand, once present in a resource-consumer-predator equilibrium, the predator can persist as long as the additional mortality is below 0.002/day, which amounts to more than five times its background mortality. For increasing values of the maximum consumer length ℓ_v that is vulnerable to predation, the range of mortality rates for which the alternative stable states can occur shrinks and disappears altogether for $\ell_v > 85$ mm.

It should be noted, though, that these changes of the parameter domain with bistability for changing R_{max}, ℓ_v, and predator mortality depend to a considerable extent on the specific choice of parameters. For example, de Roos and Persson (2002) first reported the occurrence of the emergent Allee effect using the same model for resource and size-structured consumer as analyzed here with also identical parameters. The part of that model describing predator dynamics was structurally also the same as equation (4.3) above but assumed different parameter values for the predator maximum ingestion rate, half-saturation constant, and total loss rate (in fact, the model used ignored maintenance costs altogether and only accounted for predator mortality, as it tracked predator number density instead of biomass). As a consequence of these differences in predator parameters alone, it was found that the range of mortality rates for which alternative stable states occurred increased proportionally when the maximum resource density increased. The latter result contrasts with the results shown in figure 4.11 and hence warns against drawing general conclusions on the basis of the results presented in that figure.

SIZE-STRUCTURED PREDATORS FORAGING
ON SIZE-STRUCTURED PREY

All results presented in the foregoing sections of this chapter relate to models with an unstructured predator population. These models are hence based on the simplifying assumptions that ingestion and net biomass production scale linearly with consumer body size, that there is an equally efficient biomass production through somatic growth and reproduction, and that predator mortality is size-independent (see also the discussion of the Yodzis and Innes model in chapter 3). Most important, however, these models assume that all predators forage on the same size class of consumers and do not differ in any way in their prey selectivity. In this section we will present results of a model that relaxes this latter assumption and that hence represents the fact that predators of different body sizes tend to differ in the size range of prey they forage on.

Broadly speaking, predators can be subdivided into two species categories depending on the range of prey size that they select to forage on (Peters 1983). Mammalian predators and birds of prey are "large-prey eaters," as they tend to forage on prey that on average weighs some 10 percent of their own body mass. On the other hand, piscivorous and insectivorous birds, lizards, amphibians, and fish, as well as invertebrate predators can be classified as "small-prey eaters," as they eat prey that weigh on average only 0.2 percent of their own weight. For small-prey eaters, Peters (1983, appendix VIIb) presents data showing that across many predator species the minimum ratio between the body weight of prey and predator equals $4.8 \cdot 10^{-5}$, while the maximum of this ratio equals $7.4 \cdot 10^{-2}$. In between these two limits there is, for a given predator body weight, a near-normal distribution of the logarithm of the prey weight with a variance that is virtually independent of predator size. The ratio between prey and predator body weight at which the optimum of the log-normal distribution occurs is estimated by Peters as $1.9 \cdot 10^{-3}$. If we now assume that body weight is proportional to the cubed body length, the minimum, optimum, and maximum ratio between the body length of prey and predator can be estimated (with some rounding off) to equal 0.05, 0.1, and 0.4, respectively.

For our analysis of the interaction between size-structured predators and prey we will model the dynamics of prey (or consumers; these terms we use interchangeably) with the size-structured population model based on the Kooijman-Metz energy budget model that was introduced in chapter 3. The only difference will be the fact that individuals are exposed to predation mortality on top of background mortality. Dynamics of the predator population are modeled by the same type of size-structured population model, except that predators acquire energy through preying on consumers instead of foraging on a basic resource. The body length of individual consumers and predators will be denoted by ℓ and l, respectively (note the difference between these two symbols!). The population size-distribution of consumers and predators will be represented by the density functions $c(t,\ell)$ and $p(t,l)$, respectively. Food availability for predators with length l depends on the consumer size distribution $c(t,\ell)$ and on the relative prey size-selectivity of the predator, which we represent by the relative prey vulnerability function $T(\ell,l)$.

Based on the discussion of prey-predator body size ratios above, we assume that $T(\ell,l)$ will be 0 for $\ell/l < \delta = 0.05$ and $\ell/l > \varepsilon = 0.4$, in which the parameters δ and ε represent the minimum and maximum ratio between prey and predator body length, respectively. Furthermore, the relative prey vulnerability $T(\ell,l)$ is assumed to equal 1 for $\ell/l = \varphi = 0.1$, in which φ represents the optimal prey-predator body length ratio. For simplicity we will not model the relative prey vulnerability with a log-normal distribution function, which according to the data is appropriate, but will assume that for a given predator length l the vulnerability $T(\ell,l)$ increases linearly from 0 for $\ell/l = \delta$ to 1 for $\ell/l = \varphi$ to thereafter decrease again linearly, reaching 0 when $\ell/l = \varepsilon$. In other words, for given predator length l the vulnerability $T(\ell,l)$ is a tentlike function of prey length ℓ. Figure 4.12 illustrates the relative prey vulnerability $T(\ell,l)$ as a function of the length of both prey and predator for default values of the minimum, optimum, and maximum prey-predator body length ratio.

Given the preceding assumptions, the amount of prey biomass that is available to a predator of length l to forage on equals:

$$C(l) = \int_{\delta l}^{\varepsilon l} T(\ell,l) w_C(\ell) c(t,\ell) \, d\ell \qquad (4.6)$$

This expression integrates the product of the consumer biomass density at body length ℓ, $w_C(\ell) c(t,\ell)$, and its vulnerability to the predator over the entire range of prey sizes from the minimum length δl to the maximum εl that the predator can forage on. The ingestion rate of prey biomass by a predator of length l is now modeled with the expression:

$$M_P \frac{C(l)}{H_P + C(l)} l^2 \qquad (4.7)$$

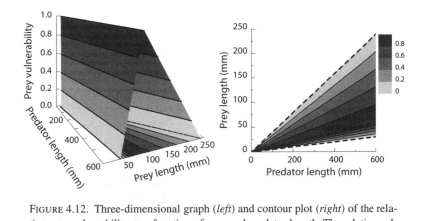

FIGURE 4.12. Three-dimensional graph (*left*) and contour plot (*right*) of the relative prey vulnerability as a function of prey and predator length. The relative vulnerability reaches a maximum of 1 for the optimum prey-predator length ratio $\varphi = 0.1$ (*right panel, middle dashed line*) and equals 0 for prey-predator length ratios below the minimum ratio $\delta = 0.05$ and above the maximum ratio $\varepsilon = 0.4$ (*right panel, lower and upper dashed lines, respectively*).

This expression is analogous to the consumer ingestion rate function (see box 3.6) except for its dependence on the prey availability $C(l)$ instead of the resource density R. From the predator ingestion rate function it can be inferred that the predation rate that a single predator with body length l imposes on prey with body length ℓ equals:

$$M_P \frac{T(\ell,l)}{H_P + C(l)} l^2$$

Taking into account the density $p(t,l)$ of these predators and the fact that the smallest and largest predators that consumers with body length ℓ are vulnerable to have body length $l = \ell/\varepsilon$ and $l = \ell/\delta$, respectively, the total predation pressure that consumers of length ℓ are exposed to equals:

$$P(\ell) = \int_{\ell/\varepsilon}^{\ell/\delta} M_P \frac{T(\ell,l)}{H_P + C(l)} l^2 p(t,l)\,dl \qquad (4.8)$$

The total mortality rate that consumers of length ℓ experience hence equals $d_C(P(\ell)) = \mu_C + P(\ell)$. This mortality rate is the only aspect in which the model for the consumer population dynamics differs from the size-structured model discussed in chapter 3 (see box 3.6). The model for the predator population dynamics is very similar to the one for the consumer, except that all life history

BOX 4.3
SIZE-STRUCTURED PREDATOR-PREY MODEL BASED ON KOOIJMAN-METZ ENERGY CHANNELING

Dynamic equations	Description
$\dfrac{\partial p(t,l)}{\partial t} + \dfrac{\partial g_P(C(l),l)p(t,l)}{\partial l} = -\mu_P p(t,l)$	Predator length-distribution dynamics
$g_P(C(l_b),l_b)p(t,l_b) = \displaystyle\int_{l_j}^{l_m} b_P(C(l),l)p(t,l)dl$	Predator total population reproduction
$\dfrac{\partial c(t,\ell)}{\partial t} + \dfrac{\partial g_C(R,\ell)c(t,\ell)}{\partial \ell} = -d_C(P(\ell))c(t,\ell)$	Consumer length-distribution dynamics
$g_C(R,\ell_b)c(t,\ell_b) = \displaystyle\int_{\ell_j}^{\ell_m} b_C(R,\ell)c(t,\ell)d\ell$	Consumer total population reproduction
$\dfrac{dR}{dt} = \rho(R_{\max} - R) - \displaystyle\int_{\ell_b}^{\ell_m} I_C(R,\ell)c(t,\ell)d\ell$	Resource biomass dynamics

Function	Expression	Description
$I_C(R,\ell)$	$M_C R/(H_C + R)\ell^2$	Resource ingestion rate by consumers
$g_C(R,\ell)$	$\gamma_C(\ell_m R/(H_C + R) - \ell)$	Consumer growth rate in length
$b_C(R,\ell)$	$\alpha_C R/(H_C + R)\ell^2$	Consumer fecundity
$d_C(P(\ell))$	$\mu_C + P(\ell)$	Consumer death rate
$w_C(\ell)$	$\beta_C \ell^3$	Consumer weight-length scaling relation
$T(\ell,l)$	$\begin{cases} \dfrac{\ell - \delta l}{(\varphi - \delta)l} & \text{for } \delta l \leq \ell \leq \varphi l \\[2mm] \dfrac{\varepsilon l - \ell}{(\varepsilon - \varphi)l} & \text{for } \varphi l \leq \ell \leq \varepsilon l \\[2mm] 0 & \text{otherwise} \end{cases}$	Consumer vulnerability for predator of length l
$C(l)$	$\displaystyle\int_{\delta l}^{\varepsilon l} T(\ell,l)w_C(\ell)c(t,\ell)d\ell$	Prey availability for predator of length l
$P(\ell)$	$\displaystyle\int_{\ell/\varepsilon}^{\ell/\delta} M_P \dfrac{T(\ell,l)}{H_P + C(l)}l^2 p(t,l)dl$	Predation mortality of consumer at length ℓ
$g_P(C(l),l)$	$\gamma_P(l_m C(l)/(H_P + C(l)) - l)$	Predator growth rate in length
$b_P(C(l),l)$	$\alpha_P C(l)/(H_P + C(l))l^2$	Predator fecundity
$w_P(l)$	$\beta_P l^3$	Predator weight-length scaling relation

functions that for the consumer depend on the resource density R, in the predator's case depend on the prey availability $C(l)$, which is different for predators of different body length. All model equations describing the dynamics of the consumer and predator size-distribution, as well as the density of the basal resource, are presented in box 4.3, which should be compared to the model presented in box 3.6 for its similarity.

Predators are assumed to be born at $l_b = 35$ mm, mature at $l_j = 300$ mm, and reach a maximum size of $l_m = 750$ mm under unlimited food conditions. These parameter values do not relate to any predator species in particular, but are chosen in order to allow that predators can feed themselves from birth by preying on consumers and hence are right after birth already big enough to consume the newborn consumers. As discussed in chapter 3, however, the maximum length in the Kooijman-Metz model is a compound parameter composed of the scaling constant M between maximum ingestion rate and squared length, the assimilation efficiency σ, the fraction of energy channeled to growth and maintenance κ, and the scaling constant ξ relating energy requirements for maintenance to the cubed length: $l_m = \sigma_P \kappa_P M_P / \xi_P$ (here we have labeled all parameters with the subscript P to indicate that they pertain to the predator). We will assume that the larger body length of the predator results from a two times higher maximum ingestion rate per unit of squared length (i.e., surface) and a 20 percent lower maintenance requirement per unit of cubed length (i.e., weight).

These two assumptions subsequently affect other compound parameters in the model, in particular the von Bertalanffy growth rate parameter γ_P and the constant α_P, relating the maximum fecundity under unlimited food conditions to the squared length. Because it is proportional to the maintenance parameter ξ_P, the growth rate γ_P for the predator is 20 percent lower than that of the consumer. The higher value for the maximum ingestion rate scaling constant M_P leads to a proportional increase in the fecundity scaling constant α_P, but this constant is reduced by the fact that newborn predators are five times longer than newborn consumers and hence weight 125 times more. Together these changes imply that the fecundity scaling constant α_P of the predator is only 1.6 percent of the corresponding parameter value of the consumer. Finally, based on the arguments presented in chapter 3 (see box 3.4) we have assumed that the half-saturation constant is not influenced by the characteristic species weight and hence is equal for consumer and predator. The above considerations have led to a default parameter set for the predator that is presented in box 4.4. The default parameters for the consumer we do not repeat in this chapter, as they were already presented in box 3.7.

Figure 4.13 shows the changes in equilibrium biomass of resource, juvenile consumers smaller and larger than 50 mm, adult consumers, juvenile predators smaller and larger than 80 mm, and adult predators with increasing values of the

BOX 4.4
DEFAULT PARAMETER VALUES FOR THE SIZE-STRUCTURED
PREDATOR IN THE MODEL OF BOX 4.3

Parameter	Default value	Unit	Description
l_b	35	mm	Length at birth
l_j	300	mm	Length at maturation
l_m	750	mm	Maximum length at unlimited food
β_P	0.009	$mg \cdot mm^{-3}$	Weight-length scaling constant
M_P	0.2	$mg \cdot day^{-1} \cdot mm^{-2}$	Maximum ingestion-length scaling constant
H_P	0.015	mg/L	Ingestion half-saturation density
γ_P	0.005	day^{-1}	Growth rate in length scaling constant
α_P	$4.8 \cdot 10^{-5}$	$day^{-1} \cdot mm^{-2}$	Maximum fecundity-length scaling constant
μ_P	0.01	day^{-1}	Background mortality rate
δ	0.05	—	Minimum prey-predator length ratio
φ	0.1	—	Optimal prey-predator length ratio
ε	0.4	—	Maximum prey-predator length ratio

Parameter values for the resource and size-structured consumer are given in box 3.7.

maximum resource density R_{max}, as well as size-independent predator mortality in excess of background mortality levels. With respect to the occurrence of alternative stable states, the bifurcation graphs resemble the graphs shown above for an unstructured predator population foraging on small consumers only (figure 4.9): over a range of values of R_{max}, as well as additional, size-independent predator mortality, a stable predator-consumer-resource equilibrium occurs as an alternative to the consumer-resource equilibrium, which is stable against invasion by the predator. The ranges of parameters with alternative stable states are, however, significantly larger. At background mortality levels, the minimum value of R_{max} for which predators can persist ($R_{max} = 0.28$ mg/L) is almost an order of magnitude lower than the smallest R_{max} value that allows for predator invasion into a consumer-resource equilibrium ($R_{max} = 2.6$ mg/L). Similarly,

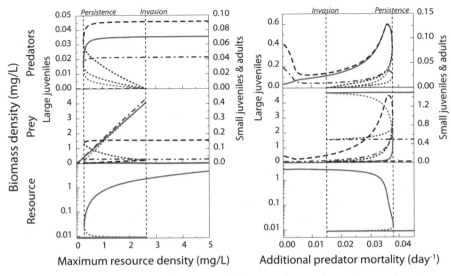

FIGURE 4.13. Changes in equilibrium biomass of basic resource (*bottom*), juvenile consumers smaller (*middle, dashed-dotted line*) and larger (*middle, solid line*) than 50 mm, adult consumers (*middle, dashed line*), juvenile predators smaller (*top, dashed-dotted line*) and larger (*top, solid line*) than 80 mm, and adult predators (*top, dashed line*) with increasing maximum resource density R_{max} (*left panels*) and predator mortality on top of background levels (*right panels*) in the size-structured consumer, size-structured predator model based on Kooijman-Metz energy channeling (see boxes 3.6 and 4.3 for model formulation). Stable equilibria are indicated with solid, dashed, or dashed-dotted lines, unstable equilibria with dotted lines. For maximum resource densities between 0.28 and 2.6 mg/L (*left panel*), as well as for additional predator mortality rates between 0.015 and 0.038 per day (*right panel*), two stable equilibria co-occur. Throughout, $R_{max} = 3.0$ mg/L (right panel only). Otherwise, default parameters apply as given in boxes 3.7 and 4.4.

for $R_{max} = 3.0$ mg/L, the maximum predator mortality in excess of background levels for which predators can persist is about twice the value above which the consumer-resource equilibrium is stable against predator invasion.

The model with a size-structured predator population, however, shows results that differ significantly from the results exhibited by all models discussed previously in this chapter (cf. figure 4.13 with figures 4.1, 4.4, 4.7, and 4.9). In particular, for increasing values of R_{max} in the model with size-structured predators, biomass densities in all size classes of consumers and predators quickly approach an asymptotic value without increasing further. Over the range of R_{max} values with alternative equilibria, slight changes can be observed in these biomass densities with increasing R_{max}, but above this range both biomass density and size distribution of the two populations are virtually independent of R_{max}, and increases in R_{max} lead only to increases in the basic resource. In the range of R_{max} values without any further changes in consumer and predator biomass, the length-age relationship of predators is strongly sigmoid, with very slow growth up to a length of 80 mm during the first five hundred days of life, rapid growth in length from 80 to 600 mm in the second five hundred days of life, and only minor increases to a maximum length around 630 mm afterward (results not shown).

Predators do not follow a von Bertalanffy growth curve as consumers do, because prey availability changes with increasing predator length and prey availability is particularly low for predators smaller than 80 mm. In equilibrium the predator population is hence controlled through a growth bottleneck at small sizes and the low survival up to a length of around 80 mm that results from it. In contrast, large juvenile and adult predators have abundant prey to forage on and hence exhibit rapid growth and fast reproduction. Consumers are also growing and reproducing at maximum rate given that resource densities in the predator-consumer-resource equilibrium (see lower panel of figure 4.13) are substantially larger than their half-saturation density ($H_C = 0.015$ mg/L). This explains why increases in R_{max} do not lead to further changes in consumer size distribution. Consumer biomass density is kept constant by the predator at a level that ensures an equilibrium; that is, a newborn predator individual on average produces a single offspring during its lifetime.

With increases in additional, size-independent predator mortality, adult predator biomass first decreases sharply until mortality is about twice its background value. Thereafter adult biomass steadily increases, reaching levels that are almost 50 percent larger than biomass density at background mortality (figure 4.13, right panel). Total juvenile predator biomass is only slightly lower than adult biomass density at background mortality and approximately equally distributed over the juvenile predator size range below and above 80 mm in length. Total juvenile predator biomass increases continuously with increasing predator mortality, reaching a maximum at roughly the same mortality level (0.036 per day) as adult predator biomass, which is close to the maximum mortality that allows for predator persistence (0.038 per day). The overall increase in juvenile predator biomass with increasing mortality masks the differences in response of small- and large-juvenile biomass. Biomass of small predators (up to 80 mm) exhibits an initial decrease with increasing mortality, similar to the decrease in adult predator biomass, and it also increases afterward. Unlike adult biomass, however, the highest density of small-predator biomass occurs at background mortality. The increase in total juvenile predator biomass hence results from an increase in biomass of juvenile predators larger than 80 mm.

The changes in biomass of juvenile and adult consumers resemble the pattern exhibited by adult predator biomass: initially juvenile and adult consumer biomass decreases with mortality and thereafter increases.

The changes in predator biomass with increasing mortality result from a complex interplay of changes in predator growth and reproduction and changes in predation mortality experienced by consumers of different sizes. At background levels of predator mortality, newborn consumers (roughly up to 15–20 mm) suffer most from predation, experiencing mortality rates that are some one hundred times larger than background mortality. While adult predator biomass decreases as a direct effect of the increase in predator mortality, small-predator biomass decreases through a combination of this higher mortality and a reduction of offspring production by the adults, who decrease in density. This decrease in small-predator biomass results, on the one hand, in a five-times lower predation mortality experienced by the newborn consumers, while on the other hand, it relaxes competition among small predators for available prey and hence disposes of the growth bottleneck that these predators experience at background mortality. The increase in somatic growth at small body sizes translates into an increase in biomass of large-juvenile predators, which more than compensates for the increased mortality that they experience.

In turn, the increase in large-juvenile predators increases the predation mortality experienced by juvenile consumers with a size between 20 and 40 mm and thereby reduces the prey availability for large-juvenile and adult predators. This decrease in prey availability translates into reduced growth rates of these larger predators; in particular, maximum predator length decreases from around 630 mm to below 500 mm. The decreased growth rate of large-juvenile predators also slows down their maturation and thereby reinforces the increase in biomass with mortality in this size range. Furthermore, the decrease in prey availability for adult predators leads to a further reduction of total reproduction than what resulted from the decrease in their density.

All these changes occur over the range where predator mortality increases from backward levels (0.01) to approximately twice that value. At twice the background mortality, predator growth in body size resembles a von Bertalanffy growth curve, although the maximum body size reached (around 500 mm) is well below the maximum size possible ($l_m = 750$ mm). Prey growth in body size as well as prey reproduction is unlimited given the high resource densities (see figure 4.13, right bottom panel). Further increases in predator mortality result in more rapid growth of predators to ever larger body sizes and significant increases in their fecundity, as a consequence of the increasing availability of prey of all sizes. The increased growth in body size is the main reason why biomass of large-juvenile and adult predators increases so strongly.

All size classes of consumer and predator increase with increasing predator mortality once the growth bottleneck for predators at small sizes has disappeared, albeit that the increase in the juvenile predator size class below 80 mm in length is small.

This remarkable increase with increasing mortality continues until juvenile and adult predator biomass and also adult consumer biomass reach their maximum, just before the predator persistence threshold (see figure 4.13). Juvenile consumer biomass continues to increase with mortality until the predator persistence threshold is reached. Over the range of additional predator mortality of 0.01 (where it is equal to background levels) to 0.036 per day (where predator biomass reaches its maximum), equilibrium resource densities decrease from a density of roughly 90 percent of its maximum ($R_{max} = 3.0$ mg/L) to about 4 percent of that value. Resource densities remain nonetheless more than an order of magnitude larger than the half-saturation density of the consumer ($H_C = 0.015$ mg/L), so consumer growth and reproduction remain unlimited by resource availability. As a consequence of the decrease in resource density, net resource productivity in equilibrium (equal to $\rho(R_{max} - \tilde{R})$, with $\tilde{R}$ the equilibrium resource density) increases almost an order of magnitude. This increased productivity forms the basis for the increases in consumer and predator biomass with increasing predator mortality.

Figure 4.14 shows for which combinations of maximum resource density and additional predator mortality a consumer-resource equilibrium as well as

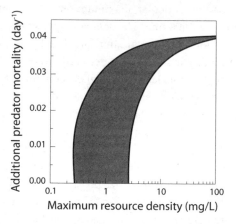

FIGURE 4.14. Range (*gray region*) of additional predator mortality rates in excess of background levels and maximum resource densities R_{max} for which both a stable resource-consumer and a stable resource-consumer-predator equilibrium occur in the size-structured consumer, size-structured predator model based on Kooijman-Metz energy channeling. Default parameters apply as given in boxes 3.7 and 4.4.

a predator-consumer-resource equilibrium occur as alternative stable states for the same parameter values. We constructed this graph by computing the type of equilibrium bifurcation graphs that are shown in figure 4.13, as a function of either R_{max} or additional predator mortality and for a range of constant values of the other parameter (that is, additional predator mortality or R_{max}, respectively). From these bifurcation graphs we inferred the combination of R_{max} and additional predator mortality at which the predator's invasion and persistence thresholds occurred. It should be noted that this procedure differs from the procedure we used to construct the analogous graphs for the stage-structured biomass model (figures 4.6 and 4.8) and the size-structured consumer-resource model with an unstructured predator population (figure 4.11), in which case we could numerically compute the parameter combination where the tangent to the equilibrium bifurcation curve is a vertical line. This latter approach is, however, technically not feasible if both the consumer and predator populations are size structured. The boundaries shown in figure 4.14 are hence not computed with the same accuracy as the analogous figures for the simpler models shown before, but the figure nonetheless provides a good indication of the parameter combinations for which we can expect alternative stable states to occur.

Figure 4.14 makes clear that bistability in the size-structured consumer, size-structured predator model mainly occurs at low maximum resource densities (i.e., low resource productivities) and is of marginal importance for maximum resource densities above $R_{max} = 50$ mg/L. This contrasts with the previously discussed models that included unstructured predators, in which an emergent Allee effect could occur at all productivities (compare figure 4.14 with the left panels of figures 4.6 and 4.8). The range of R_{max} values over which bistability occurs, however, is significantly larger than in those models with unstructured predators. Also, the bistability region in the size-structured consumer, size-structured predator model is more dependent on R_{max} than on additional predator mortality (figure 4.14). As a consequence, for low values of R_{max}, bistability occurs over a larger range of additional predator mortality levels than in the models with unstructured predators (cf. figure 4.14 with the left panels of figures 4.6 and 4.8). To summarize the results shown in figure 4.14 in an approximate way, it could be stated that predators will be able to persist in stable equilibrium with consumers and resource at resource productivities that are an order of magnitude smaller than the productivity that allows for invasion of the predator into the consumer-resource equilibrium, with the restriction that predators cannot persist at additional mortality rates that are larger than four times the background.

The bistability of a stable predator-consumer-resource equilibrium occurring next to a stable consumer-resource equilibrium for otherwise the same values of parameters is a consequence of the biomass overcompensation that occurs in the consumer population with increases in (predation) mortality. This characteristic is common to the stage-structured biomass model of predator, consumer, and resource; the size-structured consumer-resource model with unstructured predator population; and the model accounting for size structure in both consumers and predators. In contrast, the saturating relationship between consumer and predator biomass with increasing values of R_{max} and the increases in these biomasses with increasing predator mortality are consequences of the size-dependent processes in predator life history in combination with the food availability for differently sized predators. Prey availability for, and hence ingestion by, large-juvenile and adult predators is large compared with the energetic needs to cover their maintenance costs, which translates into rapid growth and high adult fecundity. The ratio between ingestion and maintenance requirements is much smaller for small predators because of the lower availability of prey.

As a consequence, a growth bottleneck for predators smaller than 80 mm develops under equilibrium conditions, which delays maturation and thereby reduces juvenile survival to such an extent that the average, per-capita fecundity (measured as total offspring production divided by the total number of predators) exactly compensates for per-capita mortality. The analysis of the size-structured consumer-resource model in chapter 3, which was based on Kooijman-Metz energetics, revealed that in equilibrium the consumer population was primarily controlled by the recruitment of large, juvenile individuals to the adult stage. Given that in that model the resource availability is the same for all consumers, it can be concluded that the basic size-scaling of ingestion and maintenance rate in the Kooijman-Metz energy budget model naturally leads to a recruitment bottleneck. The growth bottleneck for predators smaller than 80 mm in the size-structured consumer, size-structured predator model studied here is therefore not the consequence of the Kooijman-Metz energy budget model that is used to model predator energetics. Rather, it results entirely from the differences in prey availability for predators of different body sizes and leads to an equilibrium state that is mostly governed by development control. In turn, this prey availability is determined by the consumer size-distribution in combination with the function determining the vulnerability of differently sized prey to predators of a particular size (figure 4.12).

The model with size-structured predators foraging on size-structured prey does not exhibit the classical pattern of a trophic cascade, in which the densities at adjacent trophic levels show opposite trends, while densities at trophic levels that are two links apart show correlated responses with changes in productivity or mortality (Hairston, Smith, and Slobodkin 1960; Oksanen,

Fretwell, et al. 1981). Instead, owing to the size-dependent processes in predator life history, predator and consumer biomass show correlated increases with increases in mortality, which are both opposite to the trend in resource density. The increase in predator biomass with increasing mortality and the lack thereof with increasing values of R_{max} therefore not only contrast with the results from the models discussed in earlier sections of this chapter, but are also at odds with the classical theory that is based on unstructured population models.

Furthermore, these results have major implications for the exploitation of ecosystems that consist of size-structured predator species foraging on size-structured consumer species with a size-selectivity that depends on both predator and consumer body size. Such size-structured predator-prey pairs probably occur in most ecological systems, but prominent from an exploitation point of view are marine and freshwater fish stocks, in which piscivorous species such as cod, tuna, and trout are the target of intense harvesting. Harvesting does not quite correspond to an increase in size-independent predator mortality though, as it targets mostly large individuals. For this reason, in figure 4.15 we show, for two different values of R_{max}, the response in predator biomass for increases in additional mortality that targets only adult predators, which we interpret as the harvesting rate of these adults. Because the responses in predator, consumer, and resource biomass with increases in adult harvesting turn out to be

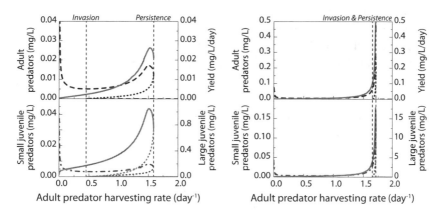

FIGURE 4.15. Changes in equilibrium biomass of juvenile predators smaller (*bottom, dashed-dotted line*) and larger (*bottom, solid line*) than 80 mm and adult predators (*top, dashed line*), as well as adult-predator harvesting yield (*top, solid line*) with increasing adult harvesting rate in the size-structured consumer, size-structured predator model based on Kooijman-Metz energy channeling for $R_{max} = 3.0$ mg/L (*left panels*) and $R_{max} = 50$ mg/L (*right panels*). Stable equilibria are indicated with solid, dashed, or dashed-dotted lines, unstable equilibria with thin dotted lines. Otherwise, default parameter values apply as given in boxes 3.7 and 4.4.

qualitatively similar to the responses with increases in size-independent predator mortality, the figure only shows the changes in biomass in the different
predator size classes. In addition, the figure shows the yield of adult predator
biomass resulting from the harvesting.

A small increase in harvesting from zero causes an almost immediate drop
in biomass of small-juvenile and adult predators, which together make up the
main part of predator biomass at background mortality (figure 4.15). Once the
harvesting rate is roughly equal to the background mortality, further increases
lead only to higher biomass densities in all predator size classes. In this aspect
the results closely resemble the results with increasing size-independent mortality. In contrast to these latter results, however, at low resource productivity
($R_{max} = 3.0$ mg/L), adult biomass never increases to levels above the density in
the absence of harvesting. At high resource productivity ($R_{max} = 50$ mg/L), biomass in all predator size classes eventually exceeds the densities at background
mortality. Irrespective of the maximum resource productivity, the harvesting
yield of adult predator biomass always increases with harvesting intensity until
it reaches a maximum at roughly the same harvesting levels where the biomass
density of predators is maximal. This maximum in yield and biomass density occurs at a mortality level close to the maximum mortality that allows for
predator persistence. At low resource productivity, bistability between a stable
predator-consumer-resource equilibrium and a stable consumer-resource equilibrium occurs over a substantial range of harvesting rates, but at high resource
productivity, the invasion and persistence threshold virtually coincide. The
closeness of the invasion and persistence thresholds at high resource productivity implies that all biomass densities increase with harvesting virtually up
to the point of extinction, given that the maximum densities and the maximum
yield occur at a mortality level between the two.

These results imply that in a scenario where harvesting intensity slowly
increases over time, all observable signals in the system continue to show
more and more positive values, as both harvesting yield and biomass densities in all predator size classes grow continuously with harvesting intensity up
to the point that leads to a collapse of the entire predator population. At low
resource productivity, some warning for this imminent collapse may present
itself shortly before the collapse, as yield and biomass start to decrease prior
to it. At high productivity, however, the collapse comes virtually unannounced,
and the population continues to become both more abundant and more productive up to the point of complete extinction (see figure 4.15). This pattern does
resemble some of the changes that have occurred in some major fish stocks (in
particular cod) that have collapsed in the last decades, as will be discussed in
more detail below. Most important, however, these results show that changes

in population biomass density and yield may be totally inappropriate indicators for the status and viability of a population, owing to the complex responses of these quantities with increases in harvesting. Because size-dependent life history processes—and, in particular, size-dependent foraging of predators on size-structured prey—is likely to be the rule rather than the exception in ecological communities, the results shed doubt on the use of abundance monitoring as a tool to protect an exploited population from overexploitation. In fact, as explained above, management based on abundance monitoring may even be an excellent approach to promote overexploitation and drive a population to extinction as quickly as possible!

EMPIRICAL EVIDENCE FOR EMERGENT ALLEE EFFECTS

Evidence for the presence of emergent Allee effects in empirical systems has surfaced as a result of overfishing of predatory fish in both marine and freshwater systems. Well-known examples are the collapse of the cod stocks in the Northwest Atlantic and in the Baltic Sea (Hutchings and Myers 1994; ICES 2006, 2007; Österblom, Casini, et al. 2006). In the Northwest Atlantic, a peak in the biomass of cod was recorded in the mid-1980s, despite an increasing fishing pressure dating from the early 1960s. However, by 1992 the cod had totally collapsed, and all stocks of cod in the Northwest Atlantic were at levels less than 15 percent of the observed maximum (Lambert and Dutil 1997; Myers, Hutchings, and Barrowman 1997). Notwithstanding a total fishing moratorium—instituted mainly as a result of the complete collapse of the fishing industry—little sign of recovery has been observed (ICES 2006; Lambert and Dutil 1997). Following the cod collapse, the biomass of prey fish, including the major prey fish of cod, capelin (*Mallotus villosus*), showed a strong increase in density, which in turn resulted in a cascading decrease in their prey, zooplankton (Carscadden, Frank, and Leggett 2001; Frank, Petrie, et al. 2005). The increase in fish prey biomass and decrease in their resource also led to a decrease in body condition and mean length of capelin (Carscadden and Frank 2002).

The Baltic cod stock increased up to the early 1980s, falling to very low levels of abundance within a few years after that peak period (figure 4.16, left panel). The cod stock stayed at these very low densities up to 2008, though some increase has been reported subsequently. At the same time as cod decreased, one of its prey fish species, herring (*Clupea harengus*), decreased, while a strong increase in the other main prey fish, sprat (*Sprattus sprattus*), occurred (figure 4.16, left panel). Concomitant with these changes in fish biomasses, a decrease in especially older cod mean weight and in sprat and herring

condition took place (Casini, Cardinale, and Hjelm 2006; ICES 2007; Österblom, Casini, et al. 2006). Furthermore, a decrease in the resource (zooplankton) of sprat and herring was recorded (Casini, Cardinale, and Hjelm 2006).

The collapses of the Northwest Atlantic and Baltic Sea cod stocks show
many similarities, including changes both in individual performance (primarily growth and body condition) as well as changes in population abundance of
predator, prey fish, and resources. For the Baltic Sea, estimates of planktivore
population fecundity furthermore show that it declined in parallel with the decrease in cod abundance, which is in line with the expectations based on the
occurrence of an emergent Allee effect (figure 4.16, right panel; van Leeuwen,
de Roos, and Persson 2008). A difference between the Northwest Atlantic and
Baltic Sea cod stocks may be that the latter has been reported to show signs
of an increase in recent years. It is, however, still too early to conclude that
this reflects a long-term recovery as a result of decreased fishing pressure. A
possible explanation for the increase in the Baltic Sea cod stock is that the
Baltic Sea, through human-induced eutrophication, is productive. From figure
4.15 (right panels) we can conclude that under these conditions a decrease in
fishing effort may indeed move the system from a consumer-resource state
to a predator-consumer-resource state (van Leeuwen, de Roos, and Persson
2008). Apart from an emergent Allee effect, several other mechanisms have

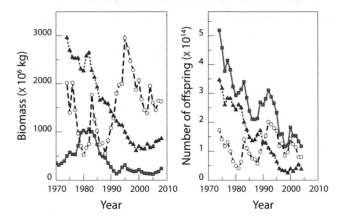

Figure 4.16. *Left*: Changes in the biomass of cod (*filled squares, solid line*),
sprat (*open circles, dashed line*), and herring (*filled triangles, dotted line*) from
1970 to 2008 in the Baltic Sea, based on International Council for the Exploration
of the Sea (ICES) estimates. *Right*: Changes in sprat population fecundity (*open
circles, dashed line*), herring population fecundity (*filled triangles, dotted line*),
and total planktivore fecundity (*filled squares, solid line*) from 1974 to 2004 in the
Baltic Sea. Data from van Leeuwen, de Roos, and Persson (2008).

been postulated to play a role in the lack of recovery of the cod stocks. These include predation of prey fish on cod eggs and competition for shared resources between planktivorous species and juvenile cod (Köster and Möllmann 2000; Möllmann, Kornilovs, et al. 2004). The potential for the latter to induce alternative stable states is something we will discuss in chapter 7. Finally, it has been proposed that the lack of cod recovery in both systems is a result of changed environmental conditions, in particular temperature and salinity (Choi, Frank, et al. 2004; Frank, Petrie, et al. 2005; Möllmann, Kornilovs, et al. 2005; Möllmann, Müller-Karulis, et al. 2008). Detailed analyses of the Baltic Sea system has, however, shown that the observed lack of cod recovery cannot be explained by environmental conditions alone (Casini, Cardinale, and Hjelm 2006; Casini, Hjelm, et al. 2009).

The strongest empirical support for the occurrence of an emergent Allee effect comes from a large-scale and long-term study of brown trout (*Salmo trutta*) and Arctic char (*Salvelinus alpinus*) in Lake Takvatn in northern Norway (Persson, Amundsen, et al. 2007). In this low-productive lake with a transparency exceeding 10 m, overharvesting of the predatory brown trout reduced their numbers to very low levels. As a consequence, Arctic char (*Salvelinus alpinus*), the prey fish for large brown trout but also a potential competitor of small brown trout for shared resources, came to totally dominate the fish community, and by 1980 trout was almost absent (figure 4.17). To improve the lake fishery, almost seven hundred thousand char were removed from the lake between 1984 and 1989. This removal was quite successful, resulting in

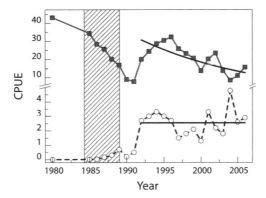

FIGURE 4.17. Changes in the abundance (catch per unit effort defined as capture per gill net per 24 hours) of brown trout (*open circles, dashed line*) and Arctic char (*closed squares, solid line*) in Lake Takvatn, 1980–2006. Heavy fishing of Arctic char took place from 1984 to 1989 (*hatched area*). Black, solid curves starting in 1991 represent trend lines. Data from Persson, Amundsen, et al. (2007).

an 80 percent decrease in char density by 1991. Subsequently, the char density rebounded to less than half its 1984 density, which was followed by a decelerating decrease toward a new steady-state density (figure 4.17). At the same time, the trout density increased significantly from 1989 to 1992 and remained steady afterward. Although char is claimed to compete with small trout for food, no evidence was found for a negative density-dependent effect of char on small brown trout (Persson, Amundsen, et al. 2007).

The Takvatn study provides several strong lines of evidence for the presence of alternative stable states induced by an emergent Allee effect. First of all, this whole-lake experiment gives particularly strong evidence for the occurrence of alternative stable states per se, as the system has remained in the present state for more than fifteen years after the termination of the char thinning, which is more than three times the generation time of brown trout in the lake (figure 4.17). Second, the growth rates in body size of char was substantially higher in postperturbation years compared with preperturbation years, especially for larger char (figure 4.18). This result can be compared with the char population in a nearby control lake, Fjellfrøsvatn, where no change in the growth rate of char was observed over the study period 1979–1999. In this lake, which has a brown trout population similar in density to the trout population in Takvatn after the char thinning, the growth rates in char body size are also similar to the char growth rates after its thinning in Takvatn (figure 4.18). Third, the size structure of the Arctic char population changed drastically as a

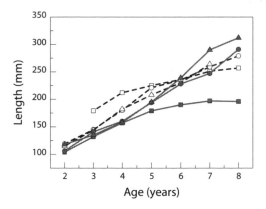

FIGURE 4.18. Individual growth curves of Arctic char in Lake Takvatn before (*closed squares, solid line*), during (*closed circles, solid line*), and after the perturbation (*closed triangles, solid line*). The individual growth curves of Arctic char in the control lake, Fjellfrøsvatn, in 1979 (*open squares, dashed line*), 1992 (*open circles, dashed line*), and 1999 (*open triangles, dashed line*) are shown for comparison. Data from Persson, Amundsen, et al. (2007).

result of the thinning. Before the manipulation, the char population was densely packed and dominated by individuals in the size range between 161 and 211 mm (figure 4.19), whereas the size distribution changed after the thinning to a reduced dominance by multiple size classes (Simpson's measure of dominance changed from 0.16 to a mean of 0.05). In comparison, the size distribution of char in the control Lake Fjellfrøsvatn did not change over the study period and was similar to that in Takvatn after the perturbation (Simpson's measure of dominance: 0.06; figure 4.19).

A critical prediction following from the hypothesis that the Takvatn system exemplifies an emergent Allee effect, is that the thinning of the char population should lead to an overcompensatory increase in abundance of prey in the size ranges that brown trout feeds on. Given their body sizes, brown trout in Takvatn will mainly feed on char with a length less than 160 mm (Persson, Amundsen, et al. 2007). The abundance of this size class in Lake Takvatn was indeed higher after the manipulation than before it (figure 4.19, upper panels). Furthermore, the abundance of this size class of Arctic char in the control Lake Fjellfrøsvatn was high throughout the entire study period (figure 4.19, lower panels). Finally, the observation that individual growth in brown trout body size was positively related to the availability of char less than 160 mm further supports the thesis that brown trout performance was limited by the availability of suitably sized prey fish (Persson, Amundsen, et al. 2007).

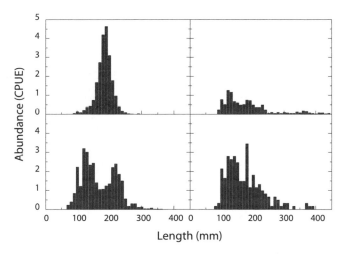

FIGURE 4.19. Size distribution of Arctic char before the thinning in 1980 (*upper left panel*) and after the thinning in 1994 (*upper right panel*) in Lake Takvatn and in the control lake, Fjellfrøsvatn, in 1992 (*lower left panel*) and in 1999 (*lower right panel*). Data from Persson, Amundsen, et al. (2007).

Overall, overfishing has revealed the potential for emergent Allee effects to play a major role in structuring fish communities. Although all three examples discussed above provide substantial evidence for the presence of emergent Allee effects, the Takvatn example undoubtedly gives the strongest evidence for the importance of stage-specific overcompensation in generating alternative stable states. Finally, in all modeling scenarios in this chapter we have assumed that all predator stages feed on the consumer, although with different efficiencies. An obvious question to address is how ontogenetic niche shifts in predator life history affect the results. This question is something we will consider in chapters 6 and 7.

CHAPTER 5

Emergent Facilitation among Predators on Size-Structured Prey

Whenever two or more consumer species compete for the same resources, their coexistence is governed by the principles of competitive exclusion (Hardin 1960; MacArthur and Levins 1967; Volterra 1928) and by niche segregation (Schoener 1974). The principle postulates that two consumer species cannot coexist in stable equilibrium when competing for a single resource. The species that manages to persist at the lowest resource density will oust its competitor. Only in cases where this superior competitor exhibits fluctuations in density—for example, because its interaction with the resource results in predator-prey oscillations—may the inferior competitor, under specific conditions, persist as well (Armstrong and McGehee 1980). On the other hand, when two consumer species compete for two different resources, coexistence is possible if the ecological niches that the two species occupy are sufficiently different. To this end, both species should be most limited by a different resource, and each species should consume proportionally more of the resource that limits its own growth most (Tilman 1982). The competitive exclusion principle and Tilman's ideas about coexistence of two consumers on two resources have become cornerstones of what is generally referred to as competition theory. At the same time, these ideas have also led to controversy, for example, about what exactly constitutes a resource and what does not. Levin (1970) extended the competitive exclusion principle by relaxing the assumption about resources to account for all "limiting factors" that can constrain population growth, but even with this extension, the identification of what represents a resource and whether consumers use different resources remains troublesome (Abrams 1988).

Prey species that exhibit distinct stage or size structure provide a good example of the difficulties surrounding the identification of resources. In the previous chapter we have focused on predators that specialize on one particular stage of a consumer species, either the juveniles or the adults. However, prey individuals are often exposed to a different suite of predators in different stages of their life (Cohen 1978; Zaret 1980). For example, marine prey fish species are exposed to predation by piscivorous fish when small, but they are hunted by marine mammals when larger (Duplisea 2005). Do these piscivorous fish and marine mammals compete for the same resource, or do they occupy different niches? On the one hand such stage-specific predators do not forage on exactly the same prey individuals, but on the other hand the prey individuals in different life history stages belong to the same species, and their dynamics are hence linked through maturation and reproduction.

Intuitively one might therefore expect that the foraging by one stage-specific predator species negatively affects the food availability for the other, leading to competition between the two. In the previous chapters we showed, however, that biomass overcompensation may occur and hence that mortality in a particular life history stage of a prey species readily leads to an increase in prey biomass in another stage, in cases where the mortality interacts with a food- or density-dependent bottleneck in the prey life history. On the basis of these results we can already hypothesize that the interaction between two predators on size- or stage-structured prey is not as straightforward as if they were competing for the same prey species and that their influence on each other might not be purely negative. Instead, in this chapter we will discuss a variety of positive interactions between predators foraging on different stages of the same prey species, which all emerge owing to the biomass overcompensation that may occur in prey life history stages in response to increased mortality. These interactions include emergent facilitation of specialist predators by generalists that forage on the same prey individuals as the specialists but in addition forage on smaller or larger prey individuals as well. Furthermore, we will show that two predators that specialize on different life-history stages of prey can facilitate each other to the extent that one predator relies on the presence of the other for its persistence. Finally, we will show that a stage-specific predator may act as a catalyst species, which promotes and in fact is necessary for the invasion of another predator species, but is subsequently outcompeted by the latter.

We will analyze these interactions between predators of stage-structured prey using the same stage-structured biomass framework as used in previous chapters. In fact, the model analyzed in the following sections is virtually the same model as analyzed in chapter 4, except for the extension to two unstructured predator populations. All functions and equations of this stage-structured prey, two-predator model are presented in box 5.1, and the default parameter set is presented in box 5.2. Dynamics of both predator populations are described by the Yodzis and Innes bioenergetics model (Yodzis and Innes 1992, equation (3.3b)), in the same manner as the biomass dynamics of predators in chapter 4. The two predator populations differ in their food selectivity, which is parameterized by the parameter ϕ_i, representing the proportion of time that

BOX 5.1

EQUATIONS AND FUNCTIONS OF THE STAGE-STRUCTURED PREY, TWO-PREDATOR BIOMASS MODEL

Dynamic equations	*Description*
$\dfrac{dR}{dt} = G(R) - \omega_J(R)J - \omega_A(R)A$	Resource biomass dynamics
$\dfrac{dJ}{dt} = v_A^+(R)A - \gamma(v_J^+, d_J)J + v_J(R)J - d_J(P_1, P_2)J$	Biomass dynamics of juvenile prey
$\dfrac{dA}{dt} = \gamma(v_J^+, d_J)J + (v_A(R) - v_A^+(R))A - d_A(P_1, P_2)A$	Biomass dynamics of adult prey
$\dfrac{dP_i}{dt} = (v_{P_i}(J,A) - \mu_{P_i})P_i$	Biomass dynamics of predator $i = 1,2$

Function	*Expression*	*Description*
$G(R)$	$\rho(R_{\max} - R)$	Intrinsic resource turnover
$\omega_J(R)$	$M_C R / (H_C + R)$	Resource intake rate by juvenile prey
$\omega_A(R)$	$q M_C R / (H_C + R)$	Resource intake rate by adult prey
$v_J(R)$	$\sigma_C \omega_J(R) - T_C$	Net energy production of juvenile prey
$v_A(R)$	$\sigma_C \omega_A(R) - T_C$	Net energy production of adult prey
$d_J(P_1, P_2)$	$\mu_C + \displaystyle\sum_{i=1,2} M_{P_i} \dfrac{\phi_i P_i}{H_{P_i} + \phi_i J + (1 - \phi_i)A}$	Mortality rate of juvenile prey
$d_A(P_1, P_2)$	$\mu_C + \displaystyle\sum_{i=1,2} M_{P_i} \dfrac{(1 - \phi_i)P_i}{H_{P_i} + \phi_i J + (1 - \phi_i)A}$	Mortality rate of adult prey
$\gamma(v_J^+, d_J)$	$(v_J^+(R) - d_J(P_1, P_2)) / (1 - z^{(1 - d_J(P_1,P_2)/v_J^+(R))})$	Maturation rate of juvenile prey
$v_{P_i}(J,A)$	$\sigma_{P_i} M_{P_i} \dfrac{\phi_i J + (1 - \phi_i)A}{H_{P_i} + \phi_i J + (1 - \phi_i)A} - T_{P_i}$	Net energy production of predators

$v_J^+(R)$ and $v_A^+(R)$ represent the value of $v_J(R)$ and $v_A(R)$, respectively, but restricted to nonnegative values. For brevity, the dependence of v_J^+ and d_J on resource and predator density, respectively, is suppressed in the maturation function $\gamma(v_J^+, d_J)$. Notice that $\gamma(v_J^+, d_J)$ is positive as long as $v_J^+(R)$ is positive. Also, $v_A(R) - v_A^+(R)$ cancels as long as $v_A(R)$ is positive and equals $v_A(R)$ otherwise. Hence, the use of $v_J^+(R)$ and $v_A^+(R)$ in the model equations implies that growth, maturation, and fecundity are positive as long as juvenile and adult net production, respectively, are positive and equal 0 otherwise.

BOX 5.2

PARAMETERS OF THE STAGE-STRUCTURED PREY,
TWO-PREDATOR BIOMASS MODEL

Parameter	Default value	Unit	Description
Resource			
ρ	0.1	day^{-1}	Resource turnover rate
R_{max}	30	mg/L	Resource maximum biomass density
Prey			
M_C	$0.1W_A^{-0.25}$	day^{-1}	Mass-specific maximum ingestion rate
H_C	3.0	mg/L	Ingestion half-saturation resource density
q	0.5, 1.0 or 2.0	—	Adult-juvenile consumer ingestion ratio
T_C	$0.01W_A^{-0.25}$	day^{-1}	Mass-specific maintenance rate
σ_C	0.5	—	Conversion efficiency
z	0.01, 0.1 or 0.5	—	Newborn-adult consumer size ratio
μ_C	$0.0015W_A^{-0.25}$	day^{-1}	Background mortality rate
Predators			
M_{P_i}	$0.1W_{P_i}^{-0.25}$	day^{-1}	Mass-specific maximum ingestion rate
H_{P_i}	3.0	mg/L	Ingestion half-saturation prey density
T_{P_i}	$0.01W_{P_i}^{-0.25}$	day^{-1}	Mass-specific maintenance rate
σ_{P_i}	0.5	—	Conversion efficiency
μ_{P_i}	$0.0015W_{P_i}^{-0.25}$	day^{-1}	Background mortality rate
ϕ_i	0.0, 0.5 or 1.0	—	Foraging preference for juveniles

W_A and W_{P_i} represent average adult body weight of consumer and predator, $i = 1, 2$, respectively.

predator individuals forage on juvenile prey. We will consider predators that specialize in preying on juvenile ($\phi_i = 1$) or adult prey ($\phi_i = 0$) or that forage equally on both prey stages ($\phi_i = 0.5$). Prey selectivity is the only difference between the predators, which are assumed to have the same characteristic adult body mass W_P. This characteristic adult body mass determines all other predator parameters following the scaling laws that were also used in previous chapters. Parameters for resource and prey dynamics are mostly the same as in chapter 4 (see box 5.2).

GENERALISTS FACILITATING SPECIALIST PREDATORS

The results presented in the previous chapter already point to the possible facilitation that specialist predators may experience from generalist predators. Figure 4.1 in the previous chapter compared the effect of generalist predators and predators specializing on juvenile prey on a prey population, which at equilibrium was mostly under reproduction control. Because the ratio between adult and juvenile maximum ingestion rate was assumed equal to $q = 0.5$, adult prey were more limited in their energetics than juvenile prey, and the adult stage constituted an energetic bottleneck. The right panel of figure 4.1 shows that predators specializing on juvenile prey can persist when the maximum resource density exceeds the predator persistence threshold, which occurs around $R_{max} = 10.5$ mg/L, whereas these predators can invade a prey-only equilibrium at maximum resource densities exceeding 16 mg/L. In equilibrium, the juvenile-specialized predator requires a juvenile prey biomass density of 0.9 mg/L. The left panel of figure 4.1 shows that in a coexistence equilibrium of prey with a generalist predator, which feeds to the same extent on juvenile and adult prey, this juvenile prey density of 0.9 mg/L already occurs at a value of R_{max} below 10 mg/L (in fact around 8 mg/L). In other words, the predator specializing on juvenile prey can invade a coexistence equilibrium of prey and generalist predators at R_{max} values that are below its own persistence threshold. The presence of the generalist predator thus expands the range of R_{max} for which the juvenile-specialized predator can persist to lower values.

Figure 5.1 illustrates the increased persistence of the juvenile-specialized predator with the invasion dynamics of both predators at $R_{max} = 10$ mg/L. Juvenile-specialized predators fail to invade a prey-only equilibrium at this resource productivity, as adult prey constitutes the largest part of prey biomass in equilibrium and the juvenile biomass density never reaches levels sufficient for predators to persist. Predator invasion leads to a short and rapid decrease in juvenile prey biomass and consequently to a decrease in adult prey biomass. The decrease in adult prey biomass is followed by overcompensation in prey total population reproduction, such that the decline in juvenile prey biomass is reversed and juvenile prey biomass increases to densities above its density in the prey-only equilibrium. This increase in juvenile prey biomass in turn slows down the decrease in biomass of the juvenile-specialized predator, but it is never sufficient to lead to an increase in predator biomass.

Following invasion, the biomass of the juvenile-specialized predator hence declines continuously. Therefore, the biomass overcompensation that occurs in the prey population at this maximum resource density is not strong enough to allow for predator persistence. In contrast, total prey biomass density in the prey-only equilibrium is sufficient for invasion of the generalist predator

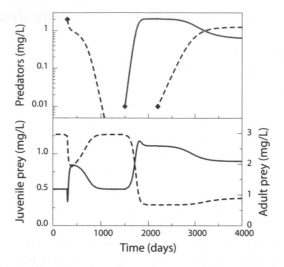

FIGURE 5.1. Invasion dynamics of generalist (*top, solid lines*) and juvenile-specialized predators (*top, dashed lines*) into an equilibrium of resource, juvenile (*bottom, solid lines*), and adult prey (*bottom, dashed lines*) for R_{max} = 10 mg/L and background predator mortality. Throughout, $q = 0.5$ and $z = 0.01$. Otherwise, default parameters apply for a prey and predators with body size $W_A = 0.0001$ and $W_P = 0.01$ gram (both predators), respectively. Generalist and specialist predators forage equally on both juvenile and adult prey ($\phi_1 = 0.5$) and exclusively on juvenile prey ($\phi_2 = 1$), respectively. Invasion of juvenile-specialized predators at $t = 300$ into the prey-only equilibrium is unsuccessful, despite the high initial density, whereas generalist predators can invade successfully even from low density ($t = 1500$). Generalist predator invasion allows for subsequent, successful invasion of specialist predators from low density ($t = 2200$).

(figure 5.1, $t = 1500$). Moreover, the generalist predator relaxes the reproduction control in the prey population, such that the equilibrium juvenile prey biomass after invasion of the generalist is much higher, whereas adult prey biomass is significantly lower. The changed biomass composition of the prey population in equilibrium with the generalist predator now allows for invasion of the juvenile-specialized predator, even from low initial densities (figure 5.1, $t = 2200$). The establishment of the juvenile-specialized predator leads to a decrease and increase in juvenile and adult prey biomass, respectively, as well as to an almost threefold reduction in generalist predator biomass. The presence of the generalist predator not only promotes invasion of the juvenile-specialized predator but is also crucial for its persistence. If for one reason or another the generalist predator goes extinct, the juvenile-specialized predator would not survive either, and the prey-only equilibrium would be the final community state. For $R_{max} = 10$ mg/L, persistence of the juvenile-specialized predator is therefore possible only in the presence of generalist predators. Despite that the generalist predator is crucial for the persistence of the community state and the juvenile-specialized predator is highly dependent on it, the latter is ultimately the most abundant predator in the community, reaching densities that are almost double the biomass density of the generalist predator.

Obviously, the generalist predator facilitates the invasion and persistence of the juvenile-specialized predator even though both predator populations forage on and compete for juvenile prey. This facilitation comes about because the mortality imposed by the generalist predator indirectly leads to an increase in equilibrium biomass of juvenile prey through biomass overcompensation. It hence emerges as a result of the interplay between the mortality inflicted by the generalist and the density-dependence in prey life history. The same interplay resulted in the positive feedback between predators and the density of their preferred prey discussed in chapter 4 that gave rise to the emergent Allee effect. Because of this analogy, we refer to this type of facilitation as "emergent facilitation."

The left panel of figure 5.2 illustrates the extent of the emergent facilitation by showing the long-term community dynamics of prey, generalist predators, and juvenile-specialized predators as a function of the maximum resource density R_{max}. Qualitatively, the long-term community dynamics are the same as shown in figure 4.1 of the previous chapter for a community of prey and generalist predators in the sense that the ultimate dynamics for a particular value of R_{max} are always uniquely determined, whereas population cycles arise for higher values of the maximum resource density. Bistability between different equilibrium states or different types of dynamics hence does not occur. For R_{max} exceeding 8 mg/L, invasion and persistence of the juvenile-specialized predator is possible, leading to stable coexistence of both predators. Comparison of figures 4.1 and 5.2 (left panels) reveals that population cycles already occur at lower values of R_{max} when the juvenile-specialized predator is present than in a community with only generalist predators. Surprisingly, the minimum density of the generalist predator during the population cycles is above its equilibrium density as a result of the interaction with the juvenile-specialized predator.

In the range of maximum resource densities R_{max}, for which the juvenile-specialized predator in isolation would exhibit an emergent Allee effect (figure 4.1, right panel), the generalist predator thus eliminates the threshold density that keeps the juvenile-specialized predator at low densities from growing, and it even allows for positive growth of the juvenile-specialized predator at values of R_{max} below this range. This last result reveals that overcompensation in juvenile prey biomass in response to increased mortality, which leads to the increased persistence of the juvenile-specialized predator, is stronger when mortality targets all prey individuals, compared with mortality that targets only juvenile prey. This is in line with the results presented in chapter 3, in which it was shown that juvenile biomass overcompensation was weakest in response to increases in juvenile mortality. For most values of R_{max} the juvenile-specialized predator is furthermore the more abundant of the two predator species, reaching densities that are more than five times the biomass density of generalists.

The right panel of figure 5.2 shows for a maximum resource density of $R_{max} = 10$ mg/L the changes in long-term community dynamics with increasing mortality of the generalist predator in excess of its background mortality. Most remarkably, increasing mortality of the generalist predator first leads to the extinction of the juvenile-specialized predator, which occurs when generalist predators experience a mortality that is roughly three and a half times their background mortality. Even though the juvenile-specialized predator

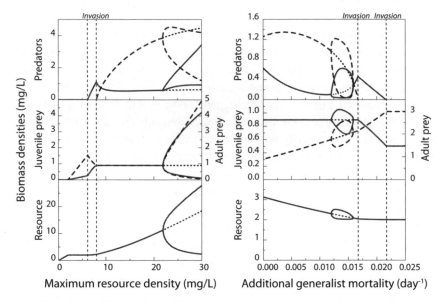

FIGURE 5.2. Changes in equilibrium biomass in a community of resource (*bottom*), juvenile (*middle, solid lines*), and adult prey (*middle, dashed lines*) and generalist (*top, solid lines*) and juvenile-specialized predators (*top, dashed lines*) at background predator mortality with increasing maximum resource density R_{max} (*left*) and for $R_{max} = 10$ mg/L with increasing mortality of generalist predators on top of background levels (*right*). Other parameters as in figure 5.1. Stable equilibria are indicated with solid or dashed lines; unstable equilibria with dotted lines. Population biomasses exhibit stable oscillations around an unstable equilibrium for maximum resource densities larger than 22 mg/L at background mortality (*left*) and for additional, generalist predator mortalities between 0.012 and 0.016 per day with $R_{max} = 10$ mg/L (*right*). Minimum and maximum biomass densities during these cycles are indicated with solid or dashed lines.

does not suffer from the additional mortality, it is hence most sensitive to it. Generalist predators only go extinct themselves when their mortality has increased to approximately four and a half times their background levels. With increasing mortality of the generalist predator, the community equilibrium destabilizes and population cycles occur at mortality levels just below the threshold where the juvenile-specialized predator goes extinct. In these population cycles both predators, but in particular the generalist, reach biomass densities close to zero.

For cases in which both predators are exposed to additional mortality, figure 5.3 shows the range of maximum resource densities R_{max} and mortality rates in excess of background mortality, for which generalist predators facilitate the persistence of juvenile-specialized predators. In essence, figure 5.3 is an extended version of figure 4.6 of the previous chapter. As was shown there, invasion of the juvenile-specialized predator into a prey-only equilibrium is always possible for high maximum resource density and low mortality. At higher maximum resource densities a higher mortality rate still allows for predator invasion, such that the predator invasion boundary is an increasing,

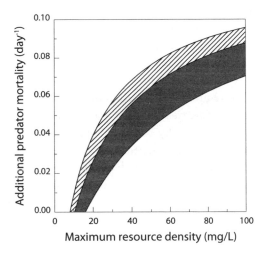

FIGURE 5.3. Range of additional predator mortality rates in excess of background levels and maximum resource densities R_{max} (*hatched region*) for which the presence of generalist predators ($\phi_1 = 0.5$) is required for persistence of juvenile-specialized predators ($\phi_2 = 1$). For parameters below this hatched region, juvenile-specialized predators can also persist in absence of generalist predators. The gray region indicates parameter values for which in the absence of generalist predators, the consumer-resource equilibria with and without juvenile-specialized predators are both stable and an emergent Allee effect occurs. Other parameters as in figure 5.1. Note that both predator species suffer from the additional mortality (cf. also with figure 4.6).

but saturating, function of maximum resource density. The range of parameters for which the juvenile-specialized predator cannot invade the prey-only equilibrium but can nonetheless persist through the emergent Allee effect borders its invasion boundary. The emergent facilitation that the juvenile-specialized predator experiences from the generalist turns the latter parameter region into a region for which the juvenile specialist can invade without suffering from the threshold density that is associated with the Allee effect. Bordering the parameter region, in which the generalist alleviates the emergent Allee effect of the juvenile-specialized predator, is a narrower parameter range for which the generalist allows for invasion and persistence of the juvenile-specialized predator, despite that the latter would never be able to persist in isolation.

Emergent facilitation occurs despite the fact that the foraging niche of the generalist predator overlaps and in fact encompasses the niche of the juvenile-specialized predator. It should, however, be noted that the generalist forages only for 50 percent of its time on juvenile prey, spending the remaining half of its foraging efforts on adult prey. This reduced effort with which the generalist forages on juvenile prey is essential: if the generalist were to forage with equal effort on juvenile prey as the juvenile specialist and in addition were to forage on adult prey, coexistence of the two predator species would never be possible, as equilibrium juvenile prey biomass would always be below the subsistence level of the juvenile-specialized predator. Equally essential for the emergent facilitation to occur is the prey stage distribution in an equilibrium of prey with generalist predators. As shown in the left panel of figure 4.1 in the previous chapter, juveniles make up a larger fraction of the prey population in such an equilibrium when R_{max} increases and dominate it for high values of R_{max}. The differences in foraging efforts by both predators on juvenile and adult prey imply that invasion of the juvenile-specialized predator is possible as soon as juvenile prey biomass equals adult prey biomass. Whether or not this equality occurs, however, depends on the values of the parameters q and z, representing the ratio between adult and juvenile prey maximum ingestion rate and the extent of the juvenile period in the prey life cycle, respectively. From chapter 3 it can be inferred that values of $q < 1$ in combination with values of z close to 1 may prevent juvenile and adult prey biomass from becoming equal with increasing R_{max}. For such parameter settings emergent facilitation of juvenile-specialized predators by generalists may hence not occur.

Emergent facilitation of predators specializing on adult prey by a generalist predator species is in principle also possible, but it does not occur at background mortality levels for the parameter values $q = 2.0$ and $z = 0.5$, which we frequently used in previous chapters as a combination resulting in a prey population under development control. Assuming that generalist predators and predators specializing on adult prey have equal maximum ingestion, maintenance, and background mortality rates, the combination $q = 2.0$ and $z = 0.5$ ensures that juveniles constitute the largest part of the prey population biomass in an equilibrium with generalist predators and that they continue to do so when R_{max} increases. For these parameters, adult prey biomass hence never reaches a density that allows for invasion of the adult-specialized predator, and generalist predators therefore outcompete the specialist predator at all values of the maximum resource density. Emergent facilitation of adult-specialized predators by generalists will therefore occur only at background mortality with smaller values of q and/or larger values of z.

The absence of emergent facilitation of adult-specialized predators by generalists at background mortality levels for $q = 2.0$ and $z = 0.5$ does not, however, preclude its occurrence at higher mortality levels of the generalist predator. In figure 4.7 in the previous chapter, it was shown that persistence of the adult-specialized predator was possible only for R_{max} values larger than 16 mg/L. Figure 5.4 illustrates how below this threshold ($R_{max} = 15$ mg/L) persistence of the adult-specialized predator may be promoted by generalist predators when the latter experience a mortality rate that is higher than background levels. For low mortality values the generalist predator outcompetes the adult-specialized predator, but at higher mortality rates the generalist requires a higher prey density to survive. The increase in adult prey biomass allows the adult-specialized predator to invade once the additional mortality of generalists exceeds 0.008/day (approximately 1.7 times background mortality). With increasing generalist mortality coexistence between prey, generalist predator, and adult-specialized predator represents the only stable community state until generalist mortality exceeds 0.04/day (~8.6 times background mortality), for which the adult-specialized predator goes extinct. Generalist predators can no longer persist for mortality rates above 0.044/day (~9.3 times background mortality). As was shown above for juvenile-specialized predators, adult-specialized predators are more sensitive to increases in generalist mortality than generalists themselves, as they go extinct before the generalist predators do, even though they are not subject to the increased mortality at all. Furthermore, adult-specialized predators are considerably more abundant than generalist predators over a substantial range of generalist mortality values, despite the fact that generalists are crucial for persistence of adult-specialized predators: because the R_{max} value is below their persistence threshold (cf. figure 4.7), adult-specialized predators will go extinct as soon as generalists, in one way or another, disappear from the community.

FACILITATION BETWEEN SPECIALIST PREDATORS

From the results presented in the previous section, it is clear that biomass overcompensation in prey populations in response to increasing mortality may give

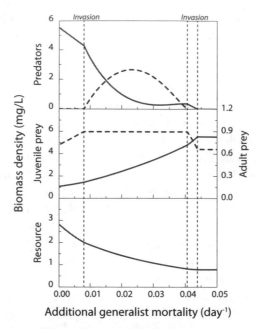

FIGURE 5.4. Changes in equilibrium biomass in a community of resource (*bottom*), juvenile (*middle, solid lines*), and adult prey (*middle, dashed lines*) and generalist (*top, solid lines*) and adult-specialized predators (*top, dashed lines*) for $R_{max} = 15$ mg/L with increasing mortality of generalist predators on top of background levels. Generalist and specialist predators forage equally on both juvenile and adult prey ($\phi_1 = 0.5$) and exclusively on adult prey ($\phi_2 = 0$), respectively. Throughout, $q = 2.0$ and $z = 0.5$. Otherwise, default parameters apply for a prey and predators with body size $W_A = 0.0001$ and $W_P = 0.01$ gram (both predators), respectively. All equilibria indicated are stable.

rise to complex interactions among the generalist and specialist predators that the prey is exposed to when the foraging niches of these predators overlap. The emergent facilitation that can occur between specialist predators that share the same prey species, but do not overlap in their foraging niche, is possibly even more complex. As will be shown below, emergent facilitation between specialist predators readily occurs and in general allows for coexistence of predators over even larger ranges of parameters than in case of a generalist and a specialist predator. However, it frequently results in bistability, with stable population cycles occurring as an alternative type of dynamics next to a stable equilibrium or next to another kind of stable population cycles.

Whether predators that specialize on juvenile prey facilitate the persistence of predators specializing on adult prey or vice versa depends on which life history process mostly limits the prey population in absence of any predation. In cases of development control, juvenile prey dominate the population in the absence of predators, and this provides the opportunity for juvenile-specialized predators to increase adult prey biomass and thereby promote persistence of predators foraging on adult prey. The reverse type of facilitation occurs in cases where the equilibrium prey population, in absence of predation, is reproduction-controlled. As shown by de Roos, Schellekens, et al. (2008a), emergent facilitation may occur when the development or reproduction control results from life history characteristics of the prey, for example, as a consequence of juvenile prey being less ($q > 1$) or more competitive foragers ($q < 1$), respectively, than adult prey. In addition, it will also occur when environmental conditions induce the development or reproduction control, such as when only juvenile or adult prey, respectively, compete for a limiting resource, with the other life stage having unlimited food supply. We will discuss in this chapter the emergent facilitation between juvenile-specialized and adult-specialized predators that occurs as a consequence of differences in food supply for juvenile and adult prey. Moreover, we consider only a single value for the ratio between size at birth and at maturation, $z = 0.1$. In this way we illustrate that emergent facilitation of adult-specialized predators by juvenile-specialized predators, as well as facilitation of juvenile-specialized predators by adult-specialized predators, may occur within the same community of prey and specialist predators, with the type of facilitation that occurs determined by whether food for juvenile or for adult prey is in abundant supply.

Figure 5.5 illustrates the emergent facilitation of predators that forage exclusively on adult prey by juvenile-specialized predators. Because it is assumed that adult prey have an unlimited food supply and only juveniles compete for the limiting resource (i.e., it is assumed that $\omega_A(R) = 0$ and $\nu_A(R) = \sigma_C M_C - T_C$, as in the bottom panels of figure 3.6), the prey population at equilibrium is under development control when predators are absent. Under these conditions, equilibrium biomass density of adult prey in the absence of predators is too low (~0.31 mg/L) for invasion of adult-specialized predators even at background mortality, as the latter need a density of 0.9 mg/L for a positive growth rate. Hence, persistence of adult-specialized predators in isolation is not possible under these conditions.

However, for mortality rates below 0.007 day^{-1}, the adult-specialized predator can invade a community of prey and juvenile-specialized predators, when the latter experience only background mortality. After invasion from low density, the biomasses of adult-specialized predators, as well as of juvenile-specialized predators, prey, and resource, smoothly and quickly approach new equilibrium levels (figure 5.6). For mortality rates between 0.002 and 0.007 day^{-1} for the adult-specialized predator this new community equilibrium is stable, with juvenile-specialized predators dominating the community (see figure 5.5). For

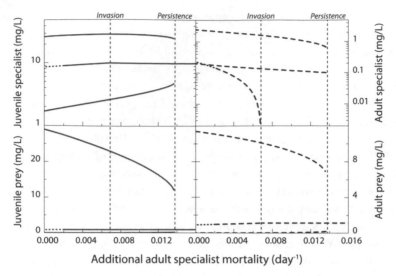

FIGURE 5.5. Changes in equilibrium biomass of juvenile-specialized preda-tors ($\phi_1 = 1$, *top-left, solid lines*), adult-specialized predators ($\phi_2 = 0$, *top-right, dashed lines*), juvenile prey (*bottom-left, solid lines*), and adult prey (*bottom-right, dashed lines*) with increasing mortality of adult-specialized predators in excess of background levels, while juvenile-specialized predators experience only background mortality. Adult-specialized predators can invade a community with only juvenile-specialized predators for mortality levels below 0.007 day^{-1}. For mortality rates below 0.014 day^{-1} (the persistence threshold of adult-specialized predators), population cycles of both predators and prey occur. Stable equilibria and minimum and maximum biomass densities of stable cycles are indicated with solid or dashed lines; unstable equilibria with dotted lines. Juvenile prey com-pete for resource ($R_{max} = 30$ mg/L), while adult prey have unlimited food supply ($\omega_A(R) = 0$ and $\nu_A(R) = \sigma_C M_C - T_C$). Throughout, $q = 1.0$ and $z = 0.1$. Otherwise, default parameters apply for prey and predators with body size $W_A = 0.0001$ and $W_P = 0.01$ gram (both predators), respectively.

mortality rates of the adult-specialized predator below 0.002 day^{-1}, however, the smooth approach to the new community equilibrium is only a transient phase. This transient may last quite long, but eventually large-amplitude cycles in prey and predator densities develop (figure 5.6). Fluctuations in predator biomass are less violent than in the biomass of juvenile and adult prey. Juvenile and adult prey biomasses fluctuate almost in synchrony and together approach densities on the order of 0.001 mg/L during the cycle minimum. Consequently, these fluctuations encompass a substantial risk of prey extinction owing to stochastic events, and hence extinction of the entire community.

The transition around a mortality rate of 0.002 day^{-1} from a stable com-munity equilibrium into population cycles represents what in technical terms

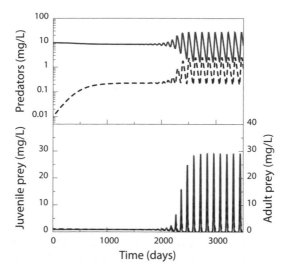

FIGURE 5.6. Invasion dynamics of adult-specialized predators ($\phi_2 = 0$, *top, dashed lines*) into an equilibrium of resource, juvenile (*bottom, solid lines*) and adult prey (*bottom, dashed lines*), and juvenile-specialized predators ($\phi_1 = 1$, *top, solid lines*) for $R_{max} = 30$ mg/L and background mortality for both predators. Other parameters as in figure 5.5.

is known as a subcritical Hopf bifurcation. The key feature of this transition is that on crossing the transition point from the side with a stable to that with an unstable equilibrium (i.e., going from high to low mortality values at 0.002 day^{-1}), the equilibrium does not give way to stable cycles that initially have (infinitely) small amplitude (such as, for example, shown in figure 5.2). Instead, cycles with large amplitude seem to arise immediately when the equilibrium turns unstable. The small-amplitude cycles do occur, but for parameter values just before the transition point (i.e., for mortality rates just above 0.002 day^{-1}). They hence form an alternative type of dynamics next to the stable community equilibrium. Furthermore, these small-amplitude cycles are unstable (and therefore not shown in figure 5.5).

Essentially, the subcritical Hopf bifurcation is no different than the transition at the invasion threshold of the juvenile-specialized predator that is shown in, for example, figure 4.4 (right panel) in the previous chapter. There, a curve or branch representing unstable equilibria originated at the transition point, where the equilibrium without predators turned unstable. As discussed in the previous chapter, this unstable equilibrium branch occurs as an alternative, unstable attractor of the dynamics next to the stable, prey-only equilibrium, up to the parameter value where the unstable equilibrium branch doubled back on itself at the turning or limit point (refer to figure 4.4, right panel). The same phenomenon occurs for the branch of unstable limit cycles that originates in the subcritical Hopf bifurcation point in figure 5.5: the unstable population cycles occur for mortality rates of the adult-specialized predator between 0.002 and 0.014 day^{-1}, at which point a turning or limit point occurs, the branch of cycles doubles back on itself, and the population cycles turn stable. The remaining part of the branch of population cycles represent a stable dynamic attractor, which the community eventually always end up in (as shown in figure 5.6). The branch of stable population cycles hence occurs for mortality rates between 0 and 0.014 day^{-1}, partly as the only stable type of dynamics (for rates up to 0.002 day^{-1}), partly as alternative stable dynamics next to a stable equilibrium with both predators present (between 0.002 and 0.007 day^{-1}) and partly as alternative stable dynamics next to a stable equilibrium with only the juvenile-specialized predator present (between 0.007 and 0.014 day^{-1}). Owing to these population cycles, persistence of the adult-specialized predator is possible over twice as large a range of mortality rates as is suggested by the mortality rate that represents the invasion threshold of the adult-specialized predator into a community of prey and juvenile-specialized predators.

Figure 5.7 shows the extent of the emergent facilitation of adult-specialized predators by predators that forage on juvenile prey in terms of the combinations of mortality rates that allow for coexistence of the two predators. Juvenile-specialized predators can invade a prey-only equilibrium and persist together with the prey as long as the additional mortality rate they experience is below 0.093 day^{-1} (roughly corresponding to twenty times its background mortality). Adult-specialized predators cannot invade a prey-only equilibrium, irrespective of the mortality they experience, nor can they persist without the juvenile-specialized predator. However, adult-specialized predators can invade an equilibrium of prey with juvenile-specialized predators for mortality rates up to 0.045 day^{-1} (corresponding to almost ten times their background mortality), depending on the mortality that juvenile-specialized predators experience. Persistence of both predators is subsequently possible either in a stable community equilibrium or while exhibiting stable population cycles.

The parameter domains leading to both types of dynamic outcomes are about the same size. The bistability between stable population cycles and a stable equilibrium with or without the adult-specialized predator, as discussed in relation to figure 5.5, occurs for a small range of parameters at low mortality rates of the juvenile-specialized predator. The emergent facilitation of the adult-specialized predator by the juvenile-specialized predator is strongest when the latter is exposed to an intermediate mortality rate. Figure 5.7 furthermore makes clear that the adult-specialized predator will always be the

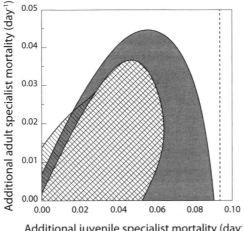

FIGURE 5.7. Ranges of additional juvenile- and adult-specialized predator mortality rate in excess of background levels for which a stable equilibrium (*gray region*) or a stable limit cycle (*hatched and cross-hatched regions*) of both predators and prey occur. Stable cycles occur as alternative attractors to stable equilibria with juvenile-specialized predators only (*hatched, white region*) or with both juvenile- and adult-specialized predators (*hatched, gray region*). In absence of juvenile-specialized predators, adult-specialized predators cannot persist, whereas juvenile-specialized predators can persist for mortality rates below 0.093 day^{-1} (*vertical, dashed line*). Adult-specialized predators can invade a stable equilibrium of juvenile-specialized predators for parameters below the solid perimeter of the gray parameter region. Juvenile prey compete for resource (R_{max} = 30 mg/L), while adult prey have unlimited food supply ($\omega_A(R) = 0$ and $v_A(R) = \sigma_C M_C - T_C$). Throughout, $q = 1.0$ and $z = 0.1$. Otherwise, default parameters apply for prey and predators with body size $W_A = 0.0001$ and $W_P = 0.01$ gram (both predators), respectively.

first species to go extinct when the mortality rate of the juvenile-specialized predator is increased, independent of the mortality that the adult-specialized predator experiences itself. This order of extinction is analogous to the earlier extinction of the specialist predator with increasing values of generalist predator mortality, which was discussed in the previous section.

In cases where food supply for juvenile prey is abundant and only adult prey compete for the limiting resource (i.e., assuming $\omega_J(R) = 0$ and $v_J(R) = \sigma_C M_C - T_C$), the prey population is under reproduction control, and adults hence make up the largest part of prey population biomass in the absence of any predator, as shown in the bottom panels of figure 3.5 in chapter 3. The juvenile biomass in a prey-only equilibrium is then too low to allow for invasion of predators that specialize on juvenile prey, even when these predators experience only background mortality. In contrast, the high adult prey biomass allows adult-specialized predators to invade a prey-only equilibrium. Note that the lack of invasion by the juvenile-specialized predator does depend on the parameter values used in this chapter, in particular the choices R_{max} = 30 mg/L, q = 1.0, and z = 0.1. Figure 4.1 in the previous chapter shows that for q = 0.5 and z = 0.01 invasion of juvenile-specialized predators would be possible for R_{max} = 30 mg/L. In fact, in the absence of the adult-specialized predator, the parameter values used in this chapter put the system in the bistable region, in which a community equilibrium of juvenile-specialized predators, prey, and resource would occur as an alternative stable state next to a stable community equilibrium with only prey and resource (cf. figure 4.1 in the previous chapter). Hence, although invasion of juvenile-specialized predators is not possible, persistence is, as long as the additional mortality experienced by the juvenile-specialized predators is below 0.024 day^{-1}.

This bistability allows the predator specializing on adult prey to play a role as a "catalyst species" for the juvenile-specialized predator: it will facilitate the invasion of juvenile-specialized predators, after which the latter outcompetes its benefactor. As shown in figure 5.8, juvenile-specialized predators cannot invade a prey-only equilibrium, even when invading at high density. Adult-specialized predators, however, can invade and will drastically change the prey biomass distribution after their establishment. This results in a juvenile prey biomass that is well above the level that is needed to support growth of the juvenile-specialized predator. The latter will hence be able to invade an equilibrium of prey and adult-specialized predators even from a low initial density. After invasion of the juvenile-specialized predator, the community exhibits some strong fluctuations in biomass density of both prey and predators, but eventually the adult-specialized predator will be driven to extinction, and a stable community equilibrium of prey with the juvenile-specialized predator results.

It should be stressed that the extinction of the adult-specialized predator is not due to stochastic events that make it go extinct when going through some low biomass densities when cycling. The extinction occurs deterministically and is an inevitable outcome of the competition between the two predators. This catalytic role of the adult-specialized predator is made possible by the emergent Allee effect that the juvenile-specialized predator experiences under these conditions: Even though it can persist when present in high biomass density, it cannot invade the stable prey-only equilibrium unless the latter is strongly perturbed. As discussed in the previous chapter, the prey-only equilibrium can be perturbed by thinning the prey population, which can lead to establishment of a top predator. Invasion of the adult-specialized predator represents a similar strong perturbation of the prey-only equilibrium. Once juvenile-specialized predators have established themselves, however, biomass densities of adult prey are too low for the adult-specialized predator to persist,

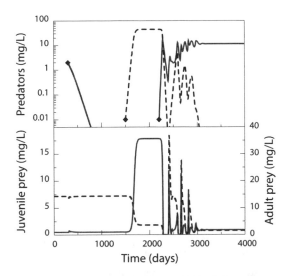

FIGURE 5.8. Invasion dynamics of juvenile-specialized ($\phi_1 = 1$, *top, solid lines*) and adult-specialized predators ($\phi_2 = 0$, *top, dashed lines*) into an equilibrium of resource, juvenile prey (*bottom, solid lines*), and adult prey (*bottom, dashed lines*) for $R_{max} = 30$ mg/L at background mortality for the juvenile-specialized predator and an additional mortality of 0.05 day^{-1} for the adult-specialized predator. Invasion of juvenile-specialized predators at $t = 300$ into the prey-only equilibrium is unsuccessful, despite a high initial density, whereas adult-specialized predators can invade successfully even from low density ($t = 1500$). Adult-specialized predator invasion allows for subsequent, successful invasion of juvenile-specialized predators from low density ($t = 2200$), upon which adult-specialized predators go extinct. Adult prey compete for resource ($R_{max} = 30$ mg/L), while juvenile prey have unlimited food supply ($\omega_J(R) = 0$ and $v_J(R) = \sigma_C M_C - T_C$). Throughout, $q = 1.0$ and $z = 0.1$. Otherwise, default parameters apply for prey and predators with body size $W_A = 0.0001$ and $W_P = 0.01$ gram (both predators), respectively.

and it is hence ousted by the juvenile-specialized predators after their invasion. Juvenile-specialized predators can be considered to have an inherent competitive advantage, as they forage on prey individuals before the latter become vulnerable to the adult-specialized predators.

Figure 5.9 shows all possible community states of resource, prey, juvenile-specialized predators, and adult-specialized predators as a function of the mortality that juvenile-specialized predators experience on top of background mortality, while adult-specialized predators experience a mortality rate of 0.05 day^{-1} on top of background mortality. For mortality rates of the juvenile-specialized predator below 0.01 day^{-1}, an equilibrium of prey and juvenile-specialized predators is the only stable outcome of the dynamics. Note,

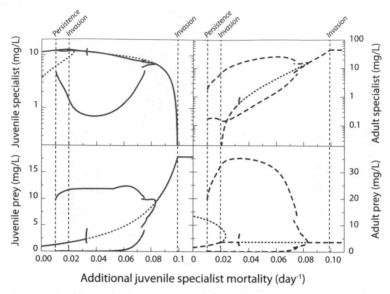

FIGURE 5.9. Changes in equilibrium biomass of juvenile-specialized ($\phi_1 = 1$, *top-left, solid lines*) and adult-specialized predators ($\phi_2 = 0$, *top-right, dashed lines*), juvenile prey (*bottom-left, solid lines*), and adult prey (*bottom-right, dashed lines*) with increasing mortality of juvenile-specialized predators in excess of background levels, while adult-specialized predators experience an additional mortality of 0.05 day^{-1}. Juvenile- and adult-specialized predators go extinct at mortality rates above 0.099 and below 0.010 day^{-1}, respectively. For mortality rates between 0.010 and 0.083 day^{-1}, population cycles of both predators and prey occur, possibly as alternative dynamics next to a stable equilibrium. Stable equilibria and minimum and maximum biomass densities during stable cycles are indicated with solid or dashed lines; unstable equilibria with dotted lines. Other parameters as in figure 5.8.

however, that this equilibrium can only be reached after the adult-specialized predator has facilitated invasion by the juvenile-specialized predator and has subsequently been outcompeted by it. Figure 5.9 does not further elaborate on the fact that juvenile-specialized predators cannot invade an equilibrium with only prey and resource, irrespective of their mortality rate, and that their invasion should first be catalyzed by adult-specialized predators. For a range of mortalities above 0.01 day^{-1}, bistability occurs between different types of stable dynamics. For mortality rates of the juvenile-specialized predator between 0.01 and 0.02 day^{-1}, the adult-specialized predator can still not invade the equilibrium with juvenile-specialized predators only. However, coexistence of both predators is possible once the adult-specialized predator has established itself, while the populations cycle with large amplitude, especially juvenile and adult prey (figure 5.9, lower panels).

For mortality rates of the juvenile-specialized predator above 0.02 day^{-1}, the adult-specialized predator can invade the equilibrium with only juvenile-specialized predators and can establish itself, leading either to a stable community equilibrium of prey with both predators or to the large-amplitude population cycles that also occur below 0.02 day^{-1}. The equilibrium and large-amplitude cycles occur next to each other as alternative stable types of dynamics up to juvenile-specialized predator mortality rates of 0.033 day^{-1}. Above that, and up to mortality rates of 0.083 day^{-1}, stable coexistence of prey and the two predators is not possible and only population cycles result. Around additional mortality rates of 0.033 and 0.075 day^{-1}, two types of population cycles, one with small and one with large amplitude, occur next to each other as alternative attractors, but the parameter ranges for which this bistability occurs are so small that we will not discuss these two types of cycles further. For high mortality rates of the juvenile-specialized predator, between 0.083 and 0.099 day^{-1}, a stable equilibrium with both predators present is the only possible outcome of the dynamics, whereas for mortality rates in excess of 0.099 day^{-1}, the juvenile-specialized predator goes extinct.

Figure 5.9 reveals how the emergent facilitation of juvenile-specialized predators by predators that forage only on adult prey may result in different types of dynamics. Figure 5.10 shows for which values of mortality rate of the juvenile-specialized and the adult-specialized predators these different types of dynamics occur, either as the only stable outcome of dynamics or as alternative attractors of the dynamics. Clearly, the presence of adult-specialized predators facilitates the invasion and persistence of the juvenile-specialized predator, despite that the latter cannot invade a prey-only equilibrium, irrespective of the mortality it experiences, and can only persist on its own for mortality rates below 0.024 day^{-1}. At mortality rates below 0.024 day^{-1} for the juvenile-specialized predator and intermediate to high mortality rates of the adult-specialized predator (above 0.02–0.03 day^{-1}), the latter tends to act as a catalyst species, facilitating its own extinction owing to the invasion of, and subsequent competition with, the juvenile-specialized predator. Persistence of the juvenile-specialized predator in coexistence with the adult-specialized predator is possible for low mortality rates of the adult-specialized predator and for mortality rates of the juvenile-specialized predator above 0.024 day^{-1}. Especially when adult-specialized predators experience intermediate levels of mortality (around 0.05 day^{-1}), the juvenile-specialized predator can endure high rates of mortality without going extinct (~ 0.1 day^{-1}, approximately twenty times background mortality). Stable population cycles occur over a large range of the parameter values, for which the two predators coexist. Sometimes these occur as an alternative type of dynamics next to a stable equilibrium with both predators or only the juvenile-specialized predator present, but for the largest part, cyclic dynamics are the unique outcome of dynamics. Coexistence of the

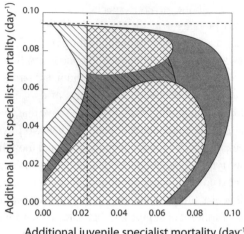

Additional juvenile specialist mortality (day^{-1})

FIGURE 5.10. Ranges of additional juvenile- and adult-specialized predator mortality rates in excess of background levels for which a stable equilibrium (*gray region*)
or stable limit cycles (*hatched and cross-hatched regions*) of both predators and
prey occur. Stable cycles occur as alternative attractors to stable equilibria with
juvenile-specialized predators only (*hatched, white region*) or with both juvenile-
and adult-specialized predators (*hatched, gray region*). Juvenile-specialized predators cannot invade a prey-resource equilibrium, but once present they can persist in
absence of adult-specialized predators for mortality rates below 0.024 day^{-1} (*vertical, dashed line*). Adult-specialized predators can persist for mortality rates below
0.095 day^{-1} (*horizontal, dashed line*). In the white parameter region to the left and
below these two dashed lines, the adult-specialized predator acts as a catalyst species for the juvenile-specialized predator. Other parameters as in figure 5.8.

two predators in stable equilibrium is mostly restricted to parameter regions
where the juvenile-specialized predator experiences high mortality rates.

MULTIPLE PREDATORS AND A SINGLE PREY

The results presented above clearly show that multiple predators can coexist in a
stable equilibrium, while all foraging on the same prey species, if the prey population is characterized by a distinct stage- or size-structure and this population
structure changes with changing density or changing food conditions. The latter
condition requires that individual development is density- or food-dependent.
The variable stage- or size-structure ensures that the prey population serves as
more than a single resource and hence supports more than a single predator. But
beyond coexistence, the results show that the predators in fact may promote

each other's population growth to the extent that one of the predator species needs its competitor in order to be able to persist at all. This aspect makes that emergent facilitation superficially resembles the indirect mutualism occurring between two predators foraging on two competing prey species (Wootton 1994). However, emergent facilitation involves only a single prey population and does not require that competition between the prey stages occurs.

Furthermore, biomass of juvenile and adult prey increase through reproduction and maturation, respectively, and thereby rely on biomass production in the other stage. Neither of the two prey stages can thus grow in isolation. Even though it functions as more than a single resource, these aspects make clear that the prey population cannot be considered equivalent to multiple, independent resources either. The similarity between emergent facilitation and previously reported types of indirect mutualism is therefore only superficial. Emergent facilitation also seems to be superficially related to the conflicting responses that prey individuals exhibit when exposed to multiple predators (Relyea 2003; Sih, Englund, and Wooster 1998), which lead to positive interactions between the predators as well. These emergent multiple predator effects (Sih , Englund, and Wooster 1998), however, occur on an individual level and a behavioral, within-generation time scale. In contrast, emergent facilitation is a community-level, multigenerational effect that does not involve any prey behavioral response and even occurs when predators forage purely exploitatively.

Theoretically, it has also been shown that emergent facilitation is not limited to two predator species foraging on the same prey species. De Roos, Schellekens, et al. (2008a) report results for a stage-structured biomass model, similar to the one analyzed in this chapter, in which four size classes of prey are distinguished (small and large juveniles and small and large adults). This size structure is shown to allow for the coexistence in stable equilibrium of up to three predator species. If juvenile prey has an unlimited food supply and all adults compete for a shared resource, the prey population in the absence of predators is governed by reproduction control and dominated by small adults. Predators that specialize on small adult prey can then most easily invade the prey-only equilibrium. Owing to biomass overcompensation, this predator invasion induces a change in size structure of the population, which increases biomass in the two juvenile prey size classes and facilitates the invasion of predators that specialize on them. Predators specializing on the smallest juvenile prey subsequently tend to outcompete a predator species that specializes on larger juvenile prey, as they forage on prey individuals preemptively, before these individuals become vulnerable to their competitors. In this setting of a prey population with four distinct size classes and reproduction control, invasion of the specialist predator on the dominant-prey stage, the small adults, therefore allows two other predator species to start competing with each other.

De Roos, Schellekens, et al. (2008a) also study the four-stage-structured prey population under conditions of development control by assuming that adult prey has an unlimited food supply and all juveniles compete for a limiting resource. In this case, the prey population in the absence of predators is dominated by small juvenile prey, and predators on these smallest prey individuals invade most readily. Invasion of this predator changes the prey size structure in such a way that biomass of large adult prey increases sufficiently for invasion of a predator species specializing on the largest prey size class. The biomass density of small-adult prey also increases but not enough to allow for invasion of a specialist predator on this size class. De Roos, Schellekens, et al. (2008a) show, however, that the invasion of a specialist predator on the large-adult prey size class leads to a further increase in the biomass of small-adult prey. The combined action of the specialist on small-juvenile prey and the specialist on large-adult prey thus allows for invasion and persistence of a specialist predator on small-adult prey. Because it is ultimately the specialist predator on large-adult prey that brings about the necessary increase in small-adult prey biomass, it can be concluded that the facilitated species has turned facilitator itself.

EXPERIMENTAL EVIDENCE

The results in this chapter, combined with the results presented by de Roos, Schellekens, et al. (2008a, supplementary information) for a prey population with a more extended population size structure, represent a "proof-of-principle" that the biomass overcompensation that occurs in prey populations with density- or food-dependent development may give rise to complex, positive interactions between predator species that forage on them. In cases of the emergent Allee effect, biomass overcompensation turns the top-down influence on a prey population exerted by a particular predator species into a positive, bottom-up effect for its own growth and persistence. Similarly, in cases of emergent facilitation, biomass overcompensation turns the top-down predation effect into a bottom-up effect that benefits another predator species. It is, however, as of yet unclear how often these indirect, positive effects of biomass overcompensation also play a role in structuring communities in nature.

Empirical evidence for the possible occurrence of emergent facilitation in natural communities comes in three different forms. Table 5.1 presents an overview of the published studies documenting these three different lines of evidence. First, in chapter 3 we have already provided direct experimental evidence for an increase in biomass in a particular population stage in response to increased mortality imposed on another stage. This type of biomass overcompensation, which has, among others, been demonstrated to occur in blowflies

(Nicholson 1957), soil mites (Cameron and Benton 2004), and fish (Schröder, Persson, and de Roos 2009), is the basic element that gives rise to emergent facilitation. We refer the reader to chapter 3 for more details and here do not discuss this type of evidence any further (see also table 5.1).

A second line of evidence for the possible occurrence of emergent facilitation in natural communities is provided by the changes in density of particular stages of a prey population in response to variation in predation mortality (see also table 5.1). Murdoch and Scott (1984) compared the structure and density of *Daphnia pulex* populations that were exposed to individuals of the backswimmer *Notonecta hoffmani* in different stages of their development. These authors found that irrespective of their developmental stage, *Notonecta* individuals preferably foraged on medium- and large-size *Daphnia*. Exposure to predation by *Notonecta* drove some *Daphnia* populations to extinction, but surviving populations maintained overall densities that were indistinguishable from the density of control populations. Populations surviving under *Notonecta* predation comprised a higher fraction of small individuals (37 percent versus 27 percent in predator-exposed and control populations, respectively) and a lower fraction of adults (19 percent versus 29 percent in predator-exposed and control populations, respectively). Taken together, these results imply that the density of small *Daphnia* increased because of predation.

Vonder Brink and Vanni (1993) carried out similar experiments with the zooplankter *Bosmina longirostris* exposed to different levels of fish predation by *Phoxinus eos*. Numerical densities of *Bosmina* were highest at low and medium fish densities and lowest when fish were absent or present in high density. Fish did significantly reduce the densities of large *Bosmina* and increased the abundance of small individuals. This overcompensation was shown to result from an increase in *Bosmina* population birth rate with increasing densities of fish. Fish thus foraged mainly on large *Bosmina*, reducing their density to such an extent that food conditions increased, leading to higher food availability and consequently a higher population birth rate. Both sets of experiments hence show that in cladoceran populations exposure to predation on large-size individuals increases the density of smaller size classes. Theoretically, this should facilitate the population growth of predators targeting these smaller individuals in a similar manner, as revealed by the models analyzed in this chapter.

The opposite type of overcompensation, predation on smaller-size individuals, which increases the density of large size classes of prey, is a common phenomenon in many fish populations (Tonn and Paszkowski 1986). Olson, Green, and Rudstam (2001) studied the changes in the population of yellow perch (*Perca flavescens*) in Canadarago Lake (U.S.A.) upon stocking the lake for six years with walleye (*Stizostedion vitreum*). Walleye is a gape-limited predator, foraging mainly on small-size prey individuals. Stocking significantly

TABLE 5.1. Experimental and Empirical Data Providing Support for the Occurrence of Emergent Facilitation or the Mechanisms Giving Rise to It

Prey species	System type	Description	Ref.
Mortality-induced increases in stage-specific prey density			
Blowflies	Laboratory	With strong adult competition, destroying 90 percent of emerging adults doubles the density of eggs, larvae, and pupae.	(1)
Blowflies	Laboratory	With strong juvenile competition, destroying 50 percent of young larvae doubles the adult density.	(1)
Soil mites	Laboratory	Harvesting eggs increases adult density if adults are superior competitors for food.	(2)
Poeciliid fish	Laboratory	Size-selective harvesting of large individuals increases densities of small and large juveniles.	(3)
Predator-induced increases in stage-specific prey density			
Daphnia pulex	Laboratory	In populations regulated through food-dependent adult fecundity, the proportion of small juveniles (< 0.8 mm) increased from 27 percent in controls to 37 percent when coexisting with the positively size-selective predator Notonecta, while total population density did not change.	(4)
Bosmina longirostris	Enclosures	Increasing fish predation from low to medium did not change total Bosmina density, but it increased densities of small individuals while decreasing densities of large individuals.	(5)
Yellow perch	Whole lake	Establishment of walleye, a gape-limited predator on juveniles, in Canadarago Lake significantly increased density of yellow perch larger than 200 mm.	(6)
Arctic char	Whole lake	Thinning of prey population led to stable recovery of predator on small prey with concomitant increases in densities of both small- and large-size prey.	(7)

Positive associations between predators with contrasting prey size preferences

Microcrustaceans	Enclosures	Planktivorous fish significantly increased abundance of juvenile microcrustaceans and predatory macroinvertebrates that forage on them.	(8)
Daphnia	Field survey	Positive correlation was found between densities of invertebrate and fish predators on small and large *Daphnia* individuals, respectively, in a range of lakes with different intensities of predation risk.	(9)
Brook stickleback	Field survey	Distribution of large, positively size-selective predators is nested within the distribution of small, negatively size-selective predators.	(10)

(1) Nicholson (1957). (2) Cameron and Benton (2004). (3) Schröder, Persson, and de Roos (2009). (4) Murdoch and Scott (1984). (5) Vonder Brink and Vanni (1993). (6) Olson, Green, and Rudstam (2001). (7) Persson, Amundsen, et al. (2007). (8) Nielsen, Hillman, et al. (2000). (9) Nielsen, Hillman, and Smith (1999); Leibold and Tessier (1991). (10) Zimmerman (2006).

increased the abundance of walleye in the lake, leading to a more than fourfold decrease in yellow perch smaller than 200 mm. In contrast, yellow perch larger than 200 mm in body length increased significantly, from being rare before the stocking of walleye to being the most abundance size class afterward. This increase was shown to result from an increased growth rate of the yellow perch from an average of 13 g/year before the stocking with walleye to an average of 25 g/year afterward. The long-term study and manipulation in Lake Takvatn, which was extensively discussed in chapter 4, provides another example of the same type of biomass overcompensation, involving Arctic char and brown trout as prey and predator, respectively. As shown in figure 4.19, the Arctic char population was dominated by a few size classes after the recovery of the brown trout population. Arctic char individuals larger than 250 mm, which were virtually absent from the lake before the brown trout recovery, represent one of these classes that made up a sizeable part of the population after the recovery. Similar to the changes in Canadarago Lake, the reestablished brown trout population in Lake Takvatn hence led to an increase in the biomass of large-size Arctic char.

The third line of evidence that provides support for the occurrence of emergent facilitation in natural communities concerns studies that do in fact involve multiple predator species (see also table 5.1). The studies with cladoceran populations discussed above led to the suggestion that emergent facilitation might be expected to occur in aquatic systems between vertebrate and invertebrate predators of zooplankton. Fish are visual hunters and typically select zooplankton of large body sizes (Brooks and Dodson 1965). In contrast, gape limitations constrain invertebrate predators of zooplankton, such as larval midges, to eat only smaller-size individuals (Campbell 1991; Zaret 1980).

Experiments in enclosures focusing on the competition between macroinvertebrates and planktivorous fish (Nielsen, Hillman, and Smith 1999; Nielsen, Hillman, et al. 2000) provide support for this expectation. Nielsen and coworkers divided the enclosures into two equal parts and added common carp-gudgeons (*Hypseleotris* spp.)—a forager on large cladocerans and adult copepods (Meredith, Matveev, and Mayes 2003)—to one half of each enclosure. The addition of the carp-gudgeons significantly increased the densities of microcrustaceans (Nielsen, Hillman, et al. 2000), in particular the densities of juvenile cladocerans, nauplii, and cyclopoid copepodites. At the same time, eight out of the fourteen predatory macroinvertebrate species also increased in the presence of fish (Nielsen, Hillman, and Smith 1999), including acarine, coleopteran, hemipteran, and dipteran predators. At least for acarine predators it has been shown experimentally that they preferably forage on small-size cladocerans (Butler and Burns 1991). This increase in eight macroinvertebrate species after the introduction of carp-gudgeons can be explained by the occurrence of emergent facilitation: the positively size-selective fish increased the densities of small, juvenile cladocerans and copepods (nauplii and copepodites) and thus promoted the population growth of the negatively size-selective macroinvertebrate predators, leading to a density increase of the latter. The remaining six species of macroinvertebrates that did not increase after introduction of carp-gudgeons included notonectids and odonates, which exhibit significant diet overlap with fish (Nielsen, Hillman, and Smith 1999) and are hence not expected to benefit from the higher densities of the smaller microcrustaceans.

Indirect evidence for emergent facilitation between vertebrate and invertebrate predators of zooplankton is further provided by a survey of the body-size patterns of coexisting *Daphnia* species in seven lakes that represented a gradient of predation intensity by *Chaoborus* and bluegill (Leibold and Tessier 1991). *Chaoborus* selectively forages on small cladocerans, whereas bluegill selects large individuals as their prey. Among the seven lakes, *Chaoborus* density was positively associated with bluegill density, as would be expected on the basis of the postulated emergent facilitation between fish and macroinvertebrates.

Based on the increases in large-size prey fish following establishment of a predator population, as discussed above for Canadarago Lake and Lake Takvatn, we can also postulate that emergent facilitation is likely to occur between predators foraging on opposing ends of the same prey fish size distribution. Piscivorous fish and marine mammals may provide an example of such predator species, as they have been shown to forage on small and large individuals of the same fish prey, respectively (Duplisea 2005). Empirical support for emergent facilitation between predators of a planktivorous prey fish species is provided by the analysis of the predator communities preying on the brook stickleback (*Culaea inconstans*) occurring at twenty-six sites across Michigan's upper peninsula (Zimmerman 2006). Small predators that were shown to be gape limited, consuming small sticklebacks only, were found to be widely distributed in contrast to large predators capable of consuming all sizes of sticklebacks, but with a presumed bias toward longer sticklebacks. These large predators always occurred only if small predators were present as well. Such a nested distribution of the large predator group is to be expected if they persisted as a result of facilitation by the small predators. In accordance with this explanation, sticklebacks were larger at sites where they coexisted only with small-predator communities than at sites where large predators also occurred.

The experimental and empirical data discussed above do not provide a direct test of the occurrence of emergent facilitation, as the studies do not show unambiguously that the persistence of one predator is facilitated by another via changes in the prey size distribution. In contrast, Huss and Nilsson (2011) carried out direct tests of the potential for emergent facilitation in a pelagic mesocosm experiment involving the predacious cladoceran zooplankton *Bythotrephes longimanus* and its prey, the herbivorous cladoceran zooplankton

Holopedium gibberum, which is also exposed to positive size-selective mortality mimicking the feeding of planktivorous fish. Owing to its relatively small size, the predation pressure of *Bythotrephes* on *Holopedium* is limited to small *Holopedium* only, because the latter is covered by a large, gelatinous sheet as adult. Based on the hypothesis that emergent facilitation occurs, fish predation was expected to primarily decrease large *Holopedium* biomass, which should lead to an overcompensatory increase in juvenile *Holopedium* biomass, if the system is regulated by reproduction control. As a consequence, fish predation should promote the performance of *Bythotrephes*. Fish predation was simulated by selectively removing large *Holopedium* at the start of the experiment from the enclosures using a plankton net. Subsequently, *Bythotrephes* was inoculated at the same densities in both control mesocosms and the mesocosms from which large *Holopedium* were removed.

Bythotrephes remained at the same density in the mesocosms where large *Holopedium* had been reduced at the start of the experiment, whereas *Bythotrephes* densities decreased to low values in the control treatments (figure 5.11, upper panel). As a result of the selective removal at the start of the experiment, large *Holopedium* were reduced to low levels. Their biomasses remained lower in harvested treatments than in control treatments at days 10 and 20 of the experiment to converge to the biomasses observed in the controls at the end of the experiment (figure 5.11, middle panel). Also juvenile *Holopedium* biomasses were lower in the harvested treatments at the start of the experiments, but at days 10 and 20 their biomasses were substantially higher in harvested treatments than in control treatments, becoming similar to control biomasses at the end of the experiment (figure 5.11, lower panel).

The study by Huss and Nilsson (2011) clearly demonstrates the potential for emergent facilitation of positive size-selective predators (fish) on *Bythotrephes* through an overcompensatory increase in juvenile *Holopedium* biomass. Analysis of this system with a dynamical fish population instead of a pulsed removal of large *Holopedium* is of course needed before definite conclusions may be drawn about the consequences of emergent facilitation for the long-term community dynamics. Still, size-selective predation of fish has invariably been shown to induce a shift in size distribution of zooplankton toward smaller sizes, making the fish-*Bythotrephes*-*Holopedium* system a strong candidate for showing coexistence based on emergent facilitation. An interesting twist to this system is that fish also prey on *Bythotrephes*, such that this system can better be characterized as an intraguild, *mutualistic* system in contrast to classical intraguild predation (IGP) systems where only the IG predator and IG prey compete for shared resource.

The foregoing discussion illustrates the potential for emergent facilitation to play a role in structuring natural communities with examples from aquatic

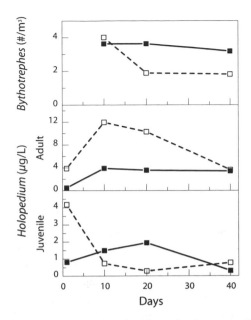

FIGURE 5.11. Changes in the densities of *Bythotrephes* (upper panel) and the biomasses of adult (middle panel) and juvenile *Holopedium* (lower panel) over forty days in control treatments (*open squares, dashed lines*) and treatments where large *Holopedium* had been harvested at the start of the experiment (*closed squares, solid lines*) in the mesocosm experiment carried out by Huss and Nilsson (2011).

systems only. Plausible arguments can be presented, however, for the occurrence of emergent facilitation in terrestrial systems as well. For example, in host-parasitoid systems, larval hosts may regularly overexploit their plant resource and suffer from intense density dependence, unless densities are suppressed by specialist parasitoids (van der Meijden and van der Veen-van Wijk 1997). Emergent facilitation could occur in systems with such strong host-density dependence, especially because most insect hosts are attacked by a range of different parasitoid species, each typically specializing on particular developmental stages of the host (Waage and Hassell 1982). Furthermore, prey species like amphibians and many insects pass through a discrete habitat shift during ontogeny and hence are naturally exposed to different suites of predators. This opens up the possibility of emergent facilitation between predators even across ecosystem boundaries (Vonesh and Osenberg 2003). A challenge for future research is therefore to consider the interaction between different predator species from the perspective of emergent facilitation to assess whether or not it does play a role.

Ontogenetic Niche Shifts

Development and growth in body size makes individual organisms change during ontogeny in a variety of ways. On the one hand, there are the changes in, for example, feeding and maintenance rate with increasing body size. These changes may lead to differences in energetics between individuals at different stages in their life history and hence give rise to the biomass overcompensation discussed in chapter 3 and its community consequences discussed in chapters 4 and 5. Such changes are mostly quantitative in nature, as they concern changes in the rate of life history processes associated with increases in body size. Individuals, however, may also experience more qualitative changes in morphology, behavior, and ecology during ontogeny. Qualitative changes are especially prominent for species that metamorphose, such as insects and amphibians. Individuals of metamorphosing species go through a complex life cycle, occupying different ecological niches and living in different habitats in different stages of their life history. In these different niches they generally use different resources as food and are exposed to different suites of predators (reviewed by Wilbur 1980). Such qualitative changes during life history are generally referred to as *ontogenetic niche shifts*.

Ontogenetic niche shifts are not unique for metamorphosing species, but are in fact also common among species that exhibit substantial growth in body size during life. In many fish species individuals are born at roughly the same size, around 1 cm in length, irrespective of whether they can eventually reach 10 or 100 cm. For example, cod individuals (*Gadus morhua*) are born at a size smaller than 1 cm in length, but can reach body lengths up to 2 meters when growing old. After hatching, these individuals start out foraging on various species of pelagic zooplankton. They extend this diet when reaching a size of 5–10 cm and gradually change to feeding in benthic environments on macro-invertebrate species. At a size of 20–30 cm, they extend their diet once again and start feeding on prey fish species, such as sprat, herring, or capelin. This shift from zooplanktivory to benthivory and eventually piscivory is characteristic for most species of piscivorous fish. Werner (1988; see also Werner and Gilliam 1984) estimated that ontogenetic niche shifts occurred ubiquitously in twenty-five out of thirty-three animal phyla, or 80 percent of all animal species, partly because of the predominance of metamorphosing species, partly because of the significant growth in body size exhibited by many species. In fact, a lack of changes in ecological niche over ontogeny may only be relevant for a few species, including mammals and birds, clonal invertebrates, filter feeders like the zooplankter *Daphnia*, scavengers, detritivores, and herbivorous insects (Rudolf and Lafferty 2011).

In theory, an ontogenetic niche shift may involve changes in a wide variety of factors, including the resources that individuals exploit in different life stages, the habitat they live in, and the predators they are exposed to. The term is, however, often used synonymously with an *ontogenetic diet shift*, which refers exclusively to a change in the resources that individuals exploit in different life history stages. Superficially, a complete shift in diet from foraging on one particular resource to foraging on a second, unrelated resource, could be interpreted as being analogous to a shift in habitat. As long as the two resources involved do not interact and hence follow dynamics independent of each other, it seems irrelevant whether the two resources occur in the same or different habitats.

However, this interpretation is confounded by a subtle problem related to the sizes of the two habitats involved: if one habitat is much larger than the other, the density-dependence within a group of individuals that moves from the smaller to the larger habitat will automatically relax. The relative size of a habitat may thus influence the dynamic consequences of an ontogenetic niche shift (de Roos, Leonardsson, et al. 2002; Persson and de Roos 2003), which is not an issue when considering ontogenetic shifts in diet only. Frequently, the sizes of the two habitats involved in an ontogenetic habitat shift are not accounted for (e.g., McCoy, Barfield, and Holt 2009), which implies that both habitats are assumed to be of equal size. One might be able to represent the difference in habitat size by a careful scaling of the model parameters, in particular the productivity of the two habitats per unit area or volume. More straightforwardly, the size differences between the two habitats can be accounted for explicitly in the modeling. For simplicity, however, we will ignore such habitat scaling issues and concentrate on ontogenetic shifts in diet, assuming that the resources that individuals switch between during ontogeny occur in the same habitat.

Implicitly we have already dealt with ontogenetic niche shifts in earlier chapters. For example, in chapter 3, we contrasted consumer-resource systems, in which juvenile and adult consumers compete for a shared resource, with systems in which either juveniles or adults have unlimited food supply. Similarly, in chapters 4 and 5 we considered predator-consumer-resource systems involving predator species that specialized on either juvenile or adult consumers.

From these analyses we can infer that an ontogenetic diet shift may promote regulation of the population in equilibrium primarily by either limited development or limited reproduction. In this chapter we will extend these analyses and consider the consequences for community structure of ontogenetic diet shifts that involve the use of different resources in different life history stages, whereby these resources are in limited supply and are hence competed for by all individuals foraging on them.

We will explore the consequences of ontogenetic diet shifts using stage-structured biomass models that account for two basic resources, a stage-structured consumer population, for which we distinguish between juveniles and adults, and up to two unstructured predator populations. The most extended model is therefore closely related to the model analyzed in the previous chapter, except for the inclusion of an additional basic resource. The equations of the full model are summarized in box 6.1, and default parameter values are listed in box 6.2. Initially we will focus on a simplified version of the full model, one that accounts only for the basic resources and the structured consumer population, and hence assumes that the predator densities P_i equal 0 in all equations shown in box 6.1.

CONSUMER-RESOURCE SYSTEMS

Werner and Gilliam (1984) were the first to extensively discuss and review ontogenetic niche shifts, although their importance was recognized much earlier (e.g., Hardy 1924). Recently there has been a revival of interest in the ubiquity of ontogenetic niche shifts and their consequences for community structure (e.g., Rudolf and Lafferty 2011; Schreiber and Rudolf 2008; Woodward and Hildrew 2002). Woodward and Hildrew quantified the diet overlap for the invertebrate predator guild in Broadstone Stream (UK). Predator body size was shown to be an important determinant of dietary overlap both between and within species. Small predators had the narrowest diets, regardless of species, and were limited to feeding on a restricted subset of the total prey size-spectrum. Niche overlap decreased as differences in body size increased both among and within species. Differently sized individuals of the same species often had less overlap in their diet than similarly sized individuals of different species. Ontogenetic shifts were hence argued to outweigh taxonomic differences. Rudolf and Lafferty (2011) quantified the average diet overlap between different life history stages of the same species in eight ecological food webs, for which sufficient information on stage-specific, consumer-resource interactions was available. Averaged over all metamorphosing species, different life history stages were shown to share on average only 0–8 percent of their resources. The empirical estimates suggested,

BOX 6.1

BIOMASS MODEL WITH TWO RESOURCES, STAGE-STRUCTURED
PREY, AND TWO UNSTRUCTURED PREDATORS

Dynamic equations	Description
$\dfrac{dR_i}{dt} = G_i(R_i) - \omega_{J,i}(R_1,R_2)J - \omega_{A,i}(R_1,R_2)A$	Dynamics of resource $i = 1,2$
$\dfrac{dJ}{dt} = v_A^+(R_1,R_2)A - \gamma(v_J^+,d_J)J + v_J(R_1,R_2)J - d_J(P_1,P_2)J$	Dynamics of juvenile prey
$\dfrac{dA}{dt} = \gamma(v_J^+,d_J)J + (v_A(R_1,R_2) - v_A^+(R_1,R_2))A - d_A(P_1,P_2)A$	Dynamics of adult prey
$\dfrac{dP_i}{dt} = (v_{P_i}(J,A) - \mu_{P_i})P_i$	Dynamics of predator $i = 1,2$

Function	Expression	Description
$G_i(R_i)$	$\rho(R_{\max,i} - R_i)$	Turnover of resource $i = 1,2$
$\omega_{J,i}(R_1,R_2)$	$M_C q_{J,i} R_i / \left(H_C + \sum\limits_{n-1,2} q_{J,n} R_n\right)$	Juvenile intake of resource i
$\omega_{A,i}(R_1,R_2)$	$M_C q_{A,i} R_i / \left(H_C + \sum\limits_{n=1,2} q_{A,n} R_n\right)$	Adult intake of resource i
$v_J(R_1,R_2)$	$\sigma_C \sum\limits_{i=1,2} \omega_{J,i}(R_1,R_2) - T_C$	Juvenile net energy production
$v_A(R_1,R_2)$	$\sigma_C \sum\limits_{i=1,2} \omega_{A,i}(R_1,R_2) - T_C$	Adult net energy production
$d_J(P_1,P_2)$	$\mu_C + \sum\limits_{i=1,2} M_{P_i} \dfrac{\phi_i P_i}{H_{P_i} + \phi_i J + (1-\phi_i)A}$	Juvenile mortality rate
$d_A(P_1,P_2)$	$\mu_C + \sum\limits_{i=1,2} M_{P_i} \dfrac{(1-\phi_i) P_i}{H_{P_i} + \phi_i J + (1-\phi_i)A}$	Adult mortality rate
$\gamma(v_J^+,d_J)$	$(v_J^+ - d_J)/(1 - z^{(1-d_J/v_J^+)})$	Juvenile maturation rate

As in previous chapters, v_J^+ and v_A^+ represent the value of v_J and v_A, respectively, but restricted to non-negative values, while, for brevity, the dependence of v_J^+ and d_J on resource and predator density, respectively, is suppressed in the maturation function $\gamma(v_J^+,d_J)$. See also the remarks following box 3.2.

however, that even for predatory species without metamorphosis, different size classes of individuals may share on average only 40 percent of their diet (see also Woodward and Hildrew 2002). Rudolf and Lafferty (2011) furthermore showed by means of model food web analysis that ontogenetic niche shifts may have a large effect on persistence and stability of ecological communities.

BOX 6.2

DEFAULT PARAMETER VALUES FOR THE STAGE-STRUCTURED
BIOMASS MODEL OF BOX 6.1

Parameter	Default value	Unit	Description
Resources			
ρ	0.1	day^{-1}	Turnover rate of resources
$R_{\max,i}$	30	mg/L	Maximum density of resource $i=1,2$
Prey			
M_C	$0.1W_A^{-0.25}$	day^{-1}	Mass-specific maximum ingestion rate
H_C	3.0	mg/L	Ingestion half-saturation resource density
$q_{J,i}$	varied	—	Juvenile foraging on resource $i=1,2$
$q_{A,i}$	varied	—	Adult foraging on resource $i=1,2$
T_C	$0.01W_A^{-0.25}$	day^{-1}	Mass-specific maintenance rate
σ_C	0.5	—	Conversion efficiency
z	0.01 or 0.2	—	Newborn-adult consumer size ratio
μ_C	$0.0015W_A^{-0.25}$	day^{-1}	Background mortality rate
Predators			
M_{P_i}	$0.1W_{P_i}^{-0.25}$	day^{-1}	Mass-specific maximum ingestion rate
H_{P_i}	3.0	mg/L	Ingestion half-saturation prey density
T_{P_i}	$0.01W_{P_i}^{-0.25}$	day^{-1}	Mass-specific maintenance rate
σ_{P_i}	0.5	—	Conversion efficiency
μ_{P_i}	$0.0015W_{P_i}^{-0.25}$	day^{-1}	Background mortality rate
ϕ_i	0.0, 0.5 or 1.0	—	Foraging preference for juveniles

W_A and W_{P_i} represent average adult body weight of consumer and predator, $i=1,2$, respectively.

As Werner and Gilliam (1984) point out, a change in diet over ontogeny can take a variety of forms, depending on the overlap in resource use between smaller- and larger-size conspecifics. The changes in diet of cod that were described above provide two examples. When switching from planktivory to benthivory, individuals gradually spend more effort foraging in the benthic environment, while simultaneously reducing the effort devoted to foraging in the pelagic on zooplankton. Such a gradual switch from one resource to another can appropriately be called an *ontogenetic diet shift* in a strict sense. In its most extreme form the individuals would switch from foraging exclusively on one resource to foraging exclusively on another resource, preventing all competitive interactions for resources between individuals in the consecutive life history stages. Another form of diet change occurs when, during ontogeny, individuals extend their diet, as exemplified by the onset of piscivory in the life history of cod. Individual cod never stop feeding on benthic macroinvertebrates but simply add prey fish to their diet when growing larger. This type is more appropriately referred to as an *ontogenetic diet broadening*. In its most extreme form this type of diet change would imply that both juvenile and adult individuals feed on a shared resource, but adults also feed on an additional, exclusive resource of their own. The extreme forms of both ontogenetic niche shifts and ontogenetic niche broadenings would imply that an individual changes its use of resources drastically and discretely at some point in its life history. Werner and Gilliam (1984) argue that such discrete changes are not limited to species that metamorphose but also readily occur among fish, reptiles, and invertebrates like midges, damselflies, and copepods.

We will analyze the community consequences of both ontogenetic niche shifts and ontogenetic niche broadenings. For simplicity we mostly focus on these changes in resource use in their most extreme form, as they represent two convenient, limiting cases of a whole spectrum of possible changes in resource use over ontogeny. To capture these various forms in our stage-structured model, we assume that the foraging of juvenile and adult consumers on resource i is described by:

$$\omega_{J,i}(R_1,R_2) = \frac{M_C q_{J,i} R_i}{\left(H_C + \sum_{n=1,2} q_{J,n} R_n\right)} \tag{6.1}$$

and

$$\omega_{A,i}(R_1,R_2) = \frac{M_C q_{A,i} R_i}{\left(H_C + \sum_{n=1,2} q_{A,n} R_n\right)}, \tag{6.2}$$

respectively. In these expressions the parameters $q_{J,i}$ and $q_{A,i}$ represent the relative effort with which juveniles and adults, respectively, forage on resource i. A complete ontogenetic niche *shift* would be represented by assuming $q_{J,1} = q_{A,2} = 1$ and $q_{J,2} = q_{A,1} = 0$, whereas a niche *broadening* would be represented by adopting $q_{J,1} = 1$, $q_{J,2} = 0$ and $q_{A,1} = q_{A,2} = 0.5$.

Figure 6.1 shows the changes in consumer and resource equilibrium densities with increasing productivities of the two basic resources for the case in which juvenile consumers forage exclusively on resource 1, while adults use

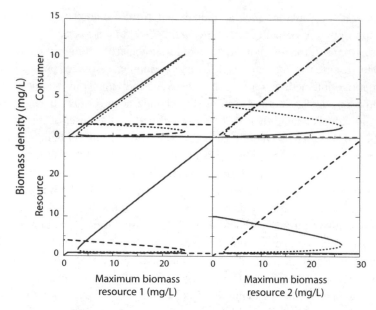

FIGURE 6.1. Changes in equilibrium biomass of resources (*bottom*) and consumers (*top*) with increasing maximum biomass density of resource 1 ($R_{max,1}$, *left*) and resource 2 ($R_{max,2}$, *right*) in the case where consumers go through a complete niche shift from resource 1 to resource 2 at maturation ($q_{J,1} = q_{A,2} = 1$ and $q_{J,2} = q_{A,1} = 0$). Stable equilibria are indicated with solid or dashed lines; unstable equilibria with dotted lines. Solid and dashed lines in bottom panels represent biomass density of resource 1 and 2, respectively; in top panels, solid and dashed lines represent juvenile and adult consumer biomass, respectively. *Left*: $R_{max,2} = 4$ mg/L; *right*: $R_{max,1} = 10$ mg/L; $z = 0.01$. Otherwise, default parameters apply as listed in box 6.2 for a consumer with body size $W_A = 0.0001$ gram.

exclusively resource 2. As in earlier chapters, the model parameters varied in this figure are the maximum biomass densities for resource 1 and 2, $R_{max,1}$ and $R_{max,2}$, respectively. The productivities of both resources scale linearly with the values of $R_{max,1}$ and $R_{max,2}$ with proportionality constant ρ. The prominent feature of both left and right panels of figure 6.1 is the occurrence of multiple, stable equilibria over broad ranges of resource productivities. For increasing values of maximum biomass of resource 1 (figure 6.1, *left*) from 0, first a consumer-resource equilibrium becomes feasible that is characterized by a dominance by juvenile individuals and high densities of resource 2. Equilibrium densities of resource 1 are low, which limits juvenile maturation, while the high densities of resource 2 translate into high total population reproduction. This particular equilibrium is obviously governed by development control.

With further increases in $R_{max,1}$, a second, stable equilibrium becomes feasible, which is characterized by a dominance of adult biomass, high densities of resource 1, and low densities of resource 2. The lowest value of $R_{max,1}$ for which this particular equilibrium occurs corresponds to a critical point that is technically known as a limit point or saddle-node bifurcation point (as also encountered in chapter 4). The equilibrium is stable for all values of $R_{max,1}$ exceeding this threshold density, which is around 3 mg/L. Owing to the low densities of resource for adult consumers in this equilibrium, it is governed by reproduction control, while the high densities of resource for juveniles translate into fast maturation. Both the development-controlled and the reproduction-controlled equilibria are stable and separated by an unstable saddle point over a considerable range of $R_{max,1}$ values, from roughly 3 mg/L to 24 mg/L. At this latter value of $R_{max,1}$, a second critical point occurs, leading to the disappearance of the development-controlled equilibrium.

Thus when resources for juveniles are in short supply relative to the adult resources, the consumer population equilibrates in a state that is development-controlled, while a reproduction-controlled equilibrium results when resource supply for adults is substantially lower than the supply for juveniles. For intermediate supply values of both resources, both types of equilibria are stable and co-occur as alternative stable states next to each other. As we will see, this basic bistability between a development-controlled and a reproduction-controlled equilibrium is the most prominent consequence of an ontogenetic niche shift for a consumer population. The right panels of figure 6.1 illustrate the occurrence of this bistability as a function of increasing productivity of the resource for adult consumers. Now, at low values of $R_{max,2}$, first the reproduction-controlled consumer-resource equilibrium becomes feasible, because the juvenile resource is in large supply relative to adult resource, whereas the development-controlled equilibrium is the only stable equilibrium for very high values of $R_{max,2}$. As before, bistability occurs at intermediate values of the adult resource. The left panel of figure 6.4 shows the range of resource productivities, in terms of $R_{max,1}$ and $R_{max,2}$, for which the bistability between a development- and reproduction-controlled equilibrium occurs. Obviously, the bistability occurs whenever the productivities of both resources are in the same range, while the region with parameters for which two alternative stable states occur expands significantly at higher levels of productivity. The region extends, however, further toward high productivities of the juvenile resource than to high values of adult resource productivity.

Schreiber and Rudolf (2008) were the first to show this bistability between an equilibrium dominated by juveniles and controlled by development and an equilibrium dominated by adults and controlled by reproduction. These authors analyzed a Lotka-Volterra-type model, in which it was assumed that juvenile and adult consumers live in different habitats and feed on their own exclusive resource. It was furthermore assumed that the two resources grow logistically, that juvenile and adult consumers forage on them following linear functional responses, while maturation and reproduction are proportional to juvenile and adult food intake, respectively. With a sufficiently high carrying capacity of the adult resource, Schreiber and Rudolf showed that at low carrying capacities of the juvenile resource, the consumer population equilibrates in a steady state dominated by juveniles, which mature slowly. Increasing juvenile resource carrying capacity first destabilizes this equilibrium, giving rise to the occurrence of population oscillations (limit cycles) and subsequently to the occurrence of a stable equilibrium dominated by adult consumers next to the stable limit cycle. With further increases in carrying capacity of the juvenile resource, the stable limit cycle disappears, resulting in an equilibrium dominated by adult consumers as the only stable state. In this adult-dominated equilibrium total population reproduction is low. At intermediate carrying capacity of the juvenile resource, variation in adult resource carrying capacity leads to an analogous pattern with a unique and stable adult-dominated equilibrium occurring at low adult resource carrying capacities. With increasing carrying capacity, bistability occurs between the adult-dominated equilibrium and an alternative, stable equilibrium dominated by juvenile consumers. Increasing the carrying capacity of the adult resource further destabilizes the latter equilibrium, leading to population limit cycles, which eventually become the sole outcome of dynamics.

Schreiber and Rudolf (2008) also showed that variations in either juvenile or adult consumer mortality could bring about the same changes in community equilibria as the changes in resource carrying capacities. They argued that the data presented by Osenberg, Mittelbach, and Wainwright (1992) on the occurrence of pumpkinseed sunfish (*Lepomis gibbosus*) provide empirical support for the occurrence of alternative stable states, dominated by either juvenile or adult sunfish. Pumpkinseed sunfish exhibit a diet shift during ontogeny from foraging on soft-bodied invertebrates as a juvenile to foraging on snails as adult. Comparing the pumpkinseed sunfish populations in three Michigan lakes, Osenberg, Mittelbach, and Wainwright (1992) indeed show that in one of these lakes the adult resource density (snails) is significantly lower and juvenile sunfish grow faster and mature more rapidly than in the other two. Data on the stage structure of the sunfish population are, however, not reported for the three lakes. Furthermore, the difference between the three lakes is attributed to the presence of small bluegill (*Lepomis macrochirus*) that compete with the juvenile sunfish for a shared resource. The environmental conditions hence differ among the three lakes, which makes the claim that the data provide support for alternative stable states *under the same environmental conditions* tenuous at best.

Guill (2009) also investigated the effects of an ontogenetic shift in habitat or diet for a consumer population that in the juvenile stage exploits a different resource than during adulthood. He modeled consumer dynamics using the same stage-structured biomass model that we have used in this and previous chapters. However, by assuming linear functional responses for juvenile and adult consumers and fast time-scale dynamics for both resources, he replaced the dependence of juvenile and adult ingestion on resource densities with a direct-density dependence, in which juvenile and adult ingestion are decreasing functions of juvenile and adult density, respectively. Similar to the results presented here as well as by Schreiber and Rudolf (2008), Guill showed bistability between a juvenile-dominated, development-controlled and an adult-dominated, reproduction-controlled equilibrium over broad ranges of resource productivities as well as consumer mortality. In fact, Guill (2009) presents a parameter plot for the occurrence of different consumer equilibria as a function of the resource productivities that is similar to the one that we show in figure 6.4 (left panel).

With equal productivity of the juvenile and adult resources, the juvenile-dominated, development-controlled equilibrium is stable for all levels of mortality for which the consumer population can persist, irrespective of whether juvenile, adult, or stage-independent mortality is varied. The adult-dominated, reproduction-controlled equilibrium occurs as an alternative equilibrium at low juvenile and adult consumer mortality. Only when resource availability for juvenile consumers is larger than for adult consumers, does the adult-dominated, reproduction-controlled equilibrium occur as the unique, stable outcome of consumer dynamics. Guill's results hence suggest that the juvenile-dominated, development-controlled equilibrium is the more likely consumer equilibrium to occur, unless juvenile consumers have an advantage over adult consumers in terms of resource availability.

In general, we can therefore conclude that, given its occurrence in both the stage-structured biomass model as well as in the Lotka-Volterra-type model analyzed by Schreiber and Rudolf (2008), an ontogenetic diet shift in consumer life history promotes bistability between a development-controlled and a reproduction-controlled consumer equilibrium under the same environmental conditions. This conclusion depends, however, on the fact that all studies discussed above have assumed an extreme form of an ontogenetic niche shift, in which consumers shift completely from exploiting one resource as a juvenile to exploiting an entirely different resource as adult.

Figure 6.2 shows that when consumers partially change their resource use, from foraging predominantly on resource 1 as a juvenile to predominantly foraging on resource 2 as an adult ($q_{J,1} = q_{A,2} = 0.75$ and $q_{J,2} = q_{A,1} = 0.25$), the bistability between the two types of consumer-resource equilibrium occurs only over a restricted range of resource productivities (figure 6.2, left panels, 18 mg/L $< R_{max,1} <$ 20 mg/L; right panels, 5.2 mg/L $< R_{max,2} <$ 5.6 mg/L). When productivity of juvenile resource is low compared with the adult resource productivity, the development-controlled equilibrium with a dominance of juvenile consumers occurs as the only stable state (figure 6.2, left panels, $R_{max,1} <$ 18 mg/L; right panels, $R_{max,2} >$ 5.6 mg/L), whereas the reproduction-controlled equilibrium with adult consumer dominance is the only stable state when adult resource productivity is relatively low (figure 6.2, left panels, $R_{max,1} >$ 20 mg/L; right panels, $R_{max,2} <$ 5.2 mg/L). Otherwise, the equilibrium characteristics are the same as described above for the case of a complete niche shift with juvenile and adult dominance coinciding with low equilibrium densities of resource 1 and 2, respectively. Figure 6.4 (middle panel) reveals that

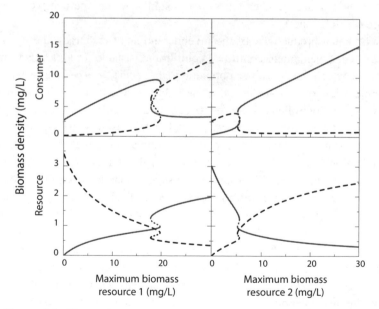

FIGURE 6.2. Changes in equilibrium resource and consumer biomasses as in figure 6.1, but here for the case where consumers go through a partial niche shift at maturation, feeding on both resources in all stages, but predominantly feeding on resource 1 and 2 as juveniles and adults, respectively ($q_{J,1} = q_{A,2} = 0.75$ and $q_{J,2} = q_{A,1} = 0.25$). *Left*: $R_{max,2} = 10$ mg/L; *right*: $R_{max,1} = 10$ mg/L; $z = 0.01$. Otherwise, default parameters apply as listed in box 6.2 for $W_A = 0.0001$ gram.

bistability indeed occurs for a very limited set of combinations of juvenile and adult resource productivity, in which juvenile resource productivity always has to exceed adult resource productivity. Figure 6.4 (middle panel) furthermore makes clear that the partial niche shift allows the consumer population to persist even in the complete absence of one of the two resources, as long as the other resource is in sufficiently large supply.

The extent of bistability between a development-controlled and a reproduction-controlled equilibrium is also limited when consumers at maturation do not shift their resource use, but extend their resource use to include resource 2 as an additional food source (figure 6.3 and figure 6.4, right panel). Our stage-structured biomass model represents this *ontogenetic niche broadening* when choosing $q_{J,1} = 1.0$, $q_{J,2} = 0.0$, and $q_{A,1} = q_{A,2} = 0.5$ for the resource selectivities of juvenile and adult consumers, respectively. At $R_{max,2} = 3$ mg/L, the development-controlled equilibrium is the only stable equilibrium as long as $R_{max,1} < 16$ mg/L, whereas the reproduction-controlled equilibrium is the only stable state for $R_{max,1} > 23$ mg/L (figure 6.3, left panels). In contrast,

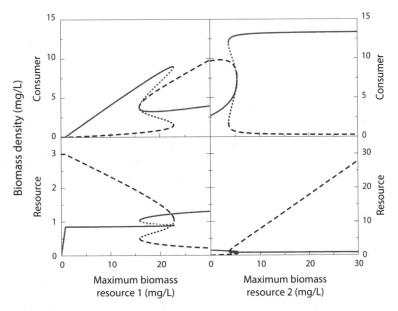

FIGURE 6.3. Changes in equilibrium resource and consumer biomasses as in figure 6.1, but here for the case where consumers broaden their diet at maturation, feeding exclusively on resource 1 as juveniles, while feeding with a reduced effort on both resources as adults ($q_{J,1} = 1.0$, $q_{J,2} = 0.0$, $q_{A,1} = q_{A,2} = 0.5$). *Left:* $R_{max,2} = 3$ mg/L; *right:* $R_{max,1} = 30$ mg/L; $z = 0.01$. Otherwise, default parameters apply as listed in box 6.2 for $W_A = 0.0001$ gram.

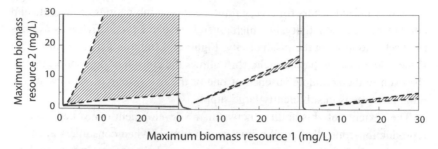

FIGURE 6.4. Combinations of maximum biomass density of resource 1 ($R_{max,1}$) and resource 2 ($R_{max,2}$) for which bistability occurs between two stable, positive consumer-resource equilibria (*hatched domain, enclosed by dashed line*). For all other parameter combinations above and/or to the right of the persistence boundary (*solid line*), a unique consumer-resource equilibrium occurs. To the right/below the hatched parameter domain, this unique equilibrium is dominated by adult biomass (reproduction-controlled equilibrium), whereas it is dominated by juvenile biomass for parameters to the left/above the hatched domain (development-controlled equilibrium). *Left*: complete ontogenetic niche shift at maturation ($q_{J,1} = q_{A,2} = 1$ and $q_{J,2} = q_{A,1} = 0$, cf. figure 6.1); *middle*: partial ontogenetic niche shift ($q_{J,1} = q_{A,2} = 0.75$ and $q_{J,2} = q_{A,1} = 0.25$, cf. figure 6.2); *right*: ontogenetic niche broadening at maturation ($q_{J,1} = 1.0$, $q_{J,2} = 0.0$, $q_{A,1} = q_{A,2} = 0.5$, cf. figure 6.3); $z = 0.01$. Otherwise, default parameters apply as listed in box 6.2 for $W_A = 0.0001$ gram.

at $R_{max,1} = 30$ mg/L, the reproduction-controlled equilibrium is the only stable state when $R_{max,2} < 3.6$ mg/L, while for $R_{max,2} > 5.3$ mg/L, only the stable development-controlled equilibrium occurs (figure 6.3, right panels).

The range of resource productivities for which both types of equilibrium are stable alternatives is limited, as is also revealed by figure 6.4 (right panel). This latter figure shows that resource 1 is essential and that a minimum productivity of it is required to allow consumers to persist, given that juvenile consumers exclusively feed on this resource. In contrast, resource 2 is not essential, as the adults also feed on resource 1 and the consumer population can persist when $R_{max,2} = 0$ as long as $R_{max,1}$ is sufficiently large. From figure 6.4 (right panel) we also conclude that the bistability between a development-controlled and reproduction-controlled equilibrium occurs only when juvenile-resource productivity is significantly higher than adult-resource productivity. The bistability region shown in the right panel of figure 6.4 is shifted to even higher values of juvenile resource productivity than in the case of a partial ontogenetic niche shift (cf. the middle panel of figure 6.4). It should also be noted that the choice of $q_{A,1} = q_{A,2} = 0.5$ for the resource selectivity of adult consumers is highly significant. This assumption implies that at maturation consumers do reduce their foraging effort on resource 1, even when resource 2 is totally absent ($R_{max,2} = 0$). The foraging effort on resource 1 by juvenile and adult consumers would only be the same if we chose $q_{A,1} = 1.0$. However, such an assumption would imply that for every non-zero value of $q_{A,2}$, adult consumers always experience higher resource availability than juveniles. As a consequence, for $q_{A,1} = 1.0$ and $q_{A,2} > 0$, the development-controlled equilibrium dominated by juvenile consumers is always the only stable consumer equilibrium state.

CONSEQUENCES FOR HIGHER TROPHIC LEVELS

Ontogenetic diet shifts may hence promote bistability between alternative equilibria that are maintained through either development or reproduction control, but the extent of the bistability clearly depends on the precise form of diet shift. Generally, a development-controlled equilibrium dominated by juvenile consumers is likely to occur under a broader range of environmental conditions than a reproduction-controlled, adult-dominated equilibrium. Irrespective of the form of the niche shift, however, changes in productivity of the juvenile or the adult resource can always induce a shift from a development-controlled to a reproduction-controlled equilibrium and hence from a dominance by juveniles to a dominance by adults (figure 6.4). In this section we will explore the consequences of this shift in equilibrium type and stage dominance for the persistence of predators that prey on one or both of the consumer life stages. We will restrict this analysis to the extreme scenario in which consumers shift from exclusively foraging on resource 1 as a juvenile to exclusively foraging on resource 2 as adult ($q_{J,1} = q_{A,2} = 1.0$, and $q_{J,2} = q_{A,1} = 0$).

EMERGENT ALLEE EFFECTS FOR GENERALIST PREDATORS

Figure 6.5 shows the changes in equilibrium biomass density in a community consisting of a generalist predator that preys with equal intensity on all consumers ($\phi = 0.5$; see the equations in box 6.1), plus juvenile and adult consumers, each with its exclusive resources, with changes in productivity of the resource for adult consumers. As the most remarkable feature, figure 6.5 makes clear that the ontogenetic diet shift in consumer life history allows for the occurrence of an emergent Allee effect for a generalist predator. At $R_{max,1} = 10$ mg/L, two stable equilibria occur for maximum biomass densities of resource 2 in the range $2.8 < R_{max,2} < 4.2$ mg/L, one predator-consumer-resource and one

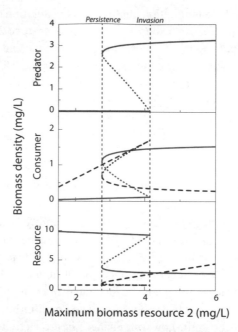

Maximum biomass resource 2 (mg/L)

FIGURE 6.5. Changes in equilibrium biomass of basic resources (*bottom*), consumer (prey, *middle*), and generalist predator ($\phi = 0.5$, *top panels*) with increasing maximum density of resource 2, $R_{max,2}$. Solid lines refer to juvenile consumer biomass or their exclusive resource 1; dashed lines to adult consumer biomass or their exclusive resource 2. Consumers experience a complete niche shift from resource 1 to resource 2 at maturation ($q_{J,1} = q_{A,2} = 1$ and $q_{J,2} = q_{A,1} = 0$). Stable equilibria are indicated with solid or dashed lines; unstable equilibria with dotted lines. $R_{max,1} = 10$ mg/L, and $z = 0.01$. Otherwise, default parameters apply as listed in box 6.2 for a prey and predator with body size $W_A = 0.0001$ and $W_P = 0.01$ gram, respectively.

consumer-resource equilibrium (figure 6.5). This contrasts with the results in chapter 4, where we found that this bistability can occur for stage-specific, but not generalist predators, if both consumer stages compete for a shared resource. The emergent Allee effect for generalist predators arises because of the ontogenetic diet shift of the consumer and in particular because of the bistability between the two different types of equilibrium that this diet shift induces.

Figure 6.1 (right panels) shows that in the absence of predators bistability occurs at $R_{max,1} = 10$ mg/L for $R_{max,2} > 2.4$ mg/L between a juvenile-dominated and adult-dominated equilibrium. In the juvenile-dominated equilibrium, juvenile consumer biomass is, moreover, higher than the biomass density that is needed for persistence of the generalist predator (cf. the juvenile biomass density of 4 mg/L in the juvenile-dominated equilibrium in the right panels of figure 6.1 with the sum of juvenile and adult consumer biomass in the predator-consumer-resource equilibrium shown in figure 6.5, which is less than 2 mg/L). In other words, the stable consumer-resource equilibrium that is dominated by juvenile consumers can be invaded for all values of $R_{max,2}$, for which it exists. The adult-dominated consumer-resource equilibrium, however, can only be invaded by generalist predators for $R_{max,2} > 4.2$ mg/L. The fact that over a range of $R_{max,2}$ values the juvenile-dominated consumer equilibrium can be invaded by predators, whereas the adult-dominated equilibrium cannot, is the reason for occurrence of the emergent Allee effect. For the lower part of this range of $R_{max,2}$ values ($2.4 < R_{max,2} < 2.8$ mg/L), predators do invade the juvenile-dominated consumer state, but subsequently change the consumer population structure to an adult-dominated state, in which they cannot persist. Hence, predator invasion leads to a flip from the juvenile-dominated to the adult-dominated consumer-resource equilibrium, followed by extinction of the predator itself. For $2.8 < R_{max,2} < 4.2$ mg/L, however, predator invasion of the juvenile-dominated consumer state leads to a stable predator-consumer-resource equilibrium, even though the adult-dominated consumer state is stable against invasion by predators.

The range of $R_{max,2}$ values over which bistability occurs between an equilibrium with and without predators is small for the default parameters used in figure 6.5. We have found, however, that this range is significantly larger when the generalist predators experience higher mortality rates (results not shown). More specifically, with an additional predator mortality equal to 0.06 per day on top of background mortality, bistability between an equilibrium with and without predators occurs for $8 < R_{max,2} < 19$ mg/L when $R_{max,1}$ equals 30 mg/L. In contrast, the bistability range is smaller for larger values of the ratio between size at birth and maturation z.

In principle, the bistability between a stable equilibrium with and without predators could also be expected to occur when varying the maximum biomass density of resource 1, $R_{max,1}$, for a given value of $R_{max,2}$. Remarkably, however, we have not found a value of $R_{max,2}$ for which increases in $R_{max,1}$ leads to the same emergent Allee effect as shown in figure 6.5. At a given value of $R_{max,2}$, increasing values of $R_{max,1}$ first allow for a stable, juvenile-dominated consumer equilibrium, followed by a range of $R_{max,1}$ values for which a stable adult-dominated equilibrium exists as well (see figure 6.1, left panels). In the left panels of figure 6.1 it can be seen that at the lowest $R_{max,1}$ value for which it exists, both the juvenile and adult consumer biomass in the adult-dominated equilibrium are considerably larger than the adult and juvenile biomass, respectively, in the juvenile-dominated equilibrium. This implies that for these values of $R_{max,1}$

the total consumer biomass density in the adult-dominated consumer equilibrium is higher than in the juvenile-dominated equilibrium. Consequently, there are indeed ranges of $R_{max,1}$ values for which the adult-dominated consumer equilibrium can be invaded by generalist predators, whereas the juvenile-dominated consumer equilibrium cannot. We have invariably found, however, that invasion of the adult-dominated state by predators eventually results in predator extinction, because the consumer population changes to the stable, juvenile-dominated equilibrium state. In these ranges of $R_{max,1}$, it is therefore not the presence of a predator species, but the mere fact that it can invade, that might cause us to observe only a juvenile-dominated consumer-resource equilibrium, despite its potential to equilibrate in an reproduction-controlled, adult-dominated equilibrium as well. At the threshold value above which the juvenile-dominated consumer equilibrium can be invaded by predators, the curve representing the predator-consumer-resource equilibria is furthermore always directed toward higher values of $R_{max,1}$, as opposed to the direction toward lower parameter values shown in figure 6.5. We thus never observed a stable predator-consumer-resource equilibrium, as long as the juvenile-dominated consumer equilibrium is stable against invasion by predators.

PERSISTENCE AND EXTINCTION OF STAGE-SPECIFIC PREDATORS

Ontogenetic diet shifts that consumers experience in life have unexpected consequences for persistence of predators that forage on only part of the consumer population. We will illustrate these consequences here by analyzing the persistence and coexistence of stage-specific predators that either forage exclusively on juvenile consumers or exclusively on adult consumers in the case in which consumers shift from exclusively foraging on resource 1 as a juvenile to exclusively foraging on resource 2 as adult ($q_{J,1} = q_{A,2} = 1.0$ and $q_{J,2} = q_{A,1} = 0$). The classic paradigm relating equilibrium densities of resource, consumers, and predators to productivity of the environment is provided by the food chain model of Oksanen, Fretwell, et al. (1981). In particular, this food chain model predicts that predator equilibrium density is positively related to resource productivity and that increases in productivity mostly benefit the highest trophic level of the food chain. In chapter 4 we already discovered that the positive relationship between predator equilibrium biomass and resource productivity may not hold for stage-specific predators at the threshold level of resource productivity at which they can invade a consumer-resource equilibrium, because predators may exhibit an emergent Allee effect and may hence be able to persist at productivities below this threshold. Nonetheless, in the stable predator-consumer-resource equilibrium, the predator biomass was found to be positively related to resource productivity as in the food chain model of Oksanen, Fretwell, et al. (1981).

Figure 6.6 shows the changes in equilibrium biomass of a predator specializing on juvenile consumers with increasing values of the maximum biomass density of resource 1 (left panels) as well as resource 2 (right panels) or, equivalently, resource productivity. Also shown are the corresponding changes in equilibrium biomass of juvenile and adult consumers, as well as their exclusive resources. These graphs are constructed using a higher value of the ratio between size at birth and maturation ($z = 0.2$) than used previously, because it allows us to illustrate the effects of both juvenile-specialized as well as adult-specialized predators using the same parameter values for consumers.

Increasing the productivity of resource 1, on which the juvenile consumers that are the main prey of the predators forage, initially leads to a change in equilibrium biomass resembling the predictions of the Oksanen, Fretwell, et al. (1981) food chain model (figure 6.6, left panels): at low productivities only the resource persists, while a stable consumer-resource equilibrium is the only stable state above the threshold resource productivity that allows for consumer persistence (at $R_{max,1} \approx 0.8$ mg/L). Juvenile-specialized predators can invade the consumer-resource equilibrium when juvenile consumer biomass is sufficient for their persistence, which occurs at the predator invasion threshold around $R_{max,1} \approx 2.7$ mg/L. At these low productivities of the juvenile resource, the consumer population at equilibrium is controlled by development and hence dominated by juvenile biomass. The alternative consumer-resource equilibrium, controlled by reproduction and dominated by adult consumers, occurs only for productivities exceeding $R_{max,1} \approx 4.3$ mg/L (see the dotted lines in the left-middle panel of figure 6.6).

Because the predators forage on the dominating stage of consumers, an emergent Allee effect does not occur, and above their invasion threshold equilibrium biomass of predators increases with productivity of resource 1. In equilibrium with predators, juvenile consumer biomass remains constant with increases in resource productivity, whereas adult consumer biomass increases. The increase in adult biomass comes about because resource availability for juveniles increases, while predators limit the competition among juveniles for the resource. Juveniles thus more rapidly mature at higher productivities of their resource, leading to the increasing adult biomass densities.

In the presence of predators, increases in juvenile resource productivity therefore lead to a consumer population that is more and more dominated by adults. This change eventually leads to a significant deviation from the pattern in the Oksanen, Fretwell, et al. (1981) food chain model, when the curve

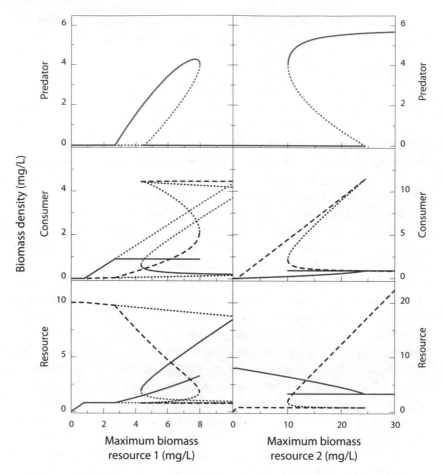

FIGURE 6.6. Changes in equilibrium biomass of basic resources (*bottom*), consumer (prey, *middle*), and juvenile-specialized predator ($\phi = 1.0$, *top panels*) with
increasing maximum density of resource 1 ($R_{max,1}$, *left*) and resource 2 ($R_{max,2}$,
right). Solid lines refer to juvenile consumer biomass or their exclusive resource
1; dashed lines to adult consumer biomass or their exclusive resource 2. Consumers experience a complete niche shift from resource 1 to resource 2 at matura-
tion ($q_{J,1} = q_{A,2} = 1$ and $q_{J,2} = q_{A,1} = 0$). Stable equilibria are indicated with solid or
dashed lines; unstable equilibria with dotted lines. $R_{max,2} = 10$ mg/L (*left panels*),
$R_{max,1} = 8$ mg/L (*right panels*), and $z = 0.2$. Otherwise, default parameters apply as
listed in box 6.2 for a prey and predator with body size $W_A = 0.0001$ and $W_P = 0.01$
gram, respectively.

representing the predator-consumer-resource equilibrium reaches a limit point and bends back toward lower parameter values at around $R_{max,1} \approx 8$ mg/L. Here a so-called saddle-node bifurcation occurs. In essence, at this critical value the equilibrium changes from being controlled by a combination of juvenile development and predation on juveniles to being controlled by adult reproduction. For all maximum resource densities for which it occurs, however, predators cannot persist in the reproduction-controlled consumer equilibrium, because it is dominated by adults and juvenile consumer biomass is too low for predator growth. The backward curve representing unstable predator-consumer-resource equilibria (saddle points) connects with the curve segment representing unstable consumer-resource equilibria (saddle points) that separates the development-controlled from the reproduction-controlled consumer-resource equilibria. As a consequence, predators go extinct and the reproduction-controlled consumer-resource equilibrium remains as the only stable state. Varying the maximum density of juvenile resource, a unique equilibrium with predators occurs for $2.7 < R_{max,1} < 4.3$ mg/L. Bistability between a predator-consumer-resource equilibrium and a reproduction-controlled consumer-only equilibrium occurs for $4.3 < R_{max,1} < 8$ mg/L, whereas the reproduction-controlled consumer-only equilibrium is the only stable community equilibrium for $R_{max,1} > 8$ mg/L. In contrast to the Oksanen, Fretwell, et al. (1981) food chain model, increases in resource productivity for its main prey does not always benefit the stage-specific predator on juveniles, but in fact drives it to extinction.

Changes in the equilibrium biomass of the juvenile-specialized predator, juvenile and adult consumers, and their exclusive resources with increasing maximum densities of the adult resource (resource 2) do not show predator extinction (figure 6.6, right panels). Rather, varying maximum adult resource density shows a bifurcation pattern that resembles the bifurcation pattern of the generalist predators shown in figure 6.5. At the maximum density of the juvenile resource of $R_{max,1} = 8$ mg/L used for the right panels of figure 6.6, bistability between a development-controlled and a reproduction-controlled consumer-only equilibrium occurs for maximum densities of the adult resource exceeding $R_{max,2} \approx 2.1$ mg/L (refer to figure 6.4, left panel), whereas the reproduction-controlled equilibrium is the only stable state below that threshold. The development-controlled consumer-only equilibrium, which is dominated by juveniles, can always be invaded by juvenile-specialized predators, because juvenile consumer biomass densities are higher (exceeding 3.1 mg/L) than what predators need for persistence (roughly 0.9 mg/L). For $R_{max,2}$ values below 10 mg/L, however, invasion by predators of the development-controlled consumer-only equilibrium leads to a switch to the reproduction-controlled consumer-only equilibrium and subsequent extinction of predators, because juvenile biomass densities in this reproduction-controlled equilibrium are too low for their persistence.

Predator invasion into the reproduction-controlled consumer-only equilibrium is possible for $R_{max,2}$ values exceeding 24 mg/L. As a function of $R_{max,2}$ the juvenile-specialized predators exhibit an emergent Allee effect, such that the curve representing predator-consumer-resource equilibria at the predator invasion threshold of $R_{max,2} \approx 24$ mg/L starts off into the direction of lower maximum resource densities. A stable predator-consumer-resource equilibrium can occur for maximum adult resource densities below that threshold (and in fact for all $R_{max,2} > 10$ mg/L), because predators induce a shift toward a consumer population state that is controlled by development, in which juvenile consumer biomass is more than sufficient for predator persistence. The equilibrium with predators is the only stable community state for $R_{max,2} > 24$ mg/L. Similar to the results presented for the generalist predator (see figure 6.5 and above), predators can thus invade one of the two alternative consumer equilibria but will only induce a shift to the alternative equilibrium and go extinct themselves for low maximum resource densities ($2.1 < R_{max,2} < 10$ mg/L). At intermediate values ($10 < R_{max,2} < 24$ mg/L), bistability between a stable predator-consumer-resource equilibrium and a stable, reproduction-controlled consumer-only equilibrium occurs, whereas higher maximum resource densities always lead to predator invasion and persistence.

Figure 6.9 (left panel) shows for which combinations of the maximum densities of the juvenile and the adult resource the predator-consumer-resource equilibrium occurs, either as the unique outcome of community dynamics or as an alternative stable state next to a consumer-only equilibrium that is controlled by consumer reproduction. As the unique outcome of dynamics, the predator-consumer-resource equilibrium occurs only over a limited range of maximum resource densities $R_{max,1}$ and $R_{max,2}$, which is significantly smaller than the parameter region for which bistability with a consumer-only equilibrium occurs. The region of parameters in which predators are driven to extinction because of a maximum density of juvenile resource that is too high (i.e., where lower values of $R_{max,1}$ do allow for predator persistence) has approximately the same extent as the region with bistability and is hence also significantly larger than the parameter region with a unique predator-consumer-resource equilibrium. Bistability and predator extinction are thus the dominating patterns over broad ranges of maximum resource densities.

The changes in equilibrium biomass of juvenile and adult consumers and their exclusive resources in the presence of a predator that forages on adult consumers only, when either maximum density of the juvenile or the adult resource increases (figure 6.7), mirror the patterns presented for a juvenile-specialized predator. More specifically, in the case of an *adult-specialized* predator, the bifurcation diagram as a function of maximum resource density for *adult* consumers is qualitatively similar to the diagram presented for *juvenile-specialized* predators as a function of maximum *juvenile* resource density (cf. right panels in figure 6.7 with left panels in figure 6.6). Similarly, equilibrium biomass changes in the case of an adult-specialized predator when varying maximum juvenile resource density, resemble the changes for juvenile-specialized predators as a function of maximum adult resource density (cf. left panels in figure 6.7 with right panels in figure 6.6). The only difference is that there are small ranges of maximum resource density for which two different types of predator-consumer-resource equilibrium occur as alternative stable states.

The left panels of figure 6.7 are constructed with a maximum resource density for adult consumers of $R_{max,2} = 6$ mg/L. For this value of $R_{max,2}$, a development-controlled and a reproduction-controlled consumer-resource equilibrium occur as alternative stable states for $R_{max,1}$ values exceeding 3.8 mg/L. Whenever it exists, the reproduction-controlled equilibrium can be invaded by adult-specialized predators, because adult consumer biomass densities in this equilibrium are larger (exceeding 2.6 mg/L) than what is required for predator persistence (≈ 0.9 mg/L). In contrast, the development-controlled consumer-only equilibrium can be invaded by adult-specialized predators only for $R_{max,1}$ values exceeding 23 mg/L (figure 6.7, left panels), when adult consumer equilibrium biomass is sufficient for predator growth. At this latter, predator-invasion threshold, the curve representing predator-consumer-resource equilibria starts off into the direction of higher values of $R_{max,1}$, but turns back toward lower parameter values at $R_{max,1} \approx 26$ mg/L. The curve bends again to higher parameter values at the limit point occurring around $R_{max,1} \approx 15$ mg/L, which corresponds to the lowest maximum adult resource density for which the predators can persist. Hence, for $3.8 < R_{max,1} < 15$ mg/L, the ultimate outcome of community dynamics is the consumer-only equilibrium that is controlled by development and dominated by juvenile consumers.

This equilibrium is unique, because the alternative, reproduction-controlled equilibrium dominated by adult consumers can be invaded by the adult-specialized predators, which induce a shift toward the development-controlled equilibrium and thereby cause their own extinction. For $15 < R_{max,1} < 23$ mg/L, bistability occurs between an equilibrium with predators, which is controlled by a combination of limited consumer reproduction and predation on adult consumers, and the development-controlled consumer-only equilibrium, followed by a small range ($23 < R_{max,1} < 26$ mg/L) with bistability between two predator-consumer-resource equilibria, one controlled by consumer reproduction and

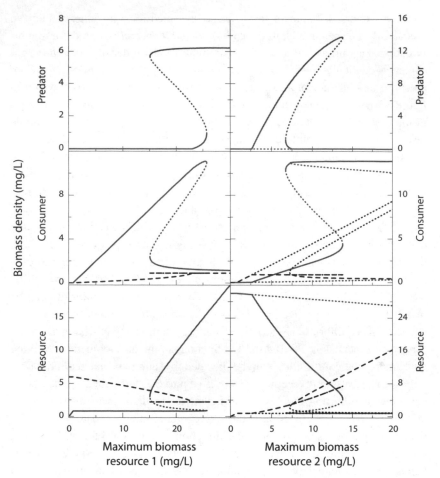

FIGURE 6.7. Changes in equilibrium biomass of basic resources (*bottom*), consumer (prey, *middle*), and adult-specialized predator ($\phi = 0.0$, *top panels*) with increasing maximum density of resource 1 ($R_{max,1}$, *left*) and resource 2 ($R_{max,2}$, *right*). Solid lines refer to juvenile consumer biomass or their exclusive resource 1; dashed lines to adult consumer biomass or their exclusive resource 2. Consumers experience a complete niche shift from resource 1 to resource 2 at maturation ($q_{J,1} = q_{A,2} = 1$ and $q_{J,2} = q_{A,1} = 0$). Stable equilibria are indicated with solid or dashed lines; unstable equilibria with dotted lines. $R_{max,2} = 6$ mg/L (*left panels*), $R_{max,1} = 30$ mg/L (*right panels*), and $z = 0.2$. Otherwise, default parameters apply as listed in box 6.2 for a prey and predator with body size $W_A = 0.0001$ and $W_P = 0.01$ gram, respectively.

predation, the other mainly by slow consumer development. For $R_{max,1}$ values exceeding 26 mg/L, the predator-consumer-resource equilibrium that is governed by a combination of consumer reproduction control and predation is the unique outcome of the community dynamics. The pattern described closely resembles the pattern for juvenile-specialized predators as presented in the right panels of figure 6.6 and extensively discussed above, except for the small range of $R_{max,1}$ values for which a predator-consumer-resource equilibrium occurs that is governed by consumer development control. It is beyond our scope to explain in detail why this type of equilibrium with predators, and hence the bistability between two different predator-consumer resource equilibria, can occur, but it is related to the fact that in a development-controlled equilibrium, an increase in adult mortality initially leads to little increase in adult consumer biomass at equilibrium (see the top-right panel in figure 3.6), whereas in a reproduction-controlled consumer equilibrium, an increase in juvenile consumer mortality translates into a much more rapid increase in juvenile equilibrium biomass (see the top-middle panel in figure 3.5).

For the maximum density of juvenile resource used to construct the graphs in the right panels of figure 6.7 ($R_{max,1} = 30$ mg/L), bistability between a development-controlled and a reproduction-controlled consumer-resource equilibrium can occur for $R_{max,2}$ values exceeding 7.2 mg/L (cf. figure 6.4, left panel). The adult-specialized predators can already invade the reproduction-controlled consumer equilibrium for lower $R_{max,2}$ values, leading to a coexistence equilibrium with increasing biomass densities of predators as well as juvenile consumers for increasing values of $R_{max,2}$. This equilibrium with predators is stable up to the predator persistence threshold that occurs at $R_{max,2} \approx 14$ mg/L, at which point the curve representing this particular equilibrium bends back toward lower values of $R_{max,2}$ in a so-called saddle-node bifurcation (figure 6.7, right panels). The continuation of the curve with decreasing densities of predator biomass and increasing juvenile consumer densities represents unstable equilibria (saddle points). The curve bends toward higher values of $R_{max,2}$ once again at $R_{max,2} \approx 6.7$ mg/L and connects at $R_{max,2} \approx 7.5$ mg/L to the curve of consumer-only equilibria in the segment that represents stable equilibria controlled by consumer development.

For all $R_{max,2}$ values exceeding 7.5 mg/L this development-controlled consumer-only equilibrium dominated by juveniles is stable against invasion by the adult-specialized predators, because equilibrium biomass densities of adult consumers are too low for predator growth. For the small range of maximum resource densities just below that threshold ($6.7 < R_{max,2} < 7.5$ mg/L), a predator-consumer-resource equilibrium occurs with low predator biomass density, which is primarily governed by development control of consumers. This particular equilibrium corresponds to the coexistence equilibrium that occurs in the left panels of figure 6.7 over a small range of $R_{max,1}$ values $(23 < R_{max,1} < 26$ mg/L). For increasing values of $R_{max,2}$, we thus find, consecutively, a range of maximum resource densities with a unique and stable predator-consumer-resource equilibrium $(2.6 < R_{max,2} < 6.7$ mg/L); a small range of bistability between two different predator-consumer-resource equilibria $(6.7 < R_{max,2} < 7.5$ mg/L); followed by a larger range of bistability between a predator-consumer-resource and a consumer-only equilibrium $(7.5 < R_{max,2} < 14$ mg/L); and a substantial range in which predators go extinct and a stable, development-controlled, consumer-only equilibrium is the unique outcome of community dynamics. Apart from the range with bistability between two equilibria with predators, this pattern is qualitatively the same as the changes in equilibrium densities for juvenile-specialized predators with increasing values of maximum density of the juvenile resource (see left panels in figure 6.6 and the corresponding discussion above).

Figure 6.9 (middle panel) shows for the case of adult-specialized predators the combinations of maximum densities for juvenile and adult resource that lead to a unique predator-consumer-resource equilibrium or to bistability between two different types of equilibria with predators or between an equilibrium with predators and a consumer-only equilibrium. Apart from the region with bistability between two different types of equilibria with predators, which separates the other two regions, this graph is a mirror image of the graph for juvenile-specialized predators shown in the left panel of figure 6.9, albeit that the parameter regions are smaller. The parameter domain leading to a unique equilibrium with adult-specialized predators is not much smaller than the corresponding domain with a unique equilibrium with juvenile-specialized predators (cf. left and middle panels in figure 6.9). However, in the case of adult-specialized predators, the parameter domain with bistability between an equilibrium with predators and a consumer-only equilibrium is significantly smaller and roughly the same size as the domain that leads to a unique predator-consumer-resource equilibrium. Compared with the other parameter regions, the region of parameters in which bistability between two different equilibria with predators occurs is negligible. Altogether, the middle panel of figure 6.9 makes clear that the parameter region, in which adult-specialized predators are driven to extinction because of high maximum densities of adult resource where lower values of $R_{max,2}$ do allow for persistence, has significantly increased in size compared with the corresponding region for juvenile-specialized predators. For adult-specialized predators, extinction is hence the dominating outcome of dynamics over broad ranges of maximum resource densities.

Mutual Facilitation between Stage-Specific Predators

The results in the previous section make clear that over considerable ranges of resource productivity stage-specific predators may not be able to persist on their own if consumers shift from using one resource during their juvenile phase to using another resource during adulthood. Either productivity of the resource that their main prey forages on is too *high* to allow any predator persistence at all; or, an equilibrium with predators present is stable, but eradication of the predator will result in a consumer-only equilibrium that is stable against their subsequent invasion, and predators are hence unable to recover from a population collapse. Combining the left and middle panels of figure 6.9, we can even deduce that there is a region of maximum resource densities for which neither the juvenile-specialized nor the adult-specialized predator can persist on its own (this particular parameter region is indicated by the hashed region in the right-most panel of figure 6.9). When both the juvenile- and adult-specialized predator are present in the community, however, they both can persist and coexist over the entire range of resource productivities for which they would otherwise run the risk of extinction.

The left panels of figure 6.8 show—for a community consisting of both juvenile- and adult-specialized predators, juvenile consumers, and adult consumers, as well as their exclusive resources—the changes in equilibrium biomass dependent on the maximum density of the juvenile consumer resource. These graphs are based on the assumption that a particular predator species will be present in the community as soon as the resource productivity allows its invasion into the community equilibrium that would occur in its absence for that particular resource productivity. Because a consumer-only equilibrium controlled by consumer development (and hence dominated by juveniles) occurs at lower values of $R_{max,1}$ than a reproduction-controlled, consumer-only equilibrium, the juvenile-specialized predator can invade the consumer-only equilibrium first. After invasion of the juvenile-specialized predator, an increase in juvenile resource productivity leads to a change in consumer population structure, because juvenile consumer biomass is kept constant by the predator, while adult consumer biomass increases. Eventually the latter increase allows for invasion of the adult-specialized predator into the community equilibrium with only juvenile-specialized predators. Further increases in juvenile resource productivity lead to increasing densities of both the juvenile- as well as the adult-specialized predator. At maximum resource densities above 15 mg/L, the equilibrium destabilizes and stable population oscillations result. The minimum biomass densities of all species during these oscillations are, however, bounded away from 0, such that neither predator runs a risk of extinction. With

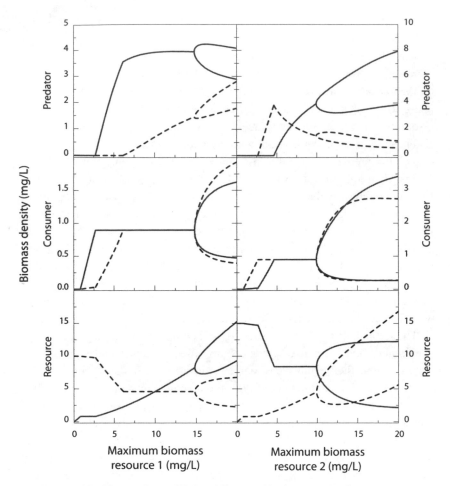

FIGURE 6.8. Changes in equilibrium biomass of basic resources (*bottom*), consumer (prey, *middle*), and juvenile-specialized ($\phi = 1.0$) and adult-specialized predator ($\phi = 0.0$, *top panels*) with increasing maximum density of resource 1 ($R_{max,1}$, *left*) and resource 2 ($R_{max,2}$, *right*). Solid lines refer to juvenile consumer biomass, their specialist predator, or their exclusive resource 1; dashed lines to adult consumer biomass, their specialist predator, or their exclusive resource 2. Consumers experience a complete niche shift from resource 1 to resource 2 at maturation ($q_{J,1} = q_{A,2} = 1$ and $q_{J,2} = q_{A,1} = 0$). For $R_{max,1} > 15$ mg/L (*left*) or $R_{max,2} > 10$ mg/L (*right*), stable oscillations around an unstable equilibrium occur. Minimum and maximum biomass densities during these cycles are indicated with thick lines. Otherwise, equilibria are stable. $R_{max,2} = 10$ mg/L (*left panels*), $R_{max,1} = 15$ mg/L (*right panels*), and $z = 0.2$. Otherwise, default parameters apply as listed in box 6.2 for a prey and predator with body size $W_A = 0.0001$ and $W_p = 0.01$ gram, respectively.

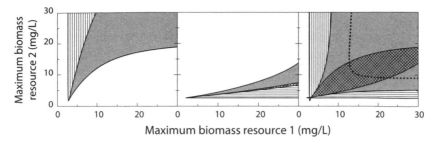

FIGURE 6.9. *Left*: combinations of maximum resource densities $R_{max,1}$ and $R_{max,2}$ for which a stable community equilibrium including juvenile-specialized predators ($\phi = 1.0$) occurs next to a stable equilibrium of only resource and consumers (*gray region*) and where it occurs as a unique outcome of community dynamics (*striped region*). *Middle*: As in the left panel, but for an equilibrium of resource, consumers, and adult-specialized predators ($\phi = 0.0$). In the additional, cross-hatched parameter region, bistability occurs between two different equilibria with adult-specialized predators present (see figure 6.7). *Right*: maximum resource densities $R_{max,1}$ and $R_{max,2}$ for which community dynamics lead to an equilibrium with only juvenile-specialized predators (*vertically striped region*), only adult-specialized predators (*horizontally striped region*), or to an equilibrium with both predators present (*gray region*). For parameter combinations in the cross-hatched area, both predators mutually facilitate each other, while for parameter combinations in the gray region above and below it, juvenile-specialized and adult-specialized predators, respectively, facilitate unilaterally their companion predator species. For parameter combinations above the dotted line, population oscillations occur. Throughout, $z = 0.2$. Otherwise, default parameters apply as listed in box 6.2 for a prey and predators with body size $W_A = 0.0001$ and $W_P = 0.01$ gram, respectively.

increasing values of the maximum juvenile resource density, we do not find any extinction of one or both of the predator species.

The right panels of figure 6.8 show the analogous pattern of changes in equilibrium biomass for a community consisting of both juvenile- and adult-specialized predators, juvenile consumers, and adult consumers, as well as their exclusive resources dependent on the maximum density of the adult consumer resource. Now, because the consumer-only equilibrium that is governed by reproduction control occurs for lower levels of $R_{max,2}$ than the development-controlled equilibrium, the adult-specialized predator is the first capable of persistence. In an equilibrium with adult-specialized predators, adult consumer biomass stays constant with further increases of the maximum adult resource density, while juvenile consumer biomass increases. Above a specific threshold of maximum adult resource density (at $R_{max,2} \approx 4.6$ mg/L), juvenile consumer

biomass is sufficiently high for juvenile-specialized predators to persist, resulting in a stable equilibrium with both juvenile- and adult-specialized predators
present. Above $R_{max,2} \approx 9.9$ mg/L, this equilibrium is unstable, and limit cycles
develop with bounded amplitude, such that neither species in the community
attains densities close to 0 during these oscillations.

The changes in equilibrium biomass when varying the maximum density
of the adult consumer resource are hence qualitatively similar to the changes
in equilibrium biomass when varying the maximum density of the juvenile
consumer resource, except that the juvenile-specialized and adult-specialized
predator have their roles exchanged (in terms of which of the two predators invades first). The only difference is that in a stable equilibrium with both predators present, an increase in maximum density of juvenile consumer resource
translates into an increase in equilibrium biomass of both predators, whereas
an increase in maximum density of adult consumer resource increases equilibrium biomass of the juvenile-specialized predator but decreases the equilibrium biomass of the adult-specialized predator. The reason for this contrasting
pattern is most likely related to the fact that juvenile-specialized predators prey
on consumer victims before adult-specialized predators do; hence, a substantial increase in biomass of the juvenile-specialized predator in the coexistence
equilibrium (as is visible in the right, but not the left panels of figure 6.8) may
limit the prey availability for adult-specialized predators.

Persistence and coexistence of both juvenile- and adult-specialized predators is hence possible for all combinations of maximum resource density that
exceed the lowest threshold values that allow for their invasion. The right
panel of figure 6.9 shows the regions of maximum resource densities with
different equilibrium states of the community including both juvenile- and
adult-specialized predators. At low maximum densities of juvenile consumer
resource, only the juvenile-specialized predator persists in a unique and stable,
single-predator-consumer-resource equilibrium. Similarly, at low maximum
densities of adult consumer resource, a stable equilibrium with consumers
and only the adult-specialized predators is the unique outcome of community
dynamics.

These two parameter regions roughly correspond to, but are somewhat
smaller than, the parameter regions that would result in a unique predator-
consumer-resource equilibrium in a community with only the juvenile-
specialized or only the adult-specialized predator. In the regions of maximum
resource densities, in which bistability between a predator-consumer-resource
equilibrium and a consumer-only equilibrium would occur in communities
with only the juvenile-specialized or only the adult-specialized predator, both
predators persist when together either in a stable equilibrium or, for higher

maximum densities of both resources, in a stable limit cycle. The same holds for the region of maximum resource densities, in which neither the juvenile-specialized predator nor the adult-specialized predator could persist when alone (hashed parameter region in the right panel of figure 6.9). In the latter parameter region, it is mutual facilitation between the juvenile-specialized and adult-specialized predator that allows for their persistence. If one of the predator species disappears from the community, the other goes extinct as well. Irrespective of which predator goes extinct, the whole guild of predators disappears, and the community collapses.

Hence, in contrast with the emergent facilitation discussed in chapter 5, the facilitation is in this case not unilateral, with one predator facilitating persistence of the other, but rather both species crucially depend on each other's presence. Unilateral facilitation of the adult-specialized predator by the juvenile-specialized predator does still occur in the parameter region, where in a community with only the juvenile-specialized predator, bistability would occur between an equilibrium with predators and a consumer-only equilibrium and where adult-specialized predators would not be able to persist at all (gray parameter region above the hashed region in the right panel of figure 6.9). In this region, extinction of the adult-specialized predator is reversible, given that the remaining community of juvenile-specialized predators, consumers, and resources can persist and can be reinvaded by it. On the other hand, extinction of the juvenile-specialized predator does not leave a viable community, as adult-specialized predators cannot persist on their own for these resource productivities. Hence, juvenile-specialized predator extinction would lead to collapse of the whole predator guild and result in a consumer-resource equilibrium.

Similarly, unilateral facilitation of the juvenile-specialized predator by the adult-specialized predator occurs in the parameter region, where bistability would occur in a community of only adult-specialized predators between a predator-consumer-resource and a consumer-only equilibrium and where juvenile-specialized predators cannot persist (gray parameter region below the hashed region in the right panel of figure 6.9). The roles of juvenile-specialized and adult-specialized predators are reversed for these resource productivities. Hence, extinction of the juvenile-specialized predator is a reversible community change, but extinction of the adult-specialized predator leads to a collapse of the predator guild and leaves a stable consumer-resource equilibrium.

In summary, the ontogenetic niche shift of the consumer, from foraging on resource 1 as a juvenile to foraging on resource 2 as adult, not only results in predator extinction with increasing resource productivity in communities with a single stage-specific predator but also allows for facilitation between predators, both mutual as well as unilateral, and hence persistence in communities with multiple stage-specific predators. Ontogenetic niche shifts in consumer life history may thus have significant consequences for the structure of communities including these consumers and their generalist or stage-specific predators.

ONTOGENETIC NICHE SHIFTS
IN PREDATOR LIFE HISTORY

Ontogenetic niche shifts characterize in particular the life history of many predator species, because predators, especially in aquatic systems, tend to grow substantially in body size throughout life. As a newborn or a small juvenile, predator individuals are often of similar body sizes as their future prey. Marine and freshwater fish provide prominent examples of these size relations. In the beginning of this chapter we mentioned that cod individuals are born with a length around 1 cm, which is approximately the same size as the size at birth of the fish species they later in life prey on. The same holds for many other, if not all, piscivores. More generally, the majority of predator species will not have a predatory lifestyle directly after birth, but will start out feeding on smaller and easier food items than the prey species they hunt for as adults. Only mammalian predators and birds of prey are an exception in this respect.

The previous sections focused on ontogenetic diet shifts in consumers that forage on different resources in different life stages. Predators are, of course, also consumers of resources, and hence it could be argued that the results presented in the previous sections hold true for predators as they do for consumers. A key difference, however, is that predators feed on a prey population that is self-renewing as opposed to on a resource, the productivity of which is independent of its own density. The main question addressed in the following sections is hence how an ontogenetic niche shift in life history affects the structure of a community when individuals change during ontogeny from feeding on a resource with constant productivity to feeding on a self-renewing resource. More specifically, we focus on a predatory life style and assume that individuals start out as a juvenile feeding on a resource with a productivity that is independent of the actual resource density and change to feeding as an adult on a self-renewing consumer population. The consumer population itself is assumed to be sustained by its own exclusive resource, the productivity of which is also independent of the actual resource density. As we will see in the following sections, a major consequence of these assumptions will be that the predator may drive the consumer to extinction, if adult predators can still feed to some extent on the alternative resource.

We will analyze these issues using two different model frameworks. We will first discuss the predictions of a stage-structured biomass model, in which both the consumer and the predator population are subdivided into juveniles and adults. Juvenile and adult consumers compete with each other for a resource that is exclusively available to them (and not to predators). Juvenile predators feed on another resource that is exclusively available to predators (but not to consumers). At maturation, predators go through a complete or a partial diet shift, such that adult predators either feed only on juvenile consumers or partly feed on the predator-exclusive resource and on juvenile consumers. Subsequently, we discuss the predictions of a model in which both consumer and predator are size-structured, with body size dynamics based on the Kooijman-Metz energy budget model. Again, all consumers compete for their own exclusive resource, while predators change from foraging on a predator-exclusive resource to preying on consumers at larger body sizes. The ontogenetic niche shift is in this case, however, more gradual with increasing body size, compared with the stage-structured biomass model.

STAGE-STRUCTURED PREDATORS FORAGING
ON STAGE-STRUCTURED CONSUMERS

The equations describing the biomass model with stage-structured consumers and stage-structured predators are presented in box 6.3, and the default parameters are presented in box 6.4. The equations for the dynamics of the two predator stages are similar to the equations that we use (and have used extensively before) to describe the dynamics of juvenile and adult consumers. Only the ingestion functions are different, as they represent that adult predators feed on both resource and consumers. In addition, predators are assumed to be larger and hence have different mass-specific maximum ingestion, maintenance, and mortality rates. Juvenile and adult consumers (or prey) compete for their own, consumer-exclusive resource, but we assume that the mass-specific, maximum ingestion rate for adult consumers is smaller than for juvenile consumers ($q_C = 0.5$; refer to chapter 3). This implies that in the absence of predators, the consumer population equilibrates in a state governed by reproduction control and is hence dominated by adults. The ratio between the size at birth and at maturation is assumed to be the same for both consumers and predators ($z_C = z_P = 0.01$). We investigate the model predictions dependent on the productivities of the consumer-exclusive and predator-exclusive resource (i.e., the maximum densities $R_{max,C}$ and $R_{max,P}$, respectively) when predators either go through a complete or through a partial niche shift at maturation. The extent

BOX 6.3

BIOMASS MODEL WITH TWO RESOURCES, STAGE-STRUCTURED PREY,
AND STAGE-STRUCTURED PREDATOR

Dynamic equations	Description
$\dfrac{dR_C}{dt} = G_C(R_C) - \omega_{C_J}(R_C)C_J - \omega_{C_A}(R_C)C_A$	Consumer-exclusive resource dynamics
$\dfrac{dR_P}{dt} = G_P(R_P) - \omega_{P_J}(R_P)P_J - \omega_{P_A}(R_P, C_J)P_A$	Predator-exclusive resource dynamics
$\dfrac{dC_J}{dt} = v_{C_A}^+(R_C)C_A - \gamma_C(v_{C_J}^+, d_{C_J})C_J + v_{C_J}(R_C)C_J - d_{C_J}(P_A)C_J$	Juvenile prey dynamics
$\dfrac{dC_A}{dt} = \gamma_C(v_{C_J}^+, d_{C_J})C_J + (v_{C_A}(R_C) - v_{C_A}^+(R_C))C_A - \mu_C C_A$	Adult prey dynamics
$\dfrac{dP_J}{dt} = v_{P_A}^+(R_P, C_J)P_A - \gamma_P(v_{P_J}^+, \mu_P)P_J + v_{P_J}(R_P)P_J - \mu_P P_J$	Juvenile predator dynamics
$\dfrac{dP_A}{dt} = \gamma_P(v_{P_J}^+, \mu_P)P_J + (v_{P_A}(R_P, C_J) - v_{P_A}^+(R_P, C_J))P_A - \mu_P P_A$	Adult predator dynamics

Function	Expression	Description
$G_i(R_i)$	$\rho(R_{\max,i} - R_i)$	Resources turnover, $i = C, P$
$\omega_{C_J}(R_C)$	$M_C R_C/(H_C + R_C)$	Juvenile consumer intake
$\omega_{C_A}(R_C)$	$q_C M_C R_C/(H_C + R_C)$	Adult consumer intake
$v_i(R_C)$	$\sigma_C \omega_i(R_C) - T_C$	Consumer net energy production, $i = C_J, C_A$
$d_{C_J}(P_A)$	$\mu_C + \dfrac{M_P(1-\phi_P)P_A}{H_P + \phi_P R_P + (1-\phi_P)C_J}$	Juvenile consumer mortality
$\gamma_C(v_{C_J}^+, d_{C_J})$	$(v_{C_J}^+ - d_{C_J})/(1 - z_C^{(1 - d_{C_J}/v_{C_J}^+)})$	Juvenile consumer maturation
$\omega_{P_J}(R_P)$	$M_P R_P/(H_P + R_P)$	Juvenile predator intake
$\omega_{P_A}(R_P, C_J)$	$\dfrac{M_P \phi_P R_P}{H_P + \phi_P R_P + (1-\phi_P)C_J}$	Adult predator intake
$v_{P_J}(R_P)$	$\sigma_P \omega_{P_J}(R_P) - T_P$	Juvenile predator net energy production
$v_{P_A}(R_P, C_J)$	$\sigma_P M_P \dfrac{\phi_P R_P + (1-\phi_P)C_J}{H_P + \phi_P R_P + (1-\phi_P)C_J} - T_P$	Adult predator net energy production
$\gamma_P(v_{P_J}^+, \mu_P)$	$(v_{P_J}^+ - \mu_P)/(1 - z_P^{(1 - \mu_P/v_{P_J}^+)})$	Juvenile predator maturation

v^+ represents the value of a function v, but restricted to non-negative values.

BOX 6.4

DEFAULT PARAMETER VALUES FOR THE STAGE-STRUCTURED
BIOMASS MODEL OF BOX 6.3

Parameter	Default value	Unit	Description
Resources			
ρ	0.1	day^{-1}	Turnover rate of resources
$R_{\max, i}$	30	mg/L	Maximum density of resource $i = C, P$
Prey			
M_C	$0.1 W_A^{-0.25}$	day^{-1}	Mass-specific maximum ingestion rate
H_C	3.0	mg/L	Ingestion half-saturation density
q_C	0.5	—	Adult-juvenile ingestion ratio
T_C	$0.01 W_A^{-0.25}$	day^{-1}	Mass-specific maintenance rate
σ_C	0.5	—	Conversion efficiency
z_C	0.01	—	Newborn-adult size ratio
μ_C	$0.0015 W_A^{-0.25}$	day^{-1}	Background mortality rate
Predators			
M_P	$0.1 W_P^{-0.25}$	day^{-1}	Mass-specific maximum ingestion rate
H_P	3.0	mg/L	Ingestion half-saturation density
T_P	$0.01 W_P^{-0.25}$	day^{-1}	Mass-specific maintenance rate
σ_P	0.5	—	Conversion efficiency
z_P	0.01	—	Newborn-adult size ratio
μ_P	$0.0015 W_P^{-0.25}$	day^{-1}	Background mortality rate
ϕ_P	0.0 or 0.25	—	Adult foraging effort on resource

W_A and W_P represent average adult body weight of consumer and predator, respectively.

of the niche shift is governed by the parameter ϕ_P. Juvenile predators forage only on the predator-exclusive resource, while adult predators are assumed to forage a fraction ϕ_P of their time on this exclusive resource and a fraction $1 - \phi_P$ of their time on juvenile consumers. Notice that predators are assumed to spend this latter fraction of their time on predation, even when the consumer is extinct.

Figure 6.10 shows the changes in equilibrium biomass of juvenile and adult predators, juvenile and adult consumers, and the two resources as a function of the maximum density of the consumer-exclusive (left panels) and predator-exclusive resource (right panels) when predators go through a complete

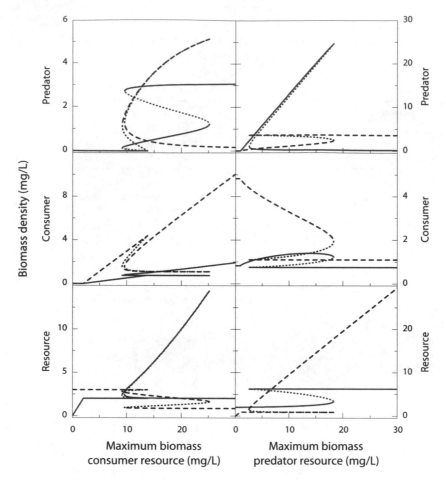

FIGURE 6.10. Changes in equilibrium biomass of predators (*top panels*), consumers (*middle*), and resources (*bottom*) with increasing maximum density of the consumer-exclusive ($R_{max,C}$, *left*) and predator-exclusive resource ($R_{max,P}$, *right*) in the biomass model with stage-structured consumers and stage-structured predators. Solid lines refer to biomass of the consumer-exclusive resource, juvenile consumer biomass, and juvenile predator biomass; dashed lines to biomass of the predator-exclusive resource, adult consumer biomass, and adult predator biomass. Stable equilibria are indicated with solid or dashed lines; unstable equilibria with dotted lines. Juvenile and adult consumers share the consumer-exclusive resource ($q_C = 0.5$), and predators experience a complete niche shift from feeding on predator-exclusive resource as a juvenile to preying on juvenile consumers as an adult ($\phi_P = 0$). $R_{max,P} = 3$ mg/L (*left panels*), $R_{max,C} = 15$ mg/L (*right panels*), $z_C = 0.01$, and $z_P = 0.01$. Otherwise, default parameters apply as listed in box 6.4 for a prey and predator with body size $W_A = 0.0001$ and $W_P = 0.01$ gram, respectively.

ontogenetic niche shift and adult predators hence forage only on juvenile consumers ($\phi_P = 0$). These bifurcation diagrams look rather complicated, mainly because they result from the interplay between two mechanisms that we have encountered before: on the one hand, the reproduction control of the consumer population in the absence of predators allows for the occurrence of an emergent Allee effect for the predators as they prey on juvenile consumers; on the other hand, the ontogenetic niche shift in predatory life history allows for the occurrence of bistability between two predator equilibrium states, one controlled by slow predator development, the other by limited predator reproduction.

When increasing the maximum consumer-exclusive resource density $R_{max,C}$ from low values the consumer-only equilibrium (including the two exclusive resources as well) is stable against predator invasion up to $R_{max,C} \approx 14$ mg/L (figure 6.10, left panels). At this predator invasion threshold, food availability for adult predators is limiting predator population growth most, and the population is hence reproduction-controlled. Because the consumer population is reproduction-controlled as well, (predation) mortality imposed on juvenile consumers leads to overcompensation in juvenile biomass and hence to a positive feedback between (adult) predator density and the availability of their prey. As a result, an emergent Allee effect occurs for the predator population such that the curve representing the predator-consumer coexistence equilibrium starts off at the predator invasion threshold in the direction of lower $R_{max,C}$ values. The initial curve segment represents unstable equilibria (saddle points) that separate the stable consumer-only equilibria from stable coexistence equilibria. The curve reaches a lower limit point, bending toward higher $R_{max,C}$ values, at $R_{max,C} \approx 9.2$ mg/L, which corresponds to the predator persistence threshold, as it is the lowest $R_{max,C}$ value that allows for a stable equilibrium with predators present. At this lowest $R_{max,C}$ value, the predator population in the stable coexistence equilibrium is governed by reproduction control and dominated by adults, because the density of juvenile consumers is limited, compared with the availability of predator-exclusive resource (cf. the middle-left and bottom-left panels in figure 6.10).

With only a slightly higher $R_{max,C}$ value, however, a second, predator-consumer coexistence equilibrium becomes feasible (at $R_{max,C} \approx 9.6$ mg/L), in which the equilibrium density of predator-exclusive resource is significantly lower. In this latter coexistence equilibrium, the predator population is hence governed by development control and dominated by juveniles. This leads to bistability between a coexistence equilibrium controlled by limited predator reproduction and another coexistence equilibrium controlled by slow predator development over a considerable range of $R_{max,C}$ values ($9.6 < R_{max,C} < 25$ mg/L). The bistability is identical to the one occurring in consumer-resource systems with increasing productivities of the resource for adult consumers, which we discussed earlier (cf. figure 6.1). The combined result, however, of the emergent Allee effect and the bistability induced by the predator ontogenetic niche shift is that, over a range of $R_{max,C}$ values ($9.6 < R_{max,C} < 14$ mg/L), three stable community equilibria occur: a consumer-only equilibrium, a coexistence equilibrium controlled by slow predator development, and a coexistence equilibrium controlled by limited predator reproduction.

With increasing maximum densities of consumer-exclusive resource, we thus find, consecutively, a region with a unique consumer-only equilibrium ($2 < R_{max,C} < 9.2$ mg/L), a minute region with bistability between a consumer-only equilibrium and a reproduction-controlled coexistence equilibrium ($9.2 < R_{max,C} < 9.6$ mg/L), the region with three stable community equilibria mentioned before ($9.6 < R_{max,C} < 14$ mg/L), a region with a stable reproduction-controlled and a stable development-controlled coexistence equilibrium ($14 < R_{max,C} < 25$ mg/L), and ultimately a unique, coexistence equilibrium controlled by slow predator development for $R_{max,C} > 25$ mg/L. The coexistence equilibrium that is controlled by limited predator reproduction is unstable, with population oscillations occurring around it for $R_{max,C} > 14.8$ mg/L. However, the amplitude of these limit cycles is invariably very small. For the value of $R_{max,P} = 3$ mg/L used to construct the left panels in figure 6.10, they are even so small that plots of the maximum and minimum biomass densities during the oscillations are graphically indistinguishable from a stable equilibrium curve. For larger values of $R_{max,P}$, the amplitude of the limit cycles is larger, but still so small that the cycles do not merit further discussion here.

The right panels of figure 6.10 reveal that the bistability between two different types of predator-consumer coexistence equilibria also shows up when varying the maximum density of the predator-exclusive resource, but that an emergent Allee effect does not occur. When increasing the maximum predator-exclusive resource density $R_{max,P}$ from 0, predators can invade the consumer-only equilibrium at $R_{max,P} \approx 0.92$ mg/L. At this predator invasion threshold, population growth of the predator is limited most by juvenile development, owing to the low density of the predator-exclusive resource. Even though the mortality imposed by adult predators on juvenile consumers leads to overcompensation in juvenile consumer biomass (see the middle-right panel in figure 6.10), the predator population as a whole does not really benefit from this increase in food availability for adult predators, as reproduction is not limiting its population growth. For this reason the overcompensation in juvenile biomass does not lead to an emergent Allee effect. At $R_{max,P}$ values just above the predator invasion threshold, the predator population is hence development-controlled and dominated by juveniles. For values of $R_{max,P}$ exceeding 2.6 mg/L, however, the second coexistence equilibrium becomes feasible, which is dominated by adult predators and governed by reproduction control of the predator. The bistability between the development-controlled and reproduction-controlled coexistence equilibrium occurs for $R_{max,P}$ values ranging from 2.6 mg/L to 18 mg/L. For higher $R_{max,P}$ values, the reproduction-controlled equilibrium is the only stable state of the community.

Irrespective of whether the maximum density of the consumer-exclusive or the predator-exclusive resource is increased, the consumer population is always dominated by adults (see middle-left and middle-right panel in figure 6.10) due to the lower mass-specific, maximum ingestion rate that we have assumed for adult consumers. In addition, increases in the maximum density of the consumer-exclusive resource only increase total predator biomass in the equilibrium in which the predator is governed by reproduction control. Once the predator-exclusive resource becomes limiting and the predator population becomes controlled by juvenile development, an increase in maximum consumer-exclusive resource density only leads to increases in juvenile and adult consumer biomass in equilibrium (middle-left panel in figure 6.10). Similarly, increases in the maximum density of the predator-exclusive resource only increase total predator biomass in the equilibrium that is governed by development control. When the availability of juvenile consumers as prey for adult predators becomes the limiting factor, further increases in maximum predator-exclusive resource density only increase the predator-exclusive resource density itself and nothing else (bottom-right panel in figure 6.10). In particular, juvenile and adult predator biomass as well as juvenile and adult consumer biomass do not respond to such further increases in maximum predator-exclusive resource density and remain constant instead.

Figure 6.12 (left panel) shows as a function of the maximum density of the consumer- and predator-exclusive resource the different regions of parameters in which the different number and types of equilibria can occur. The figure reveals that the occurrence of the emergent Allee effect is by-and-large independent of the maximum density of the predator-exclusive resource. The bistability between a stable consumer-only equilibrium and (one or two) stable coexistence equilibria that results from the emergent Allee effect occurs over a range of maximum densities of the consumer-exclusive resource, which is located below the threshold marking invasion of the predator population into the consumer-only equilibrium. The parameter region, in which two different types of coexistence equilibria occur, has qualitatively the same shape as the bistability region observed in consumer-resource systems (cf. the left panel of figure 6.4). This bistability region partly overlaps, however, with the region in which the emergent Allee effect occurs. In the overlap region, there are three stable community states, as discussed above: a consumer-only equilibrium and two coexistence equilibria. Figure 6.12 (left panel) also shows that the coexistence equilibrium that is controlled by limited predator reproduction is unstable in a large part of the region of parameters for which it occurs. As mentioned before, however, the amplitude of the resulting limit cycles is invariably small, and these cycles can hence for all practical purposes be ignored.

When predators only go through a partial niche shift at maturation, such that adult predators still forage on the predator-exclusive resource for 25 percent of their time ($\phi_P = 0.25$) and for the remainder prey on consumers, predators can persist even in the absence of consumers as long as the maximum density of the predator-exclusive resource is sufficiently high. The left panels of figure 6.11 show for such a sufficiently high value of $R_{max,P}$ the changes in equilibrium biomass of predators, consumers, and resources with increasing maximum densities of the consumer-exclusive resource. At low values of $R_{max,C}$, the predator persists on its own in an equilibrium state that is governed by reproduction control, because adult predators spend less effort to forage on the predator-exclusive resource and hence are more limited in their energetics than juveniles. Consumers can invade this predator-only equilibrium for maximum consumer-exclusive resource densities $R_{max,C}$ exceeding the consumer-invasion threshold, which occurs at $R_{max,C} \approx 3.1$ mg/L. In the coexistence equilibrium above this threshold, the predator and consumer population are both governed by reproduction control and hence dominated by adults. The consumer population is reproduction-controlled owing to the lower mass-specific maximum ingestion rate we assumed for adults. The predator population is reproduction-controlled because the equilibrium biomass of predator-exclusive resource is higher than juvenile consumer biomass and the food availability for juvenile predators therefore exceeds food availability for adult predators.

Further increases in $R_{max,C}$ lead to a region with bistability between two coexistence equilibria, in which the predator population is either governed by slow juvenile development or by limited reproduction ($9.5 < R_{max,C} < 16$ mg/L). When governed by development control, the predator population is dominated by juveniles, whereas it is dominated by adults when governed by reproduction control. For $R_{max,C}$ values beyond the bistability region, the only stable community state is the coexistence equilibrium, in which predators are governed by development control. In this latter equilibrium, juvenile predator biomass hardly increases, and adult predator biomass even decreases with further increases in maximum consumer-exclusive resource density, because of the low density of predator-exclusive resource and the consequent slow juvenile development. Furthermore, the decreasing densities of adult predators imply that consumers experience lower predation mortality and can hence sustain at a lower equilibrium biomass density of consumer-exclusive resource. As a consequence, when the predator population is governed by development control,

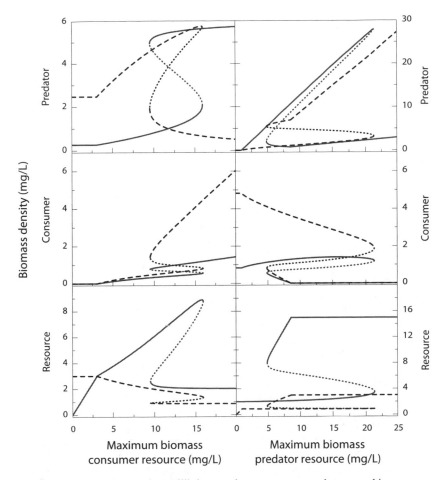

FIGURE 6.11. Changes in equilibrium predator, consumer, and resource biomass as in figure 6.10, except that predators experience a partial niche shift from feeding on predator-exclusive resource as a juvenile to feeding on resource and preying on juvenile consumers as an adult ($\phi_P = 0.25$). $R_{max,P} = 5$ mg/L (*left panels*), $R_{max,C} = 15$ mg/L (*right panels*), $q_C = 0.5$, $z_C = 0.01$, and $z_P = 0.01$. Otherwise, default parameters apply as listed in box 6.4 for a prey and predator with body size $W_A = 0.0001$ and $W_P = 0.01$ gram, respectively.

only the equilibrium biomasses of juvenile and adult consumers increase with increasing maximum densities of the consumer-exclusive resource.

The changes in equilibrium biomass of predators, consumers, and resources as a function of maximum predator-exclusive resource density $R_{max,P}$ (figure 6.11, right panels) in the case of a partial ontogenetic niche shift are

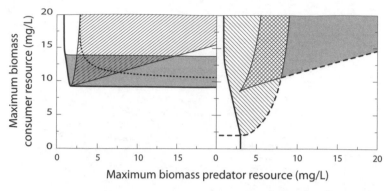

FIGURE 6.12. *Left*: combinations of maximum biomass of predator-exclusive and consumer-exclusive resource (*above/to the right of the thick solid line*) with possible persistence of stage-structured predators experiencing a complete niche shift from feeding on predator-exclusive resource as a juvenile to preying on juvenile consumers as an adult ($\phi_P = 0$). Consumers can persist for all parameters shown. In the gray parameter region, one or two stable resource-consumer-predator equilibria occur next to a stable resource-consumer state. In the hatched parameter region, two different resource-consumer-predator equilibrium states occur next to each other. For maximum resource densities above the thick dotted line, limit cycles occur around the (unstable) resource-consumer-predator equilibrium with a low juvenile and high adult predator density. Other parameters as in figure 6.10. *Right*: combinations of maximum biomass of predator-exclusive and consumer-exclusive resource (*to the right of the thick solid line*) with persistence of stage-structured predators experiencing a partial niche shift from feeding on predator-exclusive resource as a juvenile to feeding on resource and preying on juvenile consumers as an adult ($\phi_P = 0.25$). Consumers can persist for parameters above and to the left of the thick, dashed line. In the hatched parameter region, a unique, stable resource-consumer-predator equilibrium occurs. In the gray parameter region, the latter occurs next to a stable resource-predator equilibrium. In the cross-hatched parameter region, two different, stable resource-consumer-predator equilibrium states occur next to each other. Other parameters as in figure 6.11.

reminiscent of the pattern observed for a complete predator niche shift, except for the important difference that consumers go extinct at high values of $R_{\max,P}$. Above a threshold value of maximum predator-exclusive resource density, here occurring at $R_{\max,P} \approx 0.91$ mg/L, predators can invade the consumer-only equilibrium. As discussed above, an emergent Allee effect does not occur, because predator population growth is mostly limited by juvenile development and the predator population hence does not benefit from the increase in prey availability for adult predators that results from the overcompensation in juvenile consumer biomass. For $R_{\max,P}$ values exceeding 4.8 mg/L, a second stable predator-consumer coexistence equilibrium occurs, in which the predator population is controlled by limited reproduction. Because in this latter equilibrium the predator population is dominated by adults, the predation mortality imposed on juvenile consumers is high and increases with further increases of $R_{max,P}$. As a consequence, consumers cannot persist in this reproduction-controlled predator population for $R_{max,P}$ values exceeding a threshold value that occurs at $R_{max,P} \approx 8.6$ mg/L.

Above this consumer-extinction threshold, a stable predator-only equilibrium occurs, in which the population is governed by reproduction control and dominated by adult predators. The range of bistability between the two types of coexistence equilibria, one governed by development control, the other by reproduction control, is hence rather small ($4.8 < R_{max,P} < 8.6$ mg/L) for the particular value of $R_{max,C}$ used to construct the right panels in figure 6.11. Above the consumer-extinction threshold, bistability still occurs, but now between the stable coexistence equilibrium with predators governed by development control and the stable predator-only equilibrium. This bistability range is larger ($8.6 < R_{max,P} < 22$ mg/L) than the range of bistability between the two coexistence equilibria. In the coexistence equilibrium with predator development control, increases in maximum predator-exclusive resource density lead to substantial increases in biomass density of juvenile predators, but much slower increases in biomass of adult predators. The mortality that predators impose on juvenile consumers hence also increases only slowly, which allows for the consumer persistence up to the consumer-persistence threshold that occurs at $R_{max,P} \approx 22$ mg/L. In contrast, in the coexistence equilibrium with predator reproduction control, as well as in the predator-only equilibrium above the consumer-extinction threshold, increases in $R_{max,P}$ mainly increase adult predator biomass in equilibrium and decrease or only slowly increase (in the predator-only equilibrium) the equilibrium biomass of juvenile predators. The biomass of predator-exclusive resource in the predator-only equilibrium remains constant with increasing values of $R_{max,P}$, as it is the only food source sustaining the predator population.

The right panel of figure 6.12 shows, as a function of the maximum consumer-exclusive and predator-exclusive resource density, the parameter regions with different types of community equilibria for the case in which predators only partly change their diet at maturation. Broadly speaking, the parameter plane separates into two parts. At low values of the maximum consumer-exclusive resource density $R_{max,C}$ (in this case, approximately $R_{max,C} < 10$ mg/L), no bistability occurs, and the community ends up either in a stable consumer-only equilibrium, a stable coexistence equilibrium in which the predator population is controlled by slow juvenile development, or a predator-only equilibrium with the predator population controlled by limited reproduction. Which of these three community equilibria occurs and is stable mostly depends on the value of the maximum predator-exclusive resource density $R_{max,P}$ and is much less influenced by $R_{max,C}$.

In contrast, for high values of the maximum consumer-exclusive resource density $R_{max,C}$ (above $R_{max,C} \approx 10$ mg/L), two types of bistability occur: either bistability between two coexistence equilibria, one controlled by slow predator development, the other by limited predator reproduction; or, bistability between the coexistence equilibrium with predator development control and a predator-only equilibrium. Again, whether a consumer-only equilibrium, a unique coexistence equilibrium governed by predator development control, or one of the two types of bistability occurs is mainly determined by the maximum predator-exclusive resource density $R_{max,P}$ and is much less influenced by $R_{max,C}$. The right panel of figure 6.12 shows that, in particular, the region with bistability between a coexistence equilibrium with predator development control and a predator-only equilibrium is substantial and larger than the region with bistability between two types of coexistence equilibria.

One of the aspects of ontogenetic niche shifts in predator life history that has been discussed extensively in the literature is the possible occurrence of juvenile bottlenecks (Werner 1986). The fact that predator individuals start out at body sizes similar to their future prey opens up the possibility of competition between juvenile predators and consumers. Such a competition would imply that consumers and predators are engaged in a mixed interaction involving both competition and predation with juvenile predators exploiting the same niche as their future prey, while moving up to the next trophic level when growing large. Such mixed competition-predation interactions have been documented for communities involving anurans (Wilbur 1988); fish (Byström, Persson, and Wahlström 1998; Mittelbach 1983; Olson, Mittelbach, and Osenberg 1995; Persson 1988; Werner 1986); scorpions (Polis, Myers, and Holt 1989); and reptiles, crustacea, and insects (Werner and Gilliam 1984). Experimentally, a juvenile bottleneck has been shown to occur for perch (*Perca fluviatilis*, Byström, Persson, and Wahlström 1998), as a result of intense competition by its main prey, roach (*Rutilus rutilus*). In the presence of roach young-of-the-year (YOY) perch switched to feed on benthic cladocerans and chironomids, as roach depressed the free-swimming zooplankton resource. As a result, perch at the end of their first growing season were small and in poor condition and died from starvation during winter (Byström, Persson, and Whlström 1998). Olson, Mittelbach, and Osenberg (1995) showed a similar effect of YOY bluegill sunfishes on the growth of YOY largemouth bass, where bass size at the end of their first year showed a striking decline across a gradient in small bluegill density. Juvenile bottlenecks for predators have furthermore been argued to play an important role in structuring communities (Mittelbach 1983; Olson 1996; Persson 1988; Werner 1986).

We will discuss mixed interactions in the following chapter but want to point out here that the results presented above reveal that even in the absence of competition between juvenile predators and consumers, a juvenile bottleneck can occur in predator life history, which severely limits recruitment to the predatory life stage. In the coexistence equilibrium that is governed by slow predator development, juveniles dominate the predator population, and recruitment rate to the adult stage is low. This limitation of adult predator recruitment corresponds to a juvenile bottleneck, but it is induced intraspecifically, in contrast to the juvenile bottlenecks induced by interspecific competition between juvenile predators and consumers that have been discussed in the literature. The juvenile bottleneck in the coexistence equilibrium with predator development control occurs because adult predators have surplus food and hence achieve high reproductive output. The high reproduction rate subsequently leads to strong competition among juvenile predators for their limiting resource, which is not shared with consumers. Ultimately, the juvenile bottleneck in predator life history hence results from the high density of juvenile consumers that adult predators can forage on.

The right panel of figure 6.12 furthermore reveals that this juvenile bottleneck of the predator does not so much affect predator persistence, but rather promotes persistence of consumers over broad ranges of environmental productivity. More than hampering the predator, the juvenile predator bottleneck thus benefits the consumer. Without the bottleneck, recruitment to the adult predator stage would be large and predation mortality of juvenile consumers consequently high, leading to extinction of the consumer population and monopolization of the environment by the predator. This intraspecifically induced, juvenile bottleneck in predator life history is hence crucial for coexistence of consumers and predators in systems in which the predators go through a partial ontogenetic niche shift, in particular when the resource availability for consumers is high. In the following section we will return to this juvenile bottleneck and show that it also occurs and plays an important role in a more elaborate, size-structured community model as well.

SIZE-STRUCTURED PREDATORS FORAGING ON SIZE-STRUCTURED PREY

In chapter 4 we analyzed the occurrence of an emergent Allee effect in a model in which size-structured predators prey on size-structured consumers. Individual energetics of both consumer and predator individuals are modeled using the Kooijman-Metz dynamic energy budget model (Kooijman and Metz 1984; see also chapter 3 for a more detailed model description). Growth in individual length follows von Bertalanffy growth dynamics for both consumers and predators, in which the maximum attainable length is dependent on food availability. All consumers are assumed to compete for a single, basal resource, while all predators from birth onward prey on consumers. Predators of a particular body size select prey following a tent-shaped function that depends on both prey and predator body length (see figure 4.12 and box 4.3). The total food availability $C(l)$ for a predator of length l is subsequently calculated as a weighted integral over the body size-distribution of the consumer population with the tent-shaped function as weighing factor (see equation (4.6)).

Hence, the prey availability differs for predators of different sizes, as it depends on the actual size distribution of consumers, while the predation mortality faced by consumers of a particular body length varies with the size distribution of predators present (see equation (4.8)). Here we will use the same model as analyzed in chapter 4, except that we will assume that predators are born with the same body length as their prey ($\ell_b = l_b = 7$ mm, where ℓ_b and l_b refer to prey and predator length at birth, respectively), whereas in chapter 4 we assumed that predator length at birth was five times larger ($l_b = 35$ mm; see box 4.4). Assuming that basic energetics of predators do not change, the smaller length at birth entails that predator fecundity in terms of the number of offspring produced per day is 125 times larger, given that predator body mass is assumed proportional to cubed body length. The values of all other predator parameters (listed in box 6.6) are the same as used in chapter 4 (cf. box 4.4). Model parameters involved in consumer dynamics are identical to the values used in chapter 4 and are presented in box 3.7.

Because predators are born with the same length as consumers, newborn predators are too small to feed on the smallest (newborn) consumers. We therefore assume that predators start out feeding on a predator-exclusive resource, the density of which is indicated by R_p, but reduce the fraction of time that they spend feeding on this resource following an exponentially declining function of predator body length, e^{-l/l_j}. In this expression l_j refers to the length at which predators mature. For the remaining fraction of their time, $1 - e^{-l/l_j}$, predators are assumed to prey on consumers. Total food availability for a predator of length l thus equals a weighted sum of resource and prey availability:

$$F(l) = e^{-l/l_j} R_p + (1 - e^{-l/l_j}) C(l), \qquad (6.3)$$

in which $C(l)$ is the same weighted integral over the body-size distribution of the consumer population as used in chapter 4 (see equation (4.6)). Given these assumptions, the ingestion rate of a predator of length l equals:

$$M_P \frac{F(l)}{H_P + F(l)} l^2 \tag{6.4}$$

(cf. equation (4.7)), while the predation mortality experienced by a consumer with body length ℓ equals:

$$P(\ell) = \int_{\ell/\varepsilon}^{\ell/\delta} M_P \frac{(1-e^{-l/l_j})T(\ell,l)}{H_P + F(l)} l^2 p(t,l) dl \tag{6.5}$$

(cf. equation (4.8)). The total mortality rate that consumers of length ℓ experience equals $d_C(P(\ell)) = \mu_C + P(\ell)$, as in the model discussed in chapter 4.

All model equations and functions are summarized in box 6.5. They are similar to the corresponding equations in chapter 4 (cf. box 4.3), except for the exponential term representing the gradual niche shift from foraging on the predator-exclusive resource to preying on consumers, which shows up at several places in the model formulation. One additional function is presented in box 6.5 for the resource ingestion rate $I_P(R_P, F(l), l)$, with which a predator of length l feeds on the predator-exclusive resource, defined as:

$$I_P(R_P, F(l), l) = M_P e^{-l/l_j} \frac{R_P}{H_P + F(l)} l^2 \tag{6.6}$$

The only additional differential equation presented in box 6.5 describes the dynamics of the predator-exclusive resource, which equals the balance between semichemostat growth dynamics and an integral term representing the resource ingestion rate of the total predator population.

We analyze the number and types of equilibria that occur in the model for various combinations of maximum density of the consumer-exclusive and predator-exclusive resource, similar to the analysis in the previous section. It should be noted that the same cautionary remarks apply here as in chapter 4. The equilibrium biomass densities are calculated by means of the continuation methods for physiologically structured population models that are presented in more detail in section 4 of the technical appendices. These methods do not allow for simultaneous assessment of the stability properties of the computed equilibrium, in contrast to the numerical bifurcation software package MATCONT (Dhooge, Govaerts, and Yu 2003; see also box 3.5) that we used to compute equilibria of the stage-structured models presented before in this chapter. Therefore, in the bifurcation graphs presented in the remainder of this chapter we have determined the stability properties of an equilibrium on the basis of general principles from bifurcation theory (see Kuznetsov 1995, for an introduction).

BOX 6.5
PREDATOR-PREY MODEL WITH TWO RESOURCES, BASED ON KOOIJMAN-METZ ENERGETICS

Dynamic equations	*Description*
$\dfrac{\partial p(t,l)}{\partial t} + \dfrac{\partial g_P(F(l),l)\,p(t,l)}{\partial l} = -\mu_P p(t,l)$	Predator length-distribution dynamics
$g_P(F(l_b),l_b)\,p(t,l_b) = \displaystyle\int_{l_j}^{l_m} b_P(F(l),l)\,p(t,l)\,dl$	Predator population reproduction
$\dfrac{\partial c(t,\ell)}{\partial t} + \dfrac{\partial g_C(R_C,\ell)\,c(t,\ell)}{\partial \ell} = -d_C(P(\ell))\,c(t,\ell)$	Consumer length-distribution dynamics
$g_C(R_C,\ell_b)\,c(t,\ell_b) = \displaystyle\int_{\ell_j}^{\ell_m} b_C(R_C,\ell)\,c(t,\ell)\,d\ell$	Consumer population reproduction
$\dfrac{dR_P}{dt} = \rho(R_{\max,P} - R_P) - \displaystyle\int_{l_b}^{l_m} I_P(R_P,F(l),l)\,p(t,l)\,dl$	Predator-exclusive resource dynamics
$\dfrac{dR_C}{dt} = \rho(R_{\max,C} - R_C) - \displaystyle\int_{\ell_b}^{\ell_m} I_C(R_C,\ell)\,c(t,\ell)\,d\ell$	Consumer-exclusive resource dynamics

Function	*Expression*	*Description*
$I_C(R_C,\ell)$	$M_C R_C/(H_C + R_C)\,\ell^2$	Consumer resource ingestion
$g_C(R_C,\ell)$	$\gamma_C(\ell_m R_C/(H_C + R_C) - \ell)$	Consumer growth rate in length
$b_C(R_C,\ell)$	$\alpha_C R_C/(H_C + R_C)\,\ell^2$	Consumer fecundity
$d_C(P(\ell))$	$\mu_C + P(\ell)$	Consumer death rate
$w_C(\ell)$	$\beta_C \ell^3$	Consumer weight-length relation
$T(\ell,l)$	$\begin{cases} \dfrac{\ell - \delta l}{(\varphi - \delta)l} & \text{for } \delta l \le \ell < \varphi l \\[2mm] \dfrac{\varepsilon l - \ell}{(\varepsilon - \varphi)l} & \text{for } \varphi l \le \ell \le \varepsilon l \\[2mm] 0 & \text{otherwise} \end{cases}$	Vulnerability of consumer at length ℓ for predator of length l
$C(l)$	$\displaystyle\int_{\delta l}^{\varepsilon l} T(\ell,l)\,w_C(\ell)\,c(t,\ell)\,d\ell$	Prey availability for predator of length l
$F(l)$	$e^{-l/l_j}R_P + (1 - e^{-l/l_j})\,C(l)$	Food availability for predator of length l
$P(\ell)$	$M_P \displaystyle\int_{\ell/\varepsilon}^{\ell/\delta} \dfrac{(1 - e^{-l/l_j})\,T(\ell,l)}{H_P + F(l)}\,l^2 p(t,l)\,dl$	Predation mortality of consumer at length ℓ
$I_P(R_P,F(l),l)$	$M_P e^{-l/l_j}R_P/(H_P + F(l))\,l^2$	Predator resource ingestion

	(Box 6.5 continued)	
Function	*Expression*	*Description*
$g_P(F(l),l)$	$\gamma_P\big(l_m F(l)/(H_P + F(l)) - l\big)$	Predator growth rate in length
$b_P(F(l),l)$	$\alpha_P F(l)/(H_P + F(l)) l^2$	Predator fecundity
$w_P(l)$	$\beta_P l^3$	Predator weight-length relation

BOX 6.6

DEFAULT PARAMETER VALUES FOR THE SIZE-STRUCTURED PREDATOR IN THE MODEL OF BOX 6.5

Parameter	*Default value*	*Unit*	*Description*
l_b	7	mm	Length at birth
l_j	300	mm	Length at maturation
l_m	750	mm	Maximum length at unlimited food
β_P	0.009	$\text{mg} \cdot \text{mm}^{-3}$	Weight-length scaling constant
M_P	0.2	$\text{mg} \cdot \text{day}^{-1} \cdot \text{mm}^{-2}$	Maximum ingestion-length scaling constant
H_P	0.015	mg/L	Ingestion half-saturation density
γ_P	0.005	day^{-1}	Growth rate in length scaling constant
α_P	0.006	$\text{day}^{-1} \cdot \text{mm}^{-2}$	Maximum fecundity-length scaling constant
μ_P	0.01	day^{-1}	Predator background mortality rate
δ	0.05	—	Minimum prey-predator length ratio
φ	0.1	—	Optimal prey-predator length ratio
ε	0.4	—	Maximum prey-predator length ratio

Parameter values for the consumer-exclusive resource and size-structured consumer are given in box 3.7. Parameters for the predator-exclusive and consumer-exclusive resource are equal.

In particular, at so-called transcritical bifurcation or branching points, where a curve representing a coexistence equilibrium intersects a curve representing a consumer-only (or predator-only) equilibrium, the latter changes stability. The coexistence equilibrium is judged unstable if the curve starts off in the branching point toward parameter values for which the consumer-only (or predator-only) equilibrium is stable. Otherwise, the coexistence equilibrium is judged as stable. Lastly, at a so-called limit point of an equilibrium curve, where it doubles back on itself, the stability of the equilibrium changes. On the basis of these general principles we can distinguish saddle points from stable nodes and foci, but we cannot assess whether the equilibrium has become unstable through a so-called Hopf bifurcation and has given rise to limit cycles. We did not carry out any numerical simulations of population dynamics to assess the stability of computed equilibria.

Figure 6.13 shows the equilibrium biomass of predators, consumers, and resources as a function of maximum density of the consumer-exclusive (left panels) and predator-exclusive resource (right panels; cf. figures 6.10 and 6.11) for the model with size-structured predators and size-structured consumers. For both the consumer and the predator, we present the biomass densities of three different size classes: small juveniles, large juveniles, and adults (see the legend of figure 6.13 for precise details about the size ranges). In contrast to the graphs shown in figures 6.10 and 6.11, we have omitted from figure 6.13 all curve segments that represent unstable equilibria (saddle points according to the general principles discussed above). Otherwise, the graphs would have become overloaded with curves and hence too obscure. All curves shown therefore represent stable equilibria (again, according to the principles we used to assess stability). Figure 6.13 should be compared with, and indeed qualitatively resembles, figure 6.11, which showed the corresponding bifurcation graphs for the stage-structured biomass model with predators going through a partial niche shift at maturation. As a major quantitative difference, however, all axes in figure 6.13 use a logarithmic scaling, in contrast to the linearly scaled axes in figure 6.11, in order to capture the often large variations in biomass densities in this model over often large ranges of maximum resource density.

Because adult predators also feed to some extent on the predator-exclusive resource, predators can persist in the absence of consumers as long as the maximum density of predator-exclusive resource $R_{max,P}$ is sufficiently high. The left panels of figure 6.13 are constructed with such a sufficiently high $R_{max,P}$ value. Then, at low maximum densities of the consumer-exclusive resource, a unique and stable community equilibrium with only predators occurs, in which the predator population is dominated by large juveniles. Above a threshold value of maximum consumer-exclusive resource density at $R_{max,C} \approx 0.015$ mg/L (marked as "Invasion" in the left panels of figure 6.13), the consumer can successfully

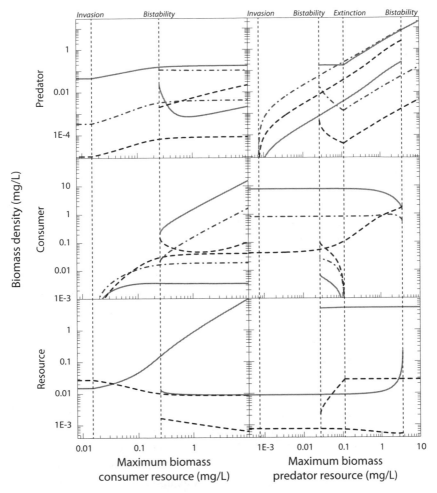

FIGURE 6.13. Changes in equilibrium biomass of consumer-exclusive (*bottom, solid line*) and predator-exclusive resource (*bottom, dashed line*), juvenile consumers smaller (*middle, dashed-dotted line*) and larger than 50 mm (*middle, solid line*), adult consumers (*middle, dashed line*), juvenile predators smaller (*top, dashed-dotted line*) and larger than 80 mm (*top, solid line*), and adult predators (*top, dashed line*) with increasing maximum density of consumer-exclusive resource $R_{max,C}$ (*left panels*) and predator-exclusive resource $R_{max,P}$ (*right panels*) in the size-structured consumer, size-structured predator model with two resources based on Kooijman-Metz energy channeling (see box 6.5 for model formulation). Only (potentially) stable equilibria are indicated. In the left panels, two different equilibria with consumers and predators present occur for $R_{max,C}$ values larger than 0.25 mg/L. In the right panels, two different equilibria with predators and consumers occur for $R_{max,P}$ values between 0.026 and 0.11 mg/L, whereas bistability between an equilibrium with and without consumers occurs for $R_{max,P}$ values between 0.11 and 3.7 mg/L. These threshold values are indicated with dashed vertical lines. $R_{max,P} = 0.05$ mg/L (*left panels*) and $R_{max,C} = 5$ mg/L (*right panels*). Otherwise, default parameters apply as given in boxes 3.7 and 6.6.

invade the predator-only equilibrium, which becomes unstable (and is hence no longer visible in figure 6.13). For $R_{max,C}$ values above the consumer-invasion threshold, the consumer population itself is dominated by small juveniles.

With further increases in $R_{max,C}$, a second threshold is crossed at $R_{max,C} \approx 0.25$ mg/L (marked as "Bistability" in the left panels of figure 6.13), above which a second, stable coexistence equilibrium becomes feasible. In this second equilibrium, the consumer population is dominated by large juveniles, whereas the predator population is dominated by small juveniles. Both coexistence equilibria, the one with many large juvenile predators and many small juvenile consumers and the other with many large juvenile consumers and many small juvenile predators, remain feasible with further increases in the maximum density of consumer-exclusive resource. Invariably, a consumer equilibrium state that is dominated by large juveniles is associated with a low equilibrium density of consumer-exclusive resource, whereas a state dominated by small juveniles coincides with high consumer-exclusive resource densities. For the predator population, the opposite holds: if the population is dominated by large juveniles, predator-exclusive resource densities are high, whereas low predator-exclusive resource densities occur when small juveniles dominate the predator population (left panels of figure 6.13).

Predators cannot persist in the absence of predator-exclusive resource, as in that case newborn predators have nothing to feed on. Hence, at very low values of the maximum predator-exclusive resource density $R_{max,P}$, the only stable and unique community equilibrium consists of a consumer population, which is in this case dominated by large juvenile individuals (right panels of figure 6.13). Above a threshold value of $R_{max,P}$, which occurs at $R_{max,P} \approx 7.5 \cdot 10^{-4}$ mg/L (marked as "Invasion" in figure 6.13, right panels), this consumer-only equilibrium becomes unstable and can be invaded by predators. After predators establish themselves in the consumer-only equilibrium at these low values of $R_{max,P}$, the predator population itself is dominated by small juveniles. With further increases in the maximum predator-exclusive resource density, a second threshold is crossed at $R_{max,P} \approx 0.026$ mg/L (left-most "Bistability" in figure 6.13, right panels), above which a second coexistence equilibrium occurs. In this latter coexistence state, the predator population is dominated by large juveniles, whereas the consumer population is dominated by small juveniles. In the coexistence equilibrium that is dominated by large juvenile consumers, increases in $R_{max,P}$ lead to proportional increases in density of all predator size classes, whereas the density of small juvenile consumers remains roughly constant, large juvenile consumers decline, and adult consumers increase with higher $R_{max,P}$ values. This particular equilibrium destabilizes and hence disappears when $R_{max,P}$ exceeds a threshold value at $R_{max,P} \approx 3.7$ mg/L (right-most

"Bistability" in figure 6.13, right panels). Above this upper bistability threshold, only a single community equilibrium remains.

In the coexistence equilibrium dominated by large juvenile predators, increases in maximum predator-exclusive resource density do not change the density of large juvenile predators, but lead to a decrease in small juvenile and adult density. Also, all consumer densities quickly decrease with increasing $R_{max,P}$ to the extent that consumers go extinct when a threshold value of $R_{max,P}$ at $R_{max,P} \approx 0.11$ mg/L is exceeded (marked "Extinction" in figure 6.13, right panels). For $R_{max,P}$ values above the consumer-extinction threshold, a stable predator-only equilibrium occurs, which cannot be invaded by consumers, owing to the high predation mortality that especially the large juvenile predators impose. The predator-only equilibrium is also the unique community equilibrium for $R_{max,P}$ values exceeding the upper bistability threshold. As was pointed out before, dominance by large (small) juvenile consumers is associated with low (high) equilibrium densities of consumer-exclusive resource, whereas low (high) equilibrium densities of predator-exclusive resource occur when small (large) juveniles dominate the predator population.

Clearly, figure 6.13 shows that there are two possible community states with coexistence of consumers and predators: one in which large juveniles dominate the consumer population, the other in which large juveniles dominate the predator population. It should be pointed out that generally the total consumer biomass also exceeds total predator biomass in equilibrium when large juveniles dominate the consumer population, whereas total predator biomass exceeds total consumer biomass when large juveniles dominate the predator population (results not shown). In fact, when large juveniles make up the largest part of consumer population biomass, their biomass exceeds the total population biomass of the predator, whereas large juvenile predator biomass is higher than total consumer biomass when large juveniles make up the main part of predator population biomass. In these two contrasting states the entire community is hence dominated by large juvenile consumers and large juvenile predators, respectively. Figures 6.14 and 6.15 present the anatomy of these two different community states in terms of consumer and predator growth curves, population size distributions, size-dependent fecundity, and mortality for a single combination of $R_{max,C}$ and $R_{max,P}$, for which both states occur as stable community equilibrium.

When large juvenile predators dominate the community, predator growth through small size ranges is rapid, but the maximum body length that predators attain is only slightly larger than the maturation size (figure 6.14, left panel). Consumers grow rapidly throughout their entire life, reaching sizes that are close to the maximum body size that they reach with surplus food. In contrast,

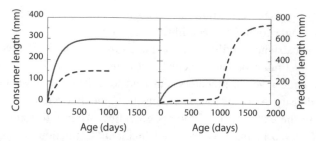

FIGURE 6.14. Growth curves of consumers (*solid line*) and predators (*dashed line*) in the two different equilibrium states that occur in the two resources, size-structured consumer, size-structured predator model based on Kooijman-Metz energy channeling for $R_{max,C} = 5.0$ and $R_{max,P} = 0.05$ mg/L (cf. figure 6.13). *Left*: equilibrium state dominated by juvenile predators larger than 80 mm ("stunted-predator state"); *right*: equilibrium state dominated by juvenile consumers larger than 50 mm ("stunted-consumer state"). Parameters as in figure 6.13.

when large juvenile consumers dominate the community, they reach maximum sizes just above their maturation length (figure 6.14, right panel). Predator growth in body size is in this state extremely slow at small body sizes, but speeds up significantly after crossing a particular size threshold and reaches maximum sizes close to what is possible with surplus food. For brevity, the two community states can hence appropriately be designated as the "stunted-predator" and the "stunted-consumer" state, respectively.

In the stunted-consumer state, the consumer size distribution is hump-shaped, while adult consumers have a low fecundity (figure 6.15, top-left panel). Newborn consumers experience a high mortality right after birth, but predation on consumers larger than 20 mm is negligible (figure 6.15, bottom-left panel), in particular because the density of intermediately sized predators is low. The predator size distribution in this stunted-consumer state is U-shaped, and adult predators have a very high fecundity (figure 6.15, top-right panel). In fact, the total population birth rate of predators in the stunted-consumer state is more than two hundred times larger than in the stunted-predator state with the same values of maximum resource densities. The bottom-right panel of figure 6.15 provides information on the food availability for predators at different body sizes. This graph shows the feeding level of predators as a function of body length, which is defined as the ratio $F(l)/(H_P + F(l))$. The feeding level is a relative measure of the predator functional response, varying from 0 in the absence of any food to 1 at very high food density. Moreover, multiplied by the maximum length at unlimited food l_m, the feeding level indicates the maximum length that predator individuals can reach given their food availability in a particular state. For

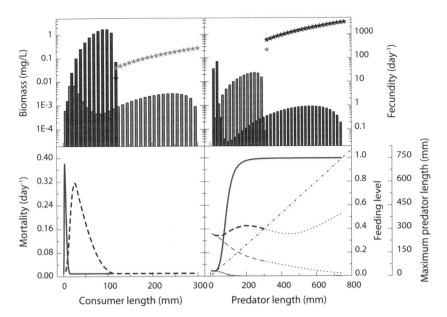

FIGURE 6.15. Characteristics of the consumer and predator population in the equilibrium dominated by juvenile consumers larger than 50 mm ("stunted-consumer") and the equilibrium dominated by juvenile predators larger than 80 mm ("stunted-predator"), which occur in the two resources, size-structured consumer, size-structured predator model based on Kooijman-Metz energy channeling for $R_{max,C} = 5.0$ and $R_{max,P} = 0.05$ mg/L (cf. figure 6.13). *Top*: biomass-size distributions (*bars*) and fecundity-size relations (*stars*) of the consumer (*left*) and predator (*right*) in the stunted-consumer (*dark gray bars and black symbols*) and stunted-predator equilibrium (*light gray bars and gray symbols*). *Bottom-left*: mortality-size relation for consumers in the stunted-consumer (*solid line*) and stunted-predator equilibrium (*dashed line*). *Bottom-right*: relation between feeding level (ingestion rate as fraction of maximum) and size for predators in the stunted-consumer (*solid line*) and stunted-predator equilibrium (*dashed line*). Thick lines represent total ingestion; thin solid or dashed lines refer to resource ingestion alone. Dotted lines represent extensions of the dashed curves to larger predator sizes in the stunted-predator equilibrium. Feeding level is proportional to maximum predator length (*right-most scale*). The diagonal, dashed-dotted line represents the 1:1 relation between predator length and maximum predator length.

comparison, the graph also shows the feeding level that predators would have on the basis of resource feeding only, which is defined as $e^{-l/l_j} R_P/(H_P + F(l))$.

Predators with a body length below 28 mm can feed only on the predator-exclusive resource, as newborn consumers are too big for them to prey on (note that the maximum prey-predator length ratio equals 0.4, see box 6.6).

This length threshold hence separates a resource-feeding niche in predator on-togeny from a predatory niche. In the stunted-consumer state, predators in the resource-feeding niche have a very low feeding level (figure 6.15, bottom-right panel), owing to the low densities of predator-exclusive resource. Predators in the resource-feeding niche are very abundant in this state, as a result of the high reproductive output by adult predators. This high density subsequently leads to intense competition and hence low resource densities. Growth throughout this niche is therefore very slow (figure 6.14, right panel). However, once predators are recruited to the predatory niche, their food availability increases dramati-cally because of the high densities of, in particular, large juvenile consumers. The feeding level of large juvenile and adult predators is close to 1, which causes them to grow rapidly to body sizes just below l_m. The predator popula-tion is in this state thus characterized by a high total population reproduction and a bottleneck in the recruitment to the predatory life stage.

In the stunted-predator equilibrium, the size distribution of the predator pop-ulation is hump-shaped (figure 6.15, top-right panel), whereas the consumer size distribution is U-shaped (figure 6.15, top-left panel). Now adult predators have a low fecundity, whereas fecundity of adult consumers is high. Predation leads to substantial mortality rates over a broad range of consumer body sizes (figure 6.15, bottom-left panel), in particular because of the high densities of large juvenile predators (in the size range of 100–300 mm). This mortality prevents the occurrence of competition among consumers for the consumer-exclusive resource. Equilibrium biomass density of the consumer-exclusive re-source is therefore high, leading to rapid consumer growth and high fecundity. Predators of all body sizes have similar feeding levels. Moreover, predator-exclusive resource makes up roughly 25 percent of the diet of even the largest predators, which implies that the availability of predator-exclusive resource is approximately the same as the availability of consumer prey. In contrast, in the stunted-consumer state, the availability of consumer prey biomass greatly exceeds the resource availability for predators in the predatory niche. Because the total population birth rate of predators is low (more than two hundred times lower than in the stunted-consumer state) and predator growth in body size is relatively rapid, the predator population is in this equilibrium state clearly governed by reproduction control.

Based on the foregoing discussion, we can conclude that the community equilibrium states, which occur in the model with size-structured consumers and size-structured predators analyzed in this section, can be classified as controlled either by slow juvenile predator development or by limited predator reproduc-tion, in a way that is very similar to the model with stage-structured consumers and predators that was analyzed in the previous section. On the basis of the growth patterns in individual life history, we also referred to the equilibrium governed by slow juvenile predator development as a stunted-consumer state, whereas the equilibrium with predator reproduction control we appropriately referred to as a stunted-predator state. Qualitatively, the predictions of both models are the same, despite the much more detailed life history of consumers and predators and the more complicated predatory interaction between consumers and predators in the size-structured model analyzed here. The main difference between the predictions of both models is that in the size-structured model, the bottleneck in juvenile development occurs at the body length where predators can start to feed on the smallest consumers, that is, where they are recruited to the predatory ontogenetic niche. In contrast, in the stage-structured model, predators are assumed to start feeding on consumers after maturation, and the bottleneck hence corresponds to a maturation bottleneck. The size-structured model, however, makes much more detailed predictions about consumer and predator life history in the two community equilibria, that is, mortality, fecundity, and food availability as a function of body length, which provide more insight into the precise mechanisms that stabilize the community in these states.

Figure 6.16 shows the occurrence of the different types of community equilibria in the model with size-structured consumers and predators dependent on the maximum densities of consumer-exclusive and predator-exclusive resource. Even at first sight, the qualitative similarity with the corresponding figure for the stage-structured model (figure 6.12, right panel) is striking. A major quantitative difference is that the maximum densities of consumer-exclusive and predator-exclusive resource are presented on a logarithmic scale in figure 6.16, as opposed to the linear scales used in figure 6.12. Otherwise, figure 6.16 shows, as does figure 6.12, that for low values of $R_{max,C}$ ($R_{max,C} < 0.1$ mg/L), either a unique consumer-only equilibrium occurs, or else a unique coexistence equilibrium governed by slow juvenile predator development (stunted-consumer state) or a unique predator-only equilibrium is established, and that the value of $R_{max,P}$ mostly determines which of these three community equilibria results. For higher values of $R_{max,C}$, the value of $R_{max,P}$ mostly determines which of the following occurs: a unique consumer-only equilibrium; a unique coexistence equilibrium governed by slow juvenile predator development (stunted-consumer state); bistability between two types of coexistence equilibria (stunted-consumer and stunted-predator state); bistability between the coexistence equilibrium governed by slow juvenile predator development (stunted-consumer state) and a stable predator-only state; or, lastly, a predator-only state as the unique outcome of community dynamics.

Taken together, the results of the last two sections on ontogenetic niche shifts in predator life history show that, on the one hand, a niche shift from

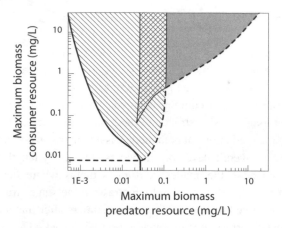

FIGURE 6.16 Combinations of maximum biomass of predator-exclusive and consumer-exclusive resource allowing for persistence of the size-structured predator (*to the right of the thick, solid line*) and the size-structured consumer (*above and to the left of the thick, dashed line*) in the size-structured consumer, size-structured predator model with two resources based on Kooijman-Metz energy channeling (see box 6.5 for model formulation). In the hatched parameter region, a unique, stable two resources-consumer-predator equilibrium occurs. In the gray parameter region, the latter occurs next to a stable two resources-predator equilibrium. In the cross-hatched parameter region, two different, stable coexistence equilibrium states occur next to each other. Other parameters as in figure 6.13.

foraging as a juvenile on a resource with constant productivity to preying as an adult on a self-renewing prey population promotes the occurrence of alternative community equilibria, in which consumers and predators coexist. These states are comparable to the states that we identified for consumer-resource systems and are governed by either development control or by reproduction control of the predator. When predators only partially change their diet over ontogeny and can hence persist in the absence of consumers, they are likely to drive consumers to extinction when the predator population is governed by reproduction control. On the other hand, when governed by development control, a bottleneck occurs in the recruitment to the ontogenetic niche in which predators prey on consumers. This bottleneck is induced intraspecifically by the high reproductive output of adult predators. Rather than hampering persistence of their own population, the bottleneck in predator development benefits persistence of the consumer population, which would otherwise be eradicated as a result of too high predation mortality. In this way, size-specific predators help their prey to persist.

Mixed Interactions

Species exhibiting ontogenetic niche shifts may be feeding at a different trophic level in different phases of their life history and can hence be considered omnivores. This will hold in particular for predator species that start to prey on other species later in life. In the previous chapter we already studied the consequences of this type of ontogenetic diet shift in predator life history for community structure. Predators were assumed to feed as juveniles on a basic resource, while switching partially or completely to preying on consumers as an adult. Juvenile predators and consumers were feeding on different resources, and the two species never overlapped in their resource use. An overlap in resource use between juvenile predators and the prey species they forage on later in life is, however, common among predator species that grow substantially in body size during ontogeny. Predator and prey are then engaged in an interaction that is not exclusively predatory or competitive, but a mixture of both. The consequences of such mixed predation-competition interactions between species are the topic of the current chapter.

Polis, Myers, and Holt (1989) coined the term intraguild predation (IGP) to describe the interaction between two species that compete for the same resource in addition to being engaged in a predator-prey relation. These authors were the first to synthesize the ubiquity of this type of interaction in the animal kingdom. The simplest community in which intraguild predation occurs includes a basic resource, a consumer species that feeds on the basic resource, and a predator species feeding on both resource and consumers. Polis, Myers, and Holt (1989) discuss many examples of such intraguild predation systems, ranging from scorpions, which eat more than fifty species of smaller arachnid and insect predators in addition to the prey of these predators, to leopards, which prey on lions, cheetahs, wild dogs, and spotted hyenas. This simple type of intraguild predation, which we will refer to as basic intraguild predation, is common in many freshwater, marine, and terrestrial food webs (Arim and Marquet 2004; Polis 1991; Polis and Strong 1996). In particular in freshwater and marine environments, predators primarily select prey on the basis of body size as opposed to taxonomic identity (Hall, Cooper, and Werner 1970;

Jennings, Pinnegar, et al. 2001), resulting in intraguild predation interactions between species.

As Polis, Myers, and Holt (1989) also pointed out, however, intraguild predation is often the result of changes in the ontogenetic niche during predator life history and related to an increase in its body size during ontogeny. This type of intraguild predation, which we shall refer to as "life history omnivory" (Pimm and Rice 1987), occurs among many generalist predators, such as spiders, scorpions, copepods, insect predators, amphibians, lizards, and fish. Polis, Myers, and Holt (1989) distinguish various types of life history omnivory, but these all share the feature that the adult, older, or larger individuals of a predator species include into their diet the juveniles of a consumer species, with which they compete for a shared resource when they are younger. When extending their diet with juvenile consumers, the adult predators may continue to feed on the shared resource or drop the latter from their diet altogether (Polis, Myers, and Holt 1989), in which case the competitive interaction and the predatory interaction between the two species are restricted to separate life stages in the predator life history. In the previous chapter we have referred to these two types of ontogenetic diet change as partial and complete ontogenetic niche shifts, respectively, but we did not consider the possibility that consumers and juvenile predators shared the resources that they were foraging. Competition between consumers and juvenile predators hence did not occur.

Intraguild predation can be considered to represent the simplest form of omnivory. Pimm and Lawton (1978) used the interaction matrix between species (i.e., the Jacobian matrix of a system of equations representing the coupled dynamics of these species) to assess the consequences of omnivory for community structure and dynamics. Their analysis suggested that communities with many omnivorous links between species are unlikely to be stable. If a stable equilibrium exists, however, omnivorous links between species may promote equilibrium stability, as perturbations of the community away from equilibrium tend to decay more rapidly in omnivorous food webs. Pimm and Rice (1987) showed that life history omnivory reduced community stability to a much lesser extent than omnivory that occurred within single life stages. Community stability was furthermore reduced if the maturation rate of juveniles varied with the availability of their food.

The theoretical results about omnivory contrast with empirical studies that have documented omnivory to be common in ecological food webs (Arim and Marquet 2004; Polis 1991; Polis and Strong 1996). This contrast has become even more pronounced since the focus of theoretical studies narrowed down to basic intraguild predation (Diehl and Feissel 2000; Holt and Polis 1997; Mylius, Klumpers, et al. 2001). Holt and Polis (1997) showed in a model for basic

IGP that for coexistence of consumers (also referred to as intraguild prey) and intraguild predators to occur at all, the intraguild prey has to be a better competitor for the basic resource than the predator. Such coexistence is, moreover, limited to environments with an intermediate productivity of the basic resource (Diehl and Feissel 2000; Holt and Polis 1997; Mylius, Klumpers, et al. 2001).

In systems with low resource productivity, competition plays the most dominant role, as consumer density is too low to be beneficial for the intraguild predator, eventually leading to predator exclusion. In contrast, in highly productive systems, intraguild predators become too abundant and hence predation too intense for the consumer to persist. Coexistence is therefore maintained only at intermediate productivity levels when competition and predation are by and large balanced (Mylius, Klumpers, et al. 2001). Given that IGP and omnivory are common in natural systems (Arim and Marquet 2004; Polis 1991; Polis and Strong 1996), many subsequent studies have focused on mechanisms that possibly promote coexistence of intraguild prey and predators, such as age-restricted predation or prey life stages invulnerable to predation (Borer 2002; Mylius, Klumpers, et al. 2001; Rudolf and Armstrong 2008); adaptive foraging behavior by intraguild predators (Krivan 2000; Krivan and Diehl 2005); spatial or temporal refuges for the intraguild prey (Amarasekare 2008; Finke and Denno 2006; Janssen, Sabelis, et al. 2007; Okuyama 2008); or additional resources for intraguild prey (Holt and Huxel 2007).

Despite the fact that intraguild predation often results as a consequence of ontogenetic niche shifts and hence corresponds to life history omnivory as opposed to basic IGP, the community consequences of this type of omnivory are known to a much lesser extent. Potential consequences have been postulated and studied experimentally. In particular, predators that grow in body size and compete as juveniles with their future prey may experience a "competitive juvenile bottleneck" when competition by prey reduces juvenile growth rates and hence limit the recruitment to the predatory life stage (Byström, Persson, and Wahlström 1998; Gilliam 1982; Persson and Brönmark 2002a; Persson and Greenberg 1990; Werner and Gilliam 1984). Juvenile bottlenecks have experimentally been shown to occur in zooplankton and fish (Byström, Persson, and Wahlström 1998; Neill 1988).

In contrast, adult predators, when abundant, may then promote juvenile development by cropping down the density of the prey individuals these juveniles compete with (Persson 1988). Together, these effects are also referred to as "cultivation/depensation effects" (Walters and Kitchell 2001). However, until now only few studies have investigated the long-term consequences of life history omnivory for community structure. Mylius, Klumpers, et al. (2001) showed that the addition of an adult prey stage that is invulnerable to predation or a juvenile predator stage that exclusively forages on basic resource did not qualitatively change the model predictions compared with the predictions for basic IGP. Life history omnivory in predator life history has even been shown to decrease the scope of coexistence between consumers and intraguild predators in cases where growth in body size is food-dependent (van de Wolfshaar, de Roos, and Persson 2006). Both these latter studies assume, however, that above a particular body size, the intraguild predators simply add intraguild prey (consumers) to their diet, while continuing to forage on basic resource at essentially the same rate. More recently, Hin, Schellekens, et al. (2011) have analyzed the consequences of predator life history omnivory for community structure in more detail, covering the entire spectrum from basic IGP to life history omnivory, in which predators exhibit a complete ontogenetic niche shift from resource feeding to preying on consumers.

In this chapter we will discuss mixed interactions from a different perspective and take as our point of departure the models for a structured prey, structured predator community with exclusive resources for both prey and predator that were analyzed in the previous chapter. For both the stage-structured biomass model and the size-structured model based on Kooijman-Metz energetics that we discussed in chapter 6, we will investigate how an increasing resource overlap between predators and consumers will change community structure and dynamics. The end point of such increasing resource overlap is a model in which consumers and predators compete for the same shared resource and adult predators feed partially or exclusively on juvenile, small-size consumers. This end point is close to the models analyzed by van de Wolfshaar, de Roos, and Persson (2006) and Hin, Schellekens, et al. (2011), and we will hence discuss our results in the light of those publications.

NICHE OVERLAP BETWEEN STAGE-STRUCTURED PREY AND PREDATORS

The key difference between the stage-structured model discussed in chapter 6 and the one analyzed in this chapter is the fact that both juvenile and adult consumers forage on a resource that is potentially shared with predators as opposed to being used exclusively by consumers. This resource will be indicated with R_S and its maximum density with $R_{\max, S}$. In the model discussed in chapter 6, juvenile predators were assumed to spend all their time foraging on predator-exclusive resource, whereas adult predators foraged on this resource for a fraction ϕ_P of their time. We will here assume that whenever predators feed on basic resources, they spend a fraction θ of that time foraging on the shared resource and a fraction $1 - \theta$ on the predator-exclusive resource. For adult predators, this implies that they forage on the shared resource for a fraction $\phi_P \theta$ of their time and for a fraction $\phi_P(1 - \theta)$ on the predator-exclusive resource. In addition, adult predators prey for a fraction $1 - \phi_P$ of their time on juvenile consumers. Obviously, for $\theta = 0$ there is no niche overlap between consumers and predators, and the model corresponds to the model analyzed in chapter 6. For $\theta = 1$ there is complete niche overlap, and the predator-exclusive resource becomes irrelevant for the community dynamics of consumers and predators. If in addition $\phi_P = 0$, predators go through a complete niche shift at maturation, and the predator hence represents a true life history omnivore. Notice that predators are assumed to spend these fractions of their time foraging on shared and predator-exclusive resource and on predation, irrespective of whether these different food sources are at all present.

The equations describing the biomass model with stage-structured consumers and stage-structured predators are presented in box 7.1, and the default parameters are presented in box 7.2. The assumptions about juvenile and adult predators feeding on two and three different food sources, respectively, complicate the model formulation. As a consequence, the functional response and net-biomass-production rate of juvenile predators depend on both the shared and predator-exclusive resource density (see the definitions of the functions $\omega_{P_J}(R_S, R_P)$, $\psi_{P_J}(R_S, R_P)$, and $\nu_{P_J}(R_S, R_P)$ in box 7.1). Analogously, the functional response and net-biomass-production rate of adult predators also depend, in addition to these two basic resource densities, on juvenile consumer biomass density, as expressed by the functions $\omega_{P_A}(R_S, R_P, C_J)$, $\psi_{P_A}(R_S, R_P, C_J)$, and $\nu_{P_A}(R_S, R_P, C_J)$ in box 7.1. Because the functional response of adult predators depends on the biomass densities of both basic resources and juvenile consumers, the predation mortality experienced by juvenile consumers, $d_{C_J}(P_A)$, is dependent on the density of the shared resource these juveniles feed on themselves. Finally, the ODE for the shared resource dynamics contains additional terms $\psi_{P_J}(R_S, R_P)P_J$ and $\psi_{P_A}(R_S, R_P, C_J)P_A$ that represent the grazing by juvenile and adult predators, respectively, on this resource.

As before, we assume that the mass-specific, maximum ingestion rate for adult consumers is smaller than for juvenile consumers ($q_C = 0.5$, refer to chapter 3). This implies that in the absence of predators, the consumer population equilibrates in a state governed by reproduction control and is hence dominated by adults. The ratio between the size at birth and at maturation is assumed to be the same for both consumers and predators ($z_C = z_P = 0.01$). Predators are assumed to be larger and hence have different mass-specific maximum ingestion, maintenance, and mortality rates. In contrast to chapter 6, however, we assume that on a mass specific basis, the maximum ingestion rate of juvenile

BOX 7.1

BIOMASS MODEL WITH TWO RESOURCES, STAGE-STRUCTURED
PREY, AND STAGE-STRUCTURED PREDATORS

Dynamic equations	Description
$\dfrac{dR_S}{dt} = G_S(R_S) - \omega_{C_J}(R_S)C_J - \omega_{C_A}(R_S)C_A$ $\qquad\qquad - \psi_{P_J}(R_S,R_P)P_J - \psi_{P_A}(R_S,R_P,C_J)P_A$	Shared resource dynamics
$\dfrac{dR_P}{dt} = G_P(R_P) - \omega_{P_J}(R_S,R_P)P_J - \omega_{P_A}(R_S,R_P,C_J)P_A$	Predator-exclusive resource dynamics
$\dfrac{dC_J}{dt} = \nu^+_{C_A}(R_S)C_A - \gamma_C(\nu^+_{C_J},d_{C_J})C_J + \nu_{C_J}(R_S)C_J - d_{C_J}(P_A)C_J$	Juvenile prey dynamics
$\dfrac{dC_A}{dt} = \gamma_C(\nu^+_{C_J},d_{C_J})C_J + (\nu_{C_A}(R_S) - \nu^+_{C_A}(R_S))C_A - \mu_C C_A$	Adult prey dynamics
$\dfrac{dP_J}{dt} = \nu^+_{P_A}(R_S,R_P,C_J)P_A - \gamma_P(\nu^+_{P_J},\mu_P)P_J + \nu_{P_J}(R_S,R_P)P_J - \mu_P P_J$	Juvenile predator dynamics
$\dfrac{dP_A}{dt} = \gamma_P(\nu^+_{P_J},\mu_P)P_J + (\nu_{P_A}(R_S,R_P,C_J) - \nu^+_{P_A}(R_S,R_P,C_J))P_A - \mu_P P_A$	Adult predator dynamics

Function	Expression	Description
$G_i(R_i)$	$\rho(R_{\max,i} - R_i)$	Resources turnover, $i = S, P$
$\omega_{C_J}(R_S)$	$M_C R_S/(H_C + R_S)$	Juvenile consumer intake
$\omega_{C_A}(R_S)$	$q_C M_C R_S/(H_C + R_S)$	Adult consumer intake
$\nu_i(R_S)$	$\sigma_C \omega_i(R_S) - T_C$	Consumer net energy production, $i = C_J, C_A$
$d_{C_J}(P_A)$	$\mu_C + \dfrac{M_{P_A}(1-\phi_P)P_A}{H_P + \phi_P(\theta R_S + (1-\theta)R_P) + (1-\phi_P)C_J}$	Juvenile consumer mortality

(Box 7.1 continued)

Function	Expression	Description
$\gamma_C(v_{C_J}^+, d_{C_J})$	$\left(v_{C_J}^+ - d_{C_J}\right)\Big/\left(1 - z_C^{(1-d_{C_J}/v_{C_J}^+)}\right)$	Juvenile consumer maturation rate
$\psi_{P_J}(R_S, R_P)$	$\dfrac{M_{P_J}\theta R_S}{H_P + \theta R_S + (1-\theta)R_P}$	Shared resource intake by juvenile predators
$\psi_{P_A}(R_S, R_P, C_J)$	$\dfrac{M_{P_A}\phi_P\theta R_S}{H_P + \phi_P(\theta R_S + (1-\theta)R_P) + (1-\phi_P)C_J}$	Shared resource intake by adult predators
$\omega_{P_J}(R_S, R_P)$	$\dfrac{M_{P_J}(1-\theta)R_P}{H_P + \theta R_S + (1-\theta)R_P}$	Juvenile predator exclusive resource intake
$\omega_{P_A}(R_S, R_P, C_J)$	$\dfrac{M_{P_A}\phi_P(1-\theta)R_P}{H_P + \phi_P(\theta R_S + (1-\theta)R_P) + (1-\phi_P)C_J}$	Adult predator exclusive resource intake
$v_{P_J}(R_S, R_P)$	$\sigma_P M_{P_J}\dfrac{\theta R_S + (1-\theta)R_P}{H_P + \theta R_S + (1-\theta)R_P} - T_P$	Juvenile predator net energy production
$v_{P_A}(R_S, R_P, C_J)$ $\quad \sigma_P M_{P_A}\dfrac{\phi_P(\theta R_S + (1-\theta)R_P) + (1-\phi_P)C_J}{H_P + \phi_P(\theta R_S + (1-\theta)R_P) + (1-\phi_P)C_J} - T_P$		Adult predator net energy production
$\gamma_P(v_{P_J}^+, \mu_P)$	$\left(v_{P_J}^+ - \mu_P\right)\Big/\left(1 - z_P^{(1-\mu_P/v_{P_J}^+)}\right)$	Juvenile predator maturation rate

v^+ represents the value of a function v, but restricted to non-negative values.

BOX 7.2

PARAMETERS OF THE STAGE-STRUCTURED BIOMASS MODEL OF BOX 7.1

Parameter	Default value	Unit	Description
Resources			
ρ	0.1	day^{-1}	Turnover rate of resources
$R_{max,i}$	30	mg/L	Maximum density of resource $i = S, P$
Prey			
M_C	$0.1W_A^{-0.25}$	day^{-1}	Mass-specific maximum ingestion rate
H_C	3.0	mg/L	Ingestion half-saturation density
q_C	0.5	—	Adult-juvenile ingestion ratio
T_C	$0.01W_A^{-0.25}$	day^{-1}	Mass-specific maintenance rate
σ_C	0.5	—	Conversion efficiency
z_C	0.01	—	Newborn-adult size ratio
μ_C	$0.0015W_A^{-0.25}$	day^{-1}	Background mortality rate
Predators			
M_{P_J}	$0.06W_P^{-0.25}$	day^{-1}	Mass-specific maximum ingestion rate by juvenile predators
M_{P_A}	$0.1W_P^{-0.25}$	day^{-1}	Mass-specific maximum ingestion rate by adult predators
H_P	3.0	mg/L	Ingestion half-saturation density
T_P	$0.01W_P^{-0.25}$	day^{-1}	Mass-specific maintenance rate
σ_P	0.5	—	Conversion efficiency
z_P	0.01	—	Newborn-adult size ratio
μ_P	$0.0015W_P^{-0.25}$	day^{-1}	Background mortality rate
θ	0.0-1.0	—	Foraging effort on shared resource
ϕ_P	0.0 or 0.25	—	Adult foraging effort on resource

W_A and W_P represent average adult body weight (in grams) of consumer and predator, respectively.

predators is lower than the maximum ingestion rate of adult predators (cf. the values $M_{P_J} = 0.06 \cdot W_P^{-0.25}$ and $M_{P_A} = 0.1 \cdot W_P^{-0.25}$ in box 7.2). This assumption implies that juvenile predators are less effective feeders on the shared resource than juvenile consumers. Juvenile consumers can just satisfy their maintenance requirements when

$$R_S = \frac{H_C T_C}{\sigma_C M_C - T_C} = 0.75 \text{ mg/L,}$$

as can be derived from the condition $v_{C_j}(R_S) = 0$.

In cases where juvenile predators forage exclusively on the shared resource ($\theta = 1$), the minimum density of shared resource at which their maintenance requirements are just satisfied is given by a similar expression (derived from $v_{P_j}(R_S, R_P) = 0$ for $\theta = 1$):

$$R_S = \frac{H_P T_P}{\sigma_P M_{P_j} - T_P} = 1.5 \text{ mg/L.}$$

The lower value for the mass-specific maximum ingestion rate of juvenile predators ($M_{P_j} = 0.06 \cdot W_P^{-0.25}$) makes that this subsistence density is higher for juvenile predators than for juvenile consumers.

We investigate the model predictions dependent on the productivities of the shared and predator-exclusive resource (i.e., the maximum densities $R_{\max,S}$ and $R_{\max,P}$, respectively), the extent of the niche overlap θ between consumers and predators, and the extent of the niche shift ϕ_P that predators go through at maturation.

Complete Niche Shifts of Adult Predators

Figure 7.1 shows the changes in equilibrium biomass of juvenile and adult predators, juvenile and adult consumers, and the two resources as a function of the maximum density of shared resource for three different levels of resource overlap between consumers and predators (from left to right: $\theta = 0$, 0.25, and 0.5, respectively). The left panels in figure 7.1 show a bifurcation diagram that is virtually identical to the diagram we encountered before in the left panels of figure 6.10, except for tiny quantitative differences that result from the assumption of a lower maximum ingestion rate of juvenile predators ($M_{P_j} = 0.06 \cdot W_P^{-0.25}$). The following discussion of this bifurcation diagram hence repeats the discussion of figure 6.10 in the previous chapter, although with less detail.

The consumer-only equilibrium is stable against predator invasion for maximum shared resource densities up to $R_{\max,S} \approx 14$ mg/L (figure 7.1, left panels). Predator population growth is at this invasion threshold limited most by food availability for adult predators, and the population is hence reproduction-controlled. Because the consumer population is reproduction-controlled as well, overcompensation in juvenile biomass occurs in response to (predation) mortality imposed on juvenile consumers, leading to a positive feedback between

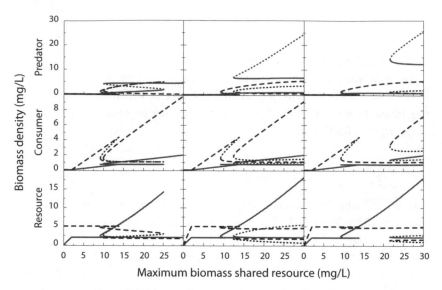

Figure 7.1. Changes in equilibrium biomass of predators (*top*), consumers (*middle*), and resources (*bottom*) with increasing maximum density of the shared resource ($R_{max,S}$) in the biomass model with stage-structured consumers and stage-structured predators, when predators experience a complete niche shift from feeding on the shared and predator-exclusive resource as a juvenile to preying exclusively on juvenile consumers as an adult ($\phi_P = 0$). Juvenile and adult consumers forage on the shared resource only ($q_C = 0.5$). Niche overlap between consumers and juvenile predators increases from left to right (*left*: $\theta = 0$; *middle*: $\theta = 0.25$; *right*: $\theta = 0.5$). Solid lines refer to biomass of shared resource, juvenile consumer, and juvenile predator biomass; dashed lines to predator-exclusive resource, adult consumer biomass, and adult predator biomass. Solid and dashed lines indicate stable equilibria; unstable equilibria are indicated with dotted lines. $R_{max,P} = 5$ mg/L, $z_C = 0.01$, and $z_P = 0.01$. Otherwise, default parameters apply as listed in box 7.2 for a prey and predator with body size $W_A = 0.0001$ and $W_P = 0.01$ gram, respectively.

(adult) predator density and the availability of their prey. As a result, the predator population exhibits an emergent Allee effect such that the curve representing the predator-consumer coexistence equilibrium starts off at the predator invasion threshold in the direction over lower $R_{max,S}$ values. The initial curve segment represents unstable equilibria (saddle points) that separate the stable consumer-only equilibria from stable coexistence equilibria. The curve reaches a lower limit point, bending toward higher $R_{max,S}$ values, at $R_{max,S} \approx 9.2$ mg/L, which corresponds to the predator persistence threshold, as it is the lowest $R_{max,S}$ value that allows for a stable equilibrium with predators present.

In the stable coexistence equilibrium at these low $R_{max,S}$ values, the predator population is still governed by reproduction control and dominated by adults, because the density of juvenile consumers is limited compared with the availability of predator-exclusive resource (cf. the middle-left and bottom-left panels in figure 7.1). For $R_{max,S}$ values exceeding 10 mg/L, however, an alternative predator-consumer coexistence equilibrium exists, in which the predator population at equilibrium is governed by development control and hence dominated by juveniles. The density of predator-exclusive resource is significantly lower in this development-controlled equilibrium. The predator-reproduction-controlled and predator-development-controlled coexistence equilibria occur as alternative stable states for $R_{max,S}$ values between 10 and 25 mg/L. In combination with the emergent Allee effect, this implies that for a range of $R_{max,S}$ values between 10 and 14 mg/L, three stable community equilibria occur: a consumer-only equilibrium, a coexistence equilibrium controlled by slow predator development, and a coexistence equilibrium controlled by limited predator reproduction. For $R_{max,S}$ values exceeding 25 mg/L, the coexistence equilibrium governed by slow predator development is the only stable state of the community. With increasing maximum densities of shared resource, we thus consecutively find a region with a unique consumer-only equilibrium ($2 < R_{max,S} < 9.2$ mg/L), a small region with bistability between a consumer-only equilibrium and a reproduction-controlled coexistence equilibrium ($9.2 < R_{max,S} < 10$ mg/L), a region with three stable community equilibria ($10 < R_{max,S} < 14$ mg/L), a region with a stable reproduction-controlled and a stable development-controlled coexistence equilibrium ($14 < R_{max,S} < 25$ mg/L), and ultimately, a unique, coexistence equilibrium controlled by slow predator development ($R_{max,S} > 25$ mg/L). As also discussed in the previous chapter, for some parameter regions, limit cycles may occur around the equilibrium controlled by limited predator reproduction, but the amplitude of these limit cycles is so small that we will not discuss them further in this chapter.

In both types of coexistence equilibria, governed by either predator development control or by predator reproduction control, the consumer population is always dominated by adults, owing to the lower mass-specific, maximum ingestion rate that we have assumed for adult consumers (figure 7.1, left-middle panel). In coexistence with predators that are governed by reproduction control, increases in the maximum density of shared resource $R_{max,S}$ hardly change the equilibrium biomass density of both juvenile and adult consumers. Only when the predator population becomes controlled by juvenile development does an increase in maximum shared resource density lead to increases in juvenile and adult consumer biomass in equilibrium as well as in shared resource biomass (left-middle and left-bottom panels in figure 7.1).

Increasing niche overlap between consumers and predators hardly affects the predator invasion and persistence thresholds. Irrespective of whether predators forage 25 percent or 50 percent of their time on the shared resource, the consumer-only equilibrium remains stable against predator invasion for $R_{max,S}$ values below 14 mg/L (cf. the left, middle, and right panels in figure 7.1). Also, the lowest $R_{max,S}$ value for which the emergent Allee effect occurs and predators can hence persist remains constant at approximately 9.2 mg/L with increasing niche overlap. As explained above, the predator population is mostly governed by limited reproduction at its invasion threshold as well as in the coexistence equilibrium that occurs as a result of the emergent Allee effect. Under these conditions, predator population growth is hardly influenced by juvenile predator development, which is rapid, owing to the high density of predator-exclusive resource ($R_P = R_{max,P} = 5$ mg/L in figure 7.1). This explains the limited influence of the extent to which juveniles forage on shared resource, given that adult predators do not forage on basic resources at all. As the only influence of the niche overlap between consumer and predators on the coexistence equilibrium in which the predator is governed by reproduction control, this state remains a stable equilibrium of the community at high values of $R_{max,S}$ even if the niche overlap is minimal. Only when niche overlap is completely absent is there an upper limit to the occurrence of the predator-reproduction-controlled coexistence equilibrium ($R_{max,S} \approx 25$ mg/L when $\theta = 0$).

Increasing niche overlap between consumers and predators therefore mostly affects the occurrence of the coexistence equilibrium in which the predator population is governed by development control. In this coexistence equilibrium, juvenile predators experience such low food availability that they are sensitive to the increased competition by consumers for the shared resource, which results from the increase in niche overlap. As a consequence, the threshold $R_{max,S}$ value above which the predator-development-controlled coexistence equilibrium occurs, increases: from 10 mg/L without niche overlap ($\theta = 0$) to 12 and 22 mg/L for 25 percent and 50 percent niche overlap, respectively ($\theta = 0.25$ and 0.5, respectively; left, middle, and right panels in figure 7.1). For 25 percent niche overlap (figure 7.1, middle panels) and increasing maximum densities of the shared resource, a region with a unique consumer-only equilibrium ($2 < R_{max,S} < 9.2$ mg/L) is hence followed by a region with bistability between a consumer-only equilibrium and a reproduction-controlled coexistence equilibrium ($9.2 < R_{max,S} < 12$ mg/L), a small region with three stable community equilibria ($12 < R_{max,S} < 14$ mg/L), and finally a region with a stable reproduction-controlled and a stable development-controlled coexistence equilibrium ($R_{max,S} > 14$ mg/L). For 50 percent niche overlap (figure 7.1, right panels), increasing maximum densities of the shared resource subsequently result in a region with a unique consumer-only equilibrium ($2 < R_{max,S} < 9.2$ mg/L), a region with bistability between a consumer-only equilibrium and a predator-reproduction-controlled coexistence equilibrium ($9.2 < R_{max,S} < 14$ mg/L), a region with a unique coexistence equilibrium governed by predator reproduction control ($14 < R_{max,S} < 22$ mg/L), and finally a region with bistability between a stable reproduction-controlled and a stable development-controlled coexistence equilibrium ($R_{max,S} > 22$ mg/L).

Figure 7.2 shows for the same three levels of resource overlap between consumers and predators (from left to right: $\theta = 0$, 0.25, and 0.5, respectively) the changes in equilibrium biomass of juvenile and adult predators, juvenile and adult consumers, and the two resources as a function of the maximum density of predator-exclusive resource, $R_{max,P}$. Without niche overlap ($\theta = 0$, figure 7.2, left panels), the bifurcation pattern is qualitatively similar to the one shown in the right panels of figure 6.10. Only minor quantitative differences occur, owing to the assumption of a lower maximum ingestion rate of juvenile predators. Predators can invade the consumer-only equilibrium for $R_{max,P}$ values exceeding 1.8 mg/L. At this predator invasion threshold, population growth of the predator is limited most by juvenile development, owing to the low density of the predator-exclusive resource. Because of the mortality imposed by adult predators on juvenile consumers, overcompensation in juvenile consumer biomass occurs (see the left-middle panel in figure 7.2), but this increase in food availability for adult predators hardly affects the predator population as a whole, as reproduction is not limiting population growth. An emergent Allee effect hence does not occur. For $R_{max,P}$ values just above the predator invasion threshold, the predator population is mostly controlled by development and dominated by juveniles. For values of $R_{max,P}$ exceeding 4.8 mg/L, however, a second coexistence equilibrium becomes feasible, which is dominated by adult predators and governed by reproduction control. The bistability between the development-controlled and reproduction-controlled coexistence equilibrium occurs for a broad range of $R_{max,P}$ values between 4.8 mg/L and 35 mg/L. For higher $R_{max,P}$ values, the reproduction-controlled equilibrium is the only stable state of the community.

In the coexistence equilibrium governed by predator development control, increases in the maximum density of the predator-exclusive resource increase juvenile, adult, and total predator biomass, as well as juvenile consumer biomass and shared resource density in equilibrium (figure 7.2, left panels). Adult consumer biomass decreases with increasing $R_{max,P}$ values, whereas the density of predator-exclusive resource remains constant. In contrast, in the coexistence equilibrium controlled by predator reproduction only, the equilibrium density of predator-exclusive resource increases with increasing $R_{max,P}$ values, because

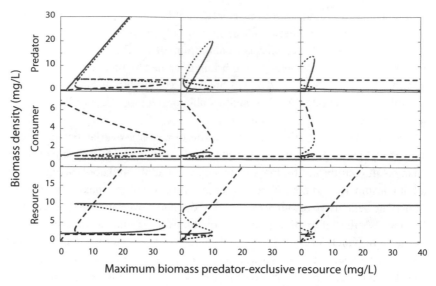

FIGURE 7.2. As in figure 7.1, but for increasing maximum density of the predator-exclusive resource ($R_{max, P}$) with $R_{max, S} = 20$ mg/L.

the availability of juvenile consumers as prey for adult predators becomes the limiting factor for predator population growth. Juvenile and adult predator biomass, juvenile and adult consumer biomass, and the density of shared resource in equilibrium stay virtually constant. In both types of coexistence equilibria, however, the consumer population is always dominated by adults and regulated by reproduction control.

Increasing niche overlap has only minor effects on the threshold value of $R_{max, P}$ above which predators can invade the consumer-resource equilibrium: for 25 percent and 50 percent niche overlap ($\theta = 0.25$ and 0.5, respectively; figure 7.2, middle and right panels), this threshold value decreases to 1.7 and 1.6 mg/L, respectively. In contrast, the threshold values of $R_{max, P}$ above which the coexistence equilibrium governed by predator reproduction control is feasible and above which it is the only stable state of the community, decrease much more strongly with increasing niche overlap. For 25 percent niche overlap between predators and consumers (figure 7.2, middle panels), the coexistence equilibrium controlled by predator reproduction already occurs for $R_{max, P}$ values exceeding 0.9 mg/L. This coexistence equilibrium is hence feasible for $R_{max, P}$ values below the threshold value that allows for invasion of predators into the consumer-only equilibrium, and bistability occurs between a consumer-only equilibrium and a coexistence equilibrium controlled by predator reproduction for $R_{max, P}$ values between 0.9 and 1.7 mg/L.

Superficially, this bistability resembles an emergent Allee effect. It is, however, not caused by overcompensation in juvenile consumer biomass in response to the increase in predation mortality, as juvenile consumer biomass is consistently lower in coexistence with a predator population governed by reproduction control. Instead, the coexistence of consumers and predators for these low $R_{max,P}$ values is based on a significant increase in the shared resource biomass in equilibrium (figure 7.2, middle-bottom panel), which comes about because of the low densities of juvenile and adult consumer biomass that occur in a predator-reproduction-controlled equilibrium. By imposing mortality on consumers, adult predators hence increase the food availability for, and facilitate the development of, juvenile predators and thus make persistence possible. Notice that adult predators do not directly crop down the adult consumers that are the main and most abundant competitors of their juveniles, but they indirectly decrease adult consumer biomass in equilibrium by limiting consumer maturation through foraging on juvenile consumers. With larger niche overlap between predators and consumers, the threshold $R_{max,P}$ value above which the coexistence equilibrium controlled by predator reproduction is feasible even drops to 0. Hence, with 50 percent niche overlap (figure 7.2, right panels), a stable coexistence equilibrium controlled by predator reproduction occurs as an alternative to a stable consumer-only equilibrium for all $R_{max,P}$ values below the predator invasion threshold at 1.6 mg/L.

The threshold $R_{max,P}$ value above which the coexistence equilibrium controlled by predator development is no longer feasible, and the coexistence equilibrium controlled by predator reproduction is the only stable community state, also decreases with increasing niche overlap between consumers and predators. With 25 percent and 50 percent niche overlap, this threshold occurs at $R_{max,P} = 11$ and 4.6 mg/L, respectively (figure 7.2, middle and right panels). However, this threshold value always occurs at a higher $R_{max,P}$ value than the threshold at which the predator can invade a consumer-only equilibrium. Qualitatively, the part of the bifurcation diagram pertaining to the consumer-only equilibrium and the coexistence equilibrium controlled by predator development does not change with increasing niche overlap; only the range of $R_{max,P}$ values decreases, over which the coexistence equilibrium controlled by predator development occurs.

With 25 percent niche overlap (figure 7.2, middle panels) between consumers and predators and increasing values of the maximum density of predator-exclusive resource, a region of $R_{max,P}$ values with the consumer-only equilibrium as unique, stable community equilibrium ($R_{max,P} < 0.9$ mg/L) is hence followed by a range of $R_{max,P}$ values ($0.9 < R_{max,P} < 1.7$ mg/L) with bistability between the consumer-only equilibrium and a coexistence equilibrium controlled by limited predator reproduction; a range of $R_{max,P}$ values ($1.7 < R_{max,P} < 11$ mg/L) with bistability between two types of coexistence equilibrium, either controlled by limited predator reproduction or by slow predator development; and finally by a range of $R_{max,P}$ values (larger than 11 mg/L), for which the coexistence equilibrium governed by predator reproduction control is the only stable community equilibrium. Qualitatively, this pattern is the same for 50 percent niche overlap (figure 7.2, right panels) except that the range with low $R_{max,P}$ values, for which the consumer-only state is the only stable community equilibrium, has disappeared and bistability occurs between this consumer-only equilibrium and the coexistence equilibrium governed by predator reproduction control up to the predator invasion threshold (at $R_{max,P} = 1.6$ mg/L).

Figure 7.3 shows for the three levels of resource overlap between consumers and predators (from left to right: $\theta = 0$, 0.25, and 0.5, respectively) the different regions of parameters in which the different number and types of equilibria can occur as a function of the maximum density of the shared and predator-exclusive resource. The left panel is very similar to the left panel of figure 6.12, showing the occurrence of the emergent Allee effect as by-and-large independent of the maximum density of the predator-exclusive resource. The bistability between a stable consumer-only equilibrium and (one or two) stable coexistence equilibria that results from the emergent Allee effect occurs over a range of maximum densities of the shared resource, which is located below the threshold marking invasion of the predator population into the consumer-only equilibrium. The two different types of coexistence equilibria occur as alternative stable states in a wedge-shaped parameter region located around the diagonal. Hence, a unique coexistence equilibrium only occurs when either the shared or the predator-exclusive resource is in significantly shorter supply than the other. This latter bistability region partly overlaps, however, with the region in which the emergent Allee effect occurs. In the overlap region there are three stable community states, as discussed above: a consumer-only equilibrium and two coexistence equilibria.

Increasing niche overlap between consumers and predators has hardly any effect on the line denoting the invasion threshold of predators into the consumer-only equilibrium (figure 7.3, middle and right panel). Nor does increasing niche overlap significantly change the parameter region in which the emergent Allee effect occurs. Increasing niche overlap mainly affects the parameter region in which two different types of coexistence equilibria occur, tilting and rotating the entire region in a counterclockwise manner. As a consequence, the changes in the number and type of equilibria occurring are more pronounced for high values of the maximum density of shared and predator-exclusive resource than for low values. Furthermore, once the left-most boundary of this parameter region, representing the threshold value of maximum resource densities above

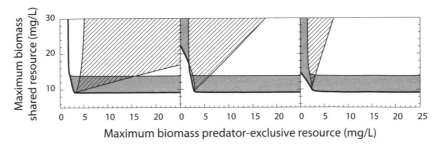

FIGURE 7.3. Combinations of maximum biomass of predator-exclusive and shared resource (*above/to the right of the thick, solid line*) with possible persistence of stage-structured predators experiencing a complete niche shift from feeding on shared and predator-exclusive resource as a juvenile to preying on juvenile consumers as an adult ($\phi_P = 0$). Niche overlap between consumers and juvenile predators increases from left to right (*left*: $\theta = 0$; *middle*: $\theta = 0.25$; *right*: $\theta = 0.5$). Consumers can persist for all parameters shown. In the gray parameter region, a stable resources-consumer-predator equilibrium occurs next to a stable resources-consumer state. In the hatched parameter region, two different resources-consumer-predator equilibrium states occur next to each other. Other parameters as in figures 7.1 and 7.2.

which the coexistence equilibrium governed by predator reproduction control is feasible, intersects with the predator-invasion boundary, a new parameter region arises (figure 7.3, middle panel, top-left gray region). In this parameter region, at low values of the maximum predator-exclusive resource density to the left of the vertical leg of the predator-invasion boundary, a stable coexistence equilibrium controlled by limited predator reproduction occurs next to the stable consumer-only equilibrium. For 50 percent niche overlap between consumers and predators (figure 7.3, right panel), this latter parameter region has even connected with the parameter region in which the emergent Allee effect occurs. Hence, for large niche overlap, the parameter region for which predators can coexist with consumers completely encompasses the parameter region for which they can invade a consumer-only equilibrium.

Predators can hence persist in coexistence with consumers even when juvenile predators only have the shared resource to feed on. Figure 7.4 shows for $R_{max,P} = 0.0$ mg/L and 50 percent niche overlap the possible equilibria at different maximum densities of the shared resource. Irrespective of the maximum shared resource density, predators can never invade a consumer-only equilibrium, as the density of shared resource in this equilibrium is too low for juvenile predators to grow and mature. Hence, the consumer outcompetes the juvenile predators. Predators and consumers can, however, coexist in stable

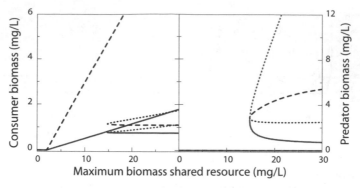

FIGURE 7.4. Changes in equilibrium biomass of consumers (*left*) and predators (*right*) with increasing maximum density of the shared resource ($R_{max,\,s}$) in the biomass model with stage-structured consumers and stage-structured predators when predator-exclusive resource is absent ($R_{max,\,P} = 0.0$ mg/L) and predators experience a complete niche shift from feeding on the shared resource as a juvenile to preying exclusively on juvenile consumers as an adult ($\phi_P = 0$). Juvenile and adult consumers forage on the shared resource only ($q_C = 0.5$). $\theta = 0.5$, which implies that juvenile predators forage for only 50 percent of their time on shared resource, despite the absence of predator-exclusive resource. Solid lines refer to juvenile consumer and predator biomass; dashed lines to adult consumer and predator biomass. Solid and dashed lines indicate stable equilibria; unstable equilibria are indicated with dotted lines. Throughout, $z_C = 0.01$ and $z_P = 0.01$. Otherwise, default parameters apply as listed in box 7.2 for a prey and predator with body size $W_A = 0.0001$ and $W_P = 0.01$ gram, respectively.

equilibrium for $R_{max,\,S}$ values exceeding 15 mg/L (in the case of 50 percent niche overlap; with larger overlap this $R_{max,\,S}$ threshold shifts to even lower values). In coexistence, the predation by adult predators is crucial for predator persistence, as it increases the food availability for, and facilitates the development of, their juveniles and thus makes persistence possible. As discussed above, however, adults do not directly crop down the adult consumers that are the main competitors of their juveniles. Instead, they limit consumer recruitment to the adult stage through foraging on juvenile consumers.

The bifurcation diagram shown in figure 7.4 resembles the results presented by Hin, Schellekens, et al. (2011), who studied the changes in community structure of a basic resource, an unstructured consumer population, and a structured predator population as a function of the extent of the ontogenetic niche shift exhibited by predators at maturation. In the case of a complete niche shift from resource feeding by juvenile predators to predation on consumers by adult predators, a consumer-only equilibrium and a coexistence equilibrium occurred as alternative stable community states. We can hence conclude that the results in figure 7.4 are independent of whether the consumer population is structured or unstructured and predators feed on all consumers equally or on juvenile consumers only.

PARTIAL NICHE SHIFTS OF ADULT PREDATORS

Figure 7.5 shows for three different levels of niche overlap between consumers and predators (from left to right: $\theta = 0$, 0.2, and 0.4, respectively) the changes in equilibrium biomass of juvenile and adult predators, juvenile and adult consumers, and the two resources as a function of the maximum density of shared resource in the case where predators only go through a partial niche shift at maturation, such that adult predators still forage on resources for 25 percent of their time ($\phi_P = 0.25$) and for the remainder prey on consumers. These diagrams represent the bifurcation pattern for a low value of the maximum predator-exclusive resource density $R_{max,P}$. At this low $R_{max,P}$ value, the changes attributable to the increasing niche overlap can be illustrated best.

However, it also implies that the bifurcation pattern without niche overlap does not resemble the pattern that we encountered in the previous chapter for the same case without niche overlap (cf. the left panels of figure 7.5 with those in figure 6.11). Without niche overlap we now find a unique, stable community equilibrium for all values of the maximum density of shared resource. For $R_{max,S}$ values below 2.6 mg/L, this is a stable predator-only equilibrium. Even though it is low, the maximum density of predator-exclusive resource is sufficiently high for predators to persist in the absence of consumers, owing to the partial niche shift that predators go through. For $R_{max,S}$ values above 2.6 mg/L, a unique coexistence equilibrium is the only stable community state, in which adults dominate the consumer population, whereas juveniles dominate the predator population, showing that the predator population is governed by development control. Increases in the maximum density of the shared resource translate into increases in both juvenile and adult consumer density, but they hardly change the equilibrium densities of juvenile and adult predators or shared and predator-exclusive resource. This coexistence equilibrium corresponds to the equilibrium state that occurs in the left panels of figure 6.11 for high values of the consumer-exclusive resource.

In contrast, the alternative coexistence equilibrium that is shown in the left panels of figure 6.11 at intermediate values of the maximum consumer-exclusive resource density does not show up in the left panels of figure 7.5 as a consequence of the low $R_{max,P}$ value. At these low $R_{max,P}$ values food availability

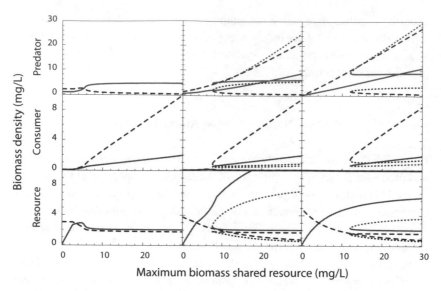

FIGURE 7.5. Changes in equilibrium biomass of predators (*top*), consumers (*middle*), and resources (*bottom*) with increasing maximum density of the shared resource ($R_{max, s}$) in the biomass model with stage-structured consumers and stage-structured predators, when predators experience a partial niche shift from feeding on the shared and predator-exclusive resource as a juvenile to feeding on these resources in addition to preying on juvenile consumers as an adult ($\phi_P = 0.25$). Juvenile and adult consumers forage on the shared resource only ($q_C = 0.5$). Niche overlap between consumers and predators increases from left to right (*left*: $\theta = 0$; *middle*: $\theta = 0.2$; *right*: $\theta = 0.4$). Solid lines refer to biomass of shared resource, juvenile consumer biomass, and juvenile predator biomass; dashed lines to predator-exclusive resource, adult consumer biomass, and adult predator biomass. Solid and dashed lines indicate stable equilibria; unstable equilibria are indicated with dotted lines. $R_{max, P} = 5$ mg/L, $z_C = 0.01$, and $z_P = 0.01$. Otherwise, default parameters apply as listed in box 7.2 for a prey and predator with body size $W_A = 0.0001$ and $W_P = 0.01$ gram, respectively.

for juvenile predators is therefore never sufficient to relax the development control and to switch to a stable equilibrium, in which the predator population is governed by reproduction control. Bistability between two coexistence equilibria, one governed by predator reproduction control and the other by predator development control, does occur for higher values of $R_{max, P}$ values, similar to the situation shown in figure 6.11, but for those parameters the influence of the niche overlap is not as illustrative.

With a niche overlap between consumers and predators of 20 percent ($\theta = 0.2$, figure 7.5, middle panel), alternative stable community equilibria do occur, in which the predator population is either governed by reproduction control or by development control. For $R_{max,S}$ values below 3.5 mg/L, a predator-only equilibrium is the only stable community equilibrium. For a range of $R_{max,S}$ values above this threshold $(3.5 < R_{max,S} < 8.7$ mg/L), the consumer can invade the predator-only equilibrium, resulting in a coexistence equilibrium in which the predator is governed by reproduction control and dominated by adults. Increases in $R_{max,S}$ translate into higher densities of both juvenile and adult predators in this reproduction-controlled equilibrium and hence into increased predation. As a consequence, consumers are excluded from this equilibrium for $R_{max,S}$ values exceeding 8.7 mg/L. At $R_{max,S}$ values above this latter threshold, a stable predator-only equilibrium once again occurs, in which the predator population is governed by reproduction control. In this predator-only equilibrium, the densities of juvenile and adult predators as well as the shared resource density increase with increasing $R_{max,S}$, whereas the equilibrium biomass density of predator-exclusive resource decreases. For $R_{max,S}$ values exceeding 7.6 mg/L, however, a second coexistence equilibrium is feasible, in which the predator population is governed by development control. This coexistence equilibrium corresponds to the state with a bottleneck in juvenile predator development, as discussed in the previous chapter, and to the state occurring at all $R_{max,S}$ values in the absence of niche overlap (cf. the left and middle panels in figure 7.5). Owing to the juvenile predator bottleneck, the consumer manages to coexist with the predator even at high maximum densities of the shared resource. As was the case in the absence of niche overlap, in this equilibrium increases in $R_{max,S}$ mainly lead to increasing densities of both juvenile and adult consumers.

In summary, for 20 percent niche overlap and increasing values of $R_{max,S}$ (figure 7.5, middle panels), a parameter region with a predator-only equilibrium as a unique, stable community state $(R_{max,S} < 3.5$ mg/L) is followed by a region with a unique coexistence equilibrium, in which the predator population is governed by reproduction control $(3.5 < R_{max,S} < 7.6$ mg/L); a small region of parameters with bistability between two different coexistence equilibria, one governed by reproduction control and the other by development control $(7.6 < R_{max,S} < 8.7$ mg/L); and eventually bistability between a predator-only equilibrium with predator reproduction control and a coexistence equilibrium with predator development control $(R_{max,S} > 8.7$ mg/L).

The changes in equilibrium biomass densities with increasing $R_{max,S}$ values and a niche overlap of 40 percent between consumers and predators are largely similar to the changes in the case of 20 percent niche overlap (cf. middle and right panels of figure 7.5), except that consumers can no longer invade the predator-only equilibrium governed by reproduction control. Irrespective of the maximum density of shared resource, a stable predator only equilibrium occurs, in which the predator population is governed by reproduction control and dominated by adults. In this predator-only equilibrium, increasing values of $R_{\text{max},S}$ translate as before into increasing densities of predators and shared resource and decreasing densities of predator-exclusive resource. For $R_{\text{max},S}$ values exceeding 12 mg/L, however, a coexistence equilibrium governed by predator development control occurs as an alternative stable community state. In this latter equilibrium, increasing values of $R_{\text{max},S}$ only increase the densities of juvenile and adult consumers. Hence, with 40 percent niche overlap ($\theta = 0.4$), only two distinct parameter regions are found: one with a unique and stable predator-only equilibrium ($R_{\text{max},S} < 12$ mg/L) and one with a predator-only equilibrium governed by predator reproduction control and a coexistence equilibrium, in which the predator population is governed by development control, as alternative stable community states ($R_{\text{max},S} > 12$ mg/L).

In the previous chapter we already pointed out that the changes in equilibrium biomass of predators, consumers, and resources as a function of maximum predator-exclusive resource density $R_{\text{max},P}$ in the case of a partial ontogenetic niche shift are reminiscent of the pattern observed for a complete predator niche shift, except that consumers go extinct at high values of $R_{\text{max},P}$. The same conclusion holds in the case of niche overlap between consumers and predators (figure 7.6). In the absence of niche overlap ($\theta = 0$, figure 7.6, left panels), predators can invade the stable consumer-only equilibrium for $R_{\text{max},P}$ values exceeding 1.8 mg/L, leading to a coexistence equilibrium in which the predator is mainly governed by limited juvenile development. For $R_{\text{max},P}$ values exceeding 9.6 mg/L, a second stable predator-consumer coexistence equilibrium occurs, in which the predator population is controlled by limited reproduction. Because in this latter equilibrium the predator population is dominated by adults, the predation mortality imposed on juvenile consumers is high and increases with further increases of $R_{\text{max},P}$.

As a consequence, consumers cannot persist with this reproduction-controlled predator population for $R_{\text{max},P}$ values exceeding a threshold value that occurs at $R_{\text{max},P} \approx 10$ mg/L. Above this consumer-extinction threshold, a stable predator-only equilibrium occurs, in which the population is governed by reproduction control and dominated by adult predators. Bistability between two different types of coexistence equilibria, one controlled by limited predator reproduction and the other by limited predator development, hence occurs over only a very small range of $R_{\text{max},P}$ values between 9.6 and 10 mg/L. Above the consumer-extinction threshold, bistability between the stable coexistence equilibrium with predators governed by development control and the stable predator-only equilibrium occurs over a much larger range of $R_{\text{max},P}$ values ($10 < R_{\text{max},P} < 33$ mg/L). In the coexistence equilibrium with predator

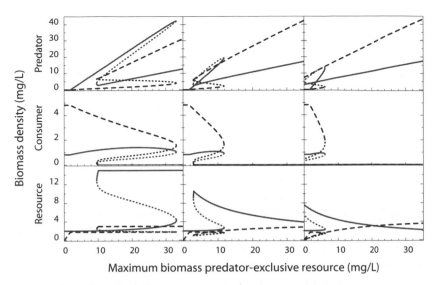

FIGURE 7.6. As in figure 7.5, but for increasing maximum density of the predator-exclusive resource ($R_{max,P}$) with $R_{max,S} = 15$ mg/L.

development control, increases in maximum predator-exclusive resource density lead to substantial increases in biomass density of juvenile predators, but much slower increases in biomass of adult predators. The mortality that predators impose on juvenile consumers hence also increases only slowly, which allows for the consumer persistence up to the consumer-persistence threshold that occurs at $R_{max,P} = 33$ mg/L. In contrast, in the coexistence equilibrium governed by predator reproduction control, as well as in the predator-only equilibrium above the consumer-extinction threshold, increases in $R_{max,P}$ mainly increase adult predator biomass in equilibrium and decrease or only slowly increase (in the predator-only equilibrium) the equilibrium biomass of juvenile predators. The biomass of predator-exclusive resource in the predator-only equilibrium remains constant with increasing values of $R_{max,P}$, as it is the only food source sustaining the predator population.

Increasing niche overlap hardly affects the threshold $R_{max,P}$ value above which predators can invade the consumer-only equilibrium: with 40 percent niche overlap it is only slightly smaller ($R_{max,P} = 1.7$ mg/L) than in the absence of niche overlap ($R_{max,P} = 1.8$ mg/L). As was also observed for the case of a complete niche shift in predator life history, increasing niche overlap between consumers and predators mainly affects the lowest $R_{max,P}$ value, for which an equilibrium with predators governed by reproduction control occurs, and the highest $R_{\text{max},P}$ value, for which an equilibrium with predators governed by development control exists. In addition, increasing niche overlap also decreases the threshold $R_{\text{max},P}$ value at which consumers are driven to extinction by the predator population that is controlled by limited reproduction (figure 7.6). In fact, the changes in this consumer-extinction threshold parallel the changes in the lowest $R_{\text{max},P}$ value for which an equilibrium with predators governed by reproduction control occurs, and the two threshold values invariably occur close to each other.

With 20 percent niche overlap between consumers and predators ($\theta = 0.2$, figure 7.6, middle panels), the coexistence equilibrium with the predator population controlled by limited reproduction occurs for $R_{\text{max},P}$ values exceeding 2.9 mg/L, while consumer density in this equilibrium becomes 0 for an only slighter larger value ($R_{\text{max},P} = 3.0$ mg/L). Hence, for $R_{\text{max},P}$ values exceeding this latter threshold, a stable predator-only equilibrium governed by reproduction control is found instead of a coexistence equilibrium. With increasing niche overlap, both these threshold $R_{\text{max},P}$ values, marking the occurrence of a coexistence equilibrium with predator reproduction control and extinction of consumers from this equilibrium, decrease, such that with 40 percent niche overlap, a coexistence equilibrium with predator reproduction control is no longer feasible and only the stable predator-only equilibrium occurs for all positive values of $R_{\text{max},P}$ (figure 7.6, right panels).

With 40 percent niche overlap, therefore, a new type of bistability occurs at low $R_{\text{max},P}$ values between a consumer-only equilibrium that is stable against invasion by predators and a stable predator-only equilibrium from which consumers are excluded owing to intense predation on juvenile consumers. Finally, the largest $R_{\text{max},P}$ value for which a coexistence equilibrium, in which the predator population is governed by development control, occurs decreases with increasing niche overlap from $R_{\text{max},P} = 33$ mg/L in the absence of niche overlap to 11.8 and 6.2 mg/L with 20 percent and 40 percent niche overlap, respectively (figure 7.6, left to right panels).

Figure 7.7 shows for the three levels of resource overlap between consumers and predators (from left to right: $\theta = 0$, 0.2, and 0.4 , respectively) the different regions of parameters in which the different number and types of equilibria can occur as a function of the maximum density of the shared and predator-exclusive resource. The left panel is qualitatively similar to the right panel of figure 6.12, showing in particular the substantial parameter region in which a stable predator-only equilibrium controlled by limited predator reproduction, and a coexistence equilibrium in which predators are controlled by limited development occur as alternative stable community states. Compared to the right panel in figure 6.12, the parameter region with bistability between two

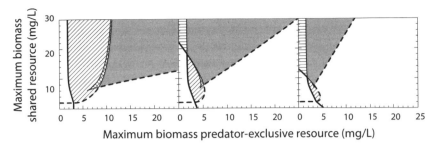

FIGURE 7.7. Combinations of maximum biomass of predator-exclusive and shared resource (*to the right of the thick, solid line*) with possible persistence of stage-structured predators experiencing a partial niche shift from feeding on shared and predator-exclusive resource as a juvenile to feeding on these resources in addition to preying on juvenile consumers as an adult ($\phi_p = 0.25$). Niche overlap between consumers and predators increases from left to right (*left*: $\theta = 0$; *middle*: $\theta = 0.2$; *right*: $\theta = 0.4$). Consumers can persist for parameters above and to the left of the thick, dashed line. In the diagonally hatched parameter region, a unique, stable resources-consumer-predator equilibrium occurs. In the gray parameter region, the latter occurs next to a stable resources-predator equilibrium. In the horizontally hatched parameter region, bistability occurs between a stable resources-consumer equilibrium and a stable resources-predator equilibrium, whereas in the cross-hatched parameter region, two different stable resources-consumer-predator equilibrium states occur next to each other. Other parameters as in figures 7.5 and 7.6.

different types of coexistence equilibria is considerably smaller, as a consequence of the assumption that juvenile predators have a smaller mass-specific maximum ingestion rate than juvenile consumers. In fact, this parameter region is so small that we will altogether ignore the bistability between two types of coexistence equilibria in the following.

By and large, the parameter space falls apart into two parts, as was also pointed out in the discussion of figure 6.12: for low values of the maximum shared resource density $R_{max,S}$ (approximately $R_{max,S} < 10$ mg/L), no bistability occurs, and the community either ends up in a stable consumer-only equilibrium, a stable coexistence equilibrium in which the predator population is controlled by slow juvenile development, or a predator-only equilibrium with the predator population controlled by limited reproduction. Which of these three community equilibria occurs and is stable mostly depends on the value of the maximum predator-exclusive resource density $R_{max,P}$. For larger values of $R_{max,S}$, the maximum predator-exclusive resource density $R_{max,P}$ mostly determines whether community dynamics lead to a stable consumer-only equilibrium or a unique and stable coexistence equilibrium controlled by limited predator development, or whether bistability occurs between this coexistence equilibrium and a stable predator-only equilibrium.

The middle and right panels of figure 7.7 show that increasing niche overlap between consumers and predators up to 40 percent hardly affects the location of the boundary marking the invasion of predators into the consumer-only equilibrium. This boundary does shift toward lower $R_{max,P}$ values, but this effect only becomes apparent with more niche overlap between consumers and predators (see below). As the most prominent effect of the increase in niche overlap, both the boundary marking the extinction of consumers from the coexistence equilibrium controlled by limited predator reproduction and the threshold marking the maximum $R_{max,P}$ value for which consumers can coexist with the predator population controlled by limited development shift to lower $R_{max,P}$ values. These changes are larger for larger values of $R_{max,S}$, which leads to the impression that the consumer-extinction boundary and the parameter region with bistability between a stable predator-only equilibrium and a stable coexistence equilibrium, in which the predator is controlled by limited development, tilt and rotate in a counterclockwise manner. Once the consumer-extinction boundary intersects with the predator-invasion boundary (figure 7.7, middle and right panel), a new parameter region, with bistability between a stable consumer-only and a stable predator-only equilibrium, arises to the left of the predator-invasion boundary and above the consumer-extinction boundary. In this parameter region, the two species hence mutually exclude each other; either the consumer drives the predator extinct by intense competition with juvenile predators, or predation by adult predators on juvenile consumers leads to extinction of the consumer population.

Predator persistence is hence possible even in the absence of a resource that they can exclusively feed on ($R_{max,P} = 0$), just as we observed in the previous section for the case where predators go through a complete niche shift at maturation. In contrast to the case of a complete niche shift, however, this persistence does not translate into coexistence with consumers when predators experience only a partial ontogenetic niche shift, because adult predators do not need consumers to be able to reproduce. Instead, with a partial ontogenetic niche shift, the consumer is excluded from any equilibrium in which predators can persist if only the shared resource is available. Figure 7.8 illustrates this for three different levels of niche overlap between consumers and predators ($\theta = 0.5$, 0.9, and 1.0, respectively). Notice that the concept of niche overlap is redundant in this case, as it is assumed that there is no predator-exclusive resource available ($R_{max,P} = 0$). Because of the partial niche shift, adult predators spend 25 percent of their time foraging on basic resources while preying on consumers during the remaining 75 percent of their time. Of the amount of

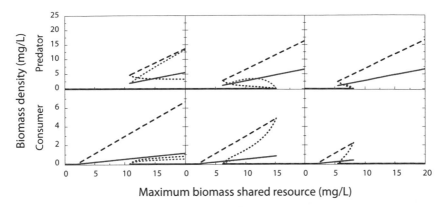

FIGURE 7.8. Changes in equilibrium biomass of predators (*top panels*) and consumers (*bottom panels*) with increasing maximum density of the shared resource ($R_{max, S}$) in the biomass model with stage-structured consumers and stage-structured predators when predator-exclusive resource is absent ($R_{max, P} = 0.0$ mg/L) and predators experience a partial niche shift from feeding on the shared resource as a juvenile to feeding on resource in addition to preying on juvenile consumers as an adult ($\phi_P = 0.25$). Juvenile and adult consumers forage on the shared resource only ($q_C = 0.5$). $\theta = 0.5$ (*left*), 0.9 (*middle*), and 1.0 (*right*), which implies that predators spend 50 percent, 90 percent, and 100 percent of their time, respectively, foraging on the shared resource, irrespective of the absence of the predator-exclusive resource. Solid lines refer to juvenile consumer and predator biomass; dashed lines to adult consumer and predator biomass. Solid and dashed lines indicate stable equilibria; unstable equilibria are indicated with dotted lines. Throughout, $z_C = 0.01$ and $z_P = 0.01$. Otherwise, default parameters apply as listed in box 7.2 for a prey and predator with body size $W_A = 0.0001$ and $W_P = 0.01$ gram, respectively.

time that juvenile and adult predators forage on basic resources, they spend only part on foraging for the shared resource, which equals 50 percent and 90 percent in the cases $\theta = 0.5$ and $\theta = 0.9$, respectively. Only for $\theta = 1.0$ do juvenile and adult predators spend all their time and 25 percent of their time, respectively, foraging on the shared resource. In the absence of the predator-exclusive resource, the parameter θ hence changes the foraging efficiency of predators on the shared resource and thereby their competitiveness relative to consumers.

The left panel of figure 7.8 shows that for $\theta = 0.5$ the foraging efficiency of predators on the shared resource is not sufficient to allow for invasion of the consumer-only equilibrium in the absence of an exclusive resource for the predator. In the stable consumer-only equilibrium, which occurs and remains stable for all values of $R_{max,S}$ we have investigated, adult consumers dominate the population. The consumer population is governed by reproduction control because of the lower maximum ingestion rate of adult consumers ($q_C = 0.5$). For $R_{max,S}$ values exceeding 11 mg/L, however, a predator-only equilibrium occurs, which is stable against consumer invasion and in which consumers hence fail to persist. This predator-only equilibrium exists for $R_{max,S}$ values below 11 mg/L as well (not shown in figure 7.8). More specifically, the predator-only equilibrium is feasible for all $R_{max,S}$ values exceeding 6 mg/L. However, for $R_{max,S}$ values between 6 and 11 mg/L, the predator density and consequently the predation pressure on consumers are too low to prevent consumer invasion. Following invasion, consumers in turn always outcompete the predators and drive them to extinction.

For $\theta = 0.9$ and $\theta = 1.0$, the foraging efficiency of predators on the shared resource is sufficiently high to allow for their invasion into the consumer-only equilibrium above some threshold value of $R_{max,S}$. For $\theta = 0.9$, this threshold occurs at $R_{max,S} = 15$ mg/L and at $R_{max,S} = 8$ mg/L for $\theta = 1.0$. Predators can invade the consumer-only equilibrium for all $R_{max,S}$ values above this threshold. Invasion by predators invariably leads to takeover and exclusion of consumers from the community, just as invasion by consumers into the predator-only equilibrium, which is possible at low $R_{max,S}$ values, leads to takeover by consumers and exclusion of predators. As a consequence, in the absence of an exclusive resource of predators, a consumer-only equilibrium is the unique, stable state of the community at low $R_{max,S}$ values, whereas for higher $R_{max,S}$ values, bistability occurs between a stable consumer-only and a stable predator-only equilibrium (figure 7.8). If the foraging efficiency of predators on the shared resource is sufficiently high, the parameter region with bistability has an upper limit, and the predator-only equilibrium is the unique, stable community state at high $R_{max,S}$ values.

This bifurcation pattern resembles the pattern found by van de Wolfshaar, de Roos, and Persson (2006), who studied an intraguild predation system consisting of a shared resource, a size-structured consumer population, and a size-structured predator population, the individuals of which preyed on consumers following an interaction function that depended on both the consumer and the predator body sizes. The model was more detailed than the model analyzed in this section and specifically tailored to the interaction between the freshwater fish species roach (*Rutilus rutilus*) and European perch (*Perca fluviatilis*). Nonetheless, model predictions were largely similar to the results presented here: increasing productivity of the shared resource resulted in a productivity range with consumer dominance and predator exclusion, followed by a productivity range with bistability between a community consisting of only consumers and an alternative community consisting only of predators, and finally a productivity range for which consumers were always excluded from the system and predators monopolized the community. Van de Wolfshaar, de Roos, and Persson (2006) also found that invasion by predators into the consumer-only equilibrium always led to takeover and exclusion of consumers, whereas consumers would take over the community and exclude predators from it for all conditions that allowed consumers to invade a community with only predators. Generally, we can hence expect that in a system with mixed competition-predation interactions between consumers and predators monopolization of the community by either consumers or predators is the rule, if foraging on basic resources alone is sufficient for persistence of predators.

NICHE OVERLAP BETWEEN SIZE-STRUCTURED
PREY AND PREDATORS

In the previous chapter we analyzed the interaction between a size-structured consumer population and a size-structured population of predators that at birth were of a similar size as the consumers. Predators started out feeding on an exclusive resource after birth but gradually switched to preying on consumers while increasing in size. More specifically, the fraction of time that predators spent feeding on basic resource followed an exponentially declining function of predator body length, e^{-l/l_j}. In this expression l_j refers to the length at which predators mature. For the remaining fraction of their time, $1 - e^{-l/l_j}$, predators are assumed to prey on consumers. To account for niche overlap between consumers and predators, we will assume in this chapter that a fraction θ of the time that predators feed on basic resources ($= e^{-l/l_j}$) is spent on foraging on the shared resource, which we will indicate by R_S. Consumers are assumed to feed only on this shared resource, which hence replaces the consumer-exclusive resource from the previous chapter (all occurrences of R_C in the model from the previous chapter are hence replaced by R_S). Predators are assumed to spend the remaining fraction $1 - \theta$ of the time that they feed on basic resources searching for the predator-exclusive resource, which will be indicated by R_P just as in the previous chapter. Total food availability for a predator of length l thus equals a weighted sum of both resources and prey availability:

$$F(l) = e^{-l/l_j}\left(\theta R_S + (1-\theta)R_p\right) + (1 - e^{-l/l_j})C(l), \qquad (7.1)$$

in which $C(l)$ is the same weighted integral over the body size distribution of the consumer population as used in chapters 4 and 6 (see equation (4.6)). This expression for the food availability for a predator of length l simplifies to its counterpart in the previous chapter (equation (6.3)) for $\theta = 0$. The parameter θ

hence scales the niche overlap between consumers and predators in the same way as it did in the stage-structured models discussed above.

The assumption about niche overlap leads to only minor changes in the formulation of the model compared with the formulation presented in the previous chapter. In addition to the altered expression above for the food availability, the expression for the feeding rate of predators on the predator-exclusive resource is changed, because the predators only forage for a fraction $1 - \theta$ of their resource feeding time on this resource:

$$I_P(R_P, F(l), l) = M_P e^{-l/l_j} \frac{(1-\theta)R_P}{H_P + F(l)} l^2.$$

Furthermore, the differential equation for the consumer-exclusive resource in the model from the previous chapter is replaced by a similar differential equation for the shared resource density. However, this differential equation includes an additional term,

$$\int_{l_b}^{l_m} I_S(R_S, F(l), l) p(t, l) \, dl,$$

which represents the feeding of predators on this shared resource. In this expression the function $I_S(R_S, F(l), l)$ represents the feeding rate of a predator of length l on the shared resource, which is defined in an analogous way as its feeding rate on predator-exclusive resource:

$$I_S(R_S, F(l), l) = M_P e^{-l/l_j} \frac{\theta R_S}{H_P + F(l)} l^2.$$

Otherwise, all model equations and model functions, summarized once again in box 7.3, are completely identical to the model presented in the previous chapter. Apart from the additional parameter θ, representing the niche overlap between consumers and predators, all model parameters are also identical to the parameters of the model presented in the previous chapter. For default values of all consumer-related and predator-related parameters, we hence refer to box 3.7 in chapter 3 and to box 6.6 in the previous chapter, respectively. We analyzed the number and types of equilibria that occur in the model for various combinations of maximum density of the shared and predator-exclusive resource, indicated by $R_{max,S}$ and $R_{max,P}$, respectively, using the same approach as presented in the previous chapter. The only other resource-related parameter, the resource turnover rate ρ, is taken equal to its default value ($\rho = 0.1$, see box 3.7 in chapter 3) for both the shared and the predator-exclusive resource. As in the previous chapter, we determined the stability properties of the computed equilibria on the basis of general principles from bifurcation theory

BOX 7.3

PREDATOR-PREY MODEL WITH TWO RESOURCES,
BASED ON KOOIJMAN-METZ ENERGETICS

Dynamic equations	Description
$\dfrac{\partial p(t,l)}{\partial t} + \dfrac{\partial g_P(F(l),l)p(t,l)}{\partial l} = -\mu_P p(t,l)$	Predator length-distribution dynamics
$g_P(F(l_b),l_b)p(t,l_b) = \displaystyle\int_{l_j}^{l_m} b_P(F(l),l)p(t,l)\,dl$	Predator population reproduction
$\dfrac{\partial c(t,\ell)}{\partial t} + \dfrac{\partial g_C(R_S,\ell)c(t,\ell)}{\partial \ell} = -d_C(P(\ell))c(t,\ell)$	Consumer length-distribution dynamics
$g_C(R_S,\ell_b)c(t,\ell_b) = \displaystyle\int_{\ell_j}^{\ell_m} b_C(R_S,\ell)c(t,\ell)\,d\ell$	Consumer population reproduction
$\dfrac{dR_P}{dt} = \rho(R_{\max,P} - R_P) - \displaystyle\int_{l_b}^{l_m} I_P(R_P,F(l),l)p(t,l)\,dl$	Predator-exclusive resource dynamics
$\dfrac{dR_S}{dt} = \rho(R_{\max,S} - R_S) - \displaystyle\int_{\ell_b}^{\ell_m} I_C(R_S,\ell)c(t,\ell)\,d\ell$ $- \displaystyle\int_{l_b}^{l_m} I_S(R_S,F(l),l)p(t,l)\,dl$	Shared resource dynamics

Function	Expression	Description
$I_C(R_S,\ell)$	$M_C R_S/(H_C + R_S)\ell^2$	Consumer resource ingestion
$g_C(R_S,\ell)$	$\gamma_C(\ell_m R_S/(H_C + R_S) - \ell)$	Consumer growth rate in length
$b_C(R_S,\ell)$	$\alpha_C R_S/(H_C + R_S)\ell^2$	Consumer fecundity
$d_C(P(\ell))$	$\mu_C + P(\ell)$	Consumer death rate
$w_C(\ell)$	$\beta_C \ell^3$	Consumer weight-length relation
$T(\ell,l)$	$\begin{cases} \dfrac{\ell - \delta l}{(\varphi - \delta)l} & \text{for } \delta l \leq \ell < \varphi l \\[2mm] \dfrac{\varepsilon l - \ell}{(\varepsilon - \varphi)l} & \text{for } \varphi l \leq \ell \leq \varepsilon l \\[2mm] 0 & \text{otherwise} \end{cases}$	Vulnerability of consumer at length ℓ for predator of length l
$C(l)$	$\displaystyle\int_{\delta l}^{\varepsilon l} T(\ell,l)w_C(\ell)c(t,\ell)\,d\ell$	Prey availability for predator of length l

(*Box 7.3 continued*)

Function	Expression	Description
$F(l)$	$e^{-l/l_j}(\theta R_S + (1-\theta)R_P) + (1-e^{-l/l_j})C(l)$	Total food availability for predator of length l
$P(\ell)$	$M_P \displaystyle\int_{l/\varepsilon}^{l/\delta} \dfrac{(1-e^{-l/l_j})T(\ell,l)}{H_P + F(l)} l^2 p(t,l)\,dl$	Predation mortality of consumer at length ℓ
$I_P(R_P,F(l),l)$	$M_P e^{-l/l_j}(1-\theta)R_P/(H_P+F(l))l^2$	Predator-exclusive resource ingestion
$I_S(R_S,F(l),l)$	$M_P e^{-l/l_j}\theta R_S/(H_P+F(l))l^2$	Shared resource ingestion by predators
$g_P(F(l),l)$	$\gamma_P\big(l_m F(l)/(H_P+F(l)) - l\big)$	Predator growth rate in length
$b_P(F(l),l)$	$\alpha_P F(l)/(H_P+F(l))l^2$	Predator fecundity
$w_P(l)$	$\beta_P l^3$	Predator weight-length relation

(see Kuznetsov 1995, for an introduction), as analytical methods for stability analysis of equilibria in a size-structured population model are not available yet. Hence, when a curve representing a coexistence equilibrium intersects a curve representing a consumer-only (or predator-only) equilibrium, the latter changes stability and the coexistence equilibrium is assumed unstable (stable) if the corresponding curve starts off in the branching point toward parameter values for which the consumer-only (or predator-only) equilibrium is stable (unstable). Second, at a (limit) point of an equilibrium curve where it doubles back on itself, the stability of the equilibrium changes from a saddle point to a stable node or vice versa.

Figure 7.9 shows the equilibrium biomass of predators, consumers and resources, as a function of maximum density of the shared (left panels) and

FIGURE 7.9. Changes in equilibrium biomass of shared (*bottom, solid line*) and predator-exclusive resource (*bottom, dashed line*), juvenile consumers smaller (*middle, dashed-dotted line*) and larger than 50 mm (*middle, solid line*), adult consumers (*middle, dashed line*), juvenile predators smaller (*top, dashed-dotted line*) and larger than 80 mm (*top, solid line*), and adult predators (*top, dashed line*) with increasing maximum density of shared resource ($R_{\text{max},S}$) (*left panels*)

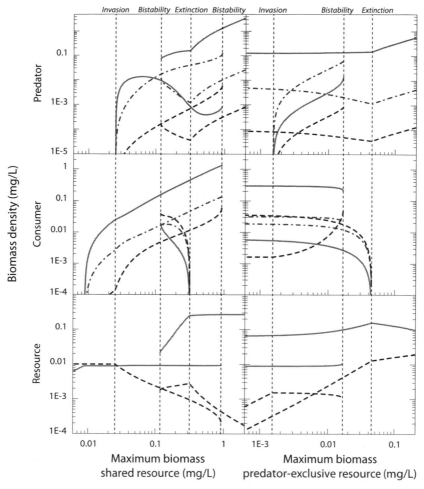

and predator-exclusive resource ($R_{\mathrm{max},\,P}$) (*right panels*) in the size-structured consumer, size-structured predator model with two resources based on Kooijman-Metz energy channeling and 10 percent foraging effort of predators on the shared resource ($\theta = 0.1$; see box 7.3 for model formulation). Only (potentially) stable equilibria are indicated. In the left panels, two different equilibria with consumers and predators present occur for ($R_{\mathrm{max},\,S}$) values between 0.12 and 0.32 mg/L, whereas bistability between an equilibrium with and without consumers occurs for $R_{\mathrm{max},\,S}$ values between 0.32 and 0.92 mg/L. In the right panels, two different equilibria with predators and consumers occur for $R_{\mathrm{max},\,P}$ values between $1.5 \cdot 10^{-3}$ and 0.017 mg/L, whereas bistability between an equilibrium with and without predators occurs for $R_{\mathrm{max},\,P}$ values below $1.5 \cdot 10^{-3}$ mg/L. These threshold values are indicated with dashed vertical lines. $R_{\mathrm{max},\,P} = 0.01$ mg/L (*left panels*) and $R_{\mathrm{max},\,S} = 0.2$ mg/L (*right panels*). Otherwise, default parameters apply, as given in boxes 3.7 and 6.6.

predator-exclusive resource (right panels) for the model with size-structured predators and size-structured consumers in the case of 10 percent niche overlap between consumers and predators ($\theta = 0.1$). The corresponding bifurcation diagrams in the absence of niche overlap ($\theta = 0$) were presented in figure 6.13 in the previous chapter, to which we refer for comparison. For both the consumer and the predator we again present the biomass densities of small juveniles, large juveniles, and adults. Only the curve segments representing stable equilibria, as determined on the basis of the principles discussed above, are shown.

The diagram in figure 7.9 looks very different from its counterpart without niche overlap (figure 6.13) in the previous chapter. We have indeed found that the change from no niche overlap to the limited niche overlap between consumers and predators ($\theta = 0.1$) used in figure 7.9 dramatically changes the arrangement of the various community equilibria, but that further increases in niche overlap ($\theta > 0.1$) have relatively little effect. Hence, we have chosen to present results for the lowest value of the niche overlap that is representative for all values of θ. Despite the differences between figure 7.9 and its counterpart without niche overlap (figure 6.13), the same four types of community equilibria can occur: either a consumer-only or a predator-only equilibrium or one of two types of coexistence equilibria, which we referred to as the stunted-consumer and stunted-predator coexistence equilibrium. These coexistence equilibria were dominated by juvenile consumers larger than 50 mm and juvenile predators larger than 80 mm, respectively. In the previous chapter we already discussed in detail the characteristics of these different types of coexistence equilibria, which we will hence not repeat here. Rather, the focus in this chapter will be on how the niche overlap between consumers and predators changes the occurrence of the four types of community equilibria and the bistability between them as a function of the maximum densities of shared and predator-exclusive resource.

Compared with the corresponding bifurcation diagram in figure 6.13 (left panels), the bifurcation diagram in figure 7.9 is constructed for a value of the maximum predator-exclusive resource density ($R_{max,P} = 0.01$ mg/L) that is not sufficient for persistence of the predators in the absence of consumers. Hence, for low values of $R_{max,S}$ in the left panels of figure 7.9, neither consumers nor predators persist. A consumer-only equilibrium occurs as the unique stable community state for $R_{max,S}$ values between $8.9 \cdot 10^{-3}$ and 0.025 mg/L. In this equilibrium, the consumer population is dominated by juvenile consumers larger than 50 mm, as is the case for all consumer-only equilibria that we encountered. For $R_{max,S}$ values exceeding 0.025 mg/L, predators can invade this consumer-only equilibrium, resulting in a coexistence equilibrium in which both the consumer and the predator population are dominated by large juveniles. The dominance of the predator population by juvenile predators larger than 80 mm results from the fact that overall food availability for predators of all body sizes is low for these values of $R_{max,S}$, such that both the consumer and the predator population are regulated in a bottom-up process by food supply.

With increasing values of $R_{max,S}$, however ($R_{max,S}$ larger than 0.1 mg/L roughly), this coexistence equilibrium changes in character to the stunted-consumer coexistence equilibrium, in which the community is dominated by juvenile consumers larger than 50 mm and the predator population is dominated by juvenile individuals smaller than 80 mm. For $R_{max,S}$ larger than 0.12 mg/L, the alternative type of coexistence equilibria becomes feasible, which is dominated by juvenile predators larger than 80 mm. Both the stunted-consumer and the stunted-predator equilibrium occur as alternative stable community states for $R_{max,S}$ values between 0.12 and 0.32 mg/L. For $R_{max,S}$ exceeding 0.32 mg/L, consumers are excluded from the stunted-predator equilibrium as a consequence of the high predation pressure that in particular the large juvenile predators impose on small consumers (see the discussion of this issue in the previous chapter). As a consequence, for $R_{max,S}$ values between 0.32 and 0.92 mg/L, bistability occurs between the stunted-consumer coexistence equilibrium and the predator-only equilibrium that resulted from extinction of consumers from the stunted-predator coexistence equilibrium. The coexistence equilibrium dominated by juvenile consumers larger than 50 mm does not exist any longer for $R_{max,S}$ larger than 0.92 mg/L. For high $R_{max,S}$ values, the predator-only equilibrium is hence the only stable community state.

The two major differences in the bifurcation diagram resulting from the niche overlap between consumers and predators are (1) that with increasing maximum density of the shared resource, consumers are driven to extinction by the stunted-predator population and (2) that the stunted-consumer coexistence equilibrium does not exist for high values of $R_{max,S}$. Without niche overlap, these two effects occurred with increases in the maximum density of the predator-exclusive resource (see the right panels of figure 6.13) but not with increases in the consumer-exclusive resource. Notice, though, that without niche overlap, the effects did occur in a qualitatively similar manner: high predation pressure by the stunted-predator population drove consumers to extinction from the stunted-predator equilibrium, whereas the stunted-consumer equilibrium ceased to exist because it reached a limit point. The fact that in the case of niche overlap predators also feed on the shared resource, even though it is to a rather limited extent, causes increases in the maximum density of the shared resource to take over the role of increases in the maximum density of the predator-exclusive resource in the absence of niche overlap.

More specifically, the extinction of consumers from the stunted-predator coexistence equilibrium results from the fact that the high maximum densities of the shared resource can support a predator population that imposes too high a predation pressure for consumers to persist, irrespective of the maximum density of the predator-exclusive resource. The disappearance of the stunted-consumer coexistence equilibrium results from the fact that the increases in maximum density of the shared resource relax the bottleneck in the development of small juvenile predators, resulting in increased recruitment to the large juvenile predator size classes and consequently to extinction of consumers. This lifting of the bottleneck in small juvenile predator development could only come about because of increases in maximum predator-exclusive resource density in the absence of niche overlap.

Compared with the changes that occur with increases in the maximum shared resource density, the changes in equilibrium biomass of resources, consumers, and predators with increasing maximum densities of the predator-exclusive resource in the case of niche overlap between consumers and predators (figure 7.9, right panels) are qualitatively more similar to the patterns observed without niche overlap (cf. the right panels of figure 6.13). For low values of the maximum density of predator-exclusive resource, a stable consumer-only equilibrium occurs that is dominated by juvenile consumers larger than 50 mm ($R_{max,P} < 1.5 \cdot 10^{-3}$ mg/L). This equilibrium can be invaded by predators for $R_{max,P}$ values exceeding $1.5 \cdot 10^{-3}$ mg/L, giving rise to a stunted-consumer coexistence equilibrium, in which the consumer population is dominated by juveniles larger than 50 mm while the predator population is dominated by juveniles smaller than 80 mm. This stunted-consumer coexistence equilibrium reaches a limit point and hence ceases to exist at $R_{max,P} = 0.017$ mg/L, which also occurred in the absence of niche overlap between consumers and predators but at a $R_{max,P}$ value that was more than two orders of magnitude higher (cf. the right panels of figures 6.13 and 7.9).

A coexistence equilibrium dominated by juvenile predators larger than 80 mm occurs over a broader range of $R_{max,P}$ values and is even possible with $R_{max,P} = 0$ mg/L. Consumers are driven to extinction from this stunted-predator equilibrium when $R_{max,P}$ is increased beyond 0.044 mg/L. This curve of stunted-predator coexistence equilibria is qualitatively similar to the corresponding curve in the absence of niche overlap. The main difference resulting from the niche overlap is that this curve extends down to zero maximum density of the predator-exclusive resource, whereas without niche overlap the stunted-predator coexistence equilibrium ceased to occur below a threshold $R_{max,P}$ value ($R_{max,P} = 0.026$ mg/L in the right panels of figure 6.13). The fact that predators also feed on the shared resource, even though to a limited extent, allows for their persistence at zero density of the predator-exclusive resource. Hence, at low maximum densities of the predator-exclusive resource, bistability occurs between a stable consumer-only equilibrium and a stunted-predator coexistence equilibrium ($R_{max,P} < 1.5 \cdot 10^{-3}$ mg/L). For $R_{max,P}$ values between $1.5 \cdot 10^{-3}$ and 0.017 mg/L, bistability occurs between two different coexistence equilibria, one dominated by large juvenile consumers, the other by large juvenile predators. This range of $R_{max,P}$ values is followed by a range in which the stunted-predator coexistence equilibrium is the unique, stable state of the community ($0.017 < R_{max,P} < 0.044$ mg/L), and ultimately by a range with a stable predator-only equilibrium ($R_{max,P} > 0.044$ mg/L) when predation pressure is too high for consumers to coexist with the stunted-predator population.

The two major changes resulting from the niche overlap between consumers and predators are therefore (1) the decrease to zero of the lowest $R_{max,P}$ value for which a stunted-predator coexistence equilibrium can occur and (2) the decrease of the highest $R_{max,P}$ value for which a stunted-consumer coexistence equilibrium can occur to below the consumer extinction threshold. In the absence of niche overlap, consumer extinction from the stunted-predator coexistence equilibrium occurred at a lower value of $R_{max,P}$ than the highest $R_{max,P}$ value for which a stunted-consumer coexistence equilibrium was feasible.

Figure 7.10 shows as a function of the maximum density of both the shared and predator-exclusive resource the parameter regions for which the different types of equilibria occur in the case of 10 percent niche overlap between consumers and predators ($\theta = 0.1$). The bifurcation diagrams presented in the left and right panels of figure 7.9 are cross-sections through this range spanned by the two parameters $R_{max,S}$ and $R_{max,P}$ for $R_{max,P} = 0.01$ mg/L and $R_{max,S} = 0.2$ mg/L, respectively. Comparison of figure 7.10 with the corresponding figure for the case without niche overlap (figure 6.16 in the previous chapter) shows that the entire top part of the diagram at high values of the maximum shared resource density has rotated in a counterclockwise direction. The boundary marking the invasion threshold of predators into the consumer-only equilibrium curves toward lower $R_{max,P}$ values for high values of $R_{max,S}$, reaching $R_{max,P} = 0$ for $R_{max,S} = 2.8$ mg/L. This is analogous to the effect of increasing niche overlap between consumers and predators in the stage-structured biomass model discussed in the previous section. In the latter model, the boundary marking invasion of the predator into the consumer-only equilibrium was hardly affected by 40 percent niche overlap between consumers and predators ($\theta - 0.4$, figure 7.7, right panel). However, figure 7.8 showed that with 90 percent and 100 percent niche overlap between consumers and predators, the invasion boundary of the predator into the consumer-only equilibrium reaches $R_{max,P} = 0$ mg/L at $R_{max,S} = 15$ and $R_{max,S} = 8$ mg/L, respectively.

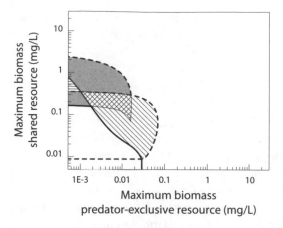

FIGURE 7.10. Combinations of maximum biomass of predator-exclusive and shared resource allowing for persistence of the size-structured predator (*to the right of the thick, black solid line*) and the size-structured consumer (*parameter region enclosed* in *the thick, black dashed line*) in the size-structured consumer, size-structured predator model with two resources based on Kooijman-Metz energy channeling and 10 percent foraging effort of predators on the shared resource ($\theta = 0.1$) (see box 7.3 for model formulation). In the diagonally hatched parameter region, a unique, stable resources-consumer-predator equilibrium occurs. In the lower and upper gray parameter regions, the latter occurs next to a stable resources-consumer equilibrium and a stable resources-predator equilibrium, respectively. In the horizontally hatched parameter region, bistability occurs between a stable resources-consumer equilibrium and a stable resources-predator equilibrium, whereas in the cross-hatched parameter region, two different stable coexistence equilibrium states occur next to each other. The thick, dark-gray solid and dashed lines represent the continuation of the boundaries marking the invasion threshold of predators into the consumer-only equilibrium and consumers into the predator-only equilibrium, respectively. The first parts of these invasion boundaries make up part of the predator and consumer persistence boundaries and are hence represented in black. Other parameters as in figure 7.9.

Also, the boundary marking the invasion of consumers into a predator-only equilibrium bends toward lower $R_{max,P}$ values at higher values of $R_{max,S}$ and reaches $R_{max,P} = 0$ for $R_{max,S} = 0.035$ mg/L. This boundary corresponds to the threshold value of $R_{max,P}$ above which consumers are driven to extinction by a stunted-predator population and the stunted-predator coexistence equilibrium ceases to exist. The effect of increasing niche overlap between consumers and predators on the consumer invasion boundary is also the same as in the stage-structured biomass model, including the fact that this effect is more pronounced than the effect of niche overlap on the predator invasion boundary. As a consequence, the two invasion boundaries intersect, resulting in a parameter region in which bistability occurs between a consumer-only and a predator-only equilibrium (the horizontally hatched region in figure 7.10).

Figure 6.16 in the previous chapter showed for the case without niche overlap that bistability occurred, either between two different coexistence equilibria or between a stunted-consumer equilibrium and a predator-only equilibrium, within a parameter domain that straddled the boundary marking the invasion of consumers into the predator-only equilibrium (i.e., the extinction boundary of consumers from the stunted-predator coexistence equilibrium). With 10 percent niche overlap between consumers and predators, the corresponding parameter domain with bistability straddles the consumer invasion boundary as well. Compared with the case without niche overlap (figure 6.16), this entire region has rotated in a counterclockwise manner as well, owing to the niche overlap between consumers and predators, and following, in fact, the change in the consumer invasion boundary. Because the changes in the parameter domain with bistability in response to the niche overlap follow the change in the consumer invasion boundary and are hence more pronounced than the changes in the boundary marking the invasion of the predator into the consumer-only equilibrium, this parameter domain separates into four parts with different types of bistability: (1) a parameter region with bistability between a consumer-only and a predator-only equilibrium (figure 7.10, horizontally hatched region), which was discussed before and which occurs where the consumer-invasion threshold occurs below and to the left of the predator-invasion boundary; (2) a parameter region with bistability between a stunted-consumer and a stunted-predator coexistence equilibrium (figure 7.10, cross-hatched region), which is enclosed by the consumer- and predator-invasion boundaries where the consumer-invasion boundary is above and to the right of the predator-invasion boundary; (3) a parameter region with bistability between a consumer-only and a stunted-predator coexistence equilibrium (figure 7.10, lower gray region), which occurs to the left and below both the consumer-invasion and the predator-invasion boundary; and finally (4) a parameter region with bistability between a stunted-consumer coexistence equilibrium and a predator-only equilibrium (figure 7.10, upper gray region), which occurs to the right and above both the consumer-invasion and the predator-invasion boundary.

It should be noted, however, that the latter parameter region disappears for very small $R_{max,P}$ values that are not shown in figure 7.10. Hence, for such small $R_{max,P}$ values, the predator-invasion boundary and the boundary marking the consumer persistence threshold (dark-gray solid line and the black dashed line, respectively, in figure 7.10) merge. As a consequence, for $R_{max,P} = 0$ mg/L with increasing maximum densities of the shared resource, first a parameter range occurs with a unique, stable consumer-only equilibrium; followed by a range with bistability between a stable consumer-only equilibrium and a stunted-predator coexistence equilibrium; a parameter range with bistability between a consumer-only and a predator-only equilibrium; and finally a parameter range with the predator-only equilibrium as the only stable community state (results not shown). The four different parameter regions with bistability all occur when the maximum biomass density of the shared resource is sufficiently high ($R_{max,S} > 0.1$ mg/L; see figure 7.10). For lower values of $R_{max,S}$, figure 7.10 to a large extent resembles the corresponding graph for the case without niche overlap (figure 6.16 in the previous chapter).

Taken together, the effects of increasing niche overlap between consumers and predators are qualitatively similar in both the stage-structured biomass model as well as in the size-structured consumer, size-structured predator model: the changes in type and number of equilibria that were observed in the absence of niche overlap for increasing maximum densities of the predator-exclusive resource can now also be observed for increasing maximum densities of the shared resource, because the niche overlap between consumers and predators implies that increases in either resource lead to an increase in food availability for predators. The key difference between the results presented for the stage-structured biomass model and for the size-structured consumer, size-structured predator model is that even a limited niche overlap of just 10 percent in the latter model produces quantitatively significant effects, which occur in the stage-structured biomass model only for a much higher level of niche overlap. This much stronger effect of niche overlap in the size-structured consumer, size-structured predator model can be explained by the more gradual ontogenetic niche shift assumed in this model, with predators starting to decrease their foraging time on resources right after birth. On average, this leads to a much smaller dependence on resource feeding than in the stage-structured biomass model and hence a stronger response to changes in resource availability.

EMPIRICAL STUDIES

There is a rich empirical literature on size-dependent, mixed interactions. Still, most of these studies have been carried out on short, within-generation time scales, which makes it difficult or even impossible to test any of the theoretical predictions that we have provided above. Three exceptions to this lack of long-term experiments are the experimental systems studied by Persson, de Roos, and Byström (2007); Montserrat, Magalhães, et al. (2008); and Schröder, Nilsson, et al. (2009). These studies involved species where the size difference between the intraguild predator and the consumer (or intraguild prey) was relatively small. As a consequence, even adult predators fed to a significant extent on the shared resource, allowing the predator to both reproduce and persist on the shared resource. Based on the theoretical results from the previous sections, these characteristics would lead us to expect little room for coexistence of IG predator and IG prey in these systems (van de Wolfshaar, de Roos, and Persson 2006). Rather, because of the significant niche overlap between consumers and predators, we expect either a consumer-only or a predator-only equilibrium as a stable community state.

Schröder, Nilsson, et al. (2009) studied the invasion success of small and large individuals of the IG predator guppy (*Poecilia reticulata*) into resident populations of the IG prey least killifish (*Heterandria formosa*) in experimental aquarium systems at three different productivity levels. The results were quite striking in that large guppies generally succeeded in invading the resident least killifish populations, whereas small guppies generally failed. Furthermore, after their successful invasion, the large guppies always excluded the resident least killifish populations, suggesting that the likelihood for coexistence between guppies and least killifish is small. The inability of small guppies to invade the resident killifish populations irrespective of productivity is logical, as small guppies cannot profit from increased system productivity by preying on the higher densities of least killifish. This contrasts with expectations from unstructured theory where each IG predator individual both preys on and competes with IG prey. Large guppies, however, immediately exposed the resident least killifish populations to a high predation pressure and hence drove them to extinction after invasion. In fact, extinction of the least killifish populations occurred within just one generation, thereby resulting in an improvement of the food conditions for the guppy offspring. For large guppies, a productivity-dependent invasion threshold is expected, although this threshold was apparently below the productivity levels used in the experiments.

Montserrat, Magalhães, et al. (2008) studied a terrestrial mite system consisting of two phytoseiid mite species, *Iphiseius degenerans* as the IG predator and *Neoseiulus cucumeris* as the IG prey, and pollen (*Typha latifolia*) as the shared resource. In this system, females of the IG prey could prey on the larvae of the IG predator; hence reciprocal intraguild predation was present. Similar to Schröder, Nilsson, et al. (2009), Montserrat, Magalhães, et al. (2008) found a strong influence of the IG prey/IG predator ratio on IG predator invasion success. Montserrat, Magalhães, et al. (2012) showed in addition that the size structure of the resident IG prey population had a strong impact on the invasion success of the IG predator. When the resident IG prey population included all size classes, the IG predator could not invade the system. In contrast, the IG predator could successfully invade when large IG prey size classes were missing. They related this result to the reciprocal nature of this IGP system. Also similar to Schröder, Nilsson, et al. (2009), coexistence was never observed, and which species dominated the system at high productivities depended on priority effects where a resident IG prey population could prevent the invasion of the IG predator and vice versa when the IG predator was the resident population.

Persson, de Roos, and Byström (2007) studied the invasion success of the fish species roach, a potential competitor and prey of European perch, in two lakes inhabited by perch only (Lakes Abborrtjärn 3 and 4, Sweden) and in two lakes containing both perch and northern pike (*Esox lucius*) (Lakes Abborrtjärn 1 and 2, Sweden). The model analyzed by van de Wolfshaar, de Roos, and Persson (2006) was specifically tailored to mimic this particular perch-roach system and, as discussed above, showed little scope for coexistence between consumer and predator. Northern pike preys on both perch and roach but has a preference for perch. In all lakes, roach were stocked using a density and size distribution representative for lakes having perch, roach, and pike. This experiment did thus not involve an invasion from low numbers, but rather from high numbers. In the two lakes lacking pike, roach were unsuccessful in invading the resident perch populations (only one successful recruitment in ten years in one of the lakes) (figure 7.11, left panel). This lack of successful recruitment

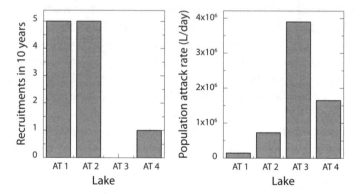

FIGURE 7.11. *Left*: number of years over a ten-year period with recorded recruitment of roach to an age of one-year-old in the Abborrtjärn Lakes (Sweden) 1 and 2 (pike present) and 3 and 4 (pike absent). *Right*: population attack rates (L/day) of the perch populations in the Abborrtjärn Lakes 1–4 on young-of-the-year roach (mean values of the last three years of the experiment). Population attack rates were calculated by multiplying estimates of size-dependent attack rates with the observed perch population size distributions in different years. Data from Persson, de Roos, and Byström (2007).

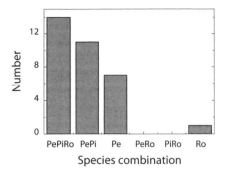

FIGURE 7.12. Species composition of the fish community in terms of the occurrence of perch (Pe), roach (Ro) and pike (Pi) in thirty Finnish lakes. Data from Sumari (1971). Note that perch co-occurs with roach only when pike is present.

can be related to the considerably higher predation pressure from perch on young-of-the-year roach in these two lakes (figure 7.11, right panel). In contrast, in the two lakes with pike, successful recruitment was observed every second year, and the roach populations were well established (figure 7.11). Further support for the idea that pike presence is a prerequisite for the coexistence of roach and perch comes from a comparative lake survey showing that perch and roach only occur together when other piscivores (mainly northern pike) are present (figure 7.12).

To conclude, all three sets of experimental systems support the theoretical expectation that the likelihood for coexistence between IG prey and IG predator is small in the absence of other mitigating factors (such as the presence of a top predator preferentially feeding on the IG predator), when both species share resources to such an extent that both can persist in isolation on the shared resource. In the case of reciprocal IGP, priority effects may furthermore have a strong impact on whether the IG predator or IG prey dominates.

Ontogenetic Niche Shifts, Predators, and Coexistence among Consumer Species

So far we have considered the effects of stage structure within consumer species on the persistence of one or several predator species (chapters 4, 5) and the effects of stage structure and ontogenetic niche shifts in predator species on predator-consumer coexistence (chapters 6, 7). In this chapter we will shift our focus to consider the effects of consumer stage structure on coexistence of different consumer species. The question of under what conditions consumers that share resources may coexist is one of the fundamental questions in ecology, going back to the classical experiments by Gause (1934) and the theoretical explorations by MacArthur and Levins (1967) and May and MacArthur (1972). Gause in particular postulated the competitive exclusion principle, stating that two consumer species cannot coexist in stable equilibrium if they share a single resource. Theoretical studies have largely supported Gause's conclusion, but in addition have suggested that coexistence of consumer species on a single resource is possible through predator mediation or under nonequilibrium conditions (Chesson 1986). For both predator-mediated coexistence and coexistence induced by temporal variation, niche differences between competitors are still assumed to be present (Chesson 2000).

In this chapter we will consider how stage structure and ontogenetic niche shifts may affect coexistence between two consumer species competing for two resources in the absence and presence of predators and how ontogenetic niche shifts may give rise to alternative stable states. More specifically, our analysis will use techniques developed within the consumer-resource framework of Tilman (1982), including consumption and renewal vectors (Schellekens, de Roos, and Persson 2010). Tilman showed that stable coexistence between consumers feeding on the same two resources is possible if each of the consumer species feeds proportionally more on the resource that limits its own growth most. Stable coexistence is, however, also affected by the form of the resource-dependent growth isoclines, which represent combinations of resource densities that lead to equal population growth of consumers. The shape of the resource-dependent growth isoclines is, in turn, determined by both differences in the nutrient quality of the resources (succinctly summarized by nutrient isoclines), as well as their distribution in space, the capacity of the consumers, and their foraging behavior (succinctly summarized by consumption constraint curves). We will show that ontogenetic niche shifts per se affect the form of the resource-dependent growth isoclines, which in turn may lead to coexistence through niche partitioning. We will also discuss how predation may promote the performance of a species undergoing ontogenetic niche shifts even in the case where it is both the inferior competitor and the preferred prey of the predator.

ONTOGENETIC NICHE SHIFTS
AND INTERSPECIFIC COMPETITION

MODEL DERIVATION

As we have discussed particularly in chapters 2 and 6, an increase in size over ontogeny is generally associated with changes in resource use, that is, ontogenetic niche shifts (Werner and Gilliam 1984). These changes can occur as an expansion of the niche, in which case larger resources are included into the diet when they become accessible, while the smallest resource size ingested remains more or less the same (niche expansion). For other organisms, however, the smaller resources are dropped from the diet when individuals increase in size, in which case the change in resource use takes the form of a distinct ontogenetic niche shift (Schellekens, de Roos, and Persson 2010). Finally, shifts in resource use may also involve a shift in habitat use, as is observed in organisms such as fish (Childress and Herrnkind 2001; Dahlgren and Eggleston 2000; Keren-Rotem, Bouskila, and Geffen 2006; Mittelbach and Osenberg 1993). In our analysis of the effects of ontogenetic niche shifts we will here focus on situations where the ontogenetic niche shifts take place as a shift in diet only (for a discussion of how to handle ontogenetic niche shifts in the form of habitat shifts, see chapter 6).

Our treatment of competition between consumers is based on the bioenergetic approach we have used in previous chapters. The reader is therefore largely referred to previous chapters and particularly chapters 3–6 for model derivations. As before, the net production of juveniles is invested into growth in body mass and maturation, whereas adult net production is invested in reproduction only. We will first consider the situation with two consumers (C, D)

competing for two resources (R_1, R_2) and thereafter also consider the effects of a shared predator on consumer coexistence. Model equations are given in box 8.1, and the default parameter values used are presented in box 8.2. However, the model specification in box 8.1 represents an extended model, which also includes a predator P foraging on both consumers. This extended model will be studied in the second part of this chapter. For the analysis in this section, we will simply assume that $P = 0$ and ignore the predator equations and functions in box 8.1. We will primarily study the case where one consumer feeds on both resources indiscriminately during its whole life span, whereas the other consumer shifts diet from one resource to the other resource at maturation. Two example organisms of this type of configuration are the cladoceran zooplankton *Daphnia* and copepod zooplankton, where the latter undergoes substantial ontogenetic niche shifts over it life period (Bern 1994; Schellekens, de Roos, and Persson 2010). We will also extend the analysis to discuss the situation in which both consumers undergo ontogenetic niche shifts.

To investigate the effects of ontogenetic niche shifts in one of the consumers, we introduce separate notations for the mass-specific consumption rates of juveniles on resource 1 $(\omega_{C_J,1})$ and resource 2 $(\omega_{C_J,2})$ and that of adults on

---

BOX 8.1
THE STAGE-STRUCTURED BIOMASS MODEL WITH TWO CONSUMERS AND ONE PREDATOR

Dynamic equations	*Description*
$$\frac{dR_1}{dt} = G_1(R_1) - \omega_{C_J,1}(R_1,R_2)C_J - \omega_{C_A,1}(R_1,R_2)C_A - \omega_{D,1}(R_1,R_2)D$$	Dynamics of resource 1
$$\frac{dR_2}{dt} = G_2(R_2) - \omega_{C_J,2}(R_1,R_2)C_J - \omega_{C_A,2}(R_1,R_2)C_A - \omega_{D,2}(R_1,R_2)D$$	Dynamics of resource 2
$$\frac{dC_J}{dt} = v_{C_A}^+(R_1,R_2)C_A - \gamma_C(v_{C_J}^+, d_C)C_J + v_{C_J}(R_1,R_2)C_J - d_C(P)C_J$$	Dynamics of juvenile niche shifter
$$\frac{dC_A}{dt} = \gamma_C(v_{C_J}^+, d_C)C_J + (v_{C_A}(R_1,R_2) - v_{C_A}^+(R_1,R_2))C_A - d_C(P)C_A$$	Dynamics of adult niche shifter
$$\frac{dD}{dt} = (v_D(R_1,R_2) - d_D(P))D$$	Dynamics of generalist
$$\frac{dP}{dt} = (v_P(C_J,C_A,D) - \mu_P)P$$	Dynamics of predator

(*Box 8.1 continued*)

Function	Expression	Description
$G_i(R_i)$	$\rho(R_{\max,i} - R_i)$	Turnover of resource $i = 1,2$
$\omega_{C_j,1}(R_1,R_2)$	$M_C(1-q_C)R_1/(H_C+(1-q_C)R_1+q_C R_2)$	Juvenile niche shifter intake of resource 1
$\omega_{C_j,2}(R_1,R_2)$	$M_C q_C R_2/(H_C+(1-q_C)R_1+q_C R_2)$	Juvenile niche shifter intake of resource 2
$\omega_{C_A,1}(R_1,R_2)$	$M_C q_C R_1/(H_C+q_C R_1+(1-q_C)R_2)$	Adult niche shifter intake of resource 1
$\omega_{C_A,2}(R_1,R_2)$	$M_C(1-q_C)R_2/(H_C+q_C R_1+(1-q_C)R_2)$	Adult niche shifter intake of resource 2
$\nu_i(R_1,R_2)$	$\sigma_C(\omega_{i,1}(R_1,R_2)+\omega_{i,2}(R_1,R_2))-T_C$	Niche shifter net energy production, $i=C_J,C_A$
$d_C(P)$	$\mu_C+M_P\dfrac{1.05\cdot P}{H_P+1.05\cdot C_J+1.05\cdot C_A+D}$	Niche shifter mortality
$\gamma_C(v_{C_J}^+,d_C)$	$(v_{C_J}^+-d_C)/\left(1-z_C^{(1-d_C/v_{C_J}^+)}\right)$	Juvenile niche shifter maturation rate
$\omega_{D,i}(R_1,R_2)$	$M_D\dfrac{0.5\cdot R_i}{H_D+0.5\cdot R_1+0.5\cdot R_2}$	Generalist intake of resource $i=1,2$
$\nu_D(R_1,R_2)$	$\sigma_D(\omega_{D,1}(R_1,R_2)+\omega_{D,2}(R_1,R_2))-T_D$	Generalist net energy production
$d_D(P)$	$\mu_D+M_P\dfrac{P}{H_P+1.05\cdot C_J+1.05\cdot C_A+D}$	Generalist mortality
$\nu_P(C_J,C_A,D)$	$\sigma_P M_P\dfrac{1.05\cdot C_J+1.05\cdot C_A+D}{H_P+1.05\cdot C_J+1.05\cdot C_A+D}-T_P$	Predator net energy production

resource 1 ($\omega_{C_A,1}$) and 2 ($\omega_{C_A,2}$) (box 8.1). We assume that juveniles spend a fraction q_C searching for resource R_2, and a fraction $1-q_C$ searching for resource R_1, while adults spend a fraction q_C of their foraging effort searching for resource R_1, and a fraction $1-q_C$ searching for resource R_2 (box 8.1). From the equations in box 8.1, it follows that each stage of the population specializes on its own exclusive resource when $q_C=0$ (R_1 for juveniles and R_2 for adults).

BOX 8.2

PARAMETERS OF THE STAGE-STRUCTURED BIOMASS MODEL OF BOX 8.1

Parameter	Default value	Unit	Description
Resources			
ρ	0.1	day^{-1}	Turnover rate of resources
$R_{max,i}$	varied	mg/L	Maximum density of resource $i = 1, 2$
Consumers			
M_C, M_D	$0.1W_A^{-0.25}$	day^{-1}	Mass-specific maximum ingestion rate
H_C, H_D	3.0	mg/L	Ingestion half-saturation density
T_C, T_D	$0.01W_A^{-0.25}$	day^{-1}	Mass-specific maintenance rate
σ_C, σ_D	0.5	—	Conversion efficiency
μ_C, μ_D	$0.0015W_A^{-0.25}$	day^{-1}	Background mortality rate
z_C	0.01	—	Newborn-adult size ratio of niche shifter
q_C	varied	—	Juvenile (adult) foraging effort on resource 2 (1)
Predators			
M_P	$0.1W_P^{-0.25}$	day^{-1}	Mass-specific maximum ingestion rate
H_P	3.0	mg/L	Ingestion half-saturation density
T_P	$0.01W_P^{-0.25}$	day^{-1}	Mass-specific maintenance rate
σ_P	0.5	—	Conversion efficiency
μ_C	$0.0015W_P^{-0.25}$	day^{-1}	Background mortality rate

W_A and W_P represent average adult body weight of consumer (0.0001 gram, the same for both consumers) and predator (0.01 gram), respectively.

For the species (*D*) feeding on the two resources without stage-specific preferences during its whole lifetime, it is assumed that every individual spends 50 percent of its time foraging on each resource (i.e., $q_D = 0.5$). As a consequence, juvenile and adult net production are the same, and the stage-structured model simplifies to the Yodzis and Innes model (box 8.1).

The only difference between the consumer populations that we will account for is the difference in resource use over the life period of one of the consumers. In all other respects, all rates are the same for both species, which means that the species are equal when $q_C = 0.5$. In this case, both species spend 50 percent of their time foraging on each resource and follow the dynamics described by the Yodzis and Innes model. In this unstructured case, the two species show neutral coexistence. By making these assumptions, we will be able to distinguish the effects of distinctive diets over ontogeny on the dynamics of the two populations from other potential factors affecting competition.

DERIVATION OF ZERO-NET-GROWTH ISOCLINES

One core element of the approach of Tilman (1982) is the zero-net-growth isocline (ZNGI). The ZNGI not only separates the combination of resources that yields positive consumer growth from those that yield negative consumer growth, but the form of the ZNGI influences the likelihood of coexistence and also the potential for alternative stable states. Following Tilman, we can calculate the minimum resource requirements for persistence for each population in the absence of the other species. From the resulting zero-net-growth isoclines (ZNGIs), potentially stable equilibria are found at the intersection(s) of the two species' ZNGIs. The ZNGI of D exhibiting no ontogenetic niche shift can be derived by setting the dynamical equation for the total biomass of D (box 8.1) to zero, leading to the following relationship between R_1 and R_2:

$$v_D(R_1,R_2) - \mu_D = \sigma_D M_D \frac{0.5 \cdot R_1 + 0.5 \cdot R_2}{H_D + 0.5 \cdot R_1 + 0.5 \cdot R_2} - T_D - \mu_D = 0. \quad (8.1)$$

This relationship yields a straight line with a slope equal to -1 in the phase plane spanned by the R_1 and R_2-axis (figure 8.1, solid, black line). The reader may here recall that this type of ZNGI is characteristic for an interaction between perfectly substitutable resources (Tilman 1982).

The derivation of the ZNGI for C, which undergoes an ontogenetic niche shift over its life span is more complicated, and we can only derive an implicit equation relating R_1 and R_2 by combining the equilibrium conditions $dC_J/dt = 0$ and $dC_A/dt = 0$. By setting $dC_A/dt = 0$, we can calculate the ratio of adult over juvenile biomass in equilibrium as $C_A/C_J = \gamma_C(v_C,\mu_C)/\mu_C$. Substitution of this ratio into the condition $dC_J/dt = 0$ yields the equation:

$$\left(v_{C_A}(R_1,R_2)\gamma_C(v_C,\mu_C) + \mu_C(v_{C_J}(R_1,R_2) - \gamma_C(v_C,\mu_C) - \mu_C)\right)C_J = 0. \quad (8.2)$$

Equilibria with positive population densities of the ontogenetic niche shifter $(C_J \neq 0)$ will therefore have to fulfill the condition:

$$v_{C_A}(R_1,R_2)\gamma_C(v_C,\mu_C) + \mu_C(v_{C_J}(R_1,R_2) - \gamma_C(v_C,\mu_C) - \mu_C) = 0 \quad \Leftrightarrow$$

$$\frac{v_{C_A}(R_1,R_2)}{\mu_C} \frac{\gamma_C(v_C,\mu_C)}{(\gamma_C(v_C,\mu_C) + \mu_C - v_{C_J}(R_1,R_2))} = 1$$

$$(8.3)$$

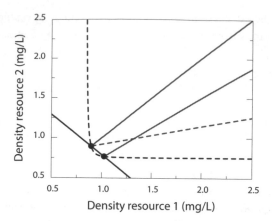

FIGURE 8.1. ZNGIs (*black*) and consumption vectors (*gray*) of consumer species *C* undergoing ontogenetic niche shifts (*dashed lines*) and *D* feeding indiscriminately on both resources over its lifetime (*solid lines*) with two possible equilibrium points. At the equilibrium point 2, the slopes of the consumption vectors of the two species are numerically indistinguishable.

This equation only depends on two unknown quantities, the resource densities R_1 and R_2 through their effect on $v_{C_J}(R_1,R_2)$ and $v_{C_A}(R_1,R_2)$. This condition can be simplified even further by replacing the maturation function $\gamma_C(v_{C_J}, \mu_C)$ with its defining expression (box 8.1), leading to the following relationship between R_1 and R_2 for ZNGI$_C$:

$$\frac{v_{C_A}(R_1,R_2)}{\mu_C} z_C^{\frac{\mu_C}{v_{C_J}(R_1,R_2)}-1} = 1 \tag{8.4}$$

If $q_C = 0$, both resources are essential for the persistence of *C*. This is reflected in the shape of the ZNGI$_C$ (figure 8.1, dashed, black curve), which resembles a ZNGI characteristic for two essential resources (Tilman 1982). Thus the introduction of an ontogenetic niche shift per se in *C* results in a change in form of the isoclines (note again that the ZNGI for $q_C = 0.5$ is a straight line) where the horizontal isocline leg of the ZNGI$_C$ reflects the constraint on population growth of *C* owing to adults requiring a minimum density of R_2 for reproduction and the vertical isocline leg of the ZNGI$_C$ reflects the constraint owing to juveniles requiring a minimum density of R_1 for growth. In other words, depending on the availabilities of R_1 and R_2, either the juvenile stage or the adult stage will form the bottleneck stage in the population.

CONSUMPTION VECTORS, SUPPLY POINT, AND RESOURCE RENEWAL

Consumption vectors represent another core element of Tilman's approach that affect coexistence patterns. Following Tilman (1982), we can first denote the point in the (R_1, R_2)-phase plane with coordinates $(R_{max,1}, R_{max,2})$ as the resource supply point. Consumption of resources by D decreases resource densities in the direction of the vector

$$\begin{pmatrix} -\omega_{D,1}(R_1, R_2) \\ -\omega_{D,2}(R_1, R_2) \end{pmatrix},$$

in which $\omega_{D,1}(R_1, R_2)$ and $\omega_{D,2}(R_1, R_2)$ are the foraging rates on resource 1 and 2, respectively. Given the expressions for $\omega_{D,1}(R_1, R_2)$ and $\omega_{D,2}(R_1, R_2)$ in terms of R_1 and R_2 (box 8.1), the direction of this consumption vector (figure 8.1, solid, gray lines) can also be expressed as

$$\begin{pmatrix} -R_1 \\ -R_2 \end{pmatrix}. \tag{8.5}$$

Counteracting resource consumption, resource supply increases resource densities in the direction of the vector

$$\begin{pmatrix} R_{max,1} - R_1 \\ R_{max,2} - R_2 \end{pmatrix}.$$

In equilibrium, the consumption rate by D is balanced by the resource supply, which occurs where the straight line connecting the origin with the resource supply point $(R_{max,1}, R_{max,2})$ intersects the ZNGI of D.

Consumption of resources by C decreases resource densities in the direction of the vector

$$\begin{pmatrix} -\omega_{C_J,1}(R_1, R_2)C_J - \omega_{C_A,1}(R_1, R_2)C_A \\ -\omega_{C_J,2}(R_1, R_2)C_J - \omega_{C_A,2}(R_1, R_2)C_A \end{pmatrix},$$

where $\omega_{C_J,1}(R_1, R_2)$ and $\omega_{C_J,2}(R_1, R_2)$ are the intake by juveniles of resource 1 and 2, respectively, and $\omega_{C_A,1}(R_1, R_2)$ and $\omega_{C_A,2}(R_1, R_2)$ are the intake by adults of resource 1 and 2, respectively. This expression makes clear that consumption of the resources depends on both the functional responses to resources as well as the population stage-structure of C.

Furthermore, by using the expression for the adult-juvenile biomass ratio in equilibrium $C_A/C_J = \gamma_C(\nu_{C_J},\mu_C)/\mu_C$, derived from $dC_A/dt = 0$, the niche shifter consumption vector can be shown to have the same direction as the vector

$$\begin{pmatrix} -\mu_C\omega_{C_{J,1}}(R_1,R_2) - \gamma_C(\nu_{C_J},\mu_C)\omega_{C_{A,1}}(R_1,R_2) \\ -\mu_C\omega_{C_{J,2}}(R_1,R_2) - \gamma_C(\nu_{C_J},\mu_C)\omega_{C_{A,2}}(R_1,R_2) \end{pmatrix}. \tag{8.6}$$

This vector is a function of the resource equilibrium densities only, such that for a given equilibrium point on the ZNGI of C with known resource densities, one can use expression (8.6) to construct a straight line through the equilibrium in the direction of this consumption vector (figure 8.1, dashed, gray lines). This line, in turn, includes all resource supply points $(R_{max,1},R_{max,2})$ that may eventually lead to this particular equilibrium point on the ZNGI of C.

COEXISTENCE PATTERNS

Using local stability analysis of coexistence equilibria, Tilman (1982) showed that two conditions are necessary for a stable equilibrium between two consumers competing for two resources: (1) each population must be limited by a different resource, and (2) each population must consume relatively more of the resource that limits its population growth rate most. We will use the same conditions to determine the potential for stable coexistence of C and D. It should be noted, though, that the different diets of juvenile and adult C imply that our dynamical system consists of five instead of the four ODEs in Tilman's model and that hence Tilman's conditions for occurrence of a stable coexistence equilibrium do not necessarily apply to the model we analyze. Nonetheless, with the numerical approach that we have used to compute model equilibria (MATCONT; Dhooge, Govaerts, and Yu 2003; see also box 3.5 in chapter 3), we have also assessed their stability properties and found that the larger number of ODEs does not invalidate Tilman's conditions, which can hence be used to draw conclusions about coexistence in the model analyzed here as well.

Intersections of the two single population ZNGIs (figure 8.1) are equilibria that potentially allow for stable coexistence of C and D (once again note that the unstructured version of the model only allows for neutral coexistence). The first (upper left) intersection (equilibrium 1 in the following) has coordinates $R_1^* = R_2^*$. At those coordinates, all non-zero consumption rates of C and D are equal $(\omega_{C_{J,1}}(R_1^*,R_2^*) = \omega_{C_{A,2}}(R_1^*,R_2^*) = \omega_{D,1}(R_1^*,R_2^*) = \omega_{D,2}(R_1^*,R_2^*))$, whereas $\omega_{C_{J,2}}(R_1^*,R_2^*) = \omega_{C_{A,1}}(R_1^*,R_2^*) = 0$ because $q_C = 0$. As a consequence, the production rates of all consumers/consumer stages are the same: $v_{C_J}(R_1^*, R_2^*) = v_{C_A}(R_1^*, R_2^*) = v_D(R_1^*, R_2^*)$. Because the production terms of juveniles and adults of C are the same at this equilibrium, the maturation rate $\gamma_C(v_{C_J}, \mu_C)$ becomes equal to $-\mu_C / \ln z_C$ (see de Roos, Schellekens, et al. 2008b). The direction of the consumption vector of C (equation (8.6)) at equilibrium 1 hence equals

$$\begin{pmatrix} \ln z_C \\ -1 \end{pmatrix}.$$

(8.7)

This vector will rotate clockwise if the ratio of mass at birth and maturation z_C decreases.

The direction of the consumption vector of D (equation (8.5)) at equilibrium 1 equals:

$$\begin{pmatrix} -1 \\ -1 \end{pmatrix}.$$

(8.8)

We can state that relative to D at this equilibrium 1, C is most limited by R_1 and that its consumption vector relative to that of D reduces R_1 most as long as $\ln z_C < -1$. Tilman's (1982) conditions thus suggest that stable coexistence of C and D is possible at equilibrium 1. The second (lower right) intersection of ZNGI_C and ZNGI_D, in which $R_1^* > R_2^*$, we will refer to as equilibrium 2 in the following. The consumption vector of C (equation (8.6)) at this equilibrium is graphically indistinguishable from the vector

$$\begin{pmatrix} -R_1^* \\ -R_2^* \end{pmatrix},$$

representing the consumption vector of D (figure 8.1).

The two coexistence equilibrium points divide both ZNGIs into three sections (figure 8.2, left panel) and give rise to distinct regions of resource supply that result either in coexistence or in competitive exclusion of one or the other population. For low and high values of R_1^*, the ZNGI of D is located at lower values of R_2^* than the ZNGI of C. Hence, for resource supply points between the corresponding sections of the ZNGI of C and D (figure 8.2, left panel; white areas right and above ZNGI_D, left and below ZNGI_C), only D persists, as C cannot persist even when alone. Furthermore, for resource supply points $(R_{max,1}, R_{max,2})$ that would allow for C persistence when alone (white areas right and above ZNGI_D and ZNGI_C), C will be outcompeted by D.

For coexistence of C and D at equilibrium 1, the combined consumption rate of C and D has to balance resource supply, which is only possible for resource supply points $(R_{max,1}, R_{max,2})$ below the line where $R_{max,1} = R_{max,2}$ (figure 8.2, left panel; light-gray and hatched area). Lines through the equilibrium point 1 that

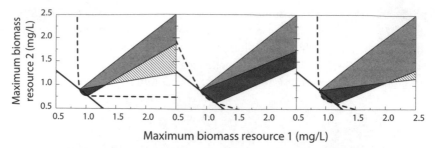

FIGURE 8.2. Coexistence patterns between C and D when $q_C = 0$ (*left panel*), when $q_C = 0.3$ (*middle panel*), and when juveniles of C forage on resource 1 only, while adults forage for 25 percent and 75 percent of their time on resource 1 and 2, respectively (*right panel*). In white areas above and to the right of the ZNGI of D, C is excluded because either the supply point is to the left and below the ZNGI of C or D competitively excludes C. In the dark-gray areas, C excludes D, and in light gray-areas, there is coexistence of C and D. In the hatched area, alternative stable states with either coexistence or only D present occur.

are spanned by the consumption vectors of C and D bound this resource supply region from below and above, respectively. Observe that for stable coexistence, the resource supply for juveniles of C (R_1) has to be larger than that for adult C (R_2), as the upper boundary is set by the consumption vector of D, which points in the direction of the diagonal $R_1 = R_2$ (vector (8.8)). The section of the ZNGI$_C$ between the two coexistence equilibrium points is located at lower values of (R_1^*, R_2^*) than the ZNGI$_D$. This results in a small region of resource supply for which only C can persist and a somewhat larger region for which C can invade an equilibrium with D only to subsequently outcompete D (figure 8.2, left panel, dark gray area). Overall, the parameter space where C can outcompete D is rather small.

Finally, for higher resource productivities, alternative stable states occur with either coexistence of C and D or only D present (figure 8.2, left panel, hatched area). The parameter space with bistability starts where the line, which originates in equilibrium 1 in the direction of the consumption vector of C toward high resource productivities, intersects with the line that starts in equilibrium 2 in the direction of the consumption vector of D (and C) toward high resource productivities. In figure 8.3 the changes in juvenile and adult biomass of C and total biomass of D are shown along a $R_{\text{max},2}$ gradient for two different values of $R_{\text{max},1}$ (1.1 and 2.0 mg/L). For the lower $R_{\text{max},1}$ value, D can invade the system at a $R_{\text{max},2}$ value around 0.7 mg/L to be excluded by C at a $R_{\text{max},2}$ value around 0.86 mg/L. C and D can coexist again for $R_{\text{max},2}$ values between 0.96 and 1.1 mg/L, while D thereafter competitively excludes C (figure 8.3, left

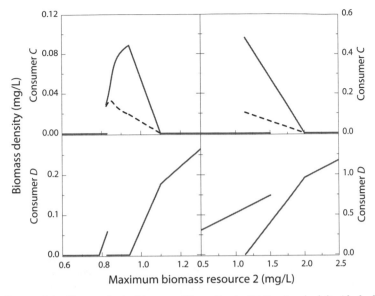

FIGURE 8.3. Changes in the biomass of juveniles (*solid lines*) and adults (*dashed lines*) of C (*upper panels*) and in total biomass of D (*lower panels*) with an increase in $R_{max,2}$ for $R_{max,1} = 1.1$ mg/L (*left panels*) and $R_{max,1} = 2.0$ mg/L (*right panels*).

panels). For the higher $R_{max,1}$ value, a $R_{max,2}$ range with bistability between a coexistence and a D-only equilibrium occurs (figure 8.3, right panels). Note that C is dominated by juveniles in biomass when present.

LESS DISTINCT DIET SHIFTS AND NICHE EXPANSION

From the above we can conclude that distinctive diets over ontogeny ($q_C = 0$), where juvenile and adult C interact with only one resource and D exploits both resources throughout its whole life allow for coexistence through niche partitioning of the two populations for certain ranges of resource supply. A necessary condition for coexistence is that resource supply for juvenile C is higher than that of adult C ($R_{max,1} > R_{max,2}$). This means that an ontogenetic diet shift can in itself provide a competitive advantage for the consumer if the supply of the adult resource is lower than the supply of the juvenile resource, but it always causes a disadvantage if the supply of the adult resource exceeds that of the juvenile resource.

What happens if the diet shift is less extreme and both stages of C exploit both resources to some extent? If $q_C > 0$, the consumption rates $\omega_{C,2}(R_1,R_2)$ and

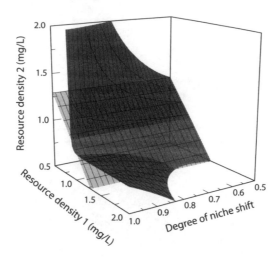

FIGURE 8.4. Changes in the ZNGI of C (*dark-gray manifold*) and the ZNGI of D (*light-gray plane*) for different values of q_C, where the degree of niche shift is identified with $(1-q_C)$. A change in q_C does not change coordinates (R_1^*, R_2^*) of equilibrium 1 (upper intersection between ZNGI$_C$ and ZNGI$_D$) but leads to a shift of equilibrium 2 to higher values of R_1^* and lower values of R_2^* with changing q_C.

$\omega_{C_A,1}(R_1, R_2)$ are larger than zero, and all functional responses will be a function of both resource densities. With increasing q_C, both juvenile and adult C divide their feeding effort more and more over both resources. As a result, the shape of the ZNGI$_C$ will gradually change from reflecting an interaction of essential to complementary, and eventually substitutable, resources with increasing q_C (figure 8.4). Ultimately, the ZNGIs and consumption vectors of both populations coincide for $q_C = 0.5$ because of the assumption that the two consumer species are energetically equivalent and completely equal when $q_C = 0.5$ (neutral coexistence).

Changes of the ZNGI$_C$ with increasing q_C potentially change the competitive outcome between C and D for a particular resource supply point. Coexistence equilibrium point 1, in which $R_1^* = R_2^*$, is independent of q_C and hence does not change coordinates (figure 8.2, middle panel; figure 8.4). As a consequence, for $q_C < 0.5$, C can potentially coexist with D in equilibrium point 1 irrespective of the value of q_C. Moreover, because both resources still equilibrate to the same density, the consumption vector of D does not change, nor do the biomass production rates of juvenile and adult C (as before equal: $\nu_{C_J}(R_1^*, R_2^*) = \nu_{C_A}(R_1^*, R_2^*)$) and the adult/juvenile biomass ratio of C at equilibrium, $C_A/C_J = \gamma_C(\nu_C, \mu_C)/\mu_C$. For $R_1^* = R_2^*$, the consumption vector (8.6) of C can be rewritten for non-zero values of q_C as:

$$\begin{pmatrix} (1-q_C)\ln z_C - q_C \\ q_C \ln z_C - (1-q_C) \end{pmatrix}. \qquad (8.9)$$

This expression shows that for increasing values of q_C, the consumption vector of C rotates counterclockwise from

$$\begin{pmatrix} \ln z_C \\ -1 \end{pmatrix}$$

at $q_C = 0$ to

$$\begin{pmatrix} -1 \\ -1 \end{pmatrix}$$

at $q_C = 0.5$. Consequently, the region of resource supply that allows for coexistence of C and D becomes smaller (figure 8.2, middle panel, gray area).

In contrast to equilibrium point 1, the location of coexistence equilibrium point 2, in which $R_1^* > R_2^*$, does change with increasing q_C such that R_1^* increases whereas R_2^* decreases (figure 8.4). As for $q_C = 0$, the consumption vectors of both populations in this equilibrium are numerically indistinguishable from each other for non-zero values of q_C as well and can hence be written as

$$\begin{pmatrix} -R_1^* \\ -R_2^* \end{pmatrix}.$$

The line through equilibrium point 2 spanned by these vectors therefore rotates clockwise with increasing q_C, owing to the changes in equilibrium resource densities. Relaxing the discrete diet shift of C thereby increases the region of resource supply points where C competitively excludes D (figure 8.2, middle panel, dark gray area). This occurs at the expense of the region of resource supply for which both populations coexist (figure 8.2, middle panel, light-gray area) and the region for which D competitively excludes C. However, irrespective of this increase in conditions with dominance by C, D will still outcompete C if the supply rate of the resource that is mostly fed upon by adults is larger ($R_{max,1} < R_{max,2}$). If q_C is only slightly smaller than 0.5, the ZNGI$_C$ crosses the ZNGI$_D$ only once at equilibrium 1, and the consumption vectors (8.5) and (8.6) in that point differ only marginally (figure 8.4). For these values of q_C, the resource supply plane is equally divided in supply regions where either C or D are competitively dominant, with C having a competitive advantage for $R_{max,1} > R_{max,2}$ and D having a competitive advantage for $R_{max,1} < R_{max,2}$.

So far, we have assumed that a single parameter q_C determines the diet of juvenile and adult C, such that consumers switch from one resource to the other at maturation and particularly that the resource selection by juveniles is the mirror image of that of adults. Making this assumption simplifies the analysis, as it involves a single parameter, but at the same time it represents only a special case of all possible forms of ontogenetic diet shifts that are present in natural systems. Another, perhaps more common situation is that an individual, when it increases in size, includes larger and larger prey items in its diet while still feeding on most of the smaller prey items as well, resulting in a niche widening (Werner and Gilliam 1984). To investigate this scenario, we will assume that juveniles only feed on R_1. In contrast, adults are assumed to feed on R_1 for 25 percent of their time and on R_2 for 75 percent of their time, such that they preferentially feed on R_2 but also include R_1 to a certain degree in their diet. Equations and functions for this slightly more general model are not presented separately, as the formulation is a straightforward extension of the model presented in box 8.1. The only difference is that the parameter q_C has a different value when it applies to juveniles ($q_C = q_J = 0$) and adults ($q_C = q_A = 0.25$). The ZNGI_C will in this case have a shape that is intermediate (figure 8.2, right panel) between the shapes of the ZNGI_C in the cases of completely exclusive (when $q_J = q_A = q_C = 0$, figure 8.2, left panel) and partially exclusive resources ($q_J = q_A = q_C = 0.3$, figure 8.2, middle panel). More specifically, the leg of the ZNGI that reflects the constraint attributable to juveniles is vertical, as for completely exclusive resources, whereas the part of the ZNGI that reflects the constraints attributable to adults is similar to that for partially exclusive resources (figure 8.2, right panel).

R_1 is thus still an essential resource for juvenile C (vertical leg of ZNGI_C), resulting in a minimum requirement of R_1^* for C to persist. In contrast, adults can compensate for a lack of R_2 with the use of R_1, enabling the persistence of C without R_2 (figure 8.2, right panel). Although the coexistence equilibrium 1 is still located at $R_1^* = R_2^*$, the wider diet of adults compared with juveniles leads to the situation in which the juvenile biomass of C is even more abundant in equilibrium 1. As a consequence, C consumes more of R_1 in coexistence equilibrium 1, rotating the consumption vector of C clockwise (cf. the slopes of the consumption vector of C in equilibrium 1, figure 8.2, left panel versus right panel). This change in consumption vector enlarges the region of resource supply with coexistence of C and D (figure 8.2, right panel, light-gray area) and diminishes the relative size of the region leading to competitive exclusion of D. At the same time, the location of coexistence equilibrium point 2, in which $R_1^* > R_2^*$, does change with ontogenetic niche widening such that R_1^* increases, whereas R_2^* decreases. This leads to an increase in the area where C excludes D and a displacement of the area with bistability to higher $R_{max,1}$ and $R_{max,2}$ values (figure 8.2, right panel). As was the case for the configuration with a single parameter q_C for both juveniles and adults assumed before, the consumption vectors of C and D in equilibrium 2 are numerically indistinguishable from each other and can again be written as

$$\begin{pmatrix} -R_1^* \\ -R_2^* \end{pmatrix}.$$

Overall, even though changes in the shape of the $ZNGI_C$ occur with an ontogenetic niche widening scenario, the conclusion still holds that the condition for coexistence between the consumers is that the resource supply for juveniles of C is higher than that for adults of C $(R_{max,1} > R_{max,2})$.

ONTOGENETIC NICHE SHIFTS IN BOTH CONSUMERS

Until now we have assumed that only one of the consumers undergoes ontogenetic shifts. Still, in many if not most situations, both consumers are expected to exhibit ontogenetic niche shifts. We will therefore in the following study two cases with ontogenetic niche shifts in both consumers. In the first scenario, we assume that both consumers undergo complete niche shifts, and in the second scenario we assume that the two consumer species differ in their extent of niche shift. As before, equations and functions for this model are not presented separately, as the formulation is similar to the model presented in box 8.1. Although both species are assumed to undergo an ontogenetic niche shift, we will continue to use the notations C and D for the consumers. The population of D is, however, now also subdivided into juvenile and adult D, indicated by D_J and D_A, respectively. The dynamics of the variables D_J and D_A are described by exactly the same set of equations as is given for C_J and C_A in box 8.1, except that the niche shift parameters, q_C and q_D, differ between the two species. In addition, the equation describing the dynamics of resource 1 (2) now contains feeding terms $\omega_{D_J,1}(R_1,R_2)$ and $\omega_{D_A,1}(R_1,R_2)$ $(\omega_{D_J,2}(R_1,R_2)$ and $\omega_{D_A,2}(R_1,R_2))$ instead of $\omega_{D,1}(R_1,R_2)$ $(\omega_{D,2}(R_1,R_2))$ to describe the feeding by juvenile and adult D, respectively, on this resource.

In the first scenario with two complete niche shifters $(q_C = q_D = 0)$, we furthermore assume that adult D is a superior competitor (tolerating the lowest resource biomass) to adult C, whereas juvenile C is a superior competitor to juvenile D. This assumption reflects an ontogenetic trade-off such that being efficient on one type of resource will have a negative effect on the species's efficiency on the other resource (Werner and Gilliam 1984). We will vary the competitiveness by assuming that the conversion efficiency of juvenile D is

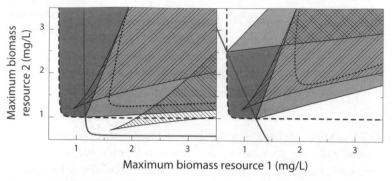

FIGURE 8.5. Coexistence patterns between C and D when both species undergo complete niche shifts (*left panel*), and when D undergoes only a partial niche shift, with juveniles feeding for 75 percent of their time on R_1 and adults for 75 percent of the time on R_2 (*right panel*). Solid and dashed lines indicate the $ZNGI_D$ and $ZNGI_C$ of C and D, respectively. In dark-gray areas above and to the right of the $ZNGI_C$, D is excluded either because the supply point is to the left of the $ZNGI_D$ or C competitively excludes D. In white areas above and to the right of the $ZNGI_D$, C is excluded because either the supply point is below the $ZNGI_C$ or D competitively excludes C. In the light-gray areas, the two species coexist. In dark-gray right-hatched (i.e., diagonally hatched) areas, two alternative stable states occur with only C present in both. In white, left-hatched (i.e., cross-diagonally hatched) areas, two alternative stable states occur with only D present in both. In light-gray, right-hatched areas, two alternative stable states occur with C present in both of them and D either present or absent, whereas in light-gray, left-hatched areas, two alternative stable states occur with D present in both of them and C either present or absent. In light-gray, cross-hatched areas, alternative stable states are present, with both species having positive biomasses in both of them. Dotted lines show the boundaries above which low amplitude cycles occur. The conversion efficiency for juveniles of D is $\sigma_{D_J} = 0.4$ and that of adults is $\sigma_{D_A} = 0.6$, and vice versa for C.

$\sigma_{D_J} = 0.4$ and that of adult D is $\sigma_{D_A} = 0.6$, and vice versa for C. From figure 8.5 it is evident that a trade-off is a necessary condition for coexistence, as the ZNGIs of the two species would otherwise not intersect. This result corresponds with the results of McCann (1998), who showed that coexistence between two species required that a competitive advantage at one stage had to be balanced by a competitive disadvantage at another stage. The consumption of resources by both consumers now depends on both the functional responses to resources as well as the population stage structure. Moreover, D and C are more limited by R_1 and R_2, respectively, and their stage structures will increase and decrease the slope of their consumption vectors above and below that of the 1:1 line, respectively, leading to an increased potential area of coexistence compared with the situation when one of the species did not exhibit any ontogenetic niche shift (figure 8.5, left panel, light gray regions, versus figure 8.2, left panel).

The scenario with complete ontogenetic niche shifts in both species also includes large areas of bistability, in which two different community equilibria occur with different stage structures in both species (figure 8.5, left panel, right-hatched and left-hatched areas). For both species, the alternative stable states with different stage structures and hence different minimum resource biomasses depend on which stage is mostly limited at equilibrium and hence forms the bottleneck for biomass turnover in the population. The reason for the occurrence of these alternative states and their relation to resource productivity was discussed in detail in chapter 6 for the one consumer case (cf. Guill 2009). Two important results from this analysis were that consumer biomass was higher and the resource levels lower when juveniles dominated the population in biomass (chapter 6; Schellekens, de Roos, and Persson 2010). Most remarkably, coexistence is possible in only one of the alternative stable states. Finally, at higher productivities, the dynamics destabilize through a Hopf bifurcation into low amplitude cycles (dotted line, figure 8.5, left panel).

In a gradient of increased productivity of R_1 for a fixed $R_{max,2}$, C (ZNGI with dashed line) will be the first species to invade because of its higher efficiency as a juvenile (figure 8.5 and figure 8.6, left panels). When invading, C is dominated by juveniles because of the low productivity of R_1. At a productivity around 1.2 mg/L, two alternative states appear, one dominated by juveniles and one dominated by adults (figure 8.5, left panel, dark-gray, right-hatched area). The alternative state, in which adults dominate the population of C, can be invaded by D after a further small increase in productivity of R_1 and leads to a region of bistability with either C and D coexisting or only C present (figure 8.5, left panel, light-gray, right-hatched area; figure 8.6, left panel). A further increase in $R_{max,1}$ leads to an area with only equilibrium coexistence (figure 8.5, left panel, light-gray area) that through a Hopf bifurcation shifts into a region with low amplitude cycles (figure 8.6, left panel). At a productivity level of R_1 around 2.6 mg/L, the system moves into another region of bistability with either juvenile or adult biomass making up the largest part of the population of D and C either present (when juveniles dominate D) or absent (when adults dominate D) (figure 8.5, left panel, light-gray, left-hatched area; figure 8.6, left panel). Finally, at a productivity level above 7 mg/L, C goes extinct owing to competition from D, and at a productivity level around 8 mg/L, the alternative stable states in D collapse into an adult-dominated state only.

In a gradient of increased productivity of R_2 for a fixed $R_{max,1}$, D is the first species to invade because of its higher efficiency as an adult (ZNGI solid line) (figure 8.5, left panel, white area; figure 8.6, right panel). Because of the low

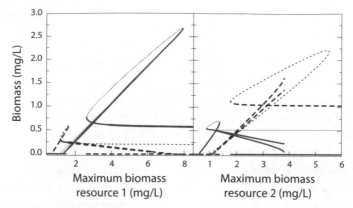

FIGURE 8.6. Changes in the total biomass of D (*solid lines*) and C (*dashed lines*) with increasing $R_{max,1}$ for a fixed $R_{max,2} = 1.5$ mg/L (*left panel*) and with increasing $R_{max,2}$ for a fixed $R_{max,1} = 2.5$ mg/L (*right panel*) when both species undergo complete niche shifts (cf. figure 8.5, left panel). Thin lines represent unstable equilibria.

productivity of R_2, however, the population is initially dominated by adults. At a productivity level of R_2 around 0.8 mg/L, alternative stable states with either juveniles or adults dominating the D population develop (figure 8.5, left panel, white, left-hatched area). From a productivity level slightly above 1 mg/L, C is either present or absent depending on whether juveniles or adults dominate D, respectively (figure 8.5, left panel, light-gray, left-hatched area; figure 8.6, right panel). After a small region with a unique and stable coexistence equilibrium, in which D is dominated by juveniles and C by adults (figure 8.5, left panel, light-gray area; figure 8.6, right panel), low amplitude dynamics in both species appear. At an $R_{max,2}$ around 1.75 mg/L, alternative stable states occur once again, in which either juveniles or adults dominate the population of C, and D is absent or present, respectively (figure 8.5, left panel, light-gray, right-hatched area; figure 8.6, right panel). Finally, at a productivity level just below 4 mg/L, D goes extinct and only one juvenile-dominated equilibrium state of C remains.

We next turn to the scenario where the two species, although both undergoing ontogenetic niche shifts, differ in the extent of the niche shift such that D undergoes only a partial niche shift ($q_D = 0.25$), whereas C undergoes a complete niche shift ($q_C = 0$). We continue to assume that the conversion efficiencies of juveniles and adults of D is $\sigma_{D_J} = 0.4$ and $\sigma_{D_A} = 0.6$, respectively, and vice versa for C. For this scenario, there are two possible equilibria similar to the situation when one species underwent an ontogenetic niche shift while the other species was a generalist (figure 8.2). However, as the two species in the present

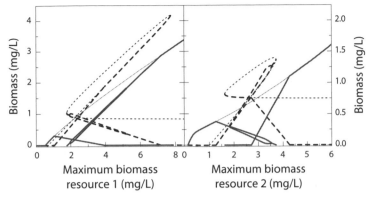

FIGURE 8.7. Changes in the total biomass of D (*solid lines*) and C (*dashed lines*) with increasing $R_{max,1}$ for a fixed $R_{max,2} = 3.0$ mg/L (*left panel*) and with increasing $R_{max,2}$ for a fixed $R_{max,1} = 2.0$ mg/L (*right panel*) when D undergoes a partial niche shift, with juveniles feeding for 75 percent of their time on R_1 and adults for 75 percent of the time on R_2, and C undergoes a complete niche shift (cf. figure 8.5, right panel). Thin lines represent unstable equilibria. Note the different values on the vertical axes in left and right panels.

case differ in their stage-specific conversion efficiencies, the slopes of the consumption vectors of the two consumer species in equilibrium 2 differ from each other (figure 8.5, right panel) (for another scenario with different slopes of the consumption vectors in equilibrium 2, see Schellekens, de Roos, and Persson 2010). As for the case with complete niche shifts in both consumers, substantial parameter areas of bistability and nonequilibrium dynamics occur.

Increasing the productivity of R_1 for a fixed $R_{max,2}$ first leads to the invasion of D in a state dominated by juveniles (figure 8.5, right panel, white area; figure 8.7, left panel). At a productivity level around 1 mg/L, C can invade the system, leading to a stable equilibrium with both species present and dominated by juveniles (figure 8.5, right panel, light-gray area; figure 8.7, left panel). With a further increase in productivity, a very narrow range with alternative states occurs, in which the population of C is either dominated by juveniles or adults and D is either present or absent, respectively (figure 8.5, right panel, light-gray, right-hatched area; figure 8.7, left panel). A slight increase in productivity shifts the system into a region with both species present at positive equilibrium biomasses in one of two alternative stable coexistence states (figure 8.5, right panel, light-gray, cross-hatched area; figure 8.7, left panel).

These coexistence equilibria differ in that C is either dominated by juveniles or by adults, whereas juveniles dominate the D population in both states. At

$R_{\text{max},1} = 2$ mg/L, the equilibrium, which is dominated by adult C, destabilizes, giving rise to small amplitude cycles. The amplitude of these cycles is, however, so small that they do not play a prominent role. Moreover, the cycles disappear again above a productivity of 5 mg/L. For a $R_{\text{max},1}$ above 4 mg/L, D is driven to extinction from the equilibrium that is dominated by juvenile C, resulting in a parameter region with bistability between a coexistence equilibrium of D and an adult-dominated C population and an equilibrium with only a juvenile-dominated C population (figure 8.5, right panel, light-gray, right-hatched area; figure 8.7, left panel). At a high $R_{\text{max},1}$ value, C is driven to extinction from the coexistence equilibrium, in which the C population is dominated by adults, owing to the competition from D. This leads to a region with bistability between an equilibrium with only C and an alternative equilibrium with only D, in which both are dominated by juveniles. For even higher $R_{\text{max},1}$ values, the C-only equilibrium ceases to exist, and the only stable equilibrium remaining is the one with only D present in a state dominated by juveniles (figure 8.7, left panel; not visible in figure 8.5).

Increasing the productivity of R_2 for a fixed $R_{\text{max},1}$ value first leads to the invasion of D dominated by juveniles (figure 8.5, right panel, white area; figure 8.7, right panel). At a $R_{\text{max},2}$ value around 1.3 mg/L, C can invade and coexist with D (figure 8.5, right panel, light-gray area; figure 8.7, right panel). In this coexistence state, the C and D populations are dominated by adults and juveniles, respectively. With a further increase in productivity until above $R_{\text{max},2} = 1.6$ mg/L, an alternative stable state occurs next to the coexistence equilibrium, in which only C is present and dominated by juveniles, whereas D is extinct (figure 8.5, right panel, light-gray, right-hatched area; figure 8.7, right panel). The dynamics of the coexistence equilibrium destabilizes at an even higher productivity into low amplitude cycles (figure 8.5, right panel, dotted line). As they are so close to the equilibrium, these cycles play a negligible role and moreover disappear at a productivity of $R_{\text{max},2} = 3.5$ mg/L. Above a productivity of $R_{\text{max},2} = 2.7$ mg/L, D can invade the juvenile-dominated C-only equilibrium, leading to two alternative equilibria with positive biomasses of C and D in both of them (figure 8.5, right panel, light-gray, cross-hatched area; figure 8.7, right panel). These coexistence equilibria differ in that one of them is dominated by adult, the other by juvenile C, whereas the D population is invariably dominated by juveniles. The coexistence equilibrium dominated by adult C disappears at a productivity level above $R_{\text{max},2} = 3.7$ mg/L, and C is driven to extinction from the other coexistence equilibrium for a productivity level above $R_{\text{max},2} = 4.3$ mg/L (figure 8.7, right panel; not visible in figure 8.5). Hence, at very high $R_{\text{max},2}$, only a single-species equilibrium of D occurs as the unique, stable community equilibrium.

Overall, complete niche shifts in both species will, if their ZNGIs intersect, lead to a broadening of the environmental conditions for which the two species may coexist. This increased coexistence potential results from the species differences in stage-specific performances affecting the slopes of the consumption vectors of both consumer species (figure 8.5 versus figure 8.2, left panel). This result resembles that of Tilman (1982) for essential resources. Notably, when the consumers do coexist, each consumer population is dominated by the stage with the lowest conversion efficiency (C by adults and D by juveniles). At the same time, the presence of stage differences in both species results in a substantial parameter space, with alternative states where either C or D is extinct in one of the states (figure 8.5). Differences in the extent of niche shifts between species obviously also leads to the potential for multiple coexistence equilibria points with different stage structures (figure 8.5, right panel, cross-hatched area).

EFFECTS OF PREDATORS ON
COEXISTENCE OF CONSUMERS

Up to now we have considered how ontogenetic niche shifts per se may result in niche partitioning among consumer species. In contemporary ecology, a major research theme has been to investigate how a predator on competing species may mediate their coexistence (Chase, Abrams, et al. 2002; Grover and Holt 1998; Holt and Lawton 1994; Leibold 1996). Two general results that have emerged from these studies are (1) that the superior consumer (maintaining itself at the lowest R^*) can tolerate the highest predator biomass and (2) that coexistence among competing consumers is possible if the predator has a preference for the superior competitor. It has also been shown that given the presence of a competition/predation trade-off among consumers, the outcome of predator presence on consumers will also depend on the productivity of the resource (Grover and Holt 1998; Holt and Lawton 1994; Leibold 1996). This occurs because the predator does not reach high enough densities at low food supply to affect consumer numbers (and hence resource levels), so the superior competitor wins. At intermediate resource supply, the predator can reduce the density of the competitively superior consumer to such an extent that increased resource levels allow the inferior competitor to coexist with the superior consumer. Finally, at high resource supply, the predator excludes the superior competitor (Grover and Holt 1998; Holt and Lawton 1994; Leibold 1996).

In the following we will analyze the effects of a predator on coexistence between competing consumer species feeding on two resources, once again assuming that one of the consumer species (C) undergoes a complete ontogenetic shift ($q_C = 0$) whereas the other (D) does not ($q_D = 0.5$). We will hence return to the model we analyzed first in this chapter and add a predator that feeds on both consumer species C and D. We will make two further assumptions. First, we assume that the ontogenetic specialist (the consumer species undergoing ontogenetic niche shifts) experiences a higher background mortality than the generalist (the consumer species not undergoing ontogenetic niche shifts), such that the ontogenetic specialist will never be able to invade an ontogenetic generalist–resources equilibrium in the absence of predation (figure 8.8, left panel). Second, we assume that the predator has a positive (5 percent) selection for the specialist consumer. Box 8.1 presents the equations and functions describing this model with two consumer and a single predator species, and box 8.2 lists its parameter values. Model equations including parameter values for the consumer species are the same as before, except that consumers in addition to their different background mortalities also experience mortality from predation. The resulting total mortality of ontogenetic specialists and generalists therefore equals (box 8.1):

$$
\begin{aligned}
d_C(P) &= \mu_C + M_P \frac{1.05 \cdot P}{(H_P + 1.05 \cdot C_J + 1.05 \cdot C_A + D)} \\
d_D(P) &= \mu_D + M_P \frac{P}{(H_P + 1.05 \cdot C_J + 1.05 \cdot C_A + D)}
\end{aligned}
\tag{8.10}
$$

The equations of the predator population (P), including predator parameters, have already been derived in chapter 6 (see also boxes 8.1 and 8.2). We continue to assume that the generalist and the niche shifter have the same adult body mass W_A, which implies that they have identical mass-specific ingestion and maintenance rates, whereas background and total mortality rates thus differ.

Given these assumptions, we conclude that the generalist consumer will always win over the specialist in the absence of predators (figure 8.8, left panel). One would intuitively think that the presence of a predator that has a preference for the specialist consumer would make the situation even more precarious for the specialist consumer. Figure 8.8 (middle panel) shows that this is not at all the case, as the addition of the predator actually allows the specialist consumer to coexist with the generalist consumer over a considerable parameter space and even outcompete the generalist consumer for certain parameter values (figure 8.8, middle panel). To understand this effect of the predator on coexistence between C and D, we first consider the effects of the predator on the ZNGI of the generalist. We showed above that the invasion boundary of the generalist consumer D competing with a niche shifter C for two resources will occur when the total resource density (the sum of R_1^* and

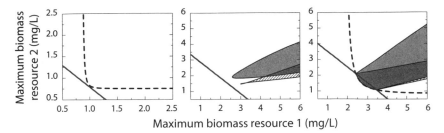

Maximum biomass resource 1 (mg/L)

FIGURE 8.8. *Left*: ZNGIs of the ontogenetic generalist (*D, solid line*) and special-ist (*C, dashed line*) consumer with a background mortality of 0.015 day^{-1} for the generalist consumer D and 0.02 day^{-1} for the specialist consumer C. *Middle*: co-existence patterns of the two consumers in the presence of a predator. The preda-tor has a 5 percent higher attack rate on C than on D (equation (8.10)). Thick, solid line: predator persistence threshold. Parameter values used for the predator are given in box 8.2. *Right*: coexistence patterns of the two consumers when the background mortality for both consumers has been increased by an additional 0.1 day^{-1} compared with the situation in the left panel. In the light-gray areas, the two consumers coexist; in dark-gray areas, C outcompetes D; and in hatched areas, bistability occurs between an equilibrium with either only C or only D pres-ent. In white areas above and to the right of the ZNGI of D, D dominates either because the resource supply is too low to allow C to invade even in the absence of D or because D outcompetes C.

R_2^*) exceeds a threshold value given by a straight line with slope -1 (figure 8.1; figure 8.8, left panel).

In the presence of a predator, D will experience an additional mortality equal to $d_D(P^*)$ (equation (8.10)). This higher mortality means that the ZNGI of D will shift to higher resource densities but will maintain the same slope (figure 8.8, right panel). This increase in equilibrium resource densities will release competition among all generalists to the same extent. In contrast, an increase in the mortality of the niche shifter increases the equilibrium density of the juvenile resource (R_1^*) more than the equilibrium density of the adult re-source (R_2^*) and thus releases competition among juvenile niche shifters more than among adults. As a consequence, the displacement of the ZNGI of diet shifters induced by increasing mortality is larger in the direction of the R_1^*-axis than in the direction of the R_2^*-axis. The predator will thus induce a change in the ratio of the two resources in equilibrium and the biomass ratio of juveniles and adults such that the consumer population after the predator invasion will be dominated by juveniles. This also means that the predator will cause an increase in net biomass production of the specialist.

This effect of predation mortality on the ZNGIs of the specialist and the generalist can be illustrated by increasing the background mortality of the generalist and specialist consumers to the same extent with 0.1 day^{-1} (figure 8.8, right panel). As before, the ZNGI of the generalist will be displaced to higher resource productivities while preserving the same slope. In contrast, the ZNGI of the specialist is substantially displaced in the R_1 direction but much less so in the R_2 direction (figure 8.8, left panel versus right panel). As a consequence, regions of coexistence, bistability, and even specialist dominance, which are absent at lower background mortalities, are now present. Note that this effect of increased mortality on consumer coexistence does not yet include any positive effect of increased specialist consumer net production on predator biomass, in turn, affecting the extinction risk of the generalist (see below).

As before, coexistence between the generalist and the specialist consumers or even the competitive extinction of the generalist by the specialist consumer will only occur if $R_{max,1} > R_{max,2}$. Besides the parameter space where only D dominates, three different regions of resource supply with different outcomes of competition (coexistence, competitive exclusion of D, and bistability between an equilibrium with either only C or only D) are present, dependent on the resource supply of R_1 and R_2. All three regions are bounded by the exclusion or invasion of one consumer population (either $C^* = 0$ or $D^* = 0$). Numerically, the boundaries where either C^* or D^* equals 0 can be continued as a function of the two resource supply values. These boundaries form curves in the $(R_{max,1}, R_{max,2})$ resource supply plane that delineate combinations of resource supply values with distinct competition outcomes (figure 8.8, middle panel).

It should be noted that the construction of these boundaries differs from the procedure we used for the consumer-resource system discussed above (i.e., figures 8.1 and 8.2). Given a particular equilibrium point on the ZNGI with known resource densities, it was possible for the consumer-resource systems to construct a straight line in the direction of the consumption vector through the equilibrium point. This line delineated all resource supply points $(R_{max,1}, R_{max,2})$ that eventually lead to this particular equilibrium point on the ZNGI. This procedure can, however, only be followed in the case of a particular, fixed mortality rate. In a predator-consumer-resources equilibrium, a change in resource supply will obviously alter the biomasses of resources and predator, and thereby the mortality rate of the niche shifter C and generalist D. Because of the change in mortality of C, however, its stage-structure will change and hence its ingestion rates of the two resources. As a consequence, the consumption vector of C changes not only because of the changed resource densities, but also as a consequence of the changed stage-structure. This precludes the use of the same approach based on consumption vectors that we used for the consumer resource system.

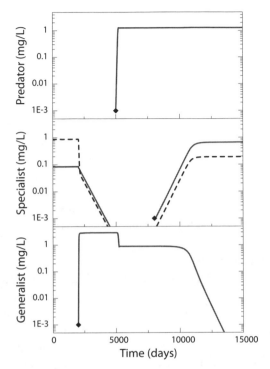

FIGURE 8.9. Time series showing that the generalist consumer excludes the specialist consumer after invasion in the absence of the predator. The subsequent invasion of the predator causes a decrease in the generalist consumer biomass and allows for the reinvasion of the specialist consumer and the subsequent extinction of the generalist consumer. Parameter values as in box 8.2, with $R_{max,1} = 6.0$ mg/L and $R_{max,2} = 2.5$ mg/L. Juvenile and adult specialist consumers are indicated with solid and dashed lines, respectively (*middle panel*).

The effect of predator presence on the competitive interactions between consumers is particularly striking in the region where D competitively excludes C in the absence of predators, while C excludes D in the presence of predators. In this region, C is, in the absence of predators, dominated by adults, and an invasion of D into a C-resources equilibrium leads to the extinction of C (figure 8.9). The productivity of the system is still high enough to allow a predator to invade, leading to a subsequent drop in biomass of D. This negative effect of the predator on the generalist consumer, in turn, allows C to reinvade with a biomass that at equilibrium is dominated by juveniles, owing to the mortality induced by the predator. These juveniles in turn make good use of the large supply of resource R_1. The presence of the predator is thus negative for the generalist through two processes: it imposes mortality on the generalist and it increases the competitiveness of the specialist. This may lead to the extinction of the generalist consumer for productivities of the two resources where the generalist would otherwise drive the specialist consumer to extinction (figure 8.9). Also note that this occurs even though the predator has a preference for the specialist consumer. As we will consider in the following, the predator-induced redistribution of the stage structure of C also boosts predator biomass, leading to an increased predation pressure on the generalist consumer, contributing to its extinction.

PREDATOR-INDUCED EFFECTS ON COEXISTENCE ALONG PRODUCTIVITY GRADIENTS

In the absence of the specialist (C), the invasion sequence of the resources, the generalist consumer (D), and the predator (P) with increasing productivity will follow the predictions of basic food chain theory (i.e., Oksanen, Fretwell, et al. 1981), because of the indiscriminate foraging of D on both resources. With two resources, the invasion boundary of the consumer will hence occur when the total resource density (the sum of R_1^* and R_2^*) exceeds a threshold value, leading to a straight line with slope -1 in the (R_1, R_2) resource plane. In the same way, the invasion boundary of the predator to invade is defined by a straight line that runs parallel with the consumer invasion boundary. The invasion sequence of the resources, C, and P with increasing productivity in the absence of D will qualitatively follow the same pattern as observed for the generalist consumer under most conditions, but we have already shown in chapter 6 that an emergent Allee effect may occur for a generalist predator that forages on an ontogenetic specialist.

The occurrence of the emergent Allee effect for the predator hints at important differences between the response of the generalist and specialist to predation. First, the predator will, as we have already discussed, change the biomass ratio of juveniles and adults such that the specialist consumer population after the predator invasion will be dominated by juveniles. This leads to an increase in net biomass production of the specialist, because of the larger supply of resource 1 ($R_{max,1} > R_{max,2}$). Second, a predator feeding on a stage-structured consumer can hence experience a nonlinear increase in biomass with increasing resource supply that will increase predation mortality and eventually lead to the exclusion of the generalists. This predator-induced shift in juvenile-adult ratio may also allow the existence of a region with bistability, which we

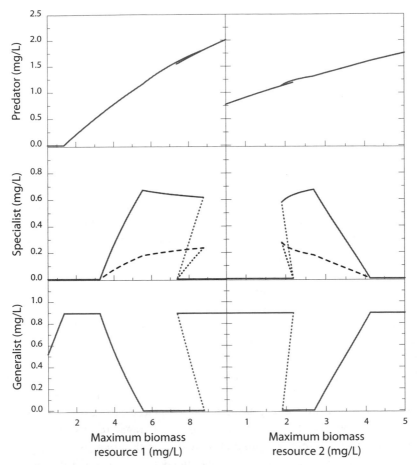

FIGURE 8.10. Bifurcation graphs of the changes in biomass of predator (*top panels*), specialist consumer (*middle panels*: juveniles, *solid line*; adults, *dashed line*), and generalist consumer (*bottom panels*) with an increase in $R_{max,1}$ for $R_{max,2} = 2.5$ mg/L (*left panels*) and with an increase in $R_{max,2}$ for $R_{max,1} = 6.0$ mg/L (*right panels*). Dotted lines indicate unstable equilibria.

discussed in detail in chapter 6 (see also Guill 2009; Schellekens, de Roos, and Persson 2010).

An increase in the resource supply of R_1 from a level where neither the predator nor the specialist consumer C can exist in the presence of the generalist consumer D, leads to a linear increase in D up to a productivity level where the predator can invade (figure 8.10, bottom left panel). Once the predator invades, its biomass increases linearly with an increase in resource productivity, whereas the biomass of D stays constant. With a further increase in productivity of R_1, C can invade with its biomass dominated by juveniles, owing to the predator-induced shift in stage distribution of C. At the same time, the biomass of D decreases. The predator-induced shift in stage distribution of C translates into a more than linear increase in predator biomass with increased productivity (figure 8.10, top left panel). D goes extinct around a resource productivity of $R_{max,1} = 5.5$ mg/L. This occurs because the predator biomass in equilibrium with the niche shifter C is higher than in the generalist-predator equilibrium for the same parameters. A further increase in the productivity of R_1 leads to the presence of an alternative stable state consisting of either the predator and the specialist consumer or the predator and the generalist consumer, where the biomass of the predator is higher in the former case owing to the predator-induced effect on the net production of C (figure 8.10, left panels). At even higher productivities, C goes extinct, and an equilibrium with the predator and D remains.

In a productivity gradient of R_2 for a specific value of $R_{max,1}$, at first only an equilibrium with the predator and the generalist D occurs, as adult C cannot survive at $R_{max,2}$ levels that are too low (figure 8.10, right panels). As before, predator biomass increases linearly while the biomass of the generalist D stays constant. At a certain $R_{max,2}$, the resource productivity is high enough to allow adult C to survive, leading to a productivity range with bistability (P coexisting with either D or C). In the equilibrium with only C, predator biomass is higher, which prevents invasion of D. With a slightly higher value of $R_{max,2}$, the equilibrium with only D can be invaded by C, which leads to the exclusion of D through the positive effect of the predator-induced increase in net production of C on predator biomass. Hence, a parameter range results with only the predator and C present at equilibrium. With a further increase in $R_{max,2}$, the predator-induced mortality on C overrides the positive effect on C net biomass production, leading to a relatively smaller increase in predator biomass with productivity. This allows for reinvasion of D and consumer coexistence (figure 8.10, right panels). Finally, at high productivities of resource R_2, D drives C to extinction.

In conclusion, in unstructured systems, predators have been shown to mediate coexistence between competitors or even to favor the inferior competitor with increasing resource supply if the predator has a preference for the superior competitor (Grover and Holt 1998; Holt and Lawton 1994; Leibold 1996). In contrast, we have found that a predator in a two-consumer system where one of the consumers (C) undergoes an ontogenetic niche shift may benefit this niche shifter even when the predator has a preference for it. This is reflected in the substantial parameter space with coexistence of C and D or with C dominating over D in the presence of the predator, where D would always dominate in the absence of the predator. This pattern results from two interacting processes: (1) the efficiency by which the niche shifter exploits and depresses the two resources will change depending on the mortality rate that it experiences as it changes its stage structure, and (2) predator biomass is boosted by the increased net production of the niche shifter that, in turn, negatively affects the generalist consumer through increased predation mortality. The increase in resource exploitation efficiency by the niche shifter induced by predation thus translates into a competitive advantage relative to generalists that does not occur in the absence of predators. This advantage, as we have found, can even counterbalance the combined effects of preferential predator feeding on diet shifters and a higher background mortality.

In chapter 4 we discussed how overcompensation in prey may lead to an emergent Allee effect, including a propensity of the predator to show catastrophic collapses in response to overexploitation. The results we have provided in this chapter suggest that predator collapses will feed back on size-structured consumer populations by decreasing the efficiency with which niche shifting consumers exploit their resources, which subsequently will enable other competing consumer species to outcompete the niche shifters and take over the community. Furthermore, it is also possible that a predator having the same preference for both stages of a stage-structured consumer can cause an increase in biomass production of both juvenile and adult consumers owing to the increase in mortality. This overcompensation occurs because the two consumer stages are not limited to the same extent by the same resource, and an increase in production of one stage does not necessarily involve a decrease in the production of the other. Therefore, predation on both consumer stages can create a positive feedback between the predator and the availability of its food that results in an emergent Allee effect, as shown in chapter 6.

PART III

ONTOGENETIC DEVELOPMENT AND
COMMUNITY DYNAMICS

CHAPTER 9

Dynamics of
Consumer-Resource Systems

In previous chapters we have in depth covered the implications of ontogenetic development for community structure. One major aspect that we have considered is how food-dependent development may give rise to alternative stable states of communities. A major insight from these chapters is that an understanding of community structure and dynamics can be gained by considering stage-dependent competitive ability and its relationship to development control versus reproduction control. In this and the following two chapters we will shift our focus to consider the effects of ontogenetic development on the dynamics of populations. Our aim is to show how the insights gained in previous chapters carry over when we want to understand the dynamics of size-structured populations. In particular, we will show that the size scaling of ecological performance (q in previous chapters) has major effects on the dynamics of both consumer-resource systems as well as systems including cannibalism (chapter 11).

In this chapter we focus on consumer-resource dynamics in systems where consumers of different sizes compete for a shared resource. We will analyze a model where we consider the implications of three important aspects of consumer life history: the explicit handling of a juvenile period leading to a delay between the time when an individual is born to when it starts to reproduce, the rate by which individual ecological processes scale with body size, and whether the rate by which the individual grows is dependent on food density or not (de Roos, Persson, and McCauley 2003). Second, we will consider the effects of different resource growth dynamics with the intention to illustrate the fundamental differences between population cycles driven by interactions between individuals of different sizes and classical predator-prey cycles driven by interactions between the consumer and the resource, also referred to as paradox of enrichment cycles (Rosenzweig and MacArthur 1963; Rosenzweig 1971) or prey escape cycles (de Roos, Metz, et al. 1990). We will also consider the empirical evidence for how common in natural systems are the type of cycles we consider in this chapter, compared with the predator-prey cycles covered in most other books. We will in this connection also to some extent cover life history processes leading to direct density dependence ("generation cycles") versus indirect density dependence ("delayed feedback cycles"), the latter found, for example, in host-parasitoid systems (Briggs, Sait, et al. 2000; Gurney and Nisbet 1985; Murdoch, Briggs, and Nisbet 2003; Nisbet and Gurney 1983). The comparison of cycles induced by life history processes and classical predator-prey cycles will include both model and empirical considerations. Specifically, we will discuss experiments with the model organism, the cladoceran zooplankton *Daphnia*, to elucidate what we do understand—and do not understand—about cycles driven by cohort inter-actions in this organism.

A SIZE-STRUCTURED POPULATION MODEL

As has been our general tenet so far in this book, we start out by formulating our model based on three basic aspects of size-dependent individual life his-tory. First, foraging intake and metabolic demands are strict functions of body size. Second, the rate by which individuals grow (develop) depends on food density. Third, small juvenile individuals allocate all their energy intake to so-matic growth, whereas adult individuals use their energy acquired for reproduc-tion only. We will characterize the individual (i) state with its size measured as body mass. Individuals are born at a size s_b and mature when reaching a size s_m. All individuals are assumed to feed on the same resource of a fixed size, whose density is R. We assume that individuals capture the resource follow-ing a Type II functional response with a half-saturation constant H_C. Maximum ingestion rate is assumed to increase linearly with body size for both juveniles and adults with proportionality constant M_C. As in previous chapters, we use q as a measure of competitive ability, where q is used as a proportionality constant such that juvenile mass-specific maximum ingestion equals M_C while that of adults equals qM_C. Adults are thus superior competitors to juveniles for $q > 1$, and vice versa for $q < 1$. Ingested food is transformed into new biomass with an assimilation efficiency σ_C. Assimilated food is first used to cover maintenance, hence we consider a net-production model (Lika and Nisbet 2000). The mainte-nance costs are assumed to scale linearly with body size with a proportionality constant T_C. Remaining energy is, as stated above, used for somatic growth by juveniles and for reproduction by adults, where the cost for each offspring equals s_b. All in all, the above assumptions lead to the following net-production rates *per unit body mass* for juveniles and adults, respectively:

$$v_J(R) = \sigma_c M_c \frac{R}{H_c + R} - T_c$$

$$v_A(R) = q\sigma_c M_c \frac{R}{H_c + R} - T_c$$

(9.1)

To handle the situation that resource levels may be lower than that required for maintenance, we assume that juvenile growth and adult reproduction equal zero when net production is negative. Individual juvenile growth rate in biomass, denoted $g(R,s)$, and individual adult birth rates in biomass, denoted $b(R, s_m)$, both restricted to non-negative values are then:

$$g(R,s) = v_J^+(R)s = \begin{cases} \left(\sigma_c M_c \dfrac{R}{H_c + R} - T_c\right)s & \text{if } R > \dfrac{H_c}{\sigma_c M_c/T_c - 1} \\ 0 & \text{otherwise} \end{cases}$$

$$b(R, s_m) = v_A^+(R)\frac{s_m}{s_b} = \begin{cases} \left(q\sigma_c M_c \dfrac{R}{H_c + R} - T_c\right)\dfrac{s_m}{s_b} & \text{if } R > \dfrac{H_c}{q\sigma_c M_c/T_c - 1} \\ 0 & \text{otherwise} \end{cases}$$

(9.2)

Note that the *number* of offspring produced is proportional to the ratio of adult size to birth size. We assume that juveniles and adults are exposed to a background mortality of μ_J and μ_A, respectively. Furthermore, when food intake is insufficient to cover maintenance requirements, juveniles will suffer from an additional starvation mortality equal to $-v_J(R)$. In a similar way, adults are assumed to experience an additional starvation mortality equal to $-v_A(R)$. The total death rates of juveniles ($d_J(R)$) and adults ($d_A(R)$), respectively, hence equal:

$$d_J(R) = \begin{cases} \mu_J & \text{if } R > \dfrac{H_c}{\sigma_c M_c/T_c - 1} \\ \mu_J - v_J(R) & \text{otherwise} \end{cases}$$

$$d_A(R) = \begin{cases} \mu_A & \text{if } R > \dfrac{H_c}{q\sigma_c M_c/T_c - 1} \\ \mu_A - v_A(R) & \text{otherwise} \end{cases}$$

(9.3)

To fully specify the model, we indicate the size distribution of juveniles over the size interval $s_b \leq s < s_m$ as $c(t,s)$ and the total number of adult consumers (size $s = s_m$) as $C(t)$. The dynamics of the consumer-resource system can be described by four equations: one describing the change in the size distribution of juveniles $c(t,s)$ as a result of individual growth and mortality; one specifying a boundary condition for $c(t,s)$, which describes the change as a result of production of offspring; one describing the dynamics of adult consumers as a balance between the maturation of juvenile individuals to the adult stage and adult mortality; and finally one describing the change in resource density as a result of resource production and foraging by consumers (box 9.1). For the resource dynamics, we will use two different renewal functions: semichemostat

---

BOX 9.1

VARIABLES, FUNCTIONS, AND POPULATION-LEVEL EQUATIONS OF THE
SIZE-STRUCTURED POPULATION MODEL WITH A SHARED RESOURCE,
WHICH FORMS THE BASIS OF THE STAGE-STRUCTURED BIOMASS MODEL

Variables	Description
R	Resource (food) density in environment
s	Individual body size
$c(t,s)$	Juvenile consumer size distribution
$C(t)$	Adult consumer density

Functions	Description
$\omega_J(R) = M_C \dfrac{R}{H_C + R}$	Juvenile mass-specific resource intake rate
$\omega_A(R) = qM_C \dfrac{R}{H_C + R}$	Adult mass-specific resource intake rate
$\nu_J(R) = \sigma_C M_C \dfrac{R}{H_C + R} - T_C$	Juvenile mass-specific, net-biomass-production rate
$\nu_A(R) = \sigma_C qM_C \dfrac{R}{H_C + R} - T_C$	Adult mass-specific, net-biomass-production rate
$g(R,s) = \begin{cases} \nu_J(R)s & \text{if } R > H_C/(\sigma_C M_C/T_C - 1) \\ 0 & \text{otherwise} \end{cases}$	Juvenile growth rate in body size
$b(R,s) = \begin{cases} \nu_A(R)s/s_b & \text{if } R > H_C/(\sigma_C qM_C/T_C - 1) \\ 0 & \text{otherwise} \end{cases}$	Adult fecundity
$d_J(R) = \begin{cases} \mu_J & \text{if } R > H_C/(\sigma_C M_C/T_C - 1) \\ \mu_J - \nu_J(R) & \text{otherwise} \end{cases}$	Juvenile mortality rate
$d_A(R) = \begin{cases} \mu_A & \text{if } R > H_C/(\sigma_C qM_C/T_C - 1) \\ \mu_A - \nu_A(R) & \text{otherwise} \end{cases}$	Adult mortality rate
$G(R) = \rho(R_{max} - R)$ or $G(R) = \rho R(1 - R/R_{max})$	Resource autonomous growth dynamics (semichemostat or logistic)

(*Box 9.1 continued*)

Population equations	Description
$\dfrac{\partial c(t,s)}{\partial t} + \dfrac{\partial g(R,s)\,c(t,s)}{\partial s} = -d_J(R)\,c(t,s)$	Change in juvenile size-distribution through growth and mortality
$g(R,s_b)\,c(t,s_b) = b(R,s_m)C$	Increase in newborn individuals through reproduction
$\dfrac{dC}{dt} = g(R,s_m)\,c(t,s_m) - d_A(R)C$	Adult consumer dynamics
$\dfrac{dR}{dt} = G(R) - \omega_J(R)\displaystyle\int_{s_b}^{s_m} s\,c(t,s)\,ds - \omega_A(R)\,s_m C$	Resource biomass dynamics

---

BOX 9.2
PARAMETERS USED IN THE SIZE-STRUCTURED CONSUMER-RESOURCE MODEL

Parameter	Default value	Unit	Description
Resource			
ρ	0.1	day^{-1}	Resource turnover rate
$R_{\max}$	100, varied	mg/L	Resource maximum biomass density
Consumer			
M_C	$0.1W_A^{-0.25}$	day^{-1}	Mass-specific maximum ingestion rate
H_C	3.0	mg/L	Ingestion half-saturation resource density
q	varied	—	Adult-juvenile consumer ingestion ratio
T_C	$0.01W_A^{-0.25}$	day^{-1}	Mass-specific maintenance rate
σ_C	0.5	—	Conversion efficiency
z	0.05	—	Newborn-adult consumer size ratio
μ_J, μ_A	$0.0015W_A^{-0.25}$	day^{-1}	Background mortality rate

W_A represents the average adult consumer body weight.

---

resource dynamics and logistic resource dynamics (box 9.1). Baseline parameter values are generally those used in chapter 3 (box 9.2).

The model derived above is a so-called physiologically structured population model (PSPM) (Metz and Diekmann 1986; de Roos 1997). For simulating its dynamics over time, we used the Escalator Boxcar Train (EBT) method, which is a numerical method specifically designed to handle the numerical integration of the equations that occur in PSPMs and which has been extensively used to study the dynamics of PSPMs under a range of different conditions including, for example, both continuous and discrete reproduction. The EBT formulation of PSPMs is explained in detail in several previous papers (de Roos 1988, 1997; de Roos, Diekmann, and Metz 1992). We see no reason to repeat this here but instead refer the readers to these original papers.

DYNAMICS WITH SEMICHEMOSTAT RESOURCE DYNAMICS

We will first study the consumer-resource model using semichemostat resource dynamics where the resource, in the absence of consumers, is described by

$$G(R) = \rho(R_{max} - R), \tag{9.4}$$

where R_{max} is the resource density in the absence of consumers and ρ is the resource turnover rate. Our choice to first consider semichemostat resource dynamics relates to the fact that we want to explore the consequences of varying stage-specific competitive ability (q) on population dynamics decoupled from the effects of logistic resource growth with the aim to unravel in more detail the mechanisms giving rise to cycles in size/stage-structured populations. As pointed out in chapter 3, semichemostat resource dynamics, in contrast to logistic resource growth, also conforms to basic principles of energy or mass conservation. Finally, semichemostat resource dynamics is a phenomenological way to handle the situation when not all resources are available to the consumers because they are either in a size refuge (i.e., some are too small to be encountered by the consumers but will become available as a result of individual growth) or in a spatial refuge (Persson, Leonardsson, et al. 1998). Overall, we will, throughout the chapter, assume that the mortality rates of juveniles and adults (μ_J, μ_A) are the same. This is done for illustrative reasons and does not entail any loss of generality (see chapter 3).

The stage-dependent competitive ability, that is, whether juveniles or adults are superior, has a profound effect on the dynamics of the consumer-resource system (figure 9.1). For low values of q ($q = 0.3 - 0.9$), the consumer-resource system exhibits population cycles. The amplitude of these cycles is high for adult and juvenile consumer biomass, whereas the amplitude for the resource is smaller (figure 9.2, left panel). Juveniles and adult consumers vary inversely to each other such that total consumer biomass and hence total consumer consumption fluctuates less, which explains why the resource fluctuates less than the two consumer stages separately (figure 9.1, upper right panel). The cycle

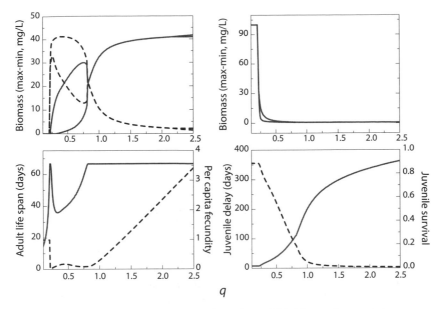

FIGURE 9.1. Changes in population dynamics and life history characteristics as a function of q. *Upper left panel*: maximum and minimum biomasses of juveniles (*solid lines*) and adults (*dashed lines*). *Upper right panel*: maximum and minimum resource biomasses. *Lower left panel*: adult life span (*solid line*) and adult fecundity (*dashed line*). *Lower right panel*: juvenile delay (*solid line*) and juvenile survival (*dashed line*). These results were obtained using long numerical simulations of the model dynamics with regular, stepwise increases as well as decreases in q. Other parameters have default values as presented in box 9.2.

is set by a dominating recruiting cohort, which depresses the resource to such a low level that adults cannot meet their energy demands, leading to zero reproduction and starvation mortality among adults. Once this dominating cohort reaches maturity, it gives rise to a new dominating cohort (figure 9.3, left panels). The life span of adults is short (figure 9.1, lower left panel; table 9.1), leading to a cycle length that is approximately equal to the time delay from birth to maturation, which is the reason why these cycles have been referred to as generation cycles (Nisbet and Gurney 1983). Because of their relation to juvenile competitive dominance, we refer to these cycles as *juvenile-driven cycles*. Other characteristics of the juvenile-driven cycles are that the resource and adult dynamics covary, hence high reproductive pulses coincide with peaks in resource densities (figure 9.2, left panel) and a short juvenile delay leading to a high juvenile survival (figure 9.1, lower right panel; table 9.1).

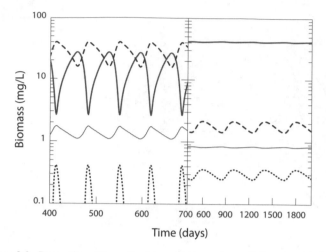

FIGURE 9.2. Dynamics of juvenile-driven (*left panel,* $q = 0.6$) and adult-driven (*right panel,* $q = 2.5$) cycles, including juvenile biomass (*thick solid line*), adult biomass (*thick dashed line*), resource (*thin solid line*), and daily biomass reproduction (*dotted line*). Other parameters have default values as presented in box 9.2.

Increasing q leads to a collapse of the cycles into a stable equilibrium in the region where juvenile and adult consumers are equal competitors (figure 9.1). This is also the situation when the size-structured model can be collapsed into Yodzis and Innes's (1992) one-stage biomass model and where cycles are a priori not expected with semichemostat resource dynamics.

With a further increase in q, cycles reappear. These cycles we will refer to as *adult-driven cycles*, because they occur when adults have a higher ingestion rate and hence competitive ability. Although the fluctuations in adult and juvenile biomasses are larger than fluctuations in the resource, the difference is smaller than for the cycles driven by juveniles (figure 9.2). In contrast to the juvenile-driven cycles, the size distribution fluctuates much less over the cycle, although distinct peaks in recruitment output are still present (figure 9.3, right panels). These cycles also have a period that is slightly longer than the juvenile delay, which can be related to the longer life span of adults (figure 9.1, lower left panel; table 9.1). Adult fecundity is also higher, despite a lower resource level related to the high consumption capacity of adults (figure 9.1, lower left panel; figure 9.2, right panel). In contrast, the low resource level leads to a long juvenile delay and, as a result, a low juvenile survival (figure 9.1, lower right panel; figure 9.2). Finally, reproduction peaks covary with peaks in adult biomass but are out of phase with peaks in resource levels, which also sets adult-driven cycles apart from the juvenile-driven cycles (figure 9.2). Interestingly,

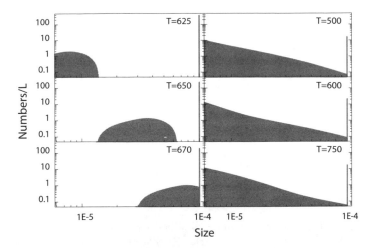

FIGURE 9.3. Changes in size distributions of juvenile-driven (*left panels*, $q = 0.6$) and adult-driven (*right panels*, $q = 2.5$) cycles over time. Other parameters have default values as presented in box 9.2.

TABLE 9.1. Some Qualitative Characteristics of Juvenile- and Adult-Driven Cycles for Semichemostat Resource Dynamics

Aspect	Juvenile-driven cycles ($q < 1$)	Adult-driven cycles ($q > 1$)
Amplitude	Large	Moderate
Period-delay ratio	≈ 1	≈ 1
Size distribution	One dominant cohort	Relatively constant
Juvenile-adult ratio	Low	High
Food density	High	Low
Adult fecundity	Low	High
Adult life span	Short	Long
Adult mortality	High	Low
Fecundity*-resource	In phase	Out of phase
Juvenile growth	Fast	Slow
Juvenile delay	Short	Long
Juvenile survival	High	Low

* Population fecundity

this means that adult per capita fecundity is at its lowest when the number of reproducing adults reaches its maximum. Despite this, the total reproduction is high in adult-driven cycles (table 9.1).

The characteristics of the two types of cohort cycles for q smaller and larger than 1, respectively, are summarized in table 9.1. For both types of cycles,

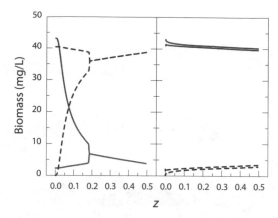

FIGURE 9.4. Effects of increasing z on juvenile-driven (*left panel*, $q = 0.6$) and adult-driven (*right panel*, $q = 2.5$) cycles. For all cases, maximum and minimum values of juveniles (*solid lines*) and adults (*dashed lines*) are given. These results were obtained using long numerical simulations of the model dynamics with regular, stepwise increases as well as decreases in z. Other parameters have default values as presented in box 9.2.

juveniles exert the major predation pressure on the resource, owing to their higher density. Overall, the patterns observed for population dynamics for different q values show that the effects of varying stage-dependent competitive ability, in previous chapters discussed in terms of development versus reproduction control, carry over into effects on population dynamics. As expected, the stage that forms the bottleneck also dominates the population in biomass, in correspondence to what we discussed in chapter 3 for the stage-structured biomass model (figure 9.2). Furthermore, as we also discussed in chapter 3, the observed fixed-point dynamics for $q = 1$ represents the symmetric borderline case that for consumer-resource dynamics separates two broad domains of conditions leading to either juvenile-driven or adult-driven cycles. The ratio of birth size to maturation size (z) was in chapter 3 shown to have substantial effects on the form of overcompensation present. This ratio also affects the likelihood for the juvenile-driven cycles to occur (figure 9.4, left), which can be related to the fact that increasing z leads to a decrease in the juvenile delay. In effect, juvenile-driven cycles are present for birth weight/maturation weight ratios below 18 percent. The fact that increasing z has no effect on the likelihood of adult-driven cycles to occur (figure 9.4, right) can on the other hand be related to the much larger juvenile delay for large q values (figure 9.1, lower right panel).

Before going into a more detailed analysis of the nature of the two types of cohort cycles, two empirical examples can be used to illustrate the oscillations

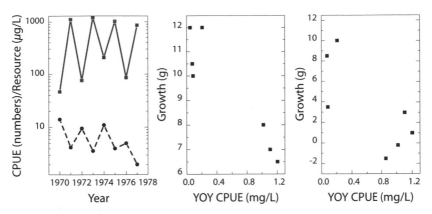

FIGURE 9.5. The dynamics of vendace in Lake Bolmen, Sweden, from 1970 through 1977 showing a two-year cycle. *Left panel*: vendace (CPUE, i.e., catch per unit effort, *solid lines*) cycling in antiphase with the resource zooplankton (biomass, *dashed lines*). *Middle panel*: growth of YOY vendace versus their own density. *Right panel*: growth of maturing (one-year-old) vendace versus the density of YOY vendace. Based on data from Hamrin and Persson (1986).

induced by cohort interactions. The first example is the zooplanktivorous fish vendace (*Coregonus albula*) (Hamrin and Persson 1986), and the second example is the cladoceran zooplankton *Daphnia* (McCauley and Murdoch 1987, 1990; Murdoch and McCauley 1985). In an eight-year study of the population dynamics of vendace, Hamrin and Persson analyzed the mechanisms behind the two-year cycle observed in this species. They first showed that the shared resource zooplankton was heavily depressed by strong recruiting cohorts of young-of-the year (YOY) vendace appearing every second year (figure 9.5, left panel). Second, the depression of zooplankton levels reduced their own growth rate (figure 9.5, middle panel). Third, the depression of zooplankton also resulted in decreased growth of maturing one-year-old vendace, as their summer growth was negatively related to YOY vendace abundance (figure 9.5, right panel). Fourth, the cycle length equaled the time from birth to maturation, the hallmark of generation cycles.

Moreover, Hamrin and Persson (1986) showed that strong YOY cohorts of vendace caused an increased mortality among adult vendace (but not vice versa). The dynamics of this system was characterized by a sequence of coupled events with a strong offspring cohort negatively affecting the growth of maturing one-year-old individuals, leading to a low reproductive output of these individuals the next year. In turn, this low reproductive output entailed that the one-year-old cohort in that year was exposed to low intercohort competition from the small YOY cohort in that year and thereby could allocate more energy to growth and reproductive tissue, thus leading to strong reproductive output the next year. As a result, the dominating strong cohort gave rise to another dominating strong cohort. To conclude, the dynamics of vendace show all the characteristics of a cohort cycle and also show that the cohort cycle is driven by competitively superior small juveniles.

Adult individuals were continuously present in the vendace population. This example thus does not represent a single-generation or single-cohort cycle in the classical sense, where adults appear only during a short time interval around reproduction and hardly overlap with juveniles (cf. Gurney and Nisbet 1985). However, it has been shown that increased background mortality in consumer populations showing single-cohort dynamics will lead to a shortening of the cycle length, increased average food density, and coexistence of juveniles and adults without changing the nature of the cycle (de Roos and Persson 2001; Persson, Leonardsson, et al. 1998). In the vendace example, piscivorous fish were the major agents behind the high mortality on vendace. The population dynamics can nevertheless be explained as a cycle induced by competitively superior juveniles where single-generation/cohort cycles represent an end point in cases with low background mortality. In the following, we will refrain from using the term "single"-generation/cohort cycle and instead use the more general term generation/cohort cycle to include situations with both overlapping generations and nonoverlapping generations.

As we will consider in more detail at the end of this chapter, *Daphnia* has served as a model organism for experimental studies of population cycles, particularly in relation to stage-structured dynamics. In a synthesizing paper, McCauley and Murdoch (1987) demonstrated the presence of fecundity suppression by recently hatched individuals (figure 9.6). Peaks in overall *Daphnia* abundance were largely due to a cohort of young individuals following a peak in reproduction output. This dominating cohort subsequently depressed fecundity among adults, reflected in both the average number of eggs per adult and, as shown in figure 9.6, the fraction of adults. McCauley and Murdoch (1987) analyzed a number of other studies where these dominance and suppression mechanisms characteristic for cohort cycles have been observed. The cycles in *Daphnia* have in several papers been interpreted as cohort cycles driven by a dominating juvenile cohort (e.g., Murdoch, Briggs, and Nisbet 2003), but recent studies have suggested the opposite (e.g., McCauley, Nelson, and Nisbet 2008). At the end of this chapter, we will in some depth discuss the nature of the cohort cycles in *Daphnia*, concluding that the population dynamics of this organism are not completely understood.

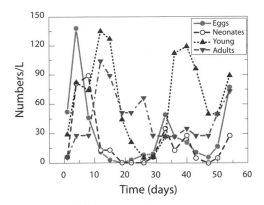

FIGURE 9.6. Example of the demography of a *Daphnia* population in a tank experiment (data from McCauley and Murdoch 1987).

Mechanisms behind Juvenile-Driven and Adult-Driven Cycles

An unstructured model in which juveniles and adults by definition do not differ in their mass-specific intake and metabolic rates will not generate cohort cycles with semichemostat resource dynamics, in correspondence with the lack of cycles for $q = 1$ with a size-independent mortality rate (e.g., Yodzis and Innes 1992). This situation represents the case of ontogenetic symmetry in energetics, where a size-structured system can be collapsed into an unstructured one. In the introduction to this chapter, we advanced three mechanisms not present in unstructured models that potentially may induce population cycles in size-structured models. First, the presence of a juvenile delay per se has in combination with a negative feedback loop been shown to be able to produce cycles (Gurney, Blythe, and Nisbet 1980). Second, food dependence in juvenile growth has been shown to lead to the possibility for variation in the juvenile delay that may also induce cycles (Gurney and Nisbet 1985). Third, the size scaling of ecological processes may have profound effects on the presence of cycles related to size-structured interactions (de Roos and Persson 2003). Gurney and Nisbet (1985) analyzed a larvae-pupae-adult (LPA) model studying the effects in four scenarios: effects through per capita larval death rate, larval maturation time, pupae survival, and adult fecundity. They found that direct effects on larvae via per capita death rate or maturation time were more likely to lead to generation-type cycles, especially when the adult period was short. In contrast, indirect effects of larval feeding success on their future pupal survival or adult fecundity led to delayed feedback cycles characterized by cycle periods two to four times the generation time. We will come back to these types of cycles later in this chapter.

In a recent contribution, Diekmann, Gyllenberg, et al. (2010) concluded that the maturation delay on its own does not lead to oscillations. More specifically, if juvenile and adult consumers differ only in that juveniles use their ingested resource for growth and adults for reproduction with fixed conversion factors, the stability properties of the structured model will exactly mimic those of an unstructured model. They further found that cycles are possible with fixed development if, near the steady state, a relatively small increase in food leads to a relatively large increase in reproduction, which is the case in net-production models (see chapter 3). However, this required that the conversion efficiency to reproduction depends rather strongly on the resource density near the equilibrium. If such cycles occur, the period is between two to four times the duration of the juvenile period.

Considering the size scaling of ecological processes, de Roos and Persson (2003) analyzed in detail the necessary conditions for the presence of juvenile-driven and adult-driven cohort cycles, respectively. The model differed, among other things, from the biomass-based model we have considered so far in assuming a Type I functional response and a constant resource growth rate. In agreement with the results presented here, we found a lack of cycles when juveniles and adults were equal competitors ($q = 1$). We subsequently contrasted model variants with food-dependent and fixed development times and found that the region with stable equilibria expands such that the equilibrium was always stable for $q > 1$ with a fixed development. In other words, with a fixed juvenile delay, adult-driven cycles are absent. In contrast, the juvenile-driven cycles present for $q < 1$ remained, with only minor changes, in the parameter region for which they occur. These results thus suggest that the cycles we observe when juveniles are competitively dominant are due to the time delay induced by a juvenile period per se. In contrast, variability in the juvenile delay, which in turn results in a large temporal variation in the survival from birth to maturation, is a crucial mechanism for the population cycles to occur when adults are competitively superior.

The fluctuations in juvenile delay in adult-driven cycles will most likely come about because the competition among juveniles is retarding their own growth. It should also be pointed out that adults have a longer life expectancy and higher fecundity in adult-driven cycles than in juvenile-driven cycles, hence adults produce more offspring that suppress food densities in adult-driven cycles, leading to an increase in the juvenile delay and a decrease in juvenile survival to maturation (see table 9.1). The conclusions about the role

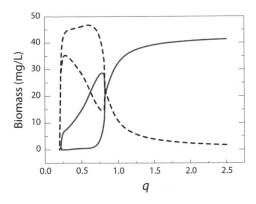

FIGURE 9.7. Maximum and minimum juvenile (*solid lines*) and adult (*dashed lines*) biomasses for varying q with a fixed maturation age. Other parameters have default values as presented in box 9.2.

of the juvenile delay, and whether or not this delay is food dependent, for juvenile- and adult-driven cycles, respectively, to occur, are supported by the characteristics of the two types of cycles (table 9.1; figures 9.1 and 9.2). The biomass-based model that we have used in this chapter also shows that adult-driven cycles are absent when a fixed juvenile period is assumed, suggesting that the conclusion about the necessity of a food-dependent development delay for adult-driven cohort cycles to occur is robust to substantial differences in model structure (figure 9.7) (see also de Roos and Persson 2003). This result is in agreement with the findings of Diekmann, Gyllenberg, et al. (2010) that a fluctuating length of the juvenile period as a result of individual juveniles' growth being negatively related to their density may on its own lead to oscillations.

For juvenile-driven cycles, the fluctuations in population reproduction have been related to (1) the fluctuations in adult density and (2) the synchrony between adult fecundity and food density (figure 9.2, left panel). Fluctuations in adult density and population reproduction are large in these cycles, and the almost perfect synchrony in the dynamics of food and adult density will also add to this large amplitude. Diekmann, Gyllenberg, et al. (2010) analyzed the special case when only juveniles exploited the resource (adult reproduction fixed) and found that cycles were promoted by a high reproduction rate and that the period is between one and two times the juvenile period, at least in a pseudo– or quasi–steady state approximation.

Using a similar approximation, de Roos and Persson (2003) investigated the importance of the almost synchronous development of food density and adult density for the occurrence of juvenile-driven cycles in a model variant in which it was assumed that the food density at any time was in pseudo–steady state with the current juvenile and adult density. The pseudo–steady state assumption might, for example, hold when resource dynamics are so rapid that the current resource density is always very close to its equilibrium density *for the current density of consumers* (note that consumers are not necessarily in equilibrium, hence the "pseudo"). As a result, resource density will be inversely related to the current feeding pressure of juvenile and adult consumers, and the indirect feedback resulting from the dynamics in food dynamics can be replaced by a direct density dependence of individual development, survival, and reproduction on the consumer population. In the model variant analyzed by de Roos and Persson (2003), the resource level was hence given by:

$$R(t) = \frac{1}{J(t) + A(t)}. \qquad (9.5)$$

The inverse relation between food and consumer density in equation (9.5) destroyed the positive correlation in time between fecundity and adult consumer density, which allowed a test of the importance of the synchrony for juvenile cycles to occur. De Roos and Persson (2003) provided a stability analysis of the equilibrium for this model variant and showed that the equilibrium was stable for all values of $q < 1$, hence juvenile driven cycles do not occur. Recently, the in-depth analysis of Diekmann, Gyllenberg, et al. (2010) supported this conclusion, as juvenile-driven cycles disappeared in cases of fast resource dynamics. In contrast, in a similar modification of the biomass-based model used in this chapter, we found that the juvenile-driven cycles remained; hence the results here seem to be sensitive to model structure, in contrast with what we found for adult-driven cycles.

To conclude, results from models of substantially different structure show that food-dependent development leading to variability in the development rate is of major importance for the adult-driven cycles to occur. In contrast, the conclusion of de Roos and Persson (2003) that an almost synchronous development of food density and adult density is vital for the juvenile-driven cohort cycles to occur is less robust and depends on the specific model structure.

DYNAMICS WITH LOGISTIC RESOURCE DYNAMICS

So far, we have considered how individual variation may generate cycles under conditions where unstructured models would not. In the following, we exchange the semichemostat resource dynamics with logistic resource growth:

$$G(R) = \rho R \left(1 - \frac{R}{R_{max}} \right) \qquad (9.6)$$

From ecological textbooks, we know that the combination of a Type II functional response and logistic growth in the resource may give rise to large amplitude cycles, particularly for parameter values with high resource productivity or a high attack rate in the predator (i.e., paradox of enrichment cycles or prey escape cycles; cf. de Roos, Metz, et al. 1990; Murdoch, Briggs, and Nisbet 2003; Rosenzweig and MacArthur 1963; Rosenzweig 1971). We will in the following discuss the conditions under which these paradox of enrichment cycles occur in size-structured consumer-resource models and, moreover, show that paradox of enrichment cycles and cohort cycles may form alternative attractors.

To start with, we consider the case where $q = 1$, that is, the condition where for semichemostat dynamics we did not find cycles, but only a stable equilibrium. At low productivities, we find the same for logistic resource growth. With increasing resource productivity (R_{max}), the stable equilibrium destabilizes and gives rise to regular cycles, the amplitude of which will increase with increasing resource productivity (figure 9.8, left). A closer inspection of these cycles reveals that adults and juveniles, in contrast to what is the case for cohort cycles discussed so far, oscillate in complete phase with each other. The in-phase oscillations of juveniles and adults is in fact a characteristic of paradox of enrichment cycles (de Roos, Metz, et al. 1990; McCauley, Nisbet, et al. 1999) (figure 9.8, right). A rapid increase in resource density causes an increased population reproduction rate, and the newborn individuals mature quickly, leading to an almost simultaneous increase in adult density. This high reproduction rate and development rate of juveniles are sustained until the resource declines to low levels because of consumption, when both reproduction and development essentially stop. Owing to the almost identical response in juvenile and adult density, the rate by which the consumer juvenile and adult population increases (or decreases) is more or less the same, such that the age and size distributions of the population remain relatively constant despite the presence of population fluctuations (cf. Roughgarden 1979). Therefore, population size structure can be ignored for $q = 1$, and the cycles present at higher resource productivities can be understood based on theory from unstructured models (Murdoch, Briggs, and Nisbet 2003; Rosenzweig and MacArthur 1963; Rosenzweig 1971).

For q smaller or larger than 1, cycles of qualitatively different types that can be separated into classical predator-prey cycles and cohort cycles will be present. We will differentiate between predator-prey cycles and cohort cycles primarily by the fact that the former have a perfect positive cross-correlation

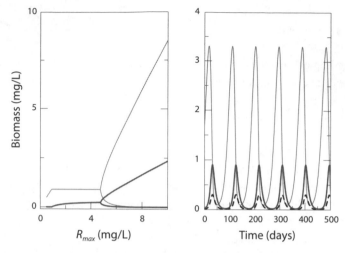

FIGURE 9.8. *Left panel*: Maximum and minimum resource (*thin solid line*) and juvenile biomass (*thick solid line*) for varying R_{max} using logistic resource growth. *Right panel*: Population oscillations in resource density (*thin solid line*), juvenile biomass (*thick solid line*), and adult biomass (*dashed thick line*) ($R_{max} = 6.0$ mg/L). Throughout, $q = 1$ was used. Note the synchronous change in juvenile and adult biomass. Other parameters have default values as presented in box 9.2.

between juveniles and adult biomasses at lag 0, whereas the latter show a lag (McCauley, Nisbet, et al. 1999). For semichemostat dynamics, we further found a period/delay ratio around 1 for both juvenile-driven and adult-driven cycles (table 9.1). We will consider whether the latter is also the case with logistic resource growth. For $q < 1$, cohort cycles of similar characteristics that we studied in detail for semichemostat resource dynamics are present up to a productivity (R_{max}) of 4.5 mg/L (figure 9.9, left panels; figure 9.10, left panel). Above this productivity, the system is characterized by high amplitude predator-prey cycles of the form shown in figure 9.8. The two types of cycles are, among other things, distinguishable through differences in juvenile-adult cross-correlations (figure 9.8 versus figure 9.10, left panel). The characteristics of the cohort cycles are with one notable exception the same as what we found for semichemostat resource dynamics (table 9.1, left, versus table 9.2, left). This exception is that the period/delay ratio is substantially larger than 1 (1.4–2.2 depending on R_{max}) for logistic resource growth, owing to the fact that the cycle period is longer (≈ 105 days) with logistic resource growth than with semichemostat dynamics (≈ 67 days) (figure 9.2, left, versus figure 9.10, left).

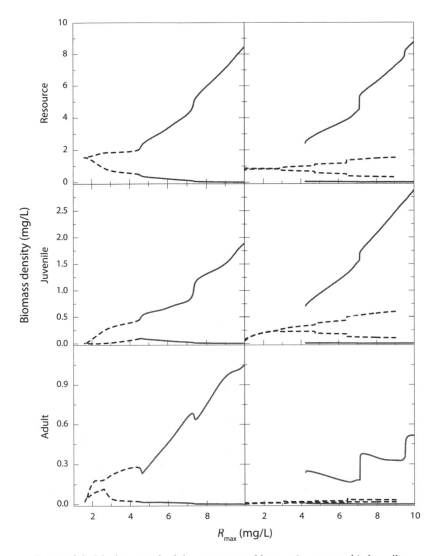

FIGURE 9.9. Maximum and minimum resource biomass (*upper panels*), juvenile biomass (*middle panels*), and adult biomass (*lower panels*) for $q = 0.6$ (*left*) and $q = 2.5$ (*right*) as a function of R_{max} using logistic resource dynamics. Dashed lines represent cohort cycles, and solid lines represent high-amplitude predator-prey cycles. Other parameters have default values as presented in box 9.2.

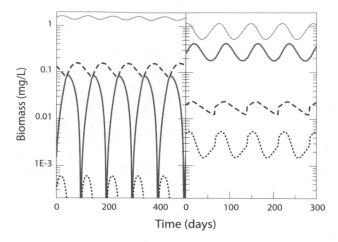

FIGURE 9.10. *Left panel*: Juvenile-driven cohort cycles showing resource (*thin solid line*), juvenile (*thick solid lines*), adult (*dashed lines*), and daily reproduction (*dotted lines*) biomasses ($q = 0.6$, $R_{max} = 2.0$ mg/L). *Right panel*: Adult-driven cohort cycles showing resource, juvenile, adult, and daily reproduction biomasses (line styles as in left panel; $q = 2.5$, $R_{max} = 6.0$ mg/L). Both cases are with logistic resource growth. Other parameters have default values as presented in box 9.2.

The longer cycle period with logistic resource growth can, in turn, be related to the fact that the resource has not fully recovered from the depletion induced by the juvenile peak in biomass when the peak in adult biomass occurs, leading to a delay in reproduction pulse until the peak in resource biomass takes place (figure 9.10, left). In contrast, for semichemostat resource dynamics, resource productivity is constant and independent of current resource density, such that recovery is fast. As a consequence, we found a perfect correlation between resource biomass, adult biomass, and reproductive output (figure 9.2).

For $q < 1$, cohort cycles do not coexist with the paradox of enrichment cycles, but rather there is a gradual development of the cycles from typical cohort cycles to typical predator-prey cycles. This gradual development includes an increase in the variability of the resource and a gradual shift in the peak in biomass of adults (figure 9.11). At low productivities, the peak in adult biomass is strongly negatively correlated with the peak in juvenile biomass. With increasing productivity, a second peak in adult biomass develops that is positively correlated with the peak in juvenile biomass, and for high productivities, the peak in adult biomass totally coincides with the peak in juvenile biomass (note that the peak in juvenile biomass in relation to the peak in resource biomass hardly changes with productivity). There is also a gradual shift in biomass dominance,

TABLE 9.2. Some Qualitative Characteristics of Juvenile-Driven
and Adult-Driven Cycles for Logistic Resource Growth

Aspect	Juvenile-driven cycles ($q < 1$)	Adult-driven cycles ($q > 1$)
Amplitude	Large	Moderate
Period-delay ratio	2.2*	0.33**
Size distribution	One dominant cohort	Relatively constant
Juvenile-adult ratio	Low	High
Food density	High	Low
Adult fecundity	Low	High
Adult life span	Shorter	Longer
Adult mortality	High	Low
Fecundity***-resource	In phase	In phase
Juvenile growth	Fast	Slow
Juvenile delay	Short	Long
Juvenile survival	High	Low

Note: The comparisons are made relative to the two different kinds of cohort cycles observed with semichemostat resource dynamics.

* for $R_{max} = 2$ mg/L; ** for $R_{max} = 6$ mg/L; *** Population fecundity.

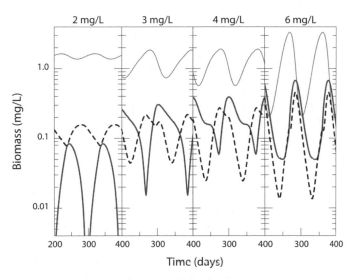

FIGURE 9.11. Population dynamics of resource (*thin solid line*), juveniles (*thick solid lines*), and adults (*dashed lines*) for different productivities (corresponding R_{max}-value shown at panel top) with $q = 0.6$ and logistic resource growth. Other parameters have default values as presented in box 9.2.

with adult dominance at low productivities and juvenile dominance at high productivities; thus, the presence of distinct juvenile-driven cohort cycles are clearly linked to adult biomass dominance (figure 9.11).

Cohort cycles and predator-prey cycles are also present for $q > 1$. The high amplitude predator-prey cycles will be present from an R_{max} of 4.2 mg/L, and the low amplitude cohort cycles from an R_{max} of 2.8 to an R_{max} of 9.1 mg/L (figure 9.9, right panels; figure 9.10 right panel). In contrast to the situation for juvenile-driven cohort cycles, these cohort cycles thus coexist with the high amplitude predator-prey cycles. As was the case for juvenile-driven cohort cycles, the low amplitude cohort cycles for $q > 1$ with logistic resource growth have many characteristics of the adult-driven cohort cycles that we discussed for semichemostat resource dynamics (figure 9.2, right panel; figure 9.10, right panel; tables 9.1 and 9.2). For example, in contrast to the juvenile-driven cycles, juveniles and adults oscillate largely in phase with each other. As we found for juvenile-driven cycles, there are also differences. First the period/delay ratio is substantially smaller than 1 with logistic resource growth, which is primarily due to a much shorter cycle length (67 time units versus 300 time units for semichemostat resource dynamics) related to the fact that the juvenile delay is shorter for logistic resource growth (on average, 230 time units versus 362 time units). A second difference between semichemostat dynamics and logistic growth is that resource biomass, adult biomass, and fecundity covary with only a small lag for logistic resource growth (table 9.1 versus table 9.2; figure 9.10, right). Although both adult reproduction and juvenile growth stop during phases with low resource densities, juveniles are more affected than adults, owing to their lower competitive ability. Because the juvenile stage duration is much longer than the cycle period, every juvenile individual experiences several occasions with retarded growth before it reaches maturation.

The circumstance that regular, adult-driven cohort cycles and paradox of enrichment cycles coexist whereas juvenile-driven cohort cycles and paradox of enrichment cycles do not may be understood by considering the characteristics of the juvenile-driven cohort cycles. For juvenile-driven cohort cycles, where juvenile and adult biomasses are out of phase with each other and adults are only present part of the time, the interplay between the oscillatory inclination of the logistic resource growth and the intrinsic cohort dynamics interfere with each other. This is less the case for adult-driven cycles, where adults and juveniles oscillate more in phase with each other and the population structure varies over time to a lesser extent. We have also already noticed the gradual shift in the adult-juvenile biomass correlation with increasing productivity from a strong negative one to a positive one (figure 9.11). The difference in the extent of interference between mechanisms promoting different kinds of cycles may also explain why adult-driven cycles always seem to be regular over their parameter range of existence, whereas juvenile-driven cycles for logistic resource growth can show a strong tendency for nonregularity (de Roos, Schellekens, et al. 2008b). Notably, the productivity level at which predator-prey cycles appear is largely independent of q (figures 9.8 and 9.9). This fact strongly suggests that the parameter region of cohort cycles is constrained by the interference with logistic resource growth, whereas the domain of predator-prey cycles is largely unaffected by intrinsic cohort dynamics. This conclusion is supported by simulation runs that vary R_{max} for semichemostat resource dynamics, which show no disappearance of cohort cycles at high productivities (see also Persson, Leonardsson, et al. 1998).

Coexistence between high-amplitude predator-prey cycles and low-amplitude cohort cycles has been demonstrated for the *Daphnia*-algal system (McCauley, Nisbet, et al. 1999). In a series of beautiful experiments, these authors demonstrated two types of dynamics, which had the characteristics of predator-prey and cohort cycles, respectively, under the same productivity conditions (figure 9.12). In one type of cycle, the amplitude of the fluctuations was much smaller, and there was also a lag between adult and juvenile peaks in density of 10–11 days (figure 9.12, left panel). In contrast, in the other type of cycle, the amplitude of the fluctuations was much higher, and adults and juveniles fluctuated in phase (figure 9.12, right panel). Both differences in characteristics of the two types of cycles are what we expect based on the

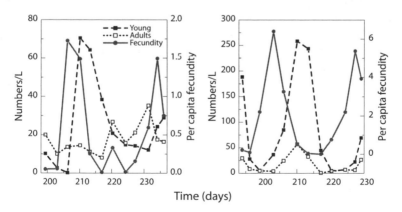

FIGURE 9.12. Alternative dynamics in the *Daphnia*–algae system. In cohort cycles (*left panel*), adults and juveniles do not oscillate in phase, whereas this is the case for predator-prey cycles (*right panel*). Note also the difference in cycle amplitude between cohort cycles and predator-prey cycles. Data from McCauley, Nisbet, et al. (1999).

theoretical insights we have provided, and the coexistence of cohort cycles and predator-prey cycles also suggests that the cohort cycles in *Daphnia* are adult-driven. We will return to this latter question later in this chapter. McCauley, Nisbet, et al. (1999) also showed that cohort cycles were more likely to occur than predator-prey cycles in the experimental systems, a fact that they related to the production of resting eggs, which under poor food conditions will temporarily decouple *Daphnia* dynamics from algal dynamics.

ARE THERE MORE TYPES OF JUVENILE-DRIVEN CYCLES?

So far we have considered two types of cycles induced by differences in stage-dependent competitive ability, for which we have coined the terms juvenile- and adult-driven cohort cycles, respectively. However, de Roos, Metz, et al. (1990) already showed that more types of juvenile-driven cycles are possible. Their physiologically structured population model was, in contrast to the model we have considered so far, based on the Kooijman-Metz (1984) energy budget model. Besides assuming a qualitatively different type of energy channeling of ingested energy, de Roos, Metz, et al. (1990) further assumed that individuals died instantaneously if ingested energy did not meet maintenance requirements (hence no starvation capacity was present) and that individuals died when reaching a fixed maximum age. As pointed out by de Roos (1997), these assumptions do not affect the characteristics of the overall dynamics, and we will hence not consider the implications of these additional assumptions here. More important, the model assumed the ingestion rate of the consumer (*Daphnia*) to be proportional to its surface area (weight$^{2/3}$), whereas the consumer's metabolic rate scaled linearly with weight. In effect, this means that we are in the range of juvenile-driven cycles ($q < 1$).

Before jumping into the possible existence of several types of juvenile-driven cycles, it is first useful to shortly highlight some insights gained by these early studies (de Roos, Metz, et al. 1990; Metz, de Roos, and van den Bosch 1988) on size-structured consumer-resource interactions that we have so far not covered. These insights relate to the fact that the changes in resource levels on food intake and fecundity in size-structured systems will involve both a direct effect on food intake and fecundity for the ambient consumer size distribution, as well as an indirect effect on food intake and fecundity mediated via the changes in the size distribution of consumers. This contrasts with unstructured models, in which only the direct effects are present. The resource-related changes in size distribution referred to as growth curve plasticity will, in turn, affect the stability properties of the system. De Roos, Metz, et al. (1990) showed that the effect of growth curve plasticity on reproduction is destabilizing, whereas its effect on consumption is stabilizing. The implications of this indirect effect of fluctuations in food availability on population dynamics may be major, which we will see in particular when we consider cannibalistic interactions (chapter 11).

We may now return to the question about more than one type of juvenile-driven cycles. De Roos, Metz, et al. (1990) showed that alternative cycles were present with semichemostat dynamics if resource productivity (R_{max}) was sufficiently high. One type of these cycles was the single-generation type cycles that we have already discussed. The other type of cycle exhibited a smaller amplitude and, importantly, the period length was substantially shorter than the juvenile delay (half of that of the single-generation cycles for default parameter values) (figure 9.13). The nature of the regulation of this latter cycle was not uncovered but was related to some type of density regulation within one generation.

The presence of two periodic attractors for size-scaling parameters where juvenile-driven cycles are expected to occur was also found by Persson, Leonardsson, et al. (1998), and a more thorough analysis of the characteristics of these alternative cycles was carried out by de Roos and Persson (2003). The latter authors found two types of cycles with very different behaviors coexisting for $q < 1$. One of these cycles is the juvenile-driven cycle that we have already considered (hereafter juvenile-driven Type I cycles), whereas the other one (hereafter juvenile-driven Type II cycles) has quite different characteristics. In

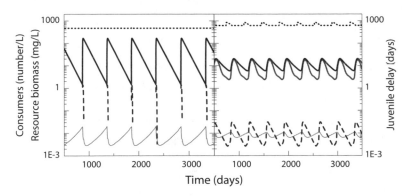

FIGURE 9.13. Juvenile-driven Type I (*left panel*) and juvenile-driven Type II (*right panel*) cycles. Numbers of juveniles (*black thick line*), adults (*dashed thick line*), and resource (*thin line*) and juvenile delay (*dotted line*) are given, and, for the juvenile-driven Type II cycles, also the numbers of juveniles < 50 mm (*gray thick line*). The Kooijman-Metz model was used (box 3.6) with a background mortality of 0.01 day^{-1}, and other parameters as listed in box 3.7.

figure 9.13, time series of the two types of juvenile-driven cycles are shown using the Kooijman-Metz model analyzed by de Roos, Metz, et al. (1990). The two types of cycles are distinguishable from each other by, among other things, the ratio between the cycle period and the juvenile delay. As we have already seen, the cycle period largely equals the juvenile delay for juvenile-driven Type I cycles with semichemostat resource dynamics. In contrast, for juvenile-driven Type II cycles, cycle length is approximately half the juvenile delay (figure 9.13). A further difference between the two cycle types is that the Type II cycles have a lower amplitude in terms of coefficient of variation in food, juvenile, and adult densities. Average food density is also higher for Type I cycles than for Type II cycles, which results in a shorter juvenile period of Type I cycles.

Comparisons of the juvenile-driven Type II cycles described by de Roos and Persson (2003) for a completely continuous model with the low amplitude, juvenile-driven cycles found by Persson, Leonardsson, et al. (1998) for a continuous model with discrete reproduction show that they have very similar characteristics, suggesting that the presence of the juvenile-driven Type II cycles is robust to major changes in model structure. In juvenile-driven Type II cycles, adults can produce another pulse of offspring when the number of individuals of the first dominating cohort has decreased sufficiently to allow the resource to recover. As a result, there are two dominant, not strongly segregated cohorts of juveniles present in the population at different stages in their development, and, furthermore, adults are always present in the system (figure 9.13). Both these characteristics clearly differentiate juvenile-driven Type II cycles from juvenile-driven Type I cycles.

OTHER SIZE-DEPENDENT
CONSUMER-RESOURCE DYNAMICS

The cycles involving a juvenile delay that we have considered so far share the common characteristic that the cycle length, with the important exception of juvenile-driven cycles with logistic growth, is equal to or smaller than the juvenile delay. Cycles with a cycle length equal to the generation time have, as mentioned above, been termed single-generation cycles, where the population is totally dominated by a single cohort that is born, grows, matures, and gives rise to another strong cohort (Gurney and Nisbet 1985). As we have also mentioned, overlapping generations do occur if background mortality is sufficiently high, which is the reason we prefer to use the more general term generation/cohort cycles.

A feature of cohort cycles is their direct density dependence in contrast to delayed density dependence found in other types of size/stage dependent interactions (Gurney and Nisbet 1985, see also below). A key element of these cycles is the formation of a dominating cohort, which suppresses the resource and thereby decreases the production of new cohorts through decreased adult fecundity and survival or through decreased survival of other immature cohorts. We have considered two examples of cohort cycles, the cladoceran *Daphnia* and the zooplanktivorous fish vendace. For vendace, an extensive empirical analysis of its dynamics has substantiated that the cycle is a juvenile-driven Type I cycle, where the cycle length equals the time from birth to maturation (Hamrin and Persson 1986). The extensive experimental studies on *Daphnia* have also shown the presence of small-amplitude, cohort-driven dynamics, although the nature of these cycles may not be completely understood, as we will consider in the next section.

Another classic example of generation cycles that has been studied extensively is the Indian meal moth (*Plodia interpunctella*) (Sait, Begon, and Thompson 1994). Although competition by itself can produce generation cycles as observed in *Plodia*, a more detailed analysis showed that if competition is assumed to be asymmetric favoring larger stages, the model produced cycles with a period of half the generation time ("half generation cycles"; Briggs, Sait, et al. 2000). The assumption that larger *Plodia* are competitively superior was empirically substantiated by the observation that strong resource limitation resulted in heavy mortality, especially among younger instars. The period/delay ratio of 0.5 observed for the *Plodia* model, with larger instars being competitively superior, is in agreement with what we have found for logistic resource growth (table 9.2). In a modeling study, Briggs, Sait, et al. (2000) showed that stage-specific cannibalism, which has been observed in *Plodia*, had to be accounted for to obtain the empirically observed cycle period and cohort structure. Essentially, stage-specific cannibalism speeded up development rate to maturation.

Another type of size/stage-dependent dynamics are delayed feedback cycles. Delayed feedback cycles are fully covered in a previous treatment of consumer-resource dynamics with a particular focus on host-parasitoid systems (Murdoch, Briggs, and Nisbet 2003), for which reason we refrain from carrying out a more extensive examination of these cycles. Instead, we restrict ourselves to pointing out some major differences between delayed feedback cycles and generation cycles. Generally speaking, delayed feedback cycles result from delayed density dependence, particularly from density dependence that is delayed across generations (Gurney and Nisbet 1985). Diekmann, Gyllenberg, et al. (2010) showed that delayed feedback cycles will occur only if there are differences between juveniles and adults in either ingestion rate or mortality. A typical scenario is the case where adult fecundity is density dependent whereas juvenile survival is density independent. In this case, a high density of adults in the current generation will affect their fecundity immediately, but owing to the density-independent juvenile survival, the density-dependent effect carries through to the number of adults recruited in the next generation (Murdoch, Briggs, and Nisbet 2003). Density-dependent fecundity thus acts with a delay. Gurney and Nisbet (1985) and Jones, Nisbet, et al. (1988) showed that the minimum period of delayed feedback cycles are at least two times the development time and, depending on adult longevity, in some cases even up to four times the development time (for a Ricker model).

A classic example of a delayed feedback cycle is presented in Nicholson's (1957) experiments with blowflies. In the case where adults competed for a fixed amount of food whereas larvae had unlimited food, adult fecundity decreased with their own density, whereas the survival during the immature period was high and invariant. The adult density at a particular time turned out to depend on the density of adults that were present at the current time minus the development delay. When larval density was high, subsequent adult density was also high, owing to the density-independent survival of immature individuals. As a result, per capita but also total population fecundity was reduced, which through immature, density-independent survival carried over to the next adult generation. A high density of adults thereby results in a low density of the next adult generation, which leads to a cycle length of at least two times the juvenile delay. Model results and empirical data have also corroborated that the cycle length for the blowfly system is slightly longer than two times the development time (Murdoch, Briggs, and Nisbet 2003).

At this stage, it is natural to ask the question: how common are cohort/ generation cycles and delayed feedback cycles in empirical systems, especially in relation to predator-prey cycles? First of all, Kendall, Prendergast, and Bjørnstad (1998) showed that population cycles occur only in around 30 percent of all populations. This analysis was based on population data from the Global Population Dynamics Database and included seven hundred long (i.e., ≥ twenty-five years) time series. Cycle incidence was found to be highest in mammals and fish populations (Kendall, Prendergast, and Bjørnstad 1998). In a subsequent analysis, Murdoch, Kendall, et al. (2002), using one hundred of the cyclic populations covering forty species from the same database, investigated the nature of the cycles with respect to whether they represented predator-prey cycles, generation cycles, or delayed feedback cycles. This was done by scaling the observed cycle length by the time to maturity. The resulting ratio for generation cycles is approximately 1, as we have concluded above, and about 2 to 4 for delayed feedback cycles.

For predator-prey cycles, the scaled cycle period was shown to be four times the juvenile delay plus two times the development time of the resource population (see Murdoch, Kendall, et al. 2002). Their analysis first showed that predator-prey cycles were restricted to specialist predators. Second, generation cycles were as common as predator-prey cycles (36 percent and 38 percent of all cycles, respectively), whereas delayed feedback cycles were present in 26 percent of the cyclic populations. Above, we found that the period/delay ratio for cohort cycles with logistic resource growth can be larger than 2 and thereby overlap with the period/delay ratio of delayed feedback cycles. These results thus show that we cannot distinguish cohort cycles from delayed feedback cycles based merely on their period/delay ratio, but that more information is needed to separate them out. Nevertheless, cycles resulting from size/stage-dependent interactions (cohort cycles and delayed feedback cycles combined) constitute the clear majority (62 percent) of all cycles observed, a figure that contrasts substantially with the poor coverage in the general ecological literature of stage/size-based cycles compared with predator-prey cycles.

DAPHNIA–ALGAE AS A MODEL SYSTEM FOR THE STUDY OF STAGE-STRUCTURED DYNAMICS

The cladoceran zooplankton *Daphnia pulex* has been the target of many studies directed toward investigating the effects of stage/size variation on the dynamics of consumer-resource systems. From the start, these studies have addressed the source and form of the cycles. In different populations of *Daphnia* and algae, different dynamics have been observed: some appeared to be stable; some *Daphnia* and algae fluctuated in joint cycles; and in some populations *Daphnia* fluctuated, whereas the algae appeared stable or almost stable (Murdoch and McCauley 1985). The cycles were further shown to be intrinsically driven rather than being a result of some external cyclic forcing (McCauley 1993). Based on data from laboratory, mesocosm, and field studies, two aspects of the dynamics—cohort dominance and fecundity suppression—were identified as important mechanisms. Furthermore, based on inspections of demographic data from laboratory and tank experiments and field studies, it was suggested that cycle length equaled generation time, as in juvenile-driven cohort cycles with semichemostat resource dynamics (McCauley and Murdoch 1987).

Based on these pieces of information, one may conclude that the dynamics observed in *Daphnia* are juvenile-driven Type I cycles, an interpretation that is reinforced by the early models of *Daphnia* (de Roos, Metz, et al. 1990; Murdoch, Briggs, and Nisbet 2003). More recent experiments in combination with new theoretical insights have, however, pointed in quite another direction: namely, that *Daphnia* dynamics have the characteristics of adult-driven cycles. Several lines of evidence supporting this conclusion are present. However, as

TABLE 9.3. Summary of the Different Experimental Evidence
for Juvenile or Adult Competitive Superiority in *Daphnia*

Aspect	Competitive superiority	Source
Individual-level experiments	Adults > juveniles	de Roos, McCauley, et al. (1997); Nisbet, McCauley, et al. (2004)
Coexistence of alternative attractors	Adults > juveniles	McCauley, Nisbet, et al. (1999); McCauley, Nelson, and Nisbet (2008)
Period/delay ratio	Adults > juveniles	McCauley, Nelson, and Nisbet (2008)
Biomass relationship	Juveniles > adults	McCauley, Nisbet, et al. (1999)
Compensation	Juveniles > adults	Nilsson, Persson, and van Kooten (2010)

we will see, there is also evidence against the thesis that *Daphnia* dynamics are adult-driven cohort cycles. To start with, as supporting evidence, de Roos, McCauley, et al. (1997) showed that parameterized models for size-dependent individual energetics of *Daphnia* (Gurney, McCauley, et al. 1990) suggest that adults are more competitive than recruits (i.e., $q > 1$) (table 9.3). Second, in a series of elegant experiments, McCauley, Nisbet, et al. (1999) and McCauley, Nelson, and Nisbet (2008) showed the presence of two alternative attractor dynamics: low-amplitude cohort cycles and predator-prey cycles (figure 9.12). As we have seen, the presence of these two alternative attractors has, in models with logistic resource growth, so far only been found for $q > 1$ but not for juvenile-driven cycles (figure 9.9) (see also de Roos, Schellekens, et al. 2008b).

In addition, Nelson, McCauley, and Nisbet (2007) and McCauley, Nelson, and Nisbet (2008) presented a model with adult competitive superiority and logistic resource growth. Competitive asymmetry in their model was not introduced as a balance between individual-level consumption capacity and metabolic costs, as we have done in this chapter, but as a size-dependent asymmetry in consumption capacity and stage-specific mortality both favoring larger stages. Their model, based on *Daphnia* parameters (McCauley, Nisbet, et al. 1996), predicted the co-occurrence of high-amplitude predator-prey cycles and low-amplitude adult-driven cohort cycles. Corresponding to the results we have provided in this chapter, they found that the low-amplitude cycles were characterized by a period/delay ratio around 0.5 (table 9.2), whereas the predator-prey cycles had a period/delay ratio around 3. In a sophisticated combination of individual-level growth

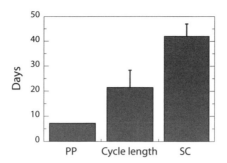

FIGURE 9.14. Juvenile stage durations occurring in cohort cycles (SC) (mean ±95 percent CL) and in predator-prey cycles (PP) (mean), and the cycle length (mean ±95 percent CL) of these cycles in the experiments of McCauley, Nelson, and Nisbet (2008). Cycle lengths of cohort cycles and predator-prey cycles were the same; see also figure 9.12 and McCauley, Nisbet, et al. (1999).

experiments and population-level experiments, McCauley, Nelson, and Nisbet (2008) could correspondingly show that the generation time in the empirically observed cohort cycle was almost twice that of the cycle length, whereas it was about one-third of the cycle length in the predator-prey cycle (figure 9.14).

So far, theoretical and experimental studies thus delineate a consistent pattern: (a) adult-driven cohort cycles and predator-prey cycles have theoretically been shown to co-occur as alternative attractors, whereas juvenile-driven cycles and predator-prey cycles have not, and (b) the cycle period/delay ratio observed in experimental systems with cohort cycles is consistent with what is expected for adult-driven cycles (table 9.3). However, there is also evidence against the supposition that the cycles in *Daphnia* populations are adult-driven cohort cycles. As we have discussed in detail in chapter 3, stage-specific competitiveness is linked to the type of control that regulates a population at equilibrium. Furthermore, de Roos, Schellekens, et al. (2008b) showed that there is a relationship between this type of control, stage-specific competitiveness, and the type of cycle present. With reproduction control, we expect juvenile-driven cohort cycles, whereas adult-driven cohort cycles are expected with development control. In terms of compensation/overcompensation, we therefore expect to see (1) a biomass dominance of juveniles and (2) compensation/overcompensation in adult biomass to harvesting, if adults are competitively superior. This is exactly the opposite of what has been observed in harvesting experiments with *Daphnia* (Nilsson, Persson, and van Kooten 2010). In these experiments, compensation was instead observed in juveniles, which points toward reproduction control. This study is not the only evidence against development control, as the data from McCauley, Nisbet, et al. (1999) show that the mean adult biomass is higher (0.148 mg/L) than the mean juvenile biomass (0.04 mg/L) in the low-amplitude cohort cycles, a result that also points toward reproduction control (table 9.3). We have already considered that it is possible to have a biomass dominance in the juvenile stage at the same time that the population maturation rate is larger than the population reproduction rate with other energy budget models (i.e., the Kooijman-Metz model, chapter 3). However, the combination of adult biomass dominance and compensatory response in juveniles is not expected on the basis of the Kooijman-Metz model at all, hence there is no theoretical support as yet for the hypothesis that biomass dominance of the adults can be combined with adult competitive superiority.

All in all, this leaves us with evidence that points in different directions regarding the type of cohort cycles in *Daphnia*. At this point, it is worthwhile to reconsider whether juvenile-driven cohort cycles in empirical populations may necessarily be generation cycles (i.e., period/delay ≈ 1), as we have shown that juvenile-driven Type II cycles have period/delay ≈ 0.5 consistent with the period/delay ratio observed in *Daphnia* (figure 9.14). For juvenile-driven Type II cycles, the shift in demography over time is also less dramatic than in the juvenile-driven Type I cycles and is closer to that of adult-driven cycles. Can juvenile-driven Type II cycles therefore be advanced as the explanation that finally unravels the true nature of the cohort cycles in *Dapnhia*? With our present knowledge we are quite hesitant to draw such a conclusion for at least three reasons. First, although the juvenile-driven Type II cycles have been observed in models of very different model structure, pointing to a substantial robustness against changes in model structure (de Roos and Persson 2003; Persson, Leonardsson, et al. 1998), they have so far not been demonstrated in the biomass-based approach that has served as a core approach for our theoretical analyses in this chapter. Second, the theoretical demonstration of the juvenile-driven Type II cycles has been restricted to models with semichemostat resource dynamics and so far has not been demonstrated with logistic resource growth. Third, experimental individual-level data actually suggest that $q > 1$, although not far from 1 (de Roos, McCauley, et al. 1997; Nisbet, McCauley, et al. 2004).

To conclude, we have learned a lot about cohort cycles from both theoretical and experimental studies on *Daphnia* dynamics, and we have no doubt progressed forward in our understanding about its dynamics. In doing this, the simultaneous use of information about individual- and population-level processes has been a major asset. We are also most likely far closer to really understanding the dynamics of this organism than we were ten years ago. Still, there are clearly elements yet to be understood before a complete knowledge of the nature of regulation in this organism can be reached.

Dynamics of Consumer-Resource Systems with Discrete Reproduction

Multiple Resources and Confronting Model Predictions with Empirical Data

So far, we considered consumer-resource dynamics under the assumption that all ecological processes such as foraging, metabolism, mortality, and reproduction take place on the same (continuous) time scale. By doing this, we could stringently link models in previous chapters that focused on community structure to fully structured models that focused on population dynamics. In this chapter, we take one step further with respect to complexity by allowing processes to take place at different time scales. More specifically, we consider models where processes like foraging, metabolism, and mortality are continuous whereas reproduction is assumed to take place as a discrete event at the start of the growth season. The assumption of discrete reproduction is relevant for many organisms living in seasonal environments (winter/summer, dry/wet seasons). The use of several time scales means that our analysis of the dynamics in this chapter will be restricted to simulations using the Escalator Boxcar Train (EBT) framework (de Roos 1988, 1997). The EBT is specifically designed to handle the numerical integration of the equations that occur in physiologically structured models using ordinary differential equations and is particularly well suited for systems with discrete reproduction (see section 8 in the technical appendices).

One advantage of the presence of discrete reproduction is that it makes it much easier to single out and follow separate age cohorts in empirical populations. First, this makes it much easier to scrutinize the mechanisms driving the observed dynamics in contrast to the problems we had in understanding *Daphnia* dynamics, as was extensively discussed in the previous chapter. Second, it allows us to present a more quantitative confrontation of model predictions with empirical data, of which we will provide two examples. This will all come with the cost that the models we will discuss in this chapter are more complex and parameter-rich than those of previous chapters. Nevertheless, as we hope to be able to convey, what we have learned about the effects of different body-size scaling, specifically the juvenile-adult scaling of foraging as reflected in the value of q in the previous chapter, will carry over to the models with discrete reproduction. Vice versa, the general understanding of the effects of different body-size scalings can also be obtained in the more parameter-rich models considered in this chapter (de Roos and Persson 2005).

In chapters 6–8 we extensively analyzed the effects of multiple resources and ontogenetic niche shifts on community structure. Here we will consider how the dynamics of consumer-resource systems are affected by the presence of more than one resource, and we will use two different scenarios: one with proportional habitat use and one with separate resources for juveniles and adults.

OVERALL MODEL CHARACTERISTICS

As in the previous chapter, we will consider a system consisting of a consumer population that is size-structured and a resource population that has a fixed size and is shared by all individuals. We assume a growth season of fixed length and assume that the resource is continuously reproducing throughout the season following semichemostat resource dynamics. The consumer is assumed to grow (or shrink, in case of starvation) continuously during the season but reproduces at discrete time instances only. The growth season of the consumer corresponds to, for example, the summer in the temperate region, and reproduction takes place at the start of the growth season (i.e., spring). We will generally ignore the winter season, which is equivalent to the situation when winter adds nothing more than a scaling down of all rates to the same extent. A winter season can readily be introduced into the dynamic formulation to, for example, investigate the effects of winter starvation on population dynamics (van de Wolfshaar, de Roos, and Persson 2008).

Through our assumption of a discrete reproduction event, the model will be a combination of a continuous dynamical system, describing growth and survival of the consumer and production and consumption of the resource during the growth season, and a discrete map describing the pulsewise reproduction of consumers every spring. This combination of a continuous dynamical system and a discrete map takes the size distribution of the consumer population and the resource density at the beginning of a season to their corresponding values at the beginning of the next season. In this case, a fixed point, or equilibrium, means that the consumer size distribution and the resource density always have the same value at a fixed time within each season, although these will naturally change during the season (sometimes referred to as a "one year cycle"; Claessen, van Oss, et al. 2002). In the same way, a stable cycle of period n is a solution that repeats itself at a fixed time every nth season.

DERIVATION OF INDIVIDUAL-LEVEL MODEL

The individual state forms the core of physiologically structured models. We therefore start out with the derivation of individual-level functions for intake, metabolism, and allocation. Our derivation of the individual-level model largely follows Persson, Leonardsson, et al. (1998) with modifications introduced by Claessen, de Roos, and Persson (2000).

We introduce two i-state variables, irreversible mass x and reversible mass y, and assume that consumer foraging, metabolism, growth, survival, and reproduction can be described as functions of these two i-state variables (Persson, Leonardsson, et al. 1998). This is done along the lines of Metz, de Roos, and van den Bosch (1988), which state that the behavior of the system should be fully determined by the values of the chosen i-state variables (plus the condition of the environment) at time t and that the values of these variables at $t + \tau$ are fully determined by their value at time t plus the intervening environmental history. In principle, we can think about a large number of relevant i-states, but we still want to keep the number of i-state variables as few as possible in order to simplify analyses and interpretations of these analyses. In irreversible mass x we include compounds like bones and organs, which cannot be starved away by the consumer. In reversible mass y we include energy reserves such as fat, muscle tissue, and gonads. These reserves may be used to cover basic metabolism during starvation. The total mass of the individual hence equals $x + y$. A similar approach is given in, for example, Broekhuizen, Gurney, et al. (1994). Observe that we for simplicity also assume that gonads are part of reversible mass.

ENERGY GAINS AND COSTS

As before, we will assume that a satiating functional response, here in the form of a Holling Type II functional response, gives an adequate description of the intake rate of the consumer as a function of resource density. Different options are possible concerning the dependency of the attack rate and handling time on i-states. In general, the attack rate and the handling time can be argued to depend on both i-state variables, because y is a measure of the condition of the individual. Here we will, however, assume that the attack rate and handling time depend only on x through the quantity:

$$w = x + q_J x. \tag{10.1}$$

We do this because functional response experiments with size-structured consumers have shown a close relationship between capture rate and body length independent of body condition (Aljetlawi, Sparrevik, and Leonardsson 2004; Finstad, Ugedal, and Berg 2006; Mittelbach 1981; Persson 1987). We call the quantity w effective body mass, where q_J stands for the maximum ratio of reversible to irreversible mass that a juvenile can have, something we will come back to below. The functional response will then be:

$$\gamma(w, R) = \frac{a(w)R}{1 + a(w)h(w)R}, \tag{10.2}$$

where $a(w)$ is the attack rate, $h(w)$ is the handling time, and R is the resource density.

For a given prey size, we will assume that the attack rate $a(w)$ is described by a hump-shaped function. Support for this assumption goes back historically to the paper by Wilson (1975). Mechanistically, the initial increase of the foraging capacity with predator size can be related to an increase in visual acuity and locomotion ability, both of which will affect the encounter with prey (Persson 1987; Peters 1983; Schoener 1969; Werner 1988; Wilson 1975). The decreasing part of the function can be related to a decreased capacity to discern small prey or to make fine-tuned maneuvers (Breck and Gitter 1983; Noakes and Godin 1988). Several models can be used to describe such a hump-shaped relationship between the attack rate and body weight. Here we have chosen the following equation for the relationship between the attack rate and effective body mass:

$$a(w) = A\left(\frac{w}{w_o} \exp\left(1 - \frac{w}{w_o}\right)\right)^{\alpha} \tag{10.3}$$

This relationship is governed by three parameters: the maximum rate, A; the body size at which this maximum rate is achieved, w_o; and a size-scaling exponent, α, which affects the rate by which the attack rate increases below and decreases above w_o. One advantage of this functional form is that it contains relatively few parameters, and each parameter represents different aspects of the hump-shaped relationship affecting the consumer-resource dynamics (see Persson, Leonardsson, et al. 1998). This functional form for the relationship between the attack rate and body mass has received considerable empirical support and has also become accepted as a general formula for size-structured predator-prey

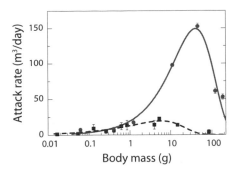

FIGURE 10.1. Attack rate of roach (*solid lines, filled circles*) and perch (*dashed lines, filled squares*) on 1.0 mm *Daphnia* as a function of their body weight. The functions were fitted to data using equation (10.3). Data points are means ±1 SD and are based on Byström and García-Berthou (1999) and Hjelm and Persson (2001).

interactions (figure 10.1) (Brose, Ehnes, et al. 2008; Byström and García-Berthou 1999; Finstad, Ugedal, and Berg 2006; Hjelm and Persson 2001).

For the handling time ($h(w)$), different expressions can be used (Claessen, de Roos, and Persson 2000; Persson, Leonardsson, et al. 1998). Calculations using fish as a model organism have shown that the main limiting process for the actual food intake is the capacity to process (digest) food. The size-dependency of digestion time can be estimated from feeding experiments where the organisms are feeding at maximum rate. We first assume that the maximum food intake (I_{max}) equals the inverse of the digestion time per unit of prey weight (i.e., $I_{max} = 1/h(w)$). We further assume that weight increment ($\Delta W(w)$) equals net ingestion ($k_e I_{max} - E_m(x, y)$, see below). Here k_e is a conversion efficiency, and $E_m(x, y)$ is the metabolic rate. Rearranging this equation and using the relationship between I_{max} and digestion time ($1/h(w)$) then yields:

$$h(w) = k_e \big/ \big(\Delta W(w) + E_m(x, y) \big) \qquad (10.4)$$

Using this relation, an allometric function for the size-dependent handling time has been shown to yield a good fit to data (Claessen, de Roos, and Persson 2000):

$$h(w) = \xi_1 w^{\xi_2} \qquad (10.5)$$

Because digestion capacity is a positive function of body size, ξ_2 takes a negative value.

The above equations complete the description of energy gain by the individual. For energy cost E_m, we will base our formulation on literature information

(Calder 1984; Lundberg and Persson 1993; Peters 1983) and assume that metabolic demands as a function of body mass $x + y$ can be described by a power function:

$$E_m(x,y) = m_1(x + y)^{m_2}, \qquad (10.6)$$

where m_1 and m_2 are positive constants. The net mass intake per unit of time $E_g(x,y,R)$ will then equal the intake per unit of time $E(w,R)$ minus metabolic demands per unit of time. Hence,

$$E_g(x,y,R) = E(w,R) - E_m(x,y), \qquad (10.7)$$

where $E(w,R)$ equals the consumption rate $\gamma(w,R)$ multiplied by the conversion factor k_e (this conversion factor takes into account prey weight, assimilation efficiency and other conversion costs).

ENERGY ALLOCATION, GROWTH, REPRODUCTION, AND SURVIVAL

If energy intake in equation (10.7) exceeds costs for metabolism, the surplus mass is invested in growth. With this assumption, we have adopted one type of energy budget model present in the literature, namely, a *net-production* energy budget model, where costs for metabolism are first paid and the remaining energy is used for somatic growth and gonad production (see Lika and Nisbet 2000). An alternative approach, *net-assimilation* energy budget models, assumes that allocation to growth of gonads has precedence and the remaining energy is used for somatic growth and metabolism. The Kooijman-Metz (1984) model that we have already used in previous chapters is one example of net-assimilation models. The most comprehensive body of theory presently exists for net-assimilation energy budget models (Kooijman 2000), but Lika and Nisbet (2000) provide an example of an energy budget model based on partitioning of net production. Theoretical modeling suggests that assumptions about energy channeling may affect predictions about toxic responses and equilibrium demography (Gurney, Middleton, et al. 1996; Nisbet, McCauley, et al. 1996). At the same time, the nature of the juvenile-driven cohort cycles that we considered in chapter 9 seems to be very robust to the different assumptions underlying net-production and net-assimilation energy budget models (observe again that only juvenile-driven cycles are expected in the Kooijman-Metz model because of its assumptions about size scalings). Furthermore, the empirical evidence in favor of one or the other energy budget model is presently meager.

Adopting a net-production model as our choice, we can subsequently derive functional relationships for how ingested energy is allocated. First, we assume that individuals mature at a fixed irreversible mass x_f. This is of course an oversimplification, but we will nevertheless make this assumption in the following with the aim of studying the effects of different size scalings on consumer-resource interactions. We will further assume that there is a maximal ratio of reversible mass to irreversible mass. Because we have limited the number of i-states to two, where gonad tissue is part of reversible mass, we will next assume that this ratio is different for juveniles and adults. We denote these ratios by q_J and q_A for juveniles and adults, respectively, where $q_A > q_J$. The maximum reversible mass of a juvenile and adult individual, respectively, hence equals:

$$y^*(x) = \begin{cases} q_J x & \text{if } x \le x_f, \\ q_A x & \text{if } x_f < x. \end{cases} \tag{10.8}$$

We assume that a fraction, $\kappa(x,y)$, of the surplus mass E_g is allocated to growth in irreversible mass and the rest into growth of reversible mass following the functions:

$$\kappa(x,y) = \begin{cases} \dfrac{1}{(1+q_J)q_J} \dfrac{y}{x} & \text{if } x \le x_f \\ \dfrac{1}{(1+q_A)q_A} \dfrac{y}{x} & \text{if } x > x_f \end{cases} \tag{10.9}$$

From equation (10.9), it follows that the individual allocates mass surplus to reversible mass at a rate $(1 - \kappa(x,y))$, which is proportional to $((1 + q_J)y^* - y)/y^*$. For $y = 0$, all mass surplus is allocated to reversible mass, whereas the mass surplus allocated to reversible mass at $y = y^*$ equals $y^*/(x + y^*) = y^*/w$. Individuals are assumed to be born with i-state (x_0, y_0), where $y_0 = q_J x_0$; that is, they are born with the maximum ratio of reversible mass to their irreversible mass (figure 10.2). x_0 and y_0 sum up to the weight at birth w_b. As long as the individual does not starve, that is, as long as $E_g \ge 0$, a juvenile individual will always have this maximum ratio of reversible versus irreversible mass and hence move along the line $y = q_J x$ in figure 10.2. When reaching the maturity size threshold $x = x_f$, it is assumed that the allocation function changes and that the individual in addition to fat reserves also allocates mass to gonads. The maximum amount of reversible mass for an adult individual is therefore proportionally larger than for juveniles and equals $q_A x$ (figure 10.2).

If the net mass intake, E_g, is negative, the individual starves, and reversible mass (but not irreversible mass) decreases (figure 10.2). We assume that the individual can withstand a certain amount of starvation, down to the point when $y = q_s x$ before this starts to increase its death rate. The rate of starving to death,

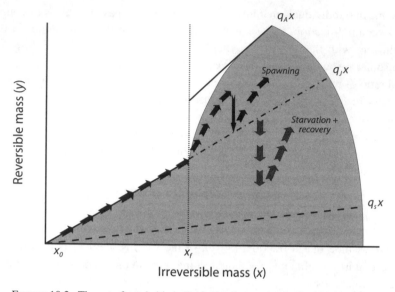

FIGURE 10.2. The set of reachable individual states, as determined by the alloca-tion of assimilated energy to irreversible (x) and reversible (y) mass. Individuals are born with an irreversible mass x_0 and a reversible mass $q_J x_0$, together summing to the weight at birth w_b. Individuals grow in mass along the line $y = q_J x$ as long as they don't starve. After maturity (reached at irreversible size x_f), the maximum amount of irreversible mass increases to $q_A x$. When adults spawn, their reversible mass drops to $q_J x$. Following spawning, mass is allocated according to the overall rule for partitioning between irreversible and reversible mass (equation (10.9)). During starvation, only reversible mass decreases, whereas during recovery after a starvation period, mass is added preferentially to restore reversible mass. Below the line $y = q_s x$ starvation will increase the mortality of individuals. Individuals can have a combination of individual-state variables anywhere in the shaded area (adapted from Persson, Leonardsson, et al. 1998).

$\mu_s(x,y)$, when the reversible mass is below $q_s x$, is assumed to increase with the proportion x/y according to the following equation:

$$\mu_s(x,y) = \begin{cases} s\left(q_s\dfrac{x}{y} - 1\right) & \text{if } \dfrac{y}{x} \leq q_s \\ 0 & \text{if } \dfrac{y}{x} > q_s \end{cases}, \tag{10.10}$$

where s is a (positive) proportionality constant. This equation ensures that death is certain before $y = 0$. Although the individual only decreases in revers-ible mass when starving, it will during recovery after a starvation period (if resource conditions improve) allocate mass to both irreversible and reversible mass, according to equation (10.9) (figure 10.2). In addition to starvation mortality, we assume that the individual may die of other causes than starvation at a rate μ_0 and that this rate is independent of the i-state. The total per capita death rate is hence the sum of the starvation and background death rates.

As we have considered above, adult individuals allocate mass to gonads and hence have a relatively higher maximum reversible mass value than juveniles. The amount $y - q_J x$ of reversible mass exceeding $q_J x$ at the end of the growth season is used for reproduction. This gonad mass is transformed into newborns on the first day of the following growth season with a conversion efficiency k_r. Hence, the number of offspring produced by a reproducing individual is given by:

$$F(x,y) = \begin{cases} k_r(y - q_J x)/w_b & \text{if } x > x_f \text{ and } y > q_J x \\ 0 & \text{otherwise} \end{cases} \tag{10.11}$$

The division by w_b, where w_b is effective offspring body mass, converts the amount of reversible mass invested into reproduction into the number of offspring produced. When an adult individual reproduces, its reversible mass is reduced to $y = q_J x$ (figure 10.2). If the reversible mass of a mature individual at the end of the growth season is less than $q_J x$, it will not reproduce. Different assumptions can be made concerning whether the mass of all newborn individuals is the same or whether an initial size distribution is present at hatching. In the context of a consumer-resource model, it has turned out that whether you use one birth size for all newborn individuals or use an initial size distribution has no influence on the population dynamics of the system (Persson, Leonardsson, et al. 1998).

Box 10.1 summarizes all equations specifying the consumer life history model discussed in detail above, and the parameters and their default values are presented in box 10.2.

THE MODEL AT THE POPULATION LEVEL

An efficient way of studying the dynamics of the system numerically is to use a formulation based on the EBT (Escalator Boxcar Train) framework (de Roos, Diekmann, and Metz 1992). Because we have assumed a pulsed reproduction, the consumer population will be naturally divided into distinct cohorts of individuals. We will further assume that individuals within a single cohort are born with the same irreversible and reversible mass (observe that we still can

BOX 10.1
Variables and Individual-Level Functions of the
Consumer-Resource Model with Discrete Reproduction

Variables	Description
R	Resource (food) density in environment
x	Irreversible mass
y	Reversible/gonad mass

Functions	Description
$w(x) = (1 + q_J)x$	Effective body mass
$a(w) = A\left(\dfrac{w}{w_o}\exp\left(1 - \dfrac{w}{w_o}\right)\right)^{\alpha}$	Size-dependent attack rate
$h(w) = \xi_1 w^{\xi_2}$	Handling (digestion) time
$\gamma(w, R) = \dfrac{a(w)R}{1 + a(w)h(w)R}$	Foraging rate
$E(w, R) = k_e \dfrac{a(w)R}{1 + a(w)h(w)R}$	Assimilation rate
$E_m(x, y) = m_1(x + y)^{m_2}$	Metabolic rate
$E_g(x, y, R) = E(w, R) - E_m(x, y)$	Net-biomass-production rate
$\kappa(x, y) = \begin{cases} \dfrac{1}{(1 + q_J)q_J}\dfrac{y}{x} & \text{if } x \le x_f \\[2mm] \dfrac{1}{(1 + q_A)q_A}\dfrac{y}{x} & \text{if } x > x_f \end{cases}$	Allocation function for net biomass production
$F(x, y) = \begin{cases} k_r(y - q_J x)/w_b & \text{if } x > x_f \text{ and } y > q_J x \\ 0 & \text{otherwise} \end{cases}$	Adult fecundity (number of offspring produced at start of growth season)
$\mu_s(x, y) = \begin{cases} s\left(q_s \dfrac{x}{y} - 1\right) & \text{if } \dfrac{y}{x} \le q_s \\[2mm] 0 & \text{if } \dfrac{y}{x} > q_s \end{cases}$	Starvation mortality rate
$\mu(x, y) = \mu_0 + \mu_s(x, y)$	Total mortality rate

For the population-level formulation, see section 8 in the technical appendices.

assume that a newborn cohort consists of several size cohorts; Persson, Leon-ardsson, et al. 1998). Individuals within a size cohort also remain identical to each other throughout their lifetime, because growth is deterministic. The consumer population hence consists of a finite number of cohorts of identical individuals.

BOX 10.2
Default Parameters for the Size-Structured
Consumer-Resource Model from Box 10.1

Parameter	Default value	Unit	Description
Resources			
r_z	0.1	day^{-1}	Pelagic resource turnover rate
K_z	3	g/m^3	Pelagic resource maximum density
r_m	0.1	day^{-1}	Benthic resource turnover rate
K_m	3	g/m^2	Benthic resource maximum density
Consumer			
μ_0	0.01	day^{-1}	Background mortality rate
μ_{nr}	0	day^{-1}	Mortality rate constant in non-refuge habitat
x_{nr}	2.0	g	Size scaling of size-dependent mortality
q_s	0.2	—	Starvation mortality threshold
s	0.2	day^{-1}	Starvation rate coefficient
w_b	0.002	g	Effective body mass at birth
x_f	4.5	g	Irreversible mass at maturation
q_J	0.75	—	Juvenile maximum condition
q_A	1.4	—	Adult maximum condition
m_1	0.03	g$^{(1-m_2)}$/day	Allometric scalar maintenance rate
m_2	0.75	—	Allometric exponent maintenance rate
α	0.6	—	Allometric exponent pelagic resource attack rate
A	30	m^3/day	Maximum pelagic resource attack rate
w_o	8	g	Optimum body size for pelagic resource foraging
α'	0.4	—	Allometric exponent benthic prey attack rate
A'	10	m^2/day	Maximum benthic prey attack rate
w'_o	20	g	Optimum body size for benthic prey foraging
ξ_1	5	day $\cdot$ g$^{-(1+\xi_2)}$	Allometric scalar of handling time function
ξ_2	−0.8	—	Allometric exponent of handling time function
k_e	0.6	—	Conversion efficiency
k_r	0.5	—	Reproductive conversion efficiency

(Box 10.2 continued)			
Other			
V	$1.0 \cdot 10^6$	m^3	Total lake volume
S	$0.5 \cdot 10^5$	m^2	Benthic bottom area
T	90	day	Length of growth season

Pelagic and benthic resources inhabit the non-refuge and refuge habitat, respectively.

Within each growth season, the resource density, the number of individuals in each consumer cohort, and their irreversible and reversible mass all change continuously. Each size cohort is represented by three ordinary differential equations (ODEs) representing the cohort's irreversible mass, reversible mass, and numbers, respectively (see section 8 in the technical appendices). Reproduction takes place at the beginning of the growth season at the discrete times $t = nT$. Each reproduction event may thus lead to the addition of one new cohort to the consumer population and thus to three new ODEs. A discrete map describes the state of the consumer population after the reproduction event as a function of its pre-reproduction state. The number of cohorts making up the consumer population is decreased if the number of individuals in a particular cohort has become negligibly small. Such cohorts are removed from the population. The EBT formulation of the population model is hence a combination of a continuous-time process, specified in terms of ordinary differential equations, and a discrete-time process, specified as a mapping (de Roos, Diekmann, and Metz 1992). A complete description of the EBT formulation for the consumer-resource system studied in this chapter and chapter 11 is given in section 8 of the technical appendices.

Notice that the model formulation represents resource densities invariably in terms of biomass densities per volume (g/m^3 or mg/L). The consumer population is, however, always expressed in terms of the total number of individuals in a lake with an arbitrarily chosen volume of $1.0 \cdot 10^6$ m^3. These choices do not affect the results in any qualitative way; they only scale the presented consumer densities. In later sections we will also consider a model variant, in which the population not only forages on a pelagic resource, but also exploits a benthic prey population, living in the littoral zone of the lake. In this particular case we will assume that the volume of the littoral zone of the lake occupies 10 percent of the entire lake volume and that the average depth of this littoral zone is 2 m. This implies that the total benthic surface area housing the additional resource population extends over $0.5 \cdot 10^5$ m^2. Densities of the benthic prey population will be expressed in terms of g/m^2. We will furthermore analyze a model variant, in which the entire lake is subdivided into a juvenile and an adult habitat part, each with its own exclusive, pelagic prey population. We will in this case vary the fraction of the entire lake that is occupied by the juvenile habitat.

CRITICAL RESOURCE DENSITY AND COHORT DYNAMICS

In previous chapters, we have shown that stage-specific competitive ability plays a crucial role for understanding both population dynamics and community structure. Specifically, we used the parameter q to scale juvenile competitive ability relative to adult competitive ability, and found that the presence and kind of cohort cycles discussed in chapter 9 was critically dependent on the value of q. Here, we will derive an expression for a *continuous* size-dependent competitive ability based on the individual-level functions for ingestion and metabolism that we have derived above. We define size-dependent competitive ability by solving for the condition when energy intake equals energy costs (Persson, Leonardsson, et al. 1998) for a non-starving individual (i.e., with $y = q_j x$). This is done by setting equation (10.7) to zero and solving for the resource density R. This yields the following expression:

$$R^* = \frac{m_1 w^{m_2}}{A\left(\dfrac{w}{w_0}\exp^{\left(1-\frac{w}{w_0}\right)}\right)^\alpha \left(k_e - m_1 w^{m_2}\xi_1 w^{\xi_2}\right)} \tag{10.12}$$

Although somewhat opaque at first sight, this expression shows that the form of the function describing the critical resource density (CRD) as a function of body size depends on the size scalings of the attack rate (equation (10.3)), handling time (equation (10.5)), and metabolic rate (equation (10.6)). Figure 10.3 shows how the critical resource density is affected by changes in the size scaling (α) for fixed size scaling of handling time and metabolic rate. For a relatively shallow slope ($\alpha = 0.60$), the resulting CRD is monotonically increasing, that is, the smaller an individual, the better competitor it is. For an intermediate slope ($\alpha = 0.80$), the curve describing CRD as a function of body size is relatively flat, and, particularly, a newborn individual has the same CRD as a just-matured individual (8 g). Finally, for a steep slope ($\alpha = 1.04$), the CRD first decreases with body size to ultimately increase again at larger sizes. Although we have chosen to vary the size scaling of the attack rate, the CRD is similarly affected by changes in the size scaling of the handling time and metabolic rate (Lundberg and Persson 1993).

The size-dependent competitive ability that we have defined through the critical resource density (equation (10.12)) obviously has its counterpart in

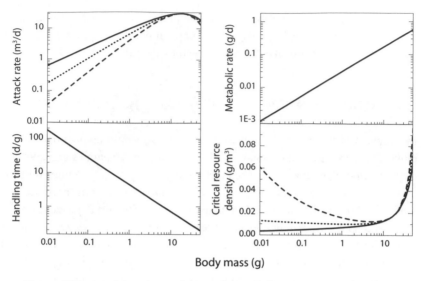

FIGURE 10.3. Attack rate (*upper left panel*), metabolic rate (*upper right panel*), handling time (*lower left panel*), and critical resource density (*lower right panel*) as a function of consumer body mass. The attack rate in the upper left panel is shown for three different α values, 0.60 (*solid line*), 0.80 (*dotted line*), and 1.04 (*dashed line*), with the resulting three different forms for the critical resource density–body weigh relationship shown in the lower right panel. Throughout, $w_o = 20$ g, and other parameters have default values as presented in box 10.2.

unstructured models. The R^* argument, where R^* is defined as the lowest resource density an organism can tolerate and still maintain itself (Tilman 1982), has often been used to study interactions among species. In particular, zero-net-growth isoclines (ZNGIs) defined for the condition of maintenance are used to determine the condition for species coexistence and extinction.

In the previous chapter, we found that the parameter q, defining whether juveniles or adults were competitively superior, largely determined the kind of dynamics observed. We have already discussed in chapter 2 ontogenetic asymmetry in terms of both the value of q as well as the form of the CRD curve. With this in mind, we would a priori expect that we can predict population dynamics on the basis of the form of the CRD curve. As we will see, this a priori expectation turns out to be completely true. For the case where smaller individuals have the lowest CRD and hence are competitively superior, the resulting dynamics is a cohort cycle driven by recruits (figure 10.4, left). In this case, when a newborn cohort is recruited to the population, it depresses the resource level to such an extent that it drives the adults to starvation extinction. This

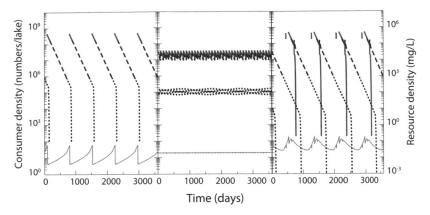

FIGURE 10.4. Time series of the population dynamics of the resource (*thin solid line*; in mg/L), $<$ one-year-old juveniles (*thick solid line*), $>$ one-year-old juveniles (*thick dashed line*), and adult consumers (*thick dotted line*) for three different α values (*left*, 0.6; *middle*, 0.8; *right*, 1.04) with semichemostat resource dynamics. Consumer densities are given in number of individuals in the entire lake ($V = 1.0 \cdot 10^6 \, \text{m}^3$). Throughout, $w_o = 20$ g, and other parameters have default values as presented in box 10.2.

complete temporal extinction of adults can be related to the imposed pulsed reproduction leading to a large number of recruits entering the system at the same time. When this dominating cohort matures, it gives rise to a new pulse of recruits, which depresses the resource, leading to starvation death of the parents. For this size scaling and with the relatively low background mortality used, the consumer population consists of only one cohort, and adults appear only during a short time period: from when the individuals of the cohort mature until they reproduce (figure 10.4, left). These cycles thus represent true single-generation (and single-cohort) cycles.

Increasing the size scaling to allow a just-matured individual to have a similar CRD as a recruiting individual (the critical resource density as a function of body size is then overall relatively flat up to maturation size, figure 10.3) leads to fixed-point dynamics (at the scale of the discrete map), where several cohorts coexist and do not outcompete each other (figure 10.4, middle). As mentioned above, these dynamics are also sometimes referred to as one-year cycles.

A further increase in the slope leads to the reappearance of large fluctuations (figure 10.4, right). These cycles do, however, differ from the cycles present for a low value for the attack rate scaling. First, the dominating cohort does not depress the resource to such an extent as to cause the adults to die of starvation, and adults reproduce for several years. In the case shown in figure 10.4

(right), individuals in the first reproductive pulse, which a dominating maturing cohort gives rise to, do all die of starvation because the resource has not
yet completely recovered. When the adult cohort reproduces a second time,
the resource is at its peak, and all recruits survive. This is also the case for the
third pulse of recruits that the adult cohort gives rise to (figure 10.4, right). In
contrast, the fourth reproductive pulse totally fails because the consumption
of the two previously recruited cohorts have now depressed the resource to a
level below the critical resource density for individuals of the size of recruiting
individuals. Because per capita consumption increases faster than the decrease
in number of individuals resulting from background mortality for a number of
years, the resource continues to decrease for several years, eventually leading
to the starvation death of adults, which have a higher CRD than the dominating
juvenile cohort (figure 10.3; figure 10.4, right).

Overall, the results of the model with discrete reproduction used in this
chapter are in line with the results of the fully structured biomass model with
continuous reproduction considered in chapter 9. As a conclusion, we can also
state that we can predict and understand the dynamics of the consumer-resource
system with pulsed recruitment based on the form of the critical resource density function. Stated in another way, the form of population dynamics observed
can readily be predicted and understood from individual-level processes. Further investigations of the model dynamics show that it is the size scalings of
the attack rate, digestion time, and the metabolic rate that lead to the different
shapes of the CRD function, which overall determines the dynamics of the system. The maximum attack rate, A (or equally the productivity of the system),
will affect the regularity and amplitude of the cycles but not the nature of the
cycle occurring (Persson, Leonardsson, et al. 1998).

In addition to the effect of α, a decreasing w_o has been shown to increase the
parameter space for cycles and particularly the juvenile-driven cycles. Below
a certain w_o value, stable fixed-point dynamics are not present any longer at
intermediate α values (Persson, Leonardsson, et al. 1998). The latter can be
related to the fact that decreasing w_o decreases the potential to have a CRD
curve that is flat over a larger size range (figure 10.3). Consequently, for a w_o
of 8 g, the dynamics with cycles driven by recruits extend to higher α values
and gradually change into regular or irregular fluctuations driven by older recruits (figure 10.5, left). In contrast, for a w_o of 20 g, the juvenile-driven cycles
collapse into fixed-point dynamics. With a further increase, these fixed-point
dynamics result in an expanding invariant loop, ending up in large-amplitude
dynamics that are largely irregular (but regular for certain values of α), with
a cycle period that equals that of juvenile-driven cycles at lower α values
(figure 10.5, right).

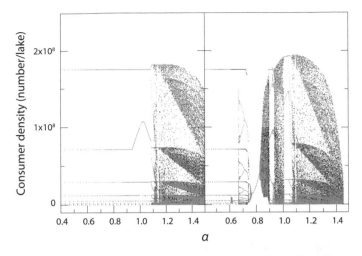

FIGURE 10.5. Bifurcation plot of consumer density (number of juveniles aged one year and older) versus the slope of the attack rate (α) for $w_o = 8$ g (*left panel*) and $w_o = 20$ g (*right panel*). The graph is compiled from consumer densities at the time of reproduction in spring (i.e., all juvenile consumers except for the offspring produced right at that moment) that were observed in numerical simulations after transient dynamics had disappeared. A regular cycle with a period of, for example, eight years in this diagram, thus shows up as eight distinct symbols at the same α value. Black symbols represent runs from low to high α values, and gray symbols represent simulations run from high to low α values. Differences between the two reveal alternative attractors. Other parameters have default values as presented in box 10.2.

The correspondence in results between what we found in chapter 9 and the results obtained in this chapter also concerns the presence of an alternative small-amplitude cycle coexisting with the juvenile-driven large-amplitude cycle (figure 9.13 and figure 10.5, right, $0.67 < \alpha < 0.78$). A closer examination shows that the two types of cycles observed for lower α values do in all major aspects resemble the Type I and II juvenile-driven cycles we discussed in chapter 9 (de Roos and Persson 2003). This similarity in types of dynamics observed with different model structures points to the robustness of the results we have found in chapters 9 and 10. Our interpretation of the existence of this similarity, despite different model structures and despite that many more parameters are included in the model with pulsed recruitment, is that the population dynamics are set by the limited number of parameters, which determine the interactions that individuals are engaged in—that is, q in chapter 9, and attack rate, handling time, and the metabolic rate affecting the form of the critical resource density function in this chapter.

The form of the critical resource demands as a function of body size can thus be used to predict the dynamics of structured consumer-resource systems. Experimental data from different planktivorous fish species show that the value of the exponent α in the scaling of attack rate with body size is in the range 0.59–0.67, leading to the conclusion that smaller individuals are in general competitively superior (Byström and Andersson 2005; Byström and García-Berthou 1999; Hjelm and Persson 2001; Mittelbach 1981; Persson and Brönmark 2002a, 2002b; Persson and de Roos 2006). Juvenile-driven cohort cycles are therefore expected for fish. In contrast, energy budget models for *Daphnia* suggest that smaller juveniles may be less competitive, in that they have a higher critical resource density than larger juveniles and small adults (see previous chapter and de Roos, McCauley, et al. 1997). Although no information on individual-level energetic performance was available, Briggs, Sait, et al. (2000) provided clear evidence that the population cycles in *Plodia* were driven by competitively superior larger stages (in combination with cannibalism). A final potential example of cycles driven by larger stages is periodic cicadas (Bulmer 1977) where preemptive competition (sensu Schoener 1983) may make larger individuals superior to smaller ones.

MULTIPLE RESOURCES AND ONTOGENETIC NICHE SHIFTS

So far, our analysis has been restricted to one consumer and one resource. Because changes in consumer size generally involve changes in resource or habitat use through ontogenetic niche shifts (Schreiber and Rudolf 2008; Werner and Gilliam 1984), it is natural to consider the effects of ontogenetic niche shifts on consumer-resource population dynamics, in the same way as we did when we investigated the effects of ontogenetic niche shifts on community structure in chapters 6–8. Here we will raise the questions of how cohort cycles are affected by the presence of multiple resources and whether the indirect coupling between cohorts via the exploitation of a shared resource is essential for cohort cycles to occur. In the previous chapter, we actually already answered the latter question when we discussed the mathematical conditions for cohort cycles to occur (Diekmann, Gyllenberg, et al. 2010). We will here engage in more detailed analyses of the characteristics of these cohort cycles for different resource productivities for juveniles and adults in order to unravel the important mechanism(s) underlying the dynamics.

Ontogenetic Niche Shifts with Shared
Resources for Juveniles and Adults

We extend the consumer-resource model with discrete reproduction to also include a second resource, representing a benthic prey population. The habitat is furthermore considered to be subdivided into a non-refuge and a refuge part, constituting 90 percent and 10 percent of the entire lake, respectively. In a lake, the refuge habitat would typically be the vegetated shore habitat and the non-refuge habitat the pelagic area (de Roos, Leonardsson, et al. 2002; Mittelbach and Chesson 1987). The additional, benthic prey population lives on the bottom surface of the refuge part of the habitat. We will assume that this benthic resource has a larger mean biomass, such that the optimal consumer body size for exploiting this resource is larger (box 10.2). Consumers thus forage on the two resources with different attack rates. Their resource ingestion is furthermore dependent on the fraction of time, referred to with the parameter p, that they spend in the refuge part of the habitat. The rate at which individual consumers now assimilate resource biomass then follows:

$$E(w,R,R') = k_e \frac{(1-p)a(w)R + pa'(w)R'}{1+(1-p)a(w)h(w)R + pa'(w)h(w)R'} \qquad (10.13)$$

In this equation, the density of the second resource is indicated with R' and the attack rate on this resource as $a'(w)$. The functional form of the attack rate is the same as used before (equation (10.3)), but with different parameters a', A', and w_o' (box 10.2). Notice that we have assumed that the handling (digestion) time $h(w)$, which is expressed in terms of the time needed to digest a certain amount of biomass, is the same for the two resources. From the above expression for the assimilation rate follow expressions for the foraging rates on resource R and R' as

$$\gamma(w,R,R') = \frac{(1-p)a(w)R}{1+(1-p)a(w)h(w)R + pa'(w)h(w)R'} \qquad (10.14)$$

and

$$\gamma'(w,R,R') = \frac{pa'(w)R'}{1+(1-p)a(w)h(w)R + pa'(w)h(w)R'}, \qquad (10.15)$$

respectively.

In addition to the random background mortality, we will also assume that consumers suffer from a mortality that is size dependent following the formula:

$$\mu_{nr} \exp\left(-\frac{x}{x_{nr}}\right) \tag{10.16}$$

In this expression μ_{nr} is a constant, and x_{nr} sets the size scaling of this additional size-dependent mortality rate. Consumers are assumed to be exposed to the size-dependent mortality only in the non-refuge (nr) habitat, whereas they experience the background mortality only in the refuge (r) habitat. The total mortality an individual experiences equals the sum of background mortality, starvation mortality, and the explicit size- and habitat-dependent predation mortality.

When all consumers use only the non-refuge habitat ($p = 0$) and hence exploit only one of the two resources, increasing the size-dependent mortality by increasing μ_{nr} leads to a shortening of the juvenile-driven cycles in a stepwise manner to lower and lower periodicity, ultimately ending up in a fixed-point dynamics before the consumer goes extinct with a further increase in mortality (figure 10.6, left panel). This shortening of the cycle length comes about because the increased mortality leads to a more rapid decline in consumer density, hence a faster recovery of the resource and a speeding up of the growth of surviving consumers. The stepwise decrease in periodicity is a result of the discrete nature of reproduction, as it can only occur at the beginning of a growth season. For values of μ_{nr} around the thresholds where the cycle period shortens, some irregular dynamics may be observed. The bifurcation pattern also shows the presence of alternative attractors. When consumers are restricted to live in the non-refuge habitat, the regular two-year cycle, for example, coexists over a significant range of μ_{nr} values with a stable fixed point (figure 10.6, left panel). Alternative attractors are furthermore likely to occur at the transitions where the cycle period shortens.

Next, we assume that individuals use the two habitats in proportion to their respective volumes ($p=0.1$, reflecting the 10 percent/90 percent subdivision of the lake into refuge (littoral) and non-refuge (pelagic) habitat) and hence exploit both resources. Increasing the size-dependent mortality in this scenario yields essentially the same results as the model with size-dependent mortality where consumers use only the non-refuge habitat (figure 10.6, right panel). Alternative attractors also occur, in this case between a regular two-year cycle and a stable fixed point and between a regular three-year cycle and either a stable fixed point or irregular, small-amplitude cycles. The robustness of the results is due to the fact that a newborn cohort rapidly controls and depresses both resources. The increase in persistence of the consumer with increasing μ_{nr} for the scenario with proportional habitat use compared with the scenario with size-dependent mortality and only one resource (figure 10.6, right versus left

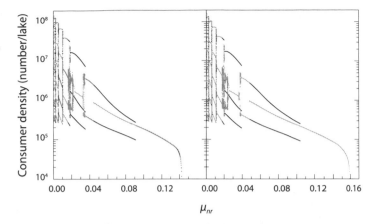

FIGURE 10.6. Bifurcation plot of the size-structured, consumer-resource model incorporating a non-refuge prey (zooplankton) and a refuge prey (macroinvertebrates). Consumers either live in the non-refuge habitat only (*left*) or use both habitats in proportion to their volume (*right*). The density of all individuals aged one year and older at the start of a growth season is plotted as a function of the constant μ_{nr}, scaling size-dependent mortality in the non-refuge habitat. A regular cycle with a period of, for example, four years in this diagram, thus shows up as four distinct symbols at the same mortality value. Transient dynamics have been discarded. Different shades of gray represent the occurrence of alternative dynamic attractors for the same parameter value. Black symbols represent runs from low to high μ_{nr} values, and gray symbols represent simulations run from high to low μ_{nr} values. Throughout, $\alpha = 0.6$, $w_o = 8\,g$, and other parameters have default values as presented in box 10.2.

panel) is a result of the fact that the consumer individuals spend a fraction of time in the habitat with lower mortality.

To conclude, the occurrence of juvenile-driven cohort cycles is not affected by the addition of a second resource per se and/or the subdivision of the habitat into refuge and non-refuge parts. Admittedly, we have used a rather specific assumption about how consumers use the different habitats. However, model scenarios using different types of flexible behavior, in which the fraction of time that consumers spend in a particular habitat depends on their intake and mortality rates in both habitats, show that the presence of cycles and the pattern with stepwise decreases in periodicity with increasing μ_{nr} remains the same. This result is largely due to the fact that recruiting juvenile cohorts rapidly deplete the refuge resource and therefore soon move out to the non-refuge habitat (for detailed analyses, see de Roos, Leonardsson, et al. 2002; Persson and de Roos 2003). The only discrepancy that occurs is that flexible behavior allows the consumer population to persist at higher values of μ_{nr}. This increased persistence is also generally associated with destabilization of the dynamics where several juvenile cohorts can be stacked in the refuge habitat owing to the high mortality risk in the non-refuge habitat, which leads to a shift from juvenile-adult competition to juvenile-juvenile competition. As we will see in the next section, this kind of dynamics is something that we also observe when juveniles and adults have separate resources and the adult habitat is much larger than the juvenile habitat.

SEPARATE RESOURCES FOR JUVENILES AND ADULTS

We next switch to consider the situation where juveniles and adults are in separate parts of the habitat and do not interact indirectly via a shared resource. In this model variant, we hence again account for two different resources, a juvenile-exclusive resource and a resource exclusive to adults, which are, however, both pelagic. The attack rates of both juvenile and adult consumers on their own exclusive resource follow the same scaling with consumer body size as in the basic model with only a single resource (equation (10.3)) with default parameters α, A, and w_o (see box 10.2) for both juveniles and adults. Juvenile and adult foraging rates, $\gamma(w, R)$ and $\gamma(w, R')$, respectively, are hence given by the same expression but depend on a different resource density (R for the juvenile-exclusive resource, R' for the adult-exclusive one). For this situation, we investigated the dynamics of the system by varying the size of the juvenile and adult habitat parts, respectively, from a juvenile/adult habitat size ratio of 0.05/0.95 to 0.95/0.05. The total volume of two habitats together equals, as before, $V = 1.0 \cdot 10^6 \text{ m}^3$. The change in juvenile/adult habitat size ratio translates into a change in the intensity of the population feedback on the two resources. An alternative perspective on this change in juvenile/adult habitat size ratio is that it changes the total productivity of the lake that is available to juveniles and adults, respectively.

For a small juvenile habitat size, adults thrive, as their resource is at its carrying capacity for most of the time (figure 10.7, left panel). When a large number of juveniles mature and adult density therefore increases, a number of strong reproductive outbursts take place that lead to a strong depression of the juvenile resource and complete recruitment failure owing to starvation among juveniles. The number of adults declines as a result of background mortality, which leads to several bursts (up to eight) of offspring that survive (figure 10.7, left panel). Several differently aged juvenile cohorts are stacked and mature at the same time, giving rise to the next peak in adult density. The first

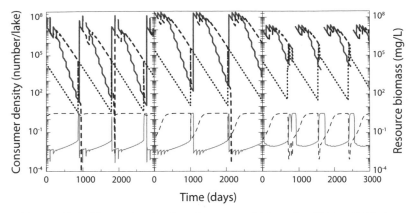

FIGURE 10.7. Time series showing < one-year-old juveniles (*thick solid line*), > one-year-old juveniles (*thick dashed line*), adults (*thick dotted line*), juvenile resource (*thin solid line*), and adult resource (*thin dashed line*) for a juvenile/ adult habitat fraction of 0.05/0.95 (*left*), 0.5/0.5 (*middle*), and 0.95/0.05 (*right*). Throughout, $w_o = 8$ g and $\mu_0 = 0.01$ in both habitats.

reproduction pulse of these newly matured adults gives rise to a strong recruiting cohort that heavily depresses the juvenile resource, leading to the death of not only the cohort itself but also older, remaining juveniles. As a result of the strong competition in the juvenile habitat, the juvenile resource is, except for a short period before a maturation pulse takes place, at a very low level (figure 10.7, left panel). We can at this point already conclude that population fluctuations of cohort type are also present when juveniles and adults do not compete for a common resource, but still are coupled via reproduction and maturation. Variation in recruitment is largely explained by variation in stage-specific mortality where a cohort by its own feeding may cause its own extinction as well as the extinction of older juvenile cohorts in the system.

Increasing the juvenile habitat fraction to 0.5 leads to a release in mortality of newborn individuals, and no complete recruitment failure is observed any more (figure 10.7, middle panel). Mortality among individuals older than one year is still present in years with strong reproduction pulses owing to the depression of the juvenile resource caused by these newly recruited individuals. Thus although starvation mortality among newly born individuals is reduced, starvation mortality is still present among juveniles as a whole. Stacking of several differently aged juvenile cohorts also occurs in this case as a result of size-dependent competitive ability. As a consequence, a maturing cohort consists of several age cohorts (figure 10.7, middle panel). Increasing the juvenile habitat fraction to 0.95 leads to a situation where the mortality of all juvenile size classes is due to background mortality only (note that slopes of decreases are the same and constant; figure 10.7 right panel). Adults produce many cohorts of offspring. Most of the differently aged cohorts mature at the same time, and when they enter the adult habitat, they cause a heavy depression of the adult resource, in turn, causing a massive death of up to 90 percent of the maturing cohort.

Overall, our analysis shows that exclusive resources for juveniles and adults lead to strong fluctuations driven by strong variation in stage-specific mortalities that will hit either juveniles or adults, dependent on the proportion each habitat constitutes. Bifurcation analyses varying the habitat size fraction show that the fluctuations are irregular over most parts of the parameter space, and only in small ranges are regular cycles observed. In the situation with a small juvenile habitat fraction, many differently aged juvenile cohorts are stacked, where smaller/younger juvenile individuals may outcompete larger/ older juvenile individuals. This dynamics, and particularly their nonregularity, resemble in many aspects the dynamics that are observed for high μ_{nr} values (see previous section) in systems with flexible habitat use involving a trade-off between energy gain and mortality risk, and where the juveniles switch instantaneously and only once during their lifetimes (de Roos, Leonardsson, et al. 2002). In a more general sense, the case with separate resources relates to the results of Diekmann, Gyllenberg, et al. (2010), with either fixed growth where net-production models may lead to cycles, owing to a delayed feedback mechanism, or with fixed reproduction where fluctuating length of the juvenile period in itself may lead to cycles. It is also noteworthy that the juvenile resource is more limiting than the adult resource for a longer time period than the reverse, even for the case with a juvenile/adult habitat size ratio of 0.95/0.05 (figure 10.7).

MODEL PREDICTIONS AND EMPIRICAL DATA

Confronting model predictions with empirical data is a fundamental but also challenging part of the scientific process. In contemporary population ecology, tests of model predictions and assumptions are generally based on statistical fits of population dynamic time series by comparing fits of different models (Ellner and Turchin 1995; Turchin and Ellner 2000; Turchin and Hanski 2001). In principle, the model selection procedure involves a successive addition of complexity, often in terms of adding nonlinear terms, where the selected model is based on some statistical measure such as the Akaiki criteria (Hilborn and Mangel 1997). This well-developed path for model testing has in contemporary ecology generally involved two major aspects: (1) model assumptions are implemented at the level of the population, and (2) model parameters are estimated from empirical data from the system under investigation, leading to interdependence between model assumptions and the dynamics of the system to be predicted.

In both these aspects, the confrontation of model predictions with data is fundamentally different for physiologically structured population models (PSPMs), a difference that can be related to the clear separation between i-states and p-states in the latter models. Because all model assumptions in PSPMs pertain to the individual level, no further assumptions are, in principle, made at the population level. The pathway for testing model predictions will hence also be different. First, the individual-level model is derived (see above equations (10.1)–(10.11)), including parameterization of the functions, a procedure that is independent of the system whose dynamics are to be predicted. As the population level is just a matter of bookkeeping, the population dynamics are subsequently predicted also largely independent of data from the system whose dynamics are to be predicted. For example, the only parameters that are not independent of the system whose dynamics are to be predicted for the consumer-resource model considered in this chapter are related to mortality (except for starvation mortality) and the productivity of the resource. Furthermore, PSPMs not only predict population-level characteristics like the presence of cycles and cycle length based on individual-level assumptions, but also the feedback of the population dynamics on individual-level rates such as individual consumption and growth rates. Physiologically structured population models thus lend themselves to a more critical confrontation of model predictions with data than unstructured population-level models, both with respect to the number of processes that can be predicted and through a higher independence between model assumptions and data to be confronted with model predictions.

Individual growth in body size at different ages has turned out to be a particularly suitable individual-level process to be predicted as a result of population feedbacks. In the previous chapter, we discussed the difficulties in sorting out the mechanisms driving the dynamics of *Daphnia* populations besides being able to define them as predator-prey or cohort cycles in general. In many respects this has been due to a difficulty in obtaining individual growth rates of differently aged individuals. A major part of the problem has been the difficulty in distinguishing different age cohorts, a problem that has only recently been resolved experimentally by keeping individuals separate but still experiencing the population feedback on the resource (McCauley, Nelson, and Nisbet 2008). With pulsed recruitment induced by seasonality, this problem largely evaporates, as age cohorts can be followed over time even under field situations.

In some organisms, like fish and trees, it may even be possible to reconstruct past growth history of an individual using the rings in scales or tree rings. This circumstance also means that population dynamics in some cases can be reconstructed backward in time (Persson, Claessen, et al. 2004). In the following, we will use two fish examples from the temperate region to show how the combination of population- and individual-level data can be used to discern the mechanisms driving dynamics.

We already in chapter 9 discussed the population cycles observed in the fish species vendace. We concluded that these cycles were juvenile-driven Type I cycles. Although it was possible to nail down the type of cycles present, the analysis was largely qualitative. Here we will take one step further to confront quantitative model predictions of individual growth rates of different age cohorts with those empirically observed. Our first example comes from a roach (*Rutilus rutilus*) population in Alderfen Broad (UK) (Cryer, Peirson, and Townsend 1986; Townsend and Perrow 1989; Townsend, Sutherland, and Perrow 1990). For this population, a two-year cycle with alternating strong and weak year-classes with a strong feedback on the resource zooplankton was documented with a density ratio of strong versus weak year-classes of 4 to 5 (figure 10.8, left). Data on adult fecundity suggested that during a growth season with strong reproduction, resource levels in the lake were suppressed to such an extent that most one-year old individuals failed to mature as two-year old individuals. In contrast, following a weak reproduction year, one-year old individuals successfully matured and managed to reproduce at an age of two years.

For roach, the individual-level model (equations (10.1)–(10.11)) has been derived and parameterized based on experimental data (for parameters used, see de Roos and Persson 2001; Hjelm and Persson 2001; Persson, Leonardsson, et al. 1998). Size scalings of foraging and metabolic demands for this species are such that smaller individuals always have a lower critical resource demand (Hjelm and Persson 2001), leading to the conclusion that cohort cycles in roach are expected to be juvenile-driven. As smaller individuals are more vulnerable to predation to size-selective predators present in Alderfen Broad, we assumed a background mortality rate that included both a size-independent and a size-dependent component (equation (10.16)). The resulting population-level dynamics in terms of numbers of individuals yield a two-year cycle with a density ratio of strong versus weak year-classes of approximately 4 to 5, the same as that observed in Alderfen Broad itself (figure 10.8, right).

It should be pointed out that the model predictions depend on the choice of the mortality parameters μ_0, μ_{nr}, and x_{nr}. The value of μ_0 was chosen on the basis of the data presented by Townsend, Sutherland, and Perrow (1990), and the values of μ_{nr} and x_{nr} were chosen such that a two-year cycle resulted.

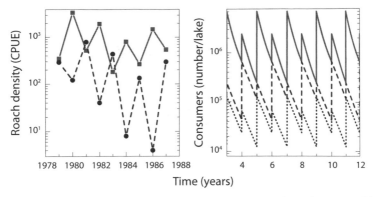

FIGURE 10.8. *Left*: variation in the number (catch per unit effort) of one-year-old (*solid lines, squares*) and two-year-old roach (*dashed lines, circles*) in Alderfen Broad. *Right*: model predictions of the number of young-of-the-year (*solid lines*), one-year-old (*dashed lines*), and two-year-old (*dotted lines*) roach in different years. Throughout, $\mu_0 = 0.014$ day^{-1}, $\mu_{nr} = 0.017$ day^{-1}, and $x_{nr} = 2.0$ g.

Model predictions are thus in these aspects not independent from the empirical observations, as the mortality parameters were tuned to obtain a qualitative match with the empirical population data. Nevertheless, given this match at the population level, we can proceed to consider to what extent the model correctly predicts the empirical observations at the individual level (individual growth rate) as a result of population feedback. Comparing predicted and observed growth rates of different age cohorts shows that the model not only correctly predicted the qualitative pattern that strong year-classes grow more slowly than weak year-classes, but also the quantitative length increase of individuals at one and two years old (figure 10.9). For older individuals (aged three and four years), a deviation is present between predictions and observations. Possible reasons for this discrepancy are sampling errors caused by the small number of older individuals caught and the fact that alternative resources used by older individuals might have influenced their growth rates.

In addition to the quantitative predictions for individual growth, the model also correctly predicts the observation that strong year-classes start to reproduce at an age of two years old, whereas a weak year-class does not start to reproduce until an age of three years old. The latter is a result of the fact that individuals of a weak year-class during their second growth season suffer from strong competition with the subsequent strong year-class. As a consequence, a strong year-class turns out to be the main contributor to the following strong year-class, as well as to the weak year-class after that. To conclude, the

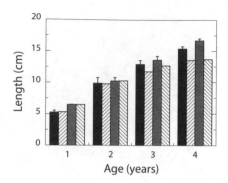

FIGURE 10.9. Length-age relations for roach in Alderfen Broad. Black and dark gray solid bars are empirical estimates for strong and weak year-classes, respectively (Townsend, Sutherland, and Perrow 1990). Risers above the solid bars indicate the range in average length observed in different years. Diagonally and cross-diagonally hatched bars represent model predictions for strong and weak year-classes, respectively.

analysis of the Alderfen Broad roach population shows a strong qualitative and quantitative agreement between predictions and observations concerning both population- and individual-level predictions.

Our second example comes from a yellow perch (*Perca flavescens*) population in Crystal Lake, Wisconsin, USA. This population exhibited a five-year cycle from the start of the study in 1981 up to 1992, after which the invading rainbow smelt (*Osmerus mordax*) took over and came to dominate the system (Beisner, Ives, and Carpenter 2003; Sanderson, Hrabik, et al. 1999) (figure 10.10). Although cannibalism is present in yellow perch, it is much less cannibalistic than its relative, the Eurasian perch (*Perca fluviatilis*) (Fulford, Rice, et al. 2006; Mittelbach and Persson 1998; Persson, de Roos, and Bertolo 2004), and the overall dynamics resemble that of cohort cycles induced by intraspecific competition (Sanderson, Hrabik, et al. 1999). Similar to the roach example, we adjusted background and size-dependent mortalities to yield a five-year cycle. Different combinations of background and size-dependent mortalities could yield a five-year cycle, and we did not have any information on background mortality for yellow perch, with the exception that overall mortality was much higher during the first year of life than in other years. We therefore derived theoretical predictions of individual growth trajectories for different combinations of background mortality and size-dependent mortality yielding a five-year cycle. Furthermore, we assumed size-dependent mortality to be present only during the first year of life. Because attack rate data for yellow perch on the zooplankton resource were available only up to a size of 40 mm

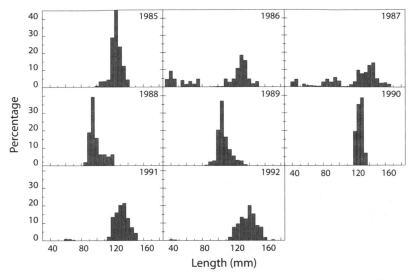

FIGURE 10.10. Development of the size distribution of yellow perch born in 1985 up to 1992 in Crystal Lake, Wisconsin. Observe that the dominating cohort born in 1985 first appears in catches in 1986. Data from Sanderson, Hrabik, et al. (1999).

(Letcher, Rice, et al. 1996) and these data did not differ substantially from that for Eurasian perch (shown in figure 10.1), we chose to use the parameter values for Eurasian perch (box 10.2).

In figure 10.11, empirical growth trajectories are shown for two strong year-classes together with theoretical predictions for two parameter combinations yielding a five-year cycle: one with a higher background mortality and lower size-dependent mortality and one with a lower background mortality and higher size-dependent mortality. The two predicted growth curves were overall very similar but differed in that the growth trajectory based on a higher background mortality yielded a slightly higher growth rate of older individuals, a result of an overall higher mortality with this scenario (figure 10.11). Predicted and observed growth trajectories showed a high correspondence up to an age of two to three years; thereafter predicted growth rates were lower than those observed. As for roach, this discrepancy between predicted and observed growth rates for older individuals is likely due to the fact that older individuals when increasing in size shifted to larger, alternative resources not accounted for in the model.

To conclude, we have established four major results in this chapter. First, we have shown that the critical resource density—analogous to q in chapter 9—can be used to predict the dynamics of size-structured consumer-resource systems.

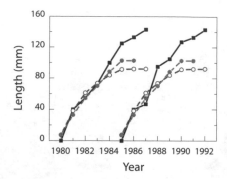

FIGURE 10.11. Observed growth trajectories (*solid lines, filled squares*) of two dominating cohorts of yellow perch born in 1980 and 1985 in Crystal Lake. Model predictions of growth rates (*dashed lines*) for a background mortality $\mu_0 = 0.007$ and a size-dependent mortality constant $\mu_{nr} = 0.046$ (*filled circles*) or a background mortality of $\mu_0 = 0.003$ and a size-dependent mortality constant of $\mu_{nr} = 0.064$ (*open circles*).

Second, we have found that the addition of an additional resource does not affect the presence and nature of cohort cycles. Third, cohort-driven fluctuations are also present when juveniles and adults do not compete for the same resource, but are only coupled via maturation and reproduction. Fourth, both the roach example and the yellow perch example illustrate how predictions of physiologically structured population models can be used to confront model predictions of individual-level growth with empirical data. We will come back to this issue in a more general discussion in chapter 12.

Cannibalism in Size-Structured Systems

In the previous two chapters, we discussed how the size scaling of foraging and metabolic rates affect the dynamics of consumer-resource systems. Using different modeling approaches, we showed that stage-dependent competitive ability (given by q or CRD) was the main predictor of population dynamics; that is, it largely set the conditions for different types of cycles to occur. The results obtained concerning consumer-resource dynamics also connected to the insights gained in chapter 3 about reproduction and development control. In this chapter, we add another intraspecific interaction on top of the consumer-resource system, namely, cannibalism. As in chapter 10, we will use a discrete-continuous population-level model based on individual-level net-production energetics to investigate the effects of cannibalism. Our focus will be on the effects of cannibalism on population dynamics related to four processes that have been discussed in the literature regarding cannibalism: effects on mortality, competition, energy gain, and the size dependence of interactions (Claessen, de Roos, and Persson 2004). As we have done in all previous chapters, we will also discuss our results in relation to development and reproduction control. We will finally relate theoretical predictions to empirical data as we did in chapter 10 and point to how relatively simple measures of cannibalistic efficiency such as gape size may be used a priori to predict population dynamics.

Cannibalistic interactions have been the focus of a vast number of theoretical and empirical studies. With respect to theoretical studies, these have focused on quite different aspects of cannibalism and have also made very different assumptions. These studies have used, on the one hand, models in which cannibalism is the only interaction incorporated and energy extraction or gain from cannibalism is not present to, on the other hand, models that both include energy gain by cannibals and competition. Owing to the diversity of models used to study cannibalism, we will first give an overview of the extent to which different studies have included different aspects of cannibalism and the different effects that cannibalism may have on population dynamics. At the same time, it should be kept in mind that we argue that an individual-based approach, which handles the consumption of the shared resource of differently sized individuals and cannibalism of larger individuals on small individuals, is to be preferred over population-level formulations, as we aim for reaching a deeper understanding of the intraspecific, mixed interaction present in many cannibalistic systems.

BACKGROUND OVERVIEW

FOUR ASPECTS OF CANNIBALISM

Cannibalism is a commonly reported interaction in ecological communities and occurs in a wide range of phyla including protozoans, mollusks, arthropods, insects, fish, amphibians, reptiles, birds, and mammals (Crowley and Hopper 1994; Elgar and Crespi 1992; Elwood 1992; Fox 1975; Polis 1981; Rudolf 2008; Smith and Reay 1991; Waddell 1992). In studies of cannibalism, a number of different aspects of cannibalism have been discussed. In particular, the size dependence of cannibal-victim interactions have been stressed in many studies (Claessen, de Roos, and Persson 2000; Claessen, van Oss, et al. 2002; Dong and DeAngelis 1998; Elgar and Crespi 1992; Fagan and Odell 1996; Orr, Murdoch, and Bence 1990; Persson, Byström, and Wahlström 2000; Rice, Crowder, and Marschall 1997; Rudolf 2008). Polis (1988), among others, also suggested that cannibalism may have a twofold advantage for the cannibal, as the mortality impact on victims (1) will reduce competition for prey that the cannibals share with victims and (2) involves an energetic gain for the cannibal from eating victims.

In a review of modeling studies on cannibalism, Claessen, de Roos, and Persson (2004) discussed the extent to which all these aspects of cannibalism had been considered in the theoretical literature: (1) the mortality impact on victims, (2) the size-based nature of cannibal-victim interactions, (3) the energy gain of cannibals from eating victims, and (4) intraspecific competition. One of these factors, cannibal-induced mortality, was found to be incorporated in all the reviewed studies, and the size-based nature of cannibalistic interactions had also been incorporated in almost all of the studies (twenty-nine out of thirty). However, sixteen of those twenty-nine studies only incorporated stage dependence in an implicit way, by assuming cannibalism to depend on age.

Concerning the third factor, energy gain, it has been a basic tenet of this book that consumption of prey, and hence biomass intake, has consequences for individual performance in terms of development and reproduction rates. Although there are examples where cannibals do not extract energy from killing victims, such as, for example, infanticide in lions (Hausfater and Hrday 1984), the energetic benefit from cannibalizing a victim is in most cases a vital process affecting individual cannibal performance (Persson, Claessen, et al. 2004; Polis 1981). The energetic extraction from cannibalism per se will also affect the interaction strength between cannibals and victims over time, as victims consumed by the cannibals are translated into growth of the cannibals, which is likely to increase their feeding and thus their cannibalistic attacks on victims. On the other hand, the mortality imposed on victims by cannibals will increase the growth of the remaining victims. As a result, the intensity of cannibalism will vary, because it is strongly coupled to the growth of both cannibal and victim individuals (Claessen, de Roos, and Persson 2000; Rice, Crowder, and Marschall 1997). Despite this fact, only half of the modeling studies reviewed by Claessen, de Roos, and Persson (2004) included energy gain. The most prominent example of the modeling studies where energy gain has not been included are the studies of *Tribolium*, an organism that in many respects has served as a model organism for cannibalism (Costantino, Desharnais, et al. 1997; Hastings and Costantino 1987).

Finally, the fact that cannibals and victims often share a common prey, the importance of which was already pointed out by Polis (1988; see also Persson 1988; Persson, Byström, and Wahlström 2000), was accounted for in seventeen of the thirty modeling studies reviewed by Claessen, de Roos, and Persson (2004). All four potential effects of cannibalism were considered in approximately one-third of the studies, of which half considered individual body size explicitly.

Effects of Cannibalism on Population Dynamics and Structure

The above overview advances four major aspects of cannibalism to potentially play a significant role in the regulation of cannibalistic populations, despite that only a minority of theoretical modeling studies have taken all of them into account. Before considering a fully size-structured model, which considers all four aspects, we will first look at different modeling attempts, where only one, or at most three, of these aspects have been considered. We will focus on the effects of cannibalism on stability and the occurrence of alternative stable equilibria (Claessen, de Roos, and Persson 2004).

A classic model of cannibalism is the Ricker model (1954), in which the relationship between the density of the present stock of fishes ($N(t)$) and that of next year's stock is given by:

$$N(t+1) = N(t)ae^{-bN(t)} \qquad\qquad (11.1)$$

The parameter a in this model has in many studies (e.g., Claessen, de Roos, and Persson 2004) been set equal to the per capita number of offspring produced, while b represents the cannibalistic intensity imposed by cannibals. However, a may also include density-independent per capita cannibalism, and similarly b may include density-dependent mechanisms other than cannibalism (Ricker 1954). In any case, stage dependence of cannibalism is in this equation only implemented in a very simple and implicit way in that cannibals have to be at least one year old and all victims are recruits. Furthermore, no term for energy gain by cannibals is present. Despite its seeming simplicity, this model can show quite complex dynamics, depending on the fecundity parameter a. For low values of a, stable fixed-point dynamics occurs, but with an increase in a, the system will destabilize, first leading to a two-point cycle, which undergoes a period doubling with further increases in a, followed by a sequence of period doublings eventually resulting in chaotic dynamics for large a values (May 1976). A study of the population dynamics of two populations of the leech *Erpobdella octoculata* in a small stream in England nicely illustrates the effect of a (Elliott 2004). One of the populations was relatively stable, whereas the other fluctuated substantially over time with a frequency of approximately four years (figure 11.1, left panel). The estimated value of the parameter a was also more than twice as high (13.43 versus 5.41) in the population exhibiting substantial fluctuations compared with the population exhibiting smaller fluctuations (figure 11.1, right panel).

This destabilizing effect of per capita fecundity also occurs in other cannibalistic models (Costantino, Desharnais, et al. 1997; Hastings and Costantino 1987; Landahl and Hansen 1975). In particular for the *Tribolium* system, there exists both modeling and experimental evidence that this system may produce complex, nonlinear dynamics, including chaotic dynamics (Costantino, Desharnais, et al. 1997). An important element of the models, which generate population cycles, is that time delays were incorporated in the form of a delay between birth and the moment individuals become cannibalistic. It is well known that such time-delayed, density-dependent processes per se may induce cycles (Gurney and Nisbet 1985).

In contrast, the studies in which competition is present as an additional density-dependent process show that cannibalism may also have a stabilizing effect (Claessen, de Roos, and Persson 2004). For example, Cushing (1991) studied a discrete-time model with a non-cannibalistic juvenile stage and a cannibalistic adult stage, which accounted for energy gain by cannibals and in which adults competed for an alternative prey (but no competition occurred

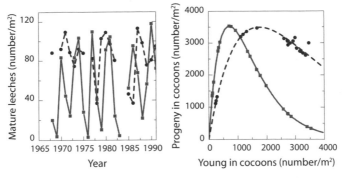

FIGURE 11.1. *Left*: fluctuations in adult leeches (*Erpobdella octoculata*) at two sites of an English stream where the population at site A (*circles, dashed lines*) shows no regular pattern, whereas the population at site B (*squares, solid lines*) shows a four-year cycle. *Right*: relationship between the number of progeny in the cocoons and young in cocoons fitted with the Ricker model in the two local populations of *Erpobdella octoculata* studied by Elliott (2004). For the population at site A, *a* was 5.41 and *b* was 0.00057, and for the population at site B, *a* was 13.43 and *a* was 0.00140. (All data provided by J. M. Elliott). Years with severe drought have been excluded (see Elliott 2004).

between stages). In the absence of cannibalism, the dynamics were characterized by the period-doubling route to chaos with increasing competition. In this case, cannibalism reduced the parameter range over which oscillations occurred, if the competition intensity among adults was sufficiently strong. Otherwise, cannibalism induced population cycles as a result of the time delay that we discussed in the previous paragraph.

Finally, Cushing (1991) argued that the interaction between positive (directly through energy gain and indirectly through reduced competition) and negative (mortality of victims) effects of cannibalism may give rise to alternative stable states. With respect to direct energy gain, energy extraction by cannibals from eating victims will increase cannibal fecundity and survival and subsequently lead to increased production of offspring. If this production of new offspring outweighs the negative effect through cannibal-induced victim mortality, cannibalism may lead to a positive feedback loop and thereby bistability (Cushing 1991; van den Bosch, de Roos, and Gabriel 1988). A special case of this bistability is the "lifeboat mechanism" that allows a cannibalistic population to persist under food conditions where a non-cannibalistic population would go extinct (Diekmann, Gyllenberg, and Metz 2003; van den Bosch, de Roos, and Gabriel 1988). The necessary ingredients for this mechanism to occur are that juveniles feed on a resource that is unavailable to the adults and the average yield in terms of new offspring produced based on cannibalizing one juvenile is higher than 1 (van den Bosch, de Roos, and Gabriel 1988).

Indirect positive effects via decreased competition may also lead to bistability. For example, consider a situation in which the individual growth of a recruiting age class and hence its size as a one-year-old individual the next year, when it cannibalizes on the following recruiting cohort, is affected by competition. If the negative effect of the number of recruits on their per capita attack rate as one-year-old cannibals outweighs the positive effect of the number of recruits on cannibal density, a negative relationship will be present between the abundance of a recruiting age class and the mortality caused by cannibals the next year. This will, in turn, result in a positive relationship between consecutive numbers of recruits and therefore a positive feedback loop, causing bistability (Fisher 1987).

A DISCRETE-CONTINUOUS MODEL FOR CANNIBALISM

With this background, we next proceed to derive an individual-based, size-structured model that incorporates all four aspects of cannibalism considered above. Because we will assume that cannibals compete with their victims for a shared prey, the resulting mixture of competition and cannibalism affecting the dynamics implies that the food web module we study can be characterized as an intraspecific, intraguild predation module (Claessen, de Roos, and Persson 2004; Polis 1991).

The cannibalism model is an extension of the consumer-resource model presented in chapter 10, which also includes cannibalistic interactions (see boxes 10.1 and 10.2 for individual-level equations and parameter values of the consumer-resource part of the model). We will assume a competitive superiority of juveniles; hence the system will exhibit juvenile-driven cohort cycles in the absence of cannibalism and will be reproduction-controlled. Adding cannibalism in a size-dependent way such that larger individuals cannibalize on smaller individuals (see below) may affect this limitation by two processes: (1) cannibals impose mortality on victims, which will affect the maturation rate both negatively (through decreased survival) as well as positively (increased growth of survivors), and (2) cannibals gain energy from cannibalizing victims, which will affect their reproduction rate. It can already here be hypothesized that cannibalism may shift a system with ontogenetic asymmetry (reproduction control) toward ontogenetic symmetry or even ontogenetic asymmetry with development control, depending on the degree of cannibal-imposed mortality and the energy gained. In contrast, in a consumer-resource system driven by adult-driven cycles (maturation control) cannibalism is likely to only increase the degree of ontogenetic asymmetry through increased maturation limitation.

In the previous chapter, we derived the individual- and population-level formulations for the consumer-resource system. What we need to do here is to add the functional relationships that describe the energy intake of cannibals and the mortality imposed by them on victims. The individual growth of an individual cannibal will then be a result of the sum of its energy gain from the shared resource and cannibalism. Although victims may be larger than cannibals, we will focus on the more general case, in which a cannibal has to be larger than its victim to cannibalize it. Following Claessen, de Roos, and Persson (2000), we assume that the length l of an individual is related to its effective body mass $w = (1 + q_J)x$ (see chapter 10) by a power function:

$$l(w) = \lambda_1 w^{\lambda_2} \tag{11.2}$$

and that positive cannibalistic attack rates are constrained between two limits of the ratio of victim (l_v) to cannibal length (l_c), which are equal to $l_v/l_c = \delta$ and ε, respectively (see box 11.1 for parameters and their values). The range between these boundaries we refer to as the cannibalistic window. The parameter

---

BOX 11.1

PARAMETER VALUES USED FOR THE FUNCTIONS
DESCRIBING THE RELATIONSHIP BETWEEN CANNIBAL
AND VICTIM IN RELATION TO THEIR SIZES

Parameter	Default value	Unit	Description
λ_1	48	mm/g$^{\lambda_2}$	Allometric length–body mass scaling constant
λ_2	0.32	—	Allometric length–body mass scaling exponent
β	0.2	m^3/day mm$^{-\sigma}$	Cannibalistic voracity scaling constant
σ	0.6	—	Cannibalistic voracity scaling exponent
δ	0.05	—	Minimum cannibal-victim length ratio
φ	0.15	—	Optimum size ratio
ε	0.45	—	Maximum cannibal-victim length ratio

See box 10.2 for non-cannibalistic parameter values.

δ sets the smallest victim size as a fraction of cannibal length below which the cannibal does not encounter victims, owing to difficulties in detecting them. In contrast, ε sets the maximum victim size as a fraction of cannibal length above which the escape ability of victims and gap constraints of the cannibal prohibit capture (Christensen 1996; Dong and DeAngelis 1998; Fagan and Odell 1996; Juanes 2003; Rice, Crowder, and Marschall 1997).

Within the boundaries set by δl_c and εl_c, respectively, we further assume that there is an optimal victim length yielding the highest attack rate for each specific cannibal size, $l_v = \varphi l_c$. A two-dimensional plot of these three linear functions for the lower and upper boundaries to cannibalism and the optimal victim-cannibal length ratio is shown in figure 11.2 (left panel), which also includes empirical data on observed cannibalism in Eurasian perch (*Perca fluviatilis*). This plot furthermore includes experimental results, in which victim size was clearly too small to be encountered by the cannibal (open circles), substantiating the presence of a lower size boundary. In this case, the inability of cannibalistic predators to encounter small victims is, in addition to their small size per se, also likely affected by the transparency of larval fish, as smaller fish larvae are more transparent than other small prey that evidently are consumed (figure 10.1 versus figure 11.2, left panel).

Within the range of cannibal and victim sizes where the cannibalistic attack rates are positive—the cannibalism window—we assume that the attack rate on differently sized victims for a specific cannibal length increases linearly with victim size, from zero at $l_v = \delta l_c$ to an optimum at $l_v = \varphi l_c$, to thereafter decrease linearly and become zero again when victim length reaches $l_v = \varepsilon l_c$ (Claessen, de Roos, and Persson 2000; Claessen, van Oss, et al. 2002). This results in a "tent" function ($T(l_v, l_c)$) of the form (see also chapter 4):

$$T(l_v, l_c) = \begin{cases} \dfrac{l_v - \delta l_c}{(\varphi - \delta)l_c} & \text{if } \delta l_c < l_v < \varphi l_c \\[2mm] \dfrac{\varepsilon l_c - l_v}{(\varepsilon - \varphi)l_c} & \text{if } \varphi l_c < l_v < \varepsilon l_c \\[2mm] 0 & \text{otherwise} \end{cases} \tag{11.3}$$

On the ridge $l_v = \varphi l_c$, representing the optimum victim length for differently sized cannibals, the attack rate is assumed to increase with cannibal length according to the power function βl_c^σ, where β represents the overall intensity of cannibalism (referred to as cannibalistic voracity in the following) and σ represents the size scaling of cannibalism (figure 11.2, right panel) (Claessen, de Roos, and Persson 2000; Claessen, van Oss, et al. 2002). Experimental support for this form of the function has been presented by, among

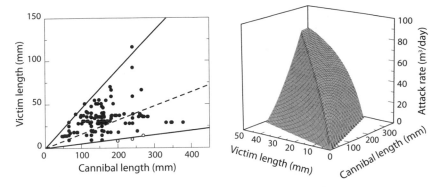

FIGURE 11.2. *Left*: lower and upper size boundaries (*solid lines*) for cannibalism to occur and victim size yielding the highest attack rate for a specific cannibal size (*dashed line*). Filled circles are empirical data representing cannibal and victim sizes, and open circles are data from experiments where the cannibal did not capture any victim (data from Persson, unpublished data; Claessen, de Roos, and Persson 2000). *Right*: three-dimensional plot of the cannibalistic attack rate (m³/day) as a function of victim and cannibal sizes. Parameter values: $\delta = 0.05$, $\varphi = 0.16$, $\varepsilon = 0.45$, $\beta = 0.4$, $\sigma = 0.6$.

others, Lundvall, Svanbäck, et al. (1999; see also Juanes 2003). The total cannibalistic attack rate ($A_c(l_v, l_c)$) of cannibals with length l_c on victims with length l_v will hence be

$$A_c(l_v, l_c) = \beta l_c^\sigma T(l_v, l_c). \tag{11.4}$$

The rate at which cannibals encounter victim biomass is subsequently given by

$$\eta_c(x_i) = \sum_j A_c(l_{v,j}, l_{c,i})(x_j + y_j) N_j, \tag{11.5}$$

where N_j stands for the number of victim individuals in the jth cohort, and $(x_j + y_j)$ refers to the total body mass of a victim in this cohort. The notations $l_{v,j}$ and $l_{v,i}$ refer to the length of the victim and the cannibal, respectively, which are related to their irreversible masses x_j and x_i following equation (11.2). The summation in the expression above is over all possible victim cohorts, that is, over all cohorts that are within the cannibalism window of a particularly sized cannibal. The total encounter rate of cannibals with food mass, both from cannibalism and from feeding on the shared resource (R), now equals

$$\eta(x_i) = \eta_R(x_i) + \eta_c(x_i), \tag{11.6}$$

where the form of the encounter rate for the shared prey $\eta_R(x_i)$ was specified in chapter 10 as $a(w_i)R$.

Given the total encounter rate of cannibals with food mass, their total intake rate is given by:

$$I(x_i) = \frac{\eta(x_i)}{1 + h(w_i)\eta(x_i)},$$ (11.7)

where $h(w_i)$ is the handling (digestion) time and $w_i = (1 + q_J)x_i$ is the effective body mass that were also specified in chapter 10. Finally, the net energy intake that can be used for somatic growth and reproduction will be equal to the total intake rate times a conversion factor k_e minus the metabolic demands, where the form of the latter was specified in chapter 10 (equation (10.6)):

$$E_g(x,y) = k_e I(x) - E_m(x,y)$$ (11.8)

The total mortality resulting from background, starvation, and cannibalism mortality is:

$$\mu(x_j,y_j) = \mu_0 + \mu_s(x_j,y_j) + \mu_c(x_j),$$ (11.9)

where the cannibalistic mortality is

$$\mu_c(x_j) = \sum_i \frac{A_c(l_{v,j}, l_{c,i}) N_i}{1 + h(w_i)\eta(x_i)}.$$ (11.10)

In these expressions, the indices i and j consistently refer to the cannibal and its victim, respectively. In the following, we will investigate the effects of the three parameters β, δ, and ε, which describe the scaling of the attack rate with cannibal length for the optimum ratio of victim/cannibal size, the lower size boundary for cannibalism, and the upper size boundary for cannibalism, respectively, on the dynamics of the cannibalistic population.

Increasing β has strong effects on the dynamics of the system (figure 11.3) (Claessen, de Roos, and Persson 2000). For low β values, cannibals are so inefficient that they impose only a nonsignificant mortality on victims, and the dynamics are characterized by the juvenile-driven cohort cycles (with a period of eight years) that we considered in the previous chapter. Increasing β first leads to a collapse of the cohort cycles to fixed-point dynamics or low-amplitude cycles. The time series of the dynamics of different stages of the cannibalistic population, resources, and individual growth trajectories of different cohorts show that cannibals reproduce every year in the region with low-amplitude dynamics but that recruits suffer from a very strong cannibalistic mortality (figure 11.4, left-middle panel). This, in turn, causes the impact of recruits on the shared resource to be relatively low, preventing cannibalistic individuals from

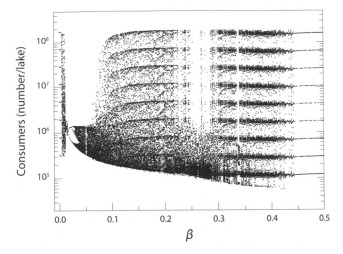

FIGURE 11.3. Bifurcation plot varying cannibalistic voracity (β). Each dot represents the total number of individuals aged one year old and older at the start of the growth season. Throughout, $\delta = 0.06$, and other parameters have default values as presented in boxes 10.2 and 11.1.

being excluded by starvation mortality (figure 11.4, left-middle and left-lower panels). Still, sufficiently many recruits survive the first year of life (all cannibalistic mortality turns out to be restricted to individuals less than one year old) to sustain the population at a relatively constant level (in figure 11.4, left-middle panel, a two-year cycle). In this parameter region, cannibals through their cannibalism prevent recruits from depleting the resource, as occurred in the juvenile-driven cohort cycles. Cannibalism thus clearly has a stabilizing effect on the dynamics in this parameter region. Also noteworthy is the fact that although cannibals impose a high mortality on victims, they gain little energy from cannibalizing them, and energetically the shared resource is their main food source, which leads to a relatively small maximum size (figure 11.4, left-upper panel). In this parameter region, cannibals thus control the number of victims but, at the same time, experience only a small energetic gain from cannibalizing.

With a further increase in cannibalistic voracity β, the low-amplitude fluctuations give rise to large-amplitude irregular fluctuations that eventually become regular cycles. These cycles at very high β values have a cycle length one year longer (nine years) than that of the cohort cycles occurring in the absence of cannibalism (figures 11.3 and 11.4, middle panels). Starting at the time when a cohort matures (in year 10, corresponding to $t = 810 - 900$ days in

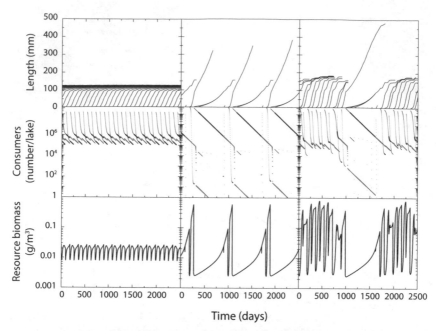

FIGURE 11.4. Time series of cannibal-driven dynamics ($\beta = 0.02$, *left panels*), dwarf-and-giant cycles ($\beta = 0.4$, *middle panels*), and mixed cannibal-driven dynamics/dwarf-and-giant cycles ($\beta = 0.2$, *right panels*). Throughout, $\delta = 0.06$, and other parameters have default values as presented in boxes 10.2 and 11.1. Upper panels show individual growth trajectories of differently aged cohorts, middle panels show number of individuals < one year old (*thin/dotted lines*), number of juveniles > 1 year old (*thick black lines*), and number of adults (*thick gray lines*), while lower panels show dynamics of the resource density (*solid line*).

figure 11.4, middle panels, given the length of the growing season of 90 days), this cohort gives rise to a pulse of recruits that decrease the shared resource only temporarily and marginally, as this pulse of recruits is totally and quickly eradicated through cannibalism by their parents (figure 11.4, middle panels). The next year another pulse of recruits is produced, which also depresses the resource for a limited time only. This recruiting cohort is heavily cannibalized, leading to a rapid recovery of the resource and a further increase in size of the mature cannibalistic cohort from year 11 to year 12 ($t = 990 - 1080$ days, figure 11.4, middle panels). The cannibalistic cohort then gives rise to a third large pulse of recruits in year 12. These recruits depress the resource, but this time for a longer time, because they are not subjected to any cannibalistic mortality. The latter results from the fact that the potential cannibals are larger in year 12 than in year 11 (figure 11.4, middle-upper and middle panels) and hence are outcompeted by the recruits in this year and starve to death before the recruits enter the cannibalistic window of the potential cannibals.

At this stage we can already see that the lower boundary for cannibalism δ can have a substantial effect on the dynamics, as it affects the timing of the onset of cannibalism relative to the resource depression induced by a recruiting cohort. The only surviving cannibals are the few individuals of the cohort born in year 11 ($t = 990$ days, figure 11.4, middle panels) that did not fall victim to the heavy cannibalism in their first year of life. These individuals cannibalize for several years on the cohort born in year 12, but owing to their low numbers, they inflict little mortality on it. In contrast, the energy they extract from cannibalism is very high, such that these cannibals grow very fast and reach "gigantic" sizes (figure 11.4, middle-upper panel). The gigantic cannibals reproduce for several years, but their low density implies that total population fecundity is low despite a high per capita fecundity. The effect of these gigantic cannibals on dynamics is hence negligible.

Once the strong cohort born in year 12 reaches maturity (at year 19, $t = 1710$ days, figure 11.4, middle panels), they give rise to a first strong recruitment pulse, and the cycle starts over again. Notice that the dominant cohort born in year 12 reaches maturity in seven years, as opposed to the eight years that it takes a dominant cohort to mature in the absence of cannibalism ($\beta = 0$). This faster development results from a lower initial cohort density than in the absence of cannibalism. Overall, the cycle is a juvenile-driven cohort cycle, similar to the cycles observed for very low or zero β values, but extended one year, owing to the presence of strong cannibalistic mortality during two years and a shortening of the juvenile period of the dominant cohort with one year, as a result of its lower density. Because of the simultaneous presence of a large cohort with slowly growing individuals and a small cohort of cannibals growing very fast, Claessen, de Roos, and Persson (2000) coined the term "dwarf-and-giant" cycle for this type of population cycle. It is noteworthy that although the cannibalistic voracity β is very high, the dynamics are for most of the time driven by cohort competition.

The dynamics of the cannibalistic system in the region with irregular, large-amplitude fluctuations are largely a mixture of periods of cannibal control, when cannibals reproduce every year but impose a high cannibalistic mortality on their offspring, and periods with dwarf-and-giant cycles (figure 11.4, right panels). The shift from the low-amplitude dynamics to the large-amplitude dynamics occurs through a series of period doublings with increasing β. In these mixed dynamics, a period with dwarf-and-giant cycles is initiated by the starvation death of most mature individuals as a consequence of a successfully recruiting cohort, which depresses the shared resource for a prolonged period of time. Notice also that the cannibals who survive a strong recruitment pulse and subsequently become giants are the smallest, and hence the most competitive, among the cannibal size classes present in the system at the moment of a major die-off of larger individuals. The fraction of time that the system is in the cannibal-controlled, low-amplitude dynamics also decreases with increasing β to eventually lead to pure dwarf-and-giant cycles (Claessen, de Roos, and Persson 2000).

The lower size boundary for cannibalism, δ, has strong effects on the dynamics. For small values of δ, cannibals can start to cannibalize on victims when the latter are very small (young), preventing any competitive effect of recruits from manifesting itself. In this parameter region, low-amplitude or fixed-point dynamics with cannibal control therefore prevails (figure 11.5). Increasing δ increases the time period during which recruits impose a competitive effect on potential cannibals, resulting in the large-amplitude dynamics with dynamics shifting between cannibal control and dwarf-and-giant cycles (for high β values only dwarf-and-giant cycles are present; see figure 11.3). A further increase in δ leads to a situation in which it takes such a long time before recruits become accessible to potential cannibals that the latter have all died of starvation owing to the sustained depression of the shared resource.

The effect of the lower size limit for cannibalism δ on population dynamics can also be investigated by varying hatching size, as a larger hatching size

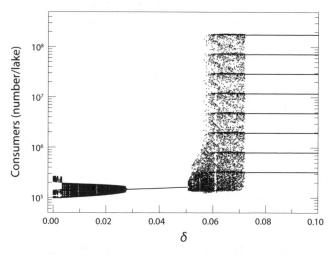

FIGURE 11.5. Bifurcation plot varying the lower size boundary δ for cannibalism to occur. Each dot represents the total number of individuals aged 1 year old and older at the start of the growth season. Other parameters have default values as presented in boxes 10.2 and 11.1.

increases the likelihood to be born in, or to grow early on into, the cannibalistic window of potential cannibals. Because increasing hatching size also
decreases the number of hatchlings (the number of hatchlings equals the mass
allocated to reproduction divided by the mass per hatchling), an increase in
hatching size will even in the absence of cannibalism lead to a decrease in
population variation, as the size of the reproductive pulses will decrease. For
example, in figure 11.6 (left panel) the coefficient of variation in population
density (numbers of individuals aged one year and older) in the absence of
cannibalism decreases from around 2 to 0.75 with an increase in hatching size
from 4 to 20 mm. This decrease in amplitude of fluctuations is associated with
a decrease in cycle length from an eleven-year cycle to a three-year cycle.

In the presence of cannibalism, but with a high minimum victim/cannibal
size ratio ($\delta = 0.1$), the coefficient of variation decreases with hatching size in
the same way as in the absence of cannibalism up to a hatching length slightly
larger than 7 mm; thereafter the coefficient of variation first increases and then
decreases to zero at a hatching length close to 10 mm (figure 11.6, left panel).
The increase in coefficient of variation between 7 and 10 mm is a result of an
extended phase with irregular large-amplitude dynamics, in which cannibals
during certain periods have control over recruiting victims, whereas the dynamics in other periods resemble more the dwarf-and-giant cycles (figure 11.4,
right panels). Decreasing δ leads to a decrease in the hatching length, at which

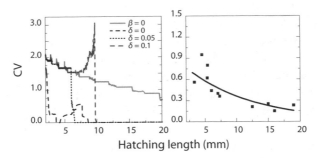

FIGURE 11.6. *Left*: coefficient of variation in the number of individuals aged
one year old and older at the start of the growth season in model time series as a
function of hatching length. Results are shown for zero cannibalism ($\beta = 0$) (*light-
gray solid line*), and for cases including cannibalism ($\beta = 0.2$) with $\delta = 0.0$ (*black
dashed line*), $\delta = 0.05$ (*black dotted line*), and $\delta = 0.1$ (*dark-gray dashed line*).
Other parameters have default values as presented in boxes 10.2 and 11.1. *Right*:
relation between hatching length and the coefficient of variation (CV) in time
series data of eleven commonly harvested cannibalistic lake fish species. Data
from van Kooten, Andersson, et al. (2010). The curve is a fit of an exponential
function to the data.

point the shift in population dynamics occurs from cohort cycles to cannibal control with fixed-point dynamics. Finally, in the scenario where cannibals have access to all victims irrespective of their hatching size ($\delta = 0$), the large-amplitude dynamics very quickly collapse to dynamics with cannibal control characterized by either fixed-point dynamics or small-amplitude cycles (figure 11.6, left panel, and figure 11.4, left panels).

Overall, we thus expect that increasing hatching size should lead to decreased variation in population density. Correspondingly, a compilation of literature data showed that cannibalistic fish species with a larger hatching size showed less variation in numbers over time (figure 11.6, right panel) (van Kooten, Andersson, et al. 2010). The reader may notice that the decrease in the coefficient of variation with increasing hatching length observed in the empirical data looks qualitatively closest to that of the effect of hatching length on population fluctuations in consumer-resource systems. However, an inclusion of only a modest level of harvesting changes the theoretical pattern for the cases with cannibalism to more closely resemble the empirically observed pattern (van Kooten, Andersson, et al. 2010).

Whereas both the cannibalistic voracity and the lower size boundary for cannibalism have major effects on the dynamics of the cannibalistic system in this model setting, the effects of the upper size boundary, ε, is smaller. For β and δ values where the dynamics are cannibal driven (i.e., competition from recruits is neutralized by intense cannibalism), variation in ε may affect the amplitude in fluctuations, but will not shift the system to other dynamic regimes (Claessen, van Oss, et al. 2002). Intuitively this may seem strange at first sight, but remember that whether small-amplitude, cannibal control or large-amplitude dynamics (pure cohort or dwarf-and-giant cycles) occurs, totally depends on high mortality of young recruits, for which the upper size boundary to cannibalism is hardly of any importance. In contrast, the upper size boundary for cannibalism has a substantial effect on the maximum size a cannibal may reach (figure 11.7). In effect, ε determines whether cannibals may reach a niche where they gain significant energy from cannibalism, that is, whether they can cannibalize on one-year-old and older cohorts.

For low ε values, the main energy extracted comes from the shared resource, whereas the small amount of energy extracted from cannibalism is solely the result of cannibalism on individuals younger than one year. Because the main energy supporting the cannibals comes from the shared resource, Claessen, van Oss, et al. (2002) referred to this niche as the shared resource niche. For low ε values, the energy balance for large cannibals (i.e., greater than 400 mm) foraging on larger victims (i.e., victims aged one year and older) is still negative, even if the cannibals had been able to reach these larger sizes (figure 11.7,

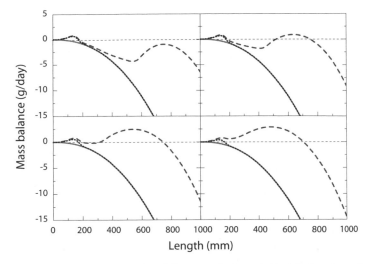

FIGURE 11.7. The size-dependent daily energy balance (g/day) during dynamics with cannibal control (see figure 11.8) for different values of the upper limit of the cannibalism window (*left-upper panel, ε = 0.2; right-upper panel, ε = 0.4; left-lower panel, ε = 0.6; right-lower panel, ε = 0.7*). The net-production rate (*dashed line*) was obtained by adding the intake rate from planktivory and piscivory to the (negative) metabolic rate (*solid line*). The distance between the dashed and solid line corresponds to the total energy intake rate. The contribution to intake from planktivory is indicated with a dotted line, and that from cannibalism is the difference between dashed and dotted lines. Throughout, $\delta = 0.0$, and other parameters have default values as presented in boxes 10.2 and 11.1.

upper-left panel). For an ε value between 0.4 and 0.6, the energy balance of large cannibals feeding in this cannibalism niche is positive, but individuals can still not reach the niche owing to the negative energy balance at intermediate cannibal sizes (figure 11.7, upper-right and lower-left panels). It is only at an ε value above 0.6 that the cannibals may break through from the shared resource niche to the cannibalism niche (figure 11.7, lower right panel). That the maximum size increases very rapidly with ε once the shared resource and the cannibalism niches have merged above the zero energy balance (figure 11.8) is a consequence of the positive energy balance of the cannibalism niche before the two niches merge and a transition from one niche to the next becomes possible (figure 11.7).

The above argumentation is based on equilibrium conditions. In the case where the system shows fluctuations, even modest recruitment pulses may result in individuals that "jump" over to the cannibalism niche (for an empirical

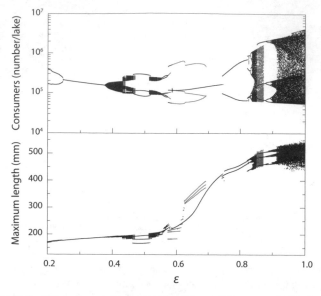

FIGURE 11.8. Bifurcation plots showing the effects of the upper size limit of the cannibalism window (ε) on population dynamics (*top panel*, total number of individuals aged one year old and older at the start of the growth season) and maximum size of individuals (*lower panel*). Gray dots are simulation runs from low to high ε values, and black dots are simulation runs from high to low ε values, revealing alternative dynamic patterns for the same parameter values. Throughout, $\delta = 0.0$, and other parameters have default values as presented in boxes 10.2 and 11.1.

example, see Byström 2006). In the parameter region where the maximum length increases rapidly with ε, bistability is also present (fixed-point dynamics versus a three-year cycle), although both dynamical regimes represent situations with cannibal control (figure 11.8). Finally and importantly, although large cannibals gain most of their energy from cannibalism, it is the smaller cannibals, who gain most of their energy from the shared resource, that exert the control over victims.

To conclude, the analysis of the major parameters of the cannibalism window shows that two of them, β and δ, have major effects on population dynamical regimes, whereas ε mainly affects the life history of individuals and the cannibal population size structure. The shifts between the major dynamical regimes (juvenile-driven cohort cycles, cannibal-driven dynamics) are strongly related to the time period during which recruiting cohorts younger than one year old are inaccessible to cannibals and can depress the density of the shared resource. This period is dependent both on the visual acuity of the cannibal per se (δ) but also on the size at hatching. The dwarf-and-giant cycles occurring at high cannibalistic voracities are also clearly related to food-dependent development of cannibalistic individuals, as the starvation of dominating cohorts after their third year of reproduction results from their increase in size in preceding years.

After investigating the effects of, in particular, the cannibalistic voracity and the lower boundary for cannibalism on population dynamics, we now return to the four factors—cannibalistic mortality, competition, energy gain, and size dependence—that we considered at the beginning of this chapter. More specifically, we focus on their relationship to different types of dynamics and whether cannibalism has a stabilizing effect or not. Overall, we have distinguished four different dynamical regimes: namely, single-cohort cycles, cannibal-driven dynamics, dwarf-and-giant dynamics, and a mixture of cannibal-driven dynamics and dwarf-and-giant cycles (table 11.1). The size dependence of cannibalistic interactions and the scaling of cannibalism with body size underlie all of these dynamical patterns. We can further conclude that cannibal control (i.e., cannibals imposing a high cannibalistic mortality on competing victims) does not coincide with energy gain; that is, when cannibals impose a high mortality, they do not gain much energy from cannibalism (cannibal-driven cycles, table 11.1). Vice versa, gigantic cannibals gain energy from their cannibalism, but their impact on mortality of victims and population fecundity are dynamically negligible. Finally, competition and energy gain can co-occur if cannibal efficiency is sufficiently high (table 11.1).

TABLE 11.1. Overview of Processes That Are Important for the Different Types of Dynamics Observed in Cannibalistic Systems

Type of dynamics	Cannibalistic mortality	Competition	Gain	Size dependence	Overall effect
Single-cohort cycles	No	Yes	No	Yes	Cycle
Cannibal-driven dynamics (CD)	Yes	No	No	Yes	Stabilization
Dwarf-and-giant cycles (DG)	No (7 years) Yes (2 years)	Yes (7 years) No (2 years)*	Yes** No	Yes Yes	No stabilization
Mixed CD and DG	Yes No	No Yes	No Yes	Yes Yes	No stabilization

*Transient competition between cannibals and victims in one of the years until the resource recovers owing to cannibalism within season.
**Until die-off of giants.

Single-cohort cycles represent the baseline dynamics of the system, induced, in the absence of cannibalism, by the competitive superiority of small individuals. Cannibalism may stabilize these dynamics if the cannibalistic voracity is sufficiently high and the lower size boundary for cannibalism is not too high (figures 11.3 and 11.4) (Claessen, de Roos, and Persson 2000; Claessen, van Oss, et al. 2002). If δ is sufficiently small, cannibal-driven dynamics are present for a wide range of β values, whereas the system destabilizes at higher β values for δ around 0.05 (Claessen, van Oss, et al. 2002). This result parallels the results of Cushing (1991), who showed that cannibalism may stabilize the dynamics when competition is present between stages (do observe, however, that he assumed competition to take place only within stages). The large-amplitude dynamics induced by a high cannibal voracity (β) shows that a high cannibalistic rate may destabilize the system, although the amplitude of this dynamics is similar to that of the single-cohort cycles present in the absence of cannibalism.

How cannibalism affects the dynamics of cannibalistic systems is thus highly influenced by β and δ. The effect of cannibalism on dynamics of the cannibalistic system can also be set in the context of whether ontogenetic asymmetry or symmetry occurs in the system. In the absence of cannibalism, the system is reproduction-controlled (ontogenetic asymmetry, juvenile-driven cohort cycles). In a gradient of increasing β for a value of $\delta = 0.0$, the dynamics of the system first shift from large-amplitude, juvenile-driven cohort cycles to damped oscillations and fixed-point dynamics and thereafter show increased population fluctuations again, although not to the same extent as in the juvenile-driven cohort cycles (figure 11.9, upper panel). Claessen, van Oss, et al. (2002) referred to all the dynamics at $\delta = 0.0$ for β values above those for which cohort cycles occur, as dynamics with cannibal control. Along this gradient of increasing β, the ratio between total population maturation and reproduction rate (both in terms of biomass), averaged over a long time period, decreases from values as high as 12 to reach a constant level around 1.4 at the β value where the dynamics stabilize to fixed-point dynamics ($\beta = 0.12$; figure 11.9, lower panel).

This value is somewhat higher than the value of unity expected for complete ontogenetic symmetry, but it should be pointed out that reproduction and hence also maturation are discrete events in the discrete-continuous, cannibalism model and that hence calculations of the *rates* of reproduction and maturation are at best approximate compared with the continuous-time models, in which we analyzed the ratio between total maturation and reproduction rate previously (chapters 3 and 9). We therefore consider the region with fixed-point dynamics ($\beta = 0.12 - 0.19$) to represent a situation with ontogenetic

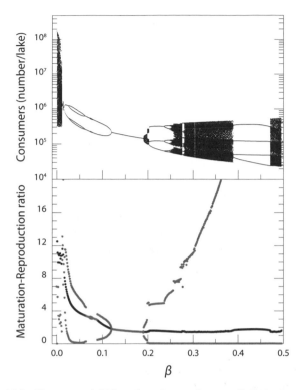

FIGURE 11.9. *Upper panel*: bifurcation plot varying cannibalistic voracity (β). Each dot represents the total number of individuals aged one year old and older at the start of the growth season. *Lower panel*: change in the ratio between the total population maturation and reproduction rate (both in terms of biomass) as a function of the cannibalistic voracity (β). Black dots represent time averages, and gray dots are maximum and minimum vales. The ratio of the total population maturation and reproduction rate was calculated by multiplying the fraction of the initial number of newborn individuals in a cohort that survived until maturation with the ratio of the effective body mass at maturation ($(1 + q_J)x_f$) and the effective body mass at birth (w_b, see box 10.2). Throughout, $\delta = 0.0$, and other parameters have default values as presented in boxes 10.2 and 11.1.

symmetry induced by a moderate intensity of cannibalism in a system where consumer-resource dynamics otherwise lead to ontogenetic asymmetry. When the dynamics destabilize again with an increase in β ($\beta > 0.19$), the average ratio between total population maturation and reproduction rate (in terms of biomass) does not decrease further. This bias of the average ratio to larger values is likely the result of the substantial variability in the ratio of population maturation and reproduction rate for these parameter values, owing to the fluctuating dynamics that involve alternating years with high reproduction and high maturation rates, respectively. The lower size limit (δ) for cannibalism to occur will also lead to a shift in regulation. An increase in δ takes the system from a situation with ontogenetic symmetry ($\beta = 0.12 - 0.19$) or ontogenetic asymmetry ($\beta > 0.19$) with development control to a situation with ontogenetic asymmetry with reproduction control (for $\delta > 0.073$, figure 11.5).

To conclude, ontogenetic asymmetry due to reproduction control is clearly associated with population cycles, as we already saw in previous chapters. A moderate intensity of cannibalism will tip this asymmetry to ontogenetic symmetry associated with fixed-point dynamics. Ontogenetic asymmetry resulting from development control caused by high cannibalism is also associated with cycles, although estimates of the maturation/reproduction rate ratio in this case do not differ from those estimated for fixed-point dynamics owing to the alternating years of high biomass maturation and reproduction rates (notice the high variation in the ratio in figure 11.9, lower panel).

EFFECTS OF HARVESTING
CANNIBALISTIC POPULATIONS

Two interrelated results that we have found so far are that (1) energy gain by cannibalism and cannibal control are mutually exclusive processes that do not occur at the same time, and (2) energy gain from cannibalism is strongly coupled to fluctuating dynamics (table 11.1). These general results have repercussions for the effects of size-selective harvesting on cannibalistic populations. Actually, the largely exclusive occurrence of either energy gain or cannibal control will turn out to be a major feature of the dynamics of harvested populations. We will continue to use the discrete-continuous model to investigate the effects of harvesting in cannibalistic populations. Van Kooten, Persson, and de Roos (2007) investigated the same issues, however, using a continuous-time model based on the Kooijman-Metz-type of individual energetics and found, by and large, results very similar to those that we will present here for the discrete-continuous model. This shows that the results obtained are not model-specific but robust to major changes in model structure. We will apply three different harvesting scenarios: one targeted at individuals 50–100 mm in size, one targeted at individuals 100–150 mm in size, and one targeted at individuals greater than 150 mm in size.

With default parameter values, fixed-point, cannibal-driven dynamics occur in the absence of harvesting (figure 11.10). Imposing a size-selective mortality on small individuals (50–100 mm) does not involve any major changes in dynamics. Only at high harvesting rates will the dynamics destabilize into

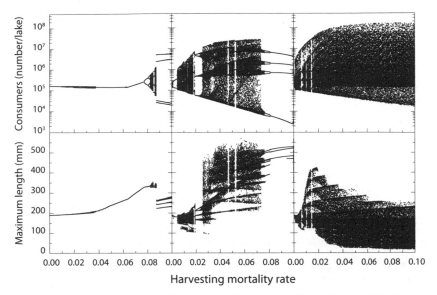

FIGURE 11.10. Bifurcation plots showing the effects of harvesting individuals with body sizes 50–100 mm (*left*), 100–150 mm (*middle*), and >150 mm (*right*). Upper panels show total number of consumers aged one year old and older at the start of the growth season, and lower panels show maximum lengths. Parameters have default values as presented in boxes 10.2 and 11.1.

a small-amplitude four-year cycle before the population goes extinct at even higher mortalities (figure 11.10, left upper panel). For this harvesting scenario, harvesting mainly involves a thinning of the population, which results in a decrease in average consumer numbers and a concomitant increase in the maximum length of individuals (figure 11.10, left panels).

Harvesting individuals in the size range of 100–150 mm has major effects on the population dynamics. Imposing harvesting first results in destabilization of the dynamics into a two-year cycle. With a further increase in harvesting, the amplitude in consumer density increases drastically and the dynamics are characterized by irregular fluctuations or regular four- or eight-year cycles until the population goes extinct at higher harvesting mortalities (figure 11.10, middle-upper panel). The maximum length that consumers can reach first increases only slightly, but then increases drastically, at a harvesting rate around 0.028 day^{-1}, from a maximum length around 200 mm to more than 500 mm (figure 11.10, middle-lower panel).

The mechanism behind this drastic increase in maximum length is revealed by inspecting the time series of numbers of differently sized individuals and individual growth trajectories. Starting with a successfully recruiting cohort, individuals of this cohort are exposed to a very low cannibalistic mortality, owing to the low numbers of cannibals. The decrease in density of this cohort is thus mainly due to background mortality. When this cohort matures, they more or less totally eradicate their first reproductive output through cannibalism (figure 11.11, middle-upper panel). After their first reproduction, the now mature individuals reach the size threshold where they become susceptible to harvesting mortality, which drastically decreases their numbers. As a result, the offspring they produce during their second reproduction event are subjected to very low cannibalistic mortality and will form the next strong cohort. This strong cohort will grow in size, mature, and give rise to the next strong cohort during their second reproductive event. While this strong cohort is growing up, however, the surviving older cannibals, their parents, grow fast and in the next year already surpass the size threshold (150 mm) above which they are no longer susceptible to harvesting (figure 11.11, middle-upper panel). These cannibals reproduce each year but their offspring are totally eradicated by the strong cohort of juveniles aged one year or older.

The population dynamics of the system when harvesting individuals with body sizes between 100 and 150 mm resemble in some respects the dwarf-and-giant cycles that we have discussed above. A closer inspection of the individual growth trajectories shows that this resemblance also concerns the appearance of the very large, gigantic individuals. Once again, starting with a successfully recruited cohort that is subjected to only low cannibalistic mortality, individuals of this cohort grow relatively slowly during their first years of life (figure 11.11; cf. middle-lower panel with left-lower panel). They impose a high cannibalistic mortality on subsequently born cohorts, but extract very little energy from cannibalizing them, and their main energy source is the shared resource. When as adults they become subjected to high harvesting mortality, they accelerate in growth as a result of the thinning. The following year, when they are no longer subjected to harvesting, they continue to grow at a high rate, because they now can cannibalize on the strong cohort born in that year (figure 11.11, middle panels). At the same time, their mortality impact on this strong cohort is minimal because of their low numbers. Eventually, the individuals of the strong cohort outgrow their cannibals, leading to zero growth among the large cannibals until they can "surf" on the next strong cohort (figure 11.11, middle lower panel).

Overall, harvesting intermediate-size individuals thus results in a destabilization of fixed-point, cannibal-driven dynamics and gives rise to the dynamics characterized by the periodic appearance of strong recruiting cohorts (a four-year cycle in figure 11.10, middle panel). The temporal appearance of these

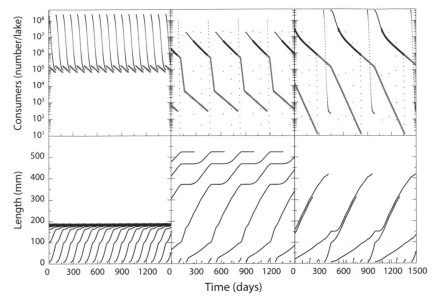

FIGURE 11.11. Shift from stable cannibal-driven dynamics (*left panels,* no harvesting) to fluctuating dynamics with cannibalistic giants when harvesting individuals with body sizes 100–150 mm (*middle panels,* harvesting mortality 0.09) and individuals with body sizes >150 mm (*right panels,* harvesting mortality 0.015). Upper panels show changes in the number of individuals < one year old (*thin dotted lines*), the number of juveniles > one year old (*thick black lines*), and adults (*thick gray lines*). Lower panels show growth trajectories of different age cohorts. Parameters have default values as presented in boxes 10.2 and 11.1.

strong cohorts is also essential for the appearance of the gigantic individuals. This result further supports the general conclusion that energy extraction and cannibal control are exclusive processes that cannot occur at the same time. Finally, over a considerable parameter range, the dynamics are characterized by large-amplitude, irregular fluctuations (figure 11.10, middle panels). This irregularity is a result of variation in time during the season when a strong cohort passes the size threshold for harvesting, that is, whether it starts to be subjected to significant harvesting mortality before or after its first reproductive output.

Harvesting individuals greater than 150 mm resembles in many respects the dynamics found when harvesting individuals in the size range 100–150 mm in that (1) harvesting induces a destabilization of the dynamics and (2) that this destabilization allows surviving, cannibalistic individuals of a strong cohort to "surf" on a following strong cohort and reach large sizes (figures 11.10 and 11.11, right panels). There are, however, also differences in a number of aspects.

First, individuals of a strong cohort will only feed on a single following strong cohort. This result is due to the fact that they never reach a size threshold above which they are no longer subjected to harvesting and are hence effectively removed from the population before they can start to feed on a second pulse of strong recruits (figure 11.11, right panels). Second, the maximum length that individuals can reach shows a peak at a harvesting rate of around 0.02 day^{-1} and thereafter decreases at higher harvesting mortalities. This result is also due to the fact that individuals never reach an upper size refuge without harvesting. Third, the dynamics are, more or less over the whole parameter region with large amplitude dynamics, characterized by irregular fluctuations, which result from the considerable variation over time in the timing of the discrete reproduction events and the moments at which individuals become subjected to harvesting. Fourth, in the example shown in figure 11.11, the cohort resulting from the first reproductive pulse of a strong cohort survives in considerable numbers, forming a cohort with high individual growth rates before the strong cohort is born the next year. The effect of this former cohort on dynamics is nevertheless negligible, and few of them reach larger sizes. Finally, for the scenario with harvesting of individuals greater than 150 mm, the rule also holds true that for cannibal individuals to gain energetically from cannibalism, they cannot control the numbers of their victims.

GIANT INDIVIDUALS: THEORY AND OBSERVATION

The temporal appearance of giant individuals is perhaps the most spectacular phenomenon observed in the dynamics discussed above. This dynamical phenomenon is a result of considering all four different aspects of cannibalism that we discussed at the start of this chapter, namely, mortality, size dependence, energy extraction, and competition with victims (table 11.1). The occurrence of giant individuals has also been documented empirically in several species, particularly fish. For example, Le Cren (1992) discussed the appearance of exceptionally large perch in Lake Windermere, England. The large size of these perch resulted from an acceleration in their growth coinciding with the appearance of particularly large numbers of small perch in different years. The most detailed study of a cannibalistic system in which the mechanisms behind temporal appearance of giants have been analyzed pertains to the Eurasian perch population in Lake Abborrtjärn 3, Sweden. Among other things, this study demonstrates, in agreement with theory, that cannibal control (high mortality) and energy gain from cannibalism are temporally exclusive processes (Persson, Byström, and Wahlström 2000; Persson, Claessen, et al. 2004).

We will address the question about the appearance of giants in relation to the broader question about the extent to which the dynamics of empirical populations can be predicted based on individual-level characteristics. For this we will consider the lower size boundary for cannibalism (δ), as this parameter is expected to have major effects on population dynamics and can, moreover, be estimated from diet analyses. We focus on three fish species, yellow perch (*Perca flavescens*), Eurasian perch, and pike (*Esox lucius*), for which information on size-dependent prey use is available and for which sufficiently long time series exist that document their dynamics in single-species communities. This combination allows for a comparison between size-dependent capacities and the type of population dynamics observed (Persson, de Roos, and Bertolo 2004). As in the previous chapter, the parameterization of the individual-level model is based on data that are independent of the data from the systems whose dynamics are to be predicted. In fact, we will use the model for Eurasian perch derived by Claessen, de Roos, and Persson (2000), which has been the baseline model of cannibalism in this chapter, including its parameterization, to predict the dynamics of all three species, while varying only the species-specific estimates for the lower size boundary of cannibalism. As a result, our contrast between predicted and observed dynamics will for yellow perch and northern pike be only semiquantitative. In other words, we will primarily make predictions about the form of individual growth trajectories and cohort dominance, not about the absolute growth rates.

Data on the life history characteristics of these three fish species show that they differ in several aspects, which are likely to affect their capacity as cannibals. Yellow perch has the highest lower size boundary for cannibalism; Eurasian perch has an intermediate lower size boundary; and pike has the lowest (table 11.2). Also, data on the upper size boundary for piscivory suggest that pike has the highest piscivorous capacity and yellow perch the lowest. The difference in size boundaries between the species is also reflected in the differences in gape sizes, suggesting that gape size may be used as a rule of thumb to predict cannibalistic population dynamics (table 11.2) (Mittelbach and Persson 1998; Persson, de Roos, and Bertolo 2004). Finally, the size at hatching is relatively similar, suggesting that this attribute is unlikely to be a significant factor affecting the dynamics of the three species differentially. In addition, other pieces of information, such as estimates of the capacity to handle differently sized prey (Mittelbach and Persson 1998), support the idea that yellow perch has the lowest, and pike the highest, capacity with respect to the size range of prey fish that can be taken.

We next proceed to predict the dynamics of the three species using the information on the lower size boundary for cannibalism. It turns out that cohort

TABLE 11.2. Life History Characteristics Influencing the
Cannibalistic Capacity of Three Different Piscivore Fish Species

Species	Minimum victim/ cannibal size ratio	Maximum victim/ cannibal size ratio	Size when switching to piscivory (mm)	Hatching size (mm)	Gape width for a 300 mm cannibal
Yellow perch *Perca flavescens*	0.08	0.4	100–170	4.8–6.0	22
Eurasian perch *Perca fluviatilis*	0.05–0.06	0.45	110–160	4.1–6.6	33
Northern pike *Esox lucius*	0.03	0.7	45–100	6.0–8.0	37

Source: Mittelbach and Persson 1998; Persson, de Roos, and Bertolo 2004.

cycles are the expected dynamics for yellow perch, where one cohort dominates the system over several years and gives rise to a new dominating cohort
when it matures (figure 11.12, left-upper panel). The empirical observations
from the yellow perch population in Crystal Lake, Wisconsin, USA, largely
correspond to this prediction, as single, strong cohorts repeatedly dominate
the population where the appearance of every new cohort is related to maturation and the subsequent die-off of the mature individuals (Sanderson, Hrabik,
et al. 1999) (figure 11.12, left lower panel, see also chapter 10). Furthermore,
evidence is present for a depression of the zooplankton resource in connection
with the appearance of a new strong cohort, hence linking the die-off of larger
individuals to intercohort competition. Despite the documented occurrence of
cannibalism in yellow perch, the dominant interaction in this system is thus
suggested to be competition leading to cohort-cycle dynamics.

For Eurasian perch parameters, the model predicts a shift between cannibal-
driven dynamics and dynamics dominated by a strong, slowly growing cohort
that is cannibalized by a low number of gigantic cannibals. These latter individuals grow very rapidly, but, as we have already discussed, impose only a
low mortality on the slowly growing, dominant cohort (figure 11.12, middle-
upper panel). Correspondingly, the observed dynamics of the Eurasian perch
population in Lake Abborrtjärn 3 has been characterized by a shift from a
phase with a dominance of cannibal-driven dynamics during which cannibals
reached a size of around 180 mm and intense cannibalism resulted in low
recruitment (1984–1993), to a phase during which older individuals died off,
recruitment was high, and surviving cannibals accelerated substantially in

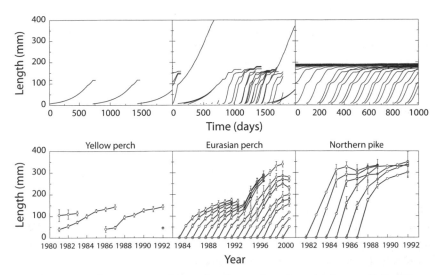

FIGURE 11.12. *Upper panels*: predicted growth trajectories in cannibalistic populations with a lower size ratio of $\delta = 0.08$ (*upper-left panel*, cohort cycle), $\delta = 0.06$ (*upper-middle panel*, mix of cannibal-driven dynamics and dwarf-and-giant cycles), and $\delta = 0.00$ (*right-upper panel*, cannibal-driven dynamics). Other parameters have default values as presented in boxes 10.2 and 11.1. *Lower panels*: growth trajectories of different age cohorts of yellow perch (*left-lower panel*), Eurasian perch (*middle-lower panel*), and northern pike (*right-lower panel*). Empirically observed growth trajectories are means ±1 SD. Data from Persson, de Roos, and Bertolo 2004.

growth (1994–1997). Subsequently, the system returned to a situation with a dominance of cannibal-driven dynamics (figure 11.12, middle-lower panel) (Persson, Byström, and Wahlström 2000; Persson, Claessen, et al. 2004). The Eurasian perch time series illustrates the alternation between a phase with cannibal control and low asymptotic sizes of cannibals and a phase with lack of cannibal control and gigantic growth among cannibals, which represents a major insight that we gained from the modeling results (table 11.3; figure 11.4). Quantitative estimates of energy intake also revealed that cannibalism amounted to a very small fraction of the energy budget of cannibals in the cannibal-driven phase but a large fraction during the giant phase (Persson, de Roos, and Bertolo 2004).

Finally, cannibal-driven dynamics were predicted for pike, including characteristics like fast growth of small individuals, small asymptotic sizes, and the successful recruitment of cohorts every year (i.e., sufficient number of individuals surviving from birth to one-year-old to sustain the population) in contrast

TABLE 11.3. Important Processes during the Two Phases of Dynamics
(Stunted and Giant) of the Eurasian Perch Population in Abborrtjärn 3

Phase	Mortality	Competition	Gain	Size dependence
Stunted	Yes	No	No	Yes
Giant	No	Yes	Yes	Yes

Source: Persson, Byström, and Wahlström 2000; Persson, Claessen, et al. 2004.

to more infrequent recruitment pulses. Data on growth trajectories of the pike populations in three Italian streams over a period of eight years, of which we present only the data from one of the populations, largely support these predictions (figure 11.12, right panels) (Persson, de Roos, and Bertolo 2004; Persson, Bertolo, and de Roos 2006). The fact that the predicted asymptotic size is substantially smaller than the observed one can be explained by the fact that only the lower size boundary was species specific, whereas all other parameter values represented default values for Eurasian perch. In particular, the upper size limit of cannibalism is substantially higher for pike than the value we used for Eurasian perch (table 11.2).

Overall, we can conclude that the individual growth trajectories of the three populations as a result of population feedbacks reflecting different types of population dynamics could be predicted quite successfully based on a priori–derived individual-level assumptions. The different types of predicted and observed population dynamics are also reflected in differences in the variation in the mean size of individuals over time, which were substantially smaller for pike (coefficient of variation, $CV = 0.07$) than for the other two species (yellow perch: $CV = 0.17$; Euarasian perch: $CV = 0.34$; Persson, de Roos, and Bertolo 2004).

The dynamics of the perch population in Abborrtjärn 3 completely corresponds to the categorization of model dynamics (table 11.1 versus table 11.3). A more in-depth analysis of the dynamics of this perch population (Persson, de Roos, et al. 2003; Persson, Claessen, et al. 2004) shows, however, that giants may actually be dynamically important. In short, during the stunted phase, cannibals overexploit their victim resource, leading to low recruitment to cannibalistic stages, which in turn leads to a decrease in the number of cannibals over time. When the cannibals have been reduced below a threshold density, a substantial fraction of young-of-the-year perch survives to one-year-olds where they are less susceptible to cannibalism. At the same time, the cannibals experience an acceleration in their growth (figure 11.12, lower-middle panel). This leads to such an increase in per capita fecundity that another strong burst of offspring is produced, which (1) outcompete the previous year's recruits and (2) survive in substantial numbers to the next year. This offspring cohort further boosts the growth of the gigantic, cannibalistic individuals. Here, giant cannibals play a key role in keeping the population in a phase with giant cannibals until their number has decreased to such an extent that their increased per capita fecundity can no longer compensate for their lower numbers. As a result, previous year's recruits experience less competition from new recruits, and hence survive and can, as two-year-old individuals, control new recruits by cannibalism, which makes the system move back to a stunted phase.

In conclusion, our analyses have shown that all aspects of cannibalistic interactions (size dependence, mortality, energy extraction, and competition) have profound effects on population dynamics, and that the different aspects play out in different ways (e.g., energy gain versus mortality, and different relationships between gain and competition) depending on population dynamics. Overall, we found that a limited number of parameters, namely, the cannibalistic voracity (β) and the lower size boundary (δ) for cannibalism, are important for determining population dynamical patterns. The impact of these parameters is related to whether ontogenetic symmetry/asymmetry occurs or not. Furthermore, the results obtained are robust to substantial changes in model structure. Finally, as we already discussed in chapter 10, physiologically structured population models form a particularly suitable framework for empirical testing, because the assumptions of the models pertain to the individual-level formulations, as opposed to population-level processes.

PART IV
EXTENSIONS AND PERSPECTIVES

Demand-Driven Systems, Model Hierarchies, and Ontogenetic Asymmetry

In chapter 2, we stated our aim to investigate the effects of ontogenetic growth on population and community dynamics from an individual-based perspective. Specifically, we used Metz and Diekmann's (1986) distinction between the individual (i) state, the population (p) state, and the condition of the environment as the point of departure for studying ecological processes (the population dynamical triad in figure 2.1). Looking back at the preceding chapters, it is apparent that all models presented in this book fit into the same individual-based framework. Depending on the questions we have raised, the complexity of the models we have used has varied from two-stage models to models with continuous size distributions and has also included different degrees of richness in details (e.g., continuous versus discrete reproduction).

Still, all models have had a clear individual-level basis. We have also shown that the simpler stage-structured models represent approximations of more complex size-structured models, in that the predictions of both modeling approaches are identical under equilibrium conditions (de Roos, Schellekens, et al. 2008b; see also section 3 of the technical appendices). The fact that simpler stage-structured models can be derived from their underlying, more complex, fully size-structured models suggests that model simplifications can be carried out within an *individual-based* framework, as opposed to through population aggregation (sensu Murdoch, Briggs, and Nisbet 2003). In this final chapter, we will discuss this question in more depth, exemplified with the consumer-resource dynamics considered in chapter 9.

In chapter 2, we further pointed out that the focus of our book was on supply-driven systems. At the same time, we indicated that the framework that we have used should be equally appropriate for studying the dynamics of demand-driven systems. In this final chapter, we will show (1) that overcompensation and cohort cycles are also found in demand-driven systems, and (2) that shifts in overcompensation patterns and cycle types can, as for supply-driven systems, be related to whether development or reproduction is more limited and controls the population at equilibrium. Furthermore, we will consider whether dynamical phenomena like cohort cycles have also been reported to occur in unicellular species, which have a limited change in size over their life cycle.

In chapter 2, we advanced as another reason in favor of individual-based models: the higher independence between model assumptions and predictions and the empirical system whose dynamics is to be predicted (including population feedbacks on individual life histories). In chapters 10 and 11, we provided several examples of this. We will in this chapter discuss this issue further in relation to structured and unstructured models.

Finally, the principles of development versus reproduction control and the concept of ontogenetic asymmetry have formed major cornerstones throughout this whole book, both when we considered community structure as well as when we considered population dynamics. Although we have already in many respects considered these topics, we will in this final chapter return to them and set them in the context of contemporary—and future—ecological theory.

DEMAND-DRIVEN SYSTEMS

The focus of our book has been on supply-driven systems. Actually, almost all dynamic energy budget (DEB) models are based on the idea that food availability determines individual food intake and the subsequent availability of energy for energy-consuming processes (e.g., Gurney, McCauley, et al. 1990; Kooijman and Metz 1984; Kooijman 1993, 2000; Lika and Nisbet 2000; Persson, Leonardsson, et al. 1998). These models hence have in common a supply-driven principle of individual energetics, in which food acquisition is always the limiting process rate in the energy budget, determining the rate of growth and reproduction. In contrast, in demand-driven models, the input results from the state and the output of the system (Kooijman 2000). In demand-driven models there is thus a feedback of energy (or any other relevant currency) demands on individual food intake. We suggested in chapter 2 that birds and mammals are organisms for which a demand-driven principle seems more appropriate than a supply-driven principle, because of their high energetic demands and their growth patterns, which are less influenced by environmental (food) conditions. Although demand-driven systems are much less explored than supply-driven systems, one example of a demand-driven model is the one developed by de Roos, Galic, and Heesterbeek (2009) for ungulate populations, where it was assumed that the growth of individuals was prescribed and under genetic control.

A question that inevitably comes to mind is the extent to which the results that we have discussed in previous chapters are also relevant for demand-driven systems. Does overcompensation also occur in demand-driven systems, or is it restricted to supply-driven systems? Do cohort cycles and the shift between juvenile-driven and adult-driven cohort cycles also occur in demand-driven systems? If the answers to these two questions are yes, this would naturally increase the generality of our results to a considerable extent. To address these questions, we will consider a demand-driven consumer-resource model where we allow the size scaling, q, of ingestion rate and fecundity to vary such that the results obtained below can be compared with the results we obtained in chapters 3 and 9.

A basic premise of demand-driven system is that energy demands are fixed, independent of current food availability, and have to be covered. The assumption of fixed demands has two basic implications. First, fixed demands require that energy reserves are needed to cover demands in the case of insufficient food supply. Second, fixed demands require that the food acquisition rate has to be limited by factors other than food density under conditions of ample food supply in order to prevent a runaway process of fat buildup. We model this limitation on food intake rate by making it dependent on the current body condition of the individual and by assuming that regulation of individual energetics is such that a target body condition is aimed for (de Roos, Galic, and Heesterbeek 2009). Individual-level functions and population equations of the basic size-structured population model, parameter values and equations, and functions defining the size-structured demand-driven model are all presented in section 9 of the technical appendices (boxes A.2 and A.3). To minimize repetition, we therefore largely restrict ourselves here to considering a schematic presentation of the energy budget of the individual.

Figure 12.1 provides a graphical overview of the energy budget of an individual organism, which is characterized by two measures of individual body size, structural mass (s) and energy reserves (E). Ingested energy is used for maintenance, growth in structural mass (which is fixed), and, in case of mature individuals, also for reproduction. If ingested energy exceeds the amount of energy needed for maintenance, structural mass growth, and reproduction, energy will also be allocated to reserves (gray solid line in figure 12.1). In contrast, if ingested energy is insufficient to meet the energy demands for maintenance, structural mass growth, and reproduction, energy reserves will be used to cover the difference between ingestion and demands (gray dashed line in figure 12.1).

The body condition of an individual is identified with the ratio of the two measures of size, E/s. Resource ingestion is assumed to follow a Type II functional response as a function of resource density. Our aim is to formulate a DEB-model as close as possible to the Kooijman-Metz energy budget model

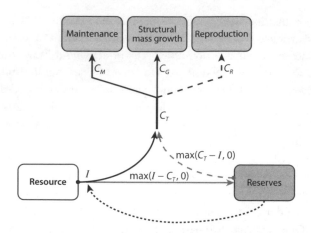

FIGURE 12.1. Schematic overview of the individual energy budget for the demand-driven model. Energy acquisition from feeding on the resource is denoted with the intake rate I, while flows of energy expenditure are characterized by different rates C. Solid black lines represent energy flows that occur in all individuals independent of their age and size. Energy investment in reproduction (*black, dashed line*) occurs only after reaching the maturation size threshold. The gray, solid line indicates energy reserve dynamics when resource intake exceeds total energy demands (anabolism), while the gray, dashed line indicates energy flow from reserves when resource intake is smaller than total energy demands (catabolism). The dotted line represents the feedback of the individual reserve state on food intake rate.

(Kooijman and Metz 1984). We therefore have assumed that the attack rate and handling time are proportional and inversely proportional, respectively, to s^q. A feedback between the body condition of an individual organism and its food ingestion rate is implemented by assuming that the attack rate is a sigmoid, decreasing function of the ratio E/s. We also assume that fecundity is an increasing function of individual body condition and individual structural mass with a scaling s^q. The latter assumption again mimics the dependence of reproduction on body size in the Kooijman-Metz model, except for the fact that in the latter model q always equals 2/3.

OVERCOMPENSATION IN DEMAND-DRIVEN SYSTEMS

Similar to the supply-driven biomass model considered in chapter 3, the demand-driven model exhibits a shift in dominance in stage-specific biomass with a change in q (figure 12.2, left panel). However, the demand-driven model

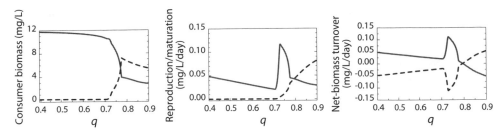

FIGURE 12.2. Juvenile (*solid lines*) and adult (*dashed lines*) biomasses (*left panel*), total population maturation (*solid lines*) and reproduction rates (*dashed lines, middle panel*), and net-biomass-turnover rates (*right panel*) of juveniles (*solid lines*) and adults (*dashed lines*) for different values of q. Net biomass turnover of both stages was computed by taking the difference between, on the one hand, the total production rate of new biomass through growth in structural mass (s) and growth in energy reserves (E), as well as through reproduction of offspring, and, on the other hand, the loss rate of biomass (structural and reserves) through mortality. Notice that in the demand-driven model the scaling with q of the attack rate, the handling time, and the fecundity leads to the shift from a positive juvenile net biomass turnover to a positive adult net biomass turnover not at $q = 1$, but rather at $q = 0.77$, compared with the biomass model in chapter 3. Otherwise, parameters have default values as presented in box A.3.

shows a pattern opposite to that of the supply-driven, stage-structured biomass model in that total juvenile biomass dominates the population biomass at low q values, whereas adults dominate the biomass at high q values, with a shifting point at a q value around 0.74. In contrast, the demand-driven model shows a similar pattern in maturation and reproduction rates to that of the supply-driven model. The adult reproduction rate is very low for low q values, starts to increase at a q value of 0.6, and reaches a maximum beyond a q value of 0.9, whereas the maturation rate decreases with increasing q (figure 12.2, middle panel).

The patterns in net biomass turnover by juveniles and adults (i.e., the balance between biomass gain through growth and reproduction and biomass loss through mortality) follow those of maturation and reproduction rates, respectively, with juvenile net biomass turnover being positive for low q values (< 0.77) and negative for high q values (> 0.77), and vice versa for the adult stage (figure 12.2, right panel). The juvenile stage as a whole is thus a net source of biomass production for low q values, whereas the adult stage is a net source of biomass production for large q values. From this we can draw the conclusion that the system is reproduction-controlled for low values of q and development-controlled for high values of q. The results obtained for the demand-driven model are thus in congruence with the results of the supply-driven model studied in chapter 3, with one important exception: changes in stage-specific biomass show a relationship with q that is opposite to that of the supply-driven model.

One factor behind this difference between the behavior of the demand-driven model and the supply-driven model is that increased resource limitation in the supply-driven model will result in decreased individual performance in growth or reproduction without necessarily affecting the mortality rate of the stage in focus. For example, high resource limitation among juveniles will lead to retarded growth, causing a biomass accumulation in this stage in a supply-driven system. In contrast, in a demand-driven system, growth in structural mass is fixed (and energy costs to cover this growth and maintenance are high), and severe resource limitation leading to depleted energy reserves will here to a larger extent affect mortality rates. As a consequence, biomass accumulation leading to biomass dominance of the most limiting stage, which occurs in supply-driven systems, is less likely in demand-driven systems, as the most limiting stage will suffer from high mortality, which produces a negative effect on the biomass of that stage. This situation is reflected in the fact that although survival throughout the juvenile period will decrease with increasing q in both the supply-driven and the demand-driven model, this takes place through different mechanisms. For the supply-driven model, the juvenile mortality rate is independent of q, and the decreased juvenile survival with increasing q is due to the longer time that the individual stays in the juvenile phase (table 9.1). In contrast, in the demand-driven model, decreased juvenile survival with increasing q is due to an increased mortality rate, as stage duration is constant and independent of q. Another important factor, which differs between demand- and supply-driven models, is that a high biomass production of juveniles will increase the total population maturation rate in demand-driven models through an increase in the amount of reserves that juveniles have when maturing, and in supply-driven models through a shortening of the juvenile period.

A detailed analysis of stage-specific responses in terms of energy reserves, condition, and resulting mortality (results not shown) also reveals that small and large juveniles differ in their response to increasing q in the size-structured, demand-driven model. For small juveniles, condition, and thereby survival, decreases monotonically with increasing q, whereas the reverse is the case for adults. For large juveniles, the pattern is more complex, as condition and survival first decrease with increasing q but thereafter increase again. This pattern in large juvenile condition and survival is also associated with the fact that the equilibrium resource biomass density first decreases with q and thereafter strongly increases.

The increase in condition of large juveniles is also associated with a strong increase in energy reserves affecting the population maturation rate. This difference in response of small and large juveniles to increasing q may first of all explain the nonmonotonic pattern in maturation rate to increasing q (figure 12.2, middle panel) and may, moreover, lead to different responses of the biomass in these two stages with increasing q. Altogether, these differences in energetics between small and large juveniles when varying q suggest that it is more appropriate to compare the responses in the demand-driven model with the responses of a three-stage biomass model or the size-structured model based on Kooijman-Metz energetics, which were also considered in chapter 3. In fact, for low q values, the results for the demand-driven model resemble the results obtained in chapter 3 for the Kooijman-Metz-based, size-structured model. In the latter model, juveniles also constitute the largest fraction of total population biomass, despite the fact that the juvenile and adult stages are a net source and net sink, respectively, of biomass production (results not shown). Within-stage differences in energetics thus break up the tight relationship between juvenile or adult biomass dominance and the source/sink status of net biomass turnover in these two stages that we found for the two-stage-structured biomass model in chapter 3.

Stage-specific biomass overcompensation also occurs in the demand-driven model. For a low q value (0.5), overcompensation in response to increased stage-independent and juvenile mortality is observed in adult biomass and small juvenile biomass. In contrast, the biomass of large juveniles, which completely dominates the total population biomass, decreases monotonically with increasing mortality (figure 12.3, upper-left and middle panels). Despite that the demand-driven model shows a reverse pattern in stage-specific biomass density with varying q, it thus shows the same pattern as the supply-driven model in response to increasing mortality, in that overcompensation is present in the stage with the lowest biomass. Overcompensation in adult biomass shows a more distinct peak at a slightly higher mortality rate for mortality targeting juveniles only than for stage-independent mortality (figure 12.3, upper-left and middle panels). For adult mortality, the pattern is reversed: adult biomass decreases monotonically, small juvenile biomass stays constant, and large (and total) juvenile biomass increases slightly with an increasing mortality rate (figure 12.3, upper-right panel).

The observed response in adult biomass to increased stage-independent and juvenile mortality in the demand-driven model for low q values can be related to the fact that increased mortality in the juvenile stage, which dominates the population in biomass at background mortality, leads to a substantial increase in resource biomass. This allows adults to attain larger body masses,

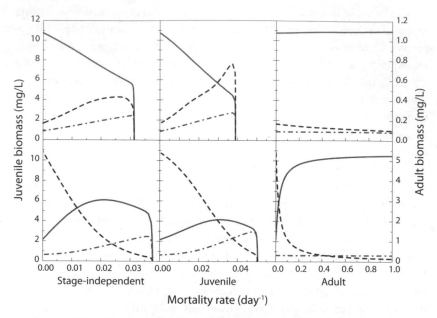

FIGURE 12.3. Changes in small-juvenile (0.002 < *s* < 0.66 gram; *dashed-dotted lines*), large-juvenile (0.66 < *s* < 7 gram; *solid lines*), and adult (7 < *s* < 30 gram; *dashed lines*) biomass with an increase in stage-independent (*left*), juvenile (*middle*), and adult (*right*) mortality for *q* = 0.5 (*upper panels*) and *q* = 0.9 (*lower panels*). Otherwise, parameters have default values as presented in box A.3.

as an increased resource level allows them to build up more reserve mass at a particular structural mass, while it also increases their probability to survive to larger sizes (i.e., larger structural masses, to which they grow at a fixed rate). Moreover, at background mortality, the energy reserves of large juveniles are relatively low, which is the reason why an increase in resource biomass has a significant positive effect on population maturation rate. This will all lead to increased population reproduction and thereby an increase in the biomass of small juveniles. In contrast, when imposing mortality on adults, the direct negative effect through the increased adult mortality will dominate, especially because the biomass dominance of juveniles will allow only a minor resource release to occur (and hence larger juveniles will still suffer from lower energy reserves).

For high *q* values (0.9), large (and total) juvenile biomass shows overcompensation with increases in stage-independent mortality and juvenile mortality as well as adult mortality (figure 12.3, lower panel). With increased adult mortality, adult biomass shows an exponentially decreasing biomass, and small juvenile biomass increases only to a negligible extent. Noteworthy, the large juvenile (and total juvenile) biomass does not show a hump in relation to increasing adult mortality, but increases asymptotically until the population goes extinct owing to the eradication of adults (figure 12.3, lower-right panel). This pattern was also observed for low q values, although the increase in juvenile biomass with increased adult mortality was in that case only slight, which can be related to the biomass dominance of juveniles at low q values. For high q values at background mortality, the system is characterized by a high adult and large-juvenile condition, a low condition of small juveniles, and relatively high resource levels. Small juveniles thereby form a survival bottleneck in the system. Increased adult mortality will release small juveniles from severe resource limitation and hence increase their survival to the large-juvenile stage, which consequently shows an increase in biomass.

In contrast, small juveniles show little response to adult harvesting, as the positive effect of resource release is counteracted by the reduced inflow resulting from low adult biomass and reproduction. For stage-independent and juvenile mortality, both small and large juveniles show overcompensation in biomass (figure 12.3, left and middle lower panels). Here a decrease in adult biomass, resulting in an increase in the resource biomass, will lead to an increased survival of small juveniles, which translates into a positive effect on both their own biomass as well as the biomass of large juveniles. The difference in the response of small-juvenile biomass to the different mortality scenarios can be related to the fact that adults, on the one hand, have a positive effect on small juveniles through reproduction but, on the other hand, have a negative effect by decreasing the resource. The response in juvenile biomass will therefore depend on the response in adult biomass, which differs between the different mortality scenarios. Interestingly, small juveniles show the strongest overcompensation with mortality that targets juveniles, although total juvenile biomass shows stronger biomass overcompensation for stage-independent and adult mortality.

Summarizing, these results show a consistent pattern in that juveniles exhibit overcompensation in biomass at high q values, whereas adults exhibit overcompensation in biomass at low q values for stage-independent and juvenile mortality. Overcompensation in biomass is always observed in the stage that has a low biomass at background mortality. Finally, the population can sustain at a much higher adult mortality than at either stage-independent or juvenile morality for both low and high q values (figure 12.3). Overall, the observed pattern in overcompensation for low q values substantially resembles the pattern observed for the Kooijman-Metz model that we considered in chapter 3. We next turn to discuss how a changing q affects the population dynamics of the demand-driven consumer-resource system.

POPULATION DYNAMICS OF THE CONSUMER-RESOURCE SYSTEM

The size-dependent competitive ability (q) affecting both consumption and fecundity has a strong effect on the population dynamics of the consumer-resource system. For values of $q < 0.74$, the system shows strong oscillations, especially in resource and the density of adults, which are not always present (figure 12.4, upper panel, lower left panel). This parameter range of q matches the range in which the population biomass at equilibrium is dominated by juveniles (figure 12.2). The dynamics are characterized by a strong pulse of newborns, which depresses the resource and causes the temporal extinction of

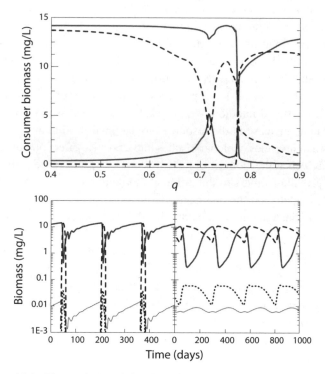

FIGURE 12.4. Changes in population dynamics in the demand-driven consumer-resource model as a function of q. *Upper panel*: maximum and minimum biomass of juveniles (*solid lines*) and adults (*dashed lines*). These results were obtained using long numerical simulations of the model dynamics with regular, stepwise increases as well as decreases in q. *Lower panels*: dynamics of juvenile-driven (*left panel*) and adult-driven (*right panel*) cycles, including juvenile biomass (*thick solid line*), adult biomass (*dashed line*), resource biomass (*thin solid line*), and total population reproduction rate in terms of biomass (*dotted line*). Otherwise, parameters have default values as presented in box A.3.

adults. Juveniles grow (at a fixed rate) in mass and simultaneously decrease in numbers, owing to background mortality. Because of the decrease in consumer numbers, the resource recovers and reaches its peak when juveniles mature, giving rise to another reproductive pulse, whereby the cycle repeats itself. In many respects, these cycles occurring for $q < 0.74$ hence resemble the juvenile-driven cycles observed in supply-driven systems (chapter 9).

Oscillations also occur for $q > 0.74$. Adults are in this case continuously present in the population (figure 12.4, upper panel, lower right panel). The range of q values where these oscillations occur also matches the region in which the population biomass at equilibrium mainly consists of adults. The main characteristics that distinguish the dynamics in this region from dynamics occurring for $q < 0.74$ (besides the fact that adults are always present in the population) are the high variability in juvenile biomass and juvenile survival. In contrast, variability in adult biomass, fecundity, and mortality is less pronounced. For $q > 0.74$, the food density is highest at the end of the cycle in adult numbers such that adult numbers and adult fecundity fluctuate out of phase (table 12.1).

Thus, the two types of cycles that we find in the demand-driven, consumer-resource model closely resemble the juvenile-driven and adult-driven cycles discussed for supply-driven systems in chapter 9 (table 12.1). This holds for all relevant characteristics except three: the juvenile-adult biomass ratio, the period delay ratio for $q > 0.74$, and the average food density (table 9.1 versus table 12.1). The difference between the supply-driven and the demand-driven

TABLE 12.1. Some Qualitative Characteristics of the Cycles for $q < 0.74$ and $q > 0.74$ in the Demand-Driven Model with Semichemostat Resource Dynamics

Aspect	$q < 0.74$	$q > 0.74$
Amplitude	Large	Moderate
Period-delay ratio	≈ 1	2.33, for $q = 0.8$
Size distribution	One dominant cohort	More constant
Juvenile-adult ratio	High	Low
Food density	Low*	High*
Adult fecundity	Low	High
Adult life span	Short	Long
Adult mortality	High	Low
Fecundity**-resource	In phase	Out of phase
Juvenile delay	Constant	Constant
Juvenile survival	High	Low

*First a slow decline to $q = 0.72$, then a rapid increase.
**Population fecundity.

systems in period-delay ratio for $q > 0.74$ is a result of the fact that maturation occurs at a fixed age in demand-driven systems (remember that individual growth was very slow for adult-driven cycles in the supply-driven system; chapter 9). For the demand-driven system, the juvenile-adult ratio is high for low q values but low for high q values, which, in turn, results in a lower average food density at low q values than at high q values, although the difference in mean food density is substantially smaller than for the supply-driven system (figure 12.3; table 12.1). The high juvenile biomass for low q values leads to a strong impact on the resource, and, in turn, a temporal extinction of adults in the demand-driven system in contrast to the supply-driven system for conditions of juvenile competitive superiority (figure 9.1 versus figure 12.3, left panel).

De Roos, Galic, and Heesterbeek (2009) studied a demand-driven model specifically tailored for ungulate (horse) populations similar to the one that we have studied here but with additional details in the energy budget model like fetal development, lactation, and seasonality. The baseline dynamics of this model was an adult-driven cycle where individuals less than three years old suffered from increased mortality. Moreover, individuals between one and two years old suffered more than foals, because the former did not profit from the energy intake from suckling (included in the model) and because as foals they had suffered from depletion of body reserves during their first winter, which spilled over on their survival probability as one-year-old individuals. De Roos, Galic, and Heesterbeek (2009) also related their model results to observations on both a horse population and Heck cattle in a natural reserve where hunting mortality is absent. For this demand-driven system, the researchers stressed that the more precarious energy balance of juvenile individuals compared with adult individuals was due to the former's high metabolic demands, which caused them to suffer to an unequal extent from resource scarcity. In other words, the form of the size-dependent critical resource density was heavily affected by the size scaling of the metabolic demands (cf. figure 10.3).

To conclude, the results regarding both overcompensation patterns at equilibrium conditions as well as population dynamics show that overcompensation and cohort cycles also occur in demand-driven systems. The relationship between the stage exhibiting biomass overcompensation and stage-specific net biomass turnover is different from what we found for the supply-driven two-stage biomass model in chapter 3 and resembles more closely the pattern observed for the size-structured model based on Kooijman-Metz energetics that we also studied in chapter 3. In contrast, the population dynamics found in the demand-driven model are overall the same as that found for the supply-driven consumer-resource model in chapter 9: low q values lead to juvenile-driven cycles, high q values lead to adult-driven cycles, and an intermediate q value (0.74) leads to fixed-point dynamics.

Overall, these results lead us to suggest that the results we have presented in chapters 3–8 regarding community structure and in chapters 9–11 regarding population dynamics are also highly relevant for demand-driven systems. An analysis of the behavior of the trophic modules that we have studied in the foregoing chapters for supply-driven systems is of course needed for demand-driven system as well in order to conclusively prove this supposition, and we see this as an important task for future research. As a final remark regarding demand-driven systems, we would like to point out that the birth mass/maturation mass ratio (z) in mammals and birds—which are the two groups that have most of the attributes of demand-driven systems—is higher than that for other organisms (e.g., poikilotherms, figure 2.4). Based on the analyses in chapters 3 and 9, we know that juvenile biomass overcompensation and juvenile-driven cohort cycles are dependent on low z values (<0.2), whereas adult biomass overcompensation and adult-driven cycles are not, and actually are even more likely to occur for higher z values. We therefore expect that the likelihood of observing biomass overcompensation and cohort cycles in demand-driven systems should be higher for development-controlled systems.

UNICELLULAR ORGANISMS

In chapter 2, we pointed out the difference between unicellular and multicellular organisms in the extent to which they undergo substantial individual growth over their life cycle. In particular, the limited extent to which the "mother" cell of a unicellular organism during mitosis is reduced to smaller "daughter" cells places substantial constraints on the extent to which dynamical patterns related to small z values (i.e., juvenile overcompensation and juvenile-driven cycles) may occur in unicellular organisms such as phytoplankton and bacteria. Therefore, we expect that adult biomass overcompensation and adult-driven cycles are more likely to occur in unicellular organisms than juvenile biomass overcompensation and juvenile-driven cycles, as we argued above for birds and mammals.

Interestingly, Massie, Blasius, et al. (2010) provide experimental and theoretical evidence for the presence of cohort cycles in a population of the phytoplankton *Chlorella vulgaris* and also other phytoplankton populations. The phytoplankton populations clearly changed their demographic structure with

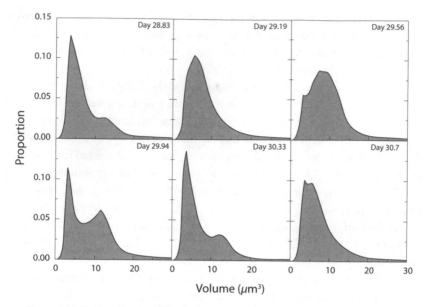

FIGURE 12.5. Size distributions in experiments with the phytoplankton *Chlorella vulgaris* along the cell cycle during one oscillation of approximately two days. Data from Massie, Blasius, et al. (2010).

the periodicity of the population cycle (figure 12.5). This phytoplankton population, consisting of unicellular units in which each cell at mitosis produces four daughter cells, was argued to provide sufficient population structure to cause these small-amplitude cycles. The relatively modest change in size structure over time and the lack of one dominant cohort show substantial resemblance to the change in size distributions observed in the adult-driven cycles in chapter 9 (figure 12.5 versus figure 9.3). This interpretation is supported by the fact that progression through the cell cycle was retarded in the nitrogen-dependent, pre-mitotic ("juvenile") phase, during which individuals born at different times where stacked. Massie, Blasius, et al. (2010) suggested that these small-amplitude cycles occurred through a synchronization between the cell cycle and the population cycles via a common nutrient pool (nitrogen). They also discussed other systems in which the cell cycle has been shown to lead to population synchronization and cohort cycles, including yeast cells and bacteria.

Studies on microcosmic unicellular organisms show that interactions among individuals within the same species are not restricted to competition for a shared resource but also include cannibalism (Holyoak and Sachdev 1998; Martinele and D'Agosto 2008; Tuffrau, Fryd-Versavel, et al. 2000). Omnivory as a result of cannibalism has also been suggested to have a stabilizing effect in microcosms including protozoans (Holyoak and Sachdev 1998). Furthermore, ciliates that undergo unequal fissions may give rise to the production of giant cells that are cannibalistic (Tuffrau, Fryd-Versavel, et al. 2000), and in mixotrophic chrysophytes, some cells have been reported to cannibalize on other sibling cells, forming giant cells two to three time larger in diameter than ordinary cells (Yubuki, Nakayama, and Inouye 2008). This documented presence of cannibalism in unicellular organisms, including the development of giant cells, suggests that the insights we gained in chapter 11 regarding the dynamical consequences of cannibalism may also be relevant for unicellular organisms.

The recognition of the presence of size structure even in populations of unicellular organisms sheds new light on why, for example, phytoplankton populations have been observed to exhibit oscillations in chemostats, where unstructured theory predicts only a steady state to be present (Massie, Blasius, et al. 2010; Pascual and Caswell 1997). Furthermore, it raises a question about under which conditions an unstructured theory may suffice to explain the population dynamical patterns in particular but also the community structure of unicellular organisms under natural (field) conditions, for which potential effects of the life (cell) cycle have traditionally been ignored (cf. Pascual and Caswell 1997).

MODEL HIERARCHIES, MODEL SIMPLIFICATIONS, AND MODEL TESTING

In chapter 9, we discussed how the dynamics of the same population may shift between dynamics that can readily be explained by unstructured theory (predator-prey cycles) and dynamics where differences in size or stage must be considered. This circumstance raises the question about links between models of different complexities and how model simplification may be carried out. Throughout this book we have used two major modeling approaches: stage-structured biomass models and fully size-structured models. In several chapters we have also discussed the extent to which both approaches yield the same or similar results. Here we will consider these modeling approaches in the context of model hierarchies and model simplifications.

STRIPPING DOWN MODEL COMPLEXITY

A long, ongoing discussion in ecology has been the difficulty in developing models that are both general and testable (cf. Holling 1966; Murdoch,

McCauley, et al. 1992; Murdoch and Nisbet 1996; Murdoch, Briggs, and Nisbet 2003). Levins (1966) first posed the fundamental problem of simultaneously achieving a high degree of generalism, realism, and precision in model building, and he argued that a particular model may reach two of these goals but not all three of them. The solution to this dilemma has generally been found in the distinction between strategic and tactical models (Holling 1966; Levins 1966; May 1974). Here strategic models have the aim to capture the "essence" of a broad range of ecological systems, whereas tactical models have the aim to quantitatively predict the dynamics of specific systems in detail, often in applied management situations.

With the aim to handle both strategic and tactical models in a common framework, Murdoch and Nisbet (Murdoch, McCauley, et al. 1992; Murdoch and Nisbet 1996; Murdoch, Briggs, and Nisbet 2003) have advocated a research strategy that is based on a hierarchy of models with different complexities. In their view, the generality of more complex models can be assessed by studying a chain of models of decreasing complexity (still embodying essential mechanisms). Specifically, they have used this hierarchic modeling approach to study parasitoid-host interactions contrasting a generic gain model for such interactions with more system-specific models, including both analytical and detailed simulation models, for the red scale–*Aphytis* interaction. Two major conclusions that Murdoch, Briggs, and Nisbet (2003) reached were that simple models were (1) essential for understanding the dynamics of complex models and (2) useful in that they serve as benchmarks for evaluating differences between them and more complex models. We will come back to these two conclusions shortly.

Murdoch, McCauley, et al. (1992) described the role of strategic models as models *"that try to reach beyond the tangle of unique detail associated with each particular ecological system to grasp and abstract the features that for those aspects of its dynamics are generic.... Generality can be obtained because, one hopes, many systems have a similar essence."* This is a true and fundamental statement about what science in general is actually all about: if there is no "essence" that is similar to many systems, but rather each system is unique, then there is no room for generalization—that is, science. We have argued that to identify the generic aspects of the dynamics and the essential ingredients of a system responsible for these aspects, a sequence of logical steps is needed, including (1) getting a detailed overview of the dynamics that the system can display, (2) finding out which features of these dynamics are generic, and (3) identifying which (essential) mechanisms are responsible for these features (de Roos and Persson 2005).

We have also strongly advocated that this sequence of steps makes it necessary to start the modeling of a population or a community at the level where the dynamics originate: the level of the *individual organism*. This choice naturally presents itself, because it is the individual that is the fundamental entity involved in all population processes such as individual growth, reproduction, and death. Moreover, the ecological interactions (such as competition and predation) that structure ecological communities are based on the performance of the individual in terms of how much it eats, its risk of being eaten, and so on. Although it is outside the scope of this book, starting at the level of the individual also allows for a direct and stringent link to study evolutionary processes incorporating ecological feedbacks and consumer-element interactions (de Roos, Boukal, and Persson 2006; Lorena, Marques, et al. 2010; Sousa, Domingos, et al. 2010).

All the model versions we have used in this book are based on individual-level assumptions. It is worthwhile to again point out that the Yodzis and Innes (1992) model that we have used as a proxy for unstructured models also represents an individual-based model, although for individuals that are characterized by ontogenetic symmetry, which can hence be faithfully represented by the average individual. In fact, even the unstructured consumer-resource models that only account for population abundance, such as Tilman's (1982) models, are based on individual-level assumptions. The demarcation line between unstructured and structured models hence does not lie in whether they are "i-state"- or "p-state"-based, as suggested by Odenbaugh (2005). Instead, the demarcation line lies in whether it is possible to reduce the i-state distribution in a consistent manner to the average individual.

Furthermore, structured models per se can vary substantially in complexity of the i-state distribution, from two-stage (juvenile, adult) models to infinite-dimensional size distributions. Importantly, the fully size-structured population model used in chapter 9, the stage-structured biomass models used extensively in chapters 3–8, and the Yodzis and Innes (1992) unstructured biomass model can be shown to form a stringent hierarchy within a purely individual-based approach (sections 2 and 3 in the technical appendices; de Roos, Schellekens, et al. 2008b). At one level, the stage-structured biomass model is derived as an approximation to the fully physiologically structured population model in such a way that all model results are completely identical under equilibrium conditions. At another level, the Yodzis and Innes (1992) model can be derived from the stage-structured model for $q = 1$ (assuming mortality rates of juveniles and adults are the same; de Roos, Schellekens, et al. 2008b). The fact that simpler models can be derived from more complex models under specific assumptions (but not the other way around) makes us propose that the fully size-structured model represents the generic model, from which simpler models can be derived for specific conditions, rather than the reverse.

The fact that the stage-structured biomass model can be reduced to a one-stage model under the limited condition of $q = 1$ (assuming that mortality rates of, and resource productivities for, the two stages are the same, see chapter 3) shows that unstructured theory represents a limiting case for the condition of *ontogenetic symmetry*. Only with ontogenetic symmetry can all individuals in a population be faithfully represented by the average individual (de Roos, Persson, and McCauley 2003; de Roos and Persson 2005). For all other q values, the system will be characterized by *ontogenetic asymmetry*, in which case, using the average individual as the basis to describe the dynamics of the system will fail to handle important elements of these dynamics owing to stage-specific bottlenecks (development control versus reproduction control). Put in another way, the concepts of ontogenetic symmetry and ontogenetic asymmetry provide us with the necessary tools to decide when and when not the average individual suffices as an appropriate *i*-state representation.

As discussed above, Murdoch and coworkers (Murdoch, McCauley, et al. 1992; Murdoch and Nisbet 1996; Murdoch, Briggs, and Nisbet 2003) used a hierarchy of linked models of different complexities to investigate the essential mechanisms underlying systems dynamics and to reveal what is lost and preserved through model simplification. However, these authors carried out this simplification within a *population-level* framework. In many contexts, such a simplification takes the form of, for example, making the recruitment rate into the juvenile stage a density-dependent function of the current adult density, as opposed to letting it be dependent on the amount of resources available for the juveniles (Murdoch, Briggs, and Nisbet 2003).

This approach contrasts with ours in that all models that we have used have an explicit individual-level interpretation, include an explicit resource dynamics, and can be stringently derived from a more complex fully size-structured individual-based model. We will use the example of the interaction between a size-structured consumer population and its unstructured resource to analyze what remains and what is lost through aggregation within an *individual-level* framework. For the fully size-structured model, we found in chapter 9 that different dynamics could be observed depending on the size-specific competitive ability (q): juvenile-driven cycles (for $q < 1$), stable, fixed-point dynamics (for $q = 1$), and adult-driven cycles (for $q > 1$) (figure 12.6, gray lines; to have symmetry at $q = 1$, background mortalities are assumed to be the same for juveniles and adults). These results are robust to substantial changes in model assumptions. For example, the results were robust to the inclusion of discrete reproduction and the possibility for somatic growth among adults (chapter 10). Furthermore, changing assumptions about resource growth dynamics still led to the same results (de Roos, Schellekens, et al. 2008b).

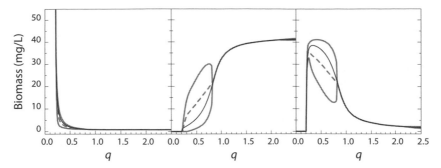

FIGURE 12.6. Biomass of resources (*left*), juveniles (*middle*), and adults (*right*) as a function of q for the two-stage-structured model (*black thin lines*) and the fully size-structured model (*gray lines*). For the fully size-structured model, minimum, maximum (*both solid lines*), and average (*dashed lines*) values are given. Parameter values are the same as in chapter 9 (box 9.2 and figure 9.1).

Taking the drastic step to reduce model complexity from an infinite-dimensional description of the consumer population to the two-dimensional two-stage biomass model first shows that the predictions of juvenile consumer, adult consumer, and resource biomasses are identical under equilibrium conditions (see figure 12.6 for q values around 1). This result naturally follows from the fact that the stage-structured model has been derived as an approximation to the fully size-structured model. Notice, though, that although the two-stage biomass model is a two-dimensional system in a technical sense, it accounts for a complete size distribution of juvenile consumers and for somatic growth of these juveniles. Second, cycles occurring for q values smaller or larger than 1 are absent. This result relates to the fact that the two-stage-structured model incorporates a distributed time delay between birth and maturation for the individual organism, whereas the fully structured model assumes that maturation takes place as a discrete event (reaching a size threshold) in the individual's life history.

The two-stage and the fully size-structured model predict the same average biomasses for the resource, juveniles, and adults for $q > 1$ (figure 12.6). We have also found this conclusion to be largely independent of the values of R_{max} and ρ that govern the semichemostat dynamics. Both parameters change the resource productivity, which equals ρR_{max}, in the same way. Increases in these parameters translate into almost proportional increases in both juvenile and adult consumer biomass, but generally do not change model dynamics in a relative sense. The equilibrium densities of resource, juvenile consumer biomass, and adult consumer biomass predicted by the stage-structured model differ somewhat more from the time-averaged values of these biomasses in the population cycles predicted by the fully size-structured model for $q < 1$ (figure 12.6). The stronger similarity between the models concerning the predictions of average stage-specific biomasses for $q > 1$ can be related to the much smaller fluctuations in the adult-driven cycles, which result from the small variation in size distribution over time during these cycles (figure 9.3). Still, despite the fact that the time-averaged biomass values are higher for juveniles and lower for adults in the size-structured model compared with the two-stage model (figure 12.6), the correspondence between time-averaged total biomass values in the size-structured model also corresponds well with that of the two-stage biomass model for $q < 1$.

The simplification of model complexity from a fully size-structured model to a two-stage consumer model clearly unravels the essential ingredients that are necessary for cohort cycles to occur. The influence of the discrete maturation event was also investigated by adding a second juvenile stage, which results in the reappearance of cycles. This reappearance of cycles that results from adding one or more juvenile stages can be explained by the fact that the presence of more than one juvenile stage introduces a finite, non-zero delay between adult reproduction and the recruitment of juveniles to the adult stage (Guill 2009). At the same time, the stage-based model captured the effects of q on stage-specific biomass dominance and the nature of regulation (development control versus reproduction control). Furthermore, we have in chapters 3–7 compared predictions of stage-structured and fully size-structured models (based on Kooijman-Metz energetics) with respect to regulatory mechanisms on community structure (i.e., emergent Allee effect versus competitive bottlenecks) and found that these mechanisms are generic and largely robust to major model simplifications. These results clearly suggest that the two-stage biomass model and its extensions may form a general modeling tool to investigate the effects of intraspecific size variation on community structure. If this is the case, our ability to efficiently study the implications of intraspecific size variation on community structure in more species-rich communities will increase tremendously.

COMPLEX VERSUS SIMPLE MODELS

Murdoch, Briggs, and Nisbet (2003) argued that simple models are essential for understanding the dynamics of complex models and useful because they serve as benchmarks for evaluating differences between them and more complex models. Within our individual-based framework, the size-structured consumer-resource system can serve as a useful example to address this question. Basically, we see model simplification as a strategic route to generalization, in which key mechanisms behind observed dynamics can be identified, including the analysis of what is lost in the model simplification process.

Historically, we can trace two major conceptual insights that have been derived about cohort cycles from model investigations. First, de Roos, Metz, et al. (1990) provided basic insights about the difference between cohort/generation cycles and predator-prey cycles (also termed prey-escape cycles). Second, Persson, Leonardsson, et al. (1998) provided basic insights about the influence of the size-scaling of competitive ability on the presence and nature of cohort cycles (fixed-point dynamics, juvenile-driven versus adult-driven cycles). The latter paper also included a mechanistic and operational derivation of intraspecific, size-dependent competitive ability (critical resource density, CRD), operational in the sense that the parameters of the CRD function can readily be estimated empirically (see chapter 10). In both cases, these insights were derived from analyses of complex, fully structured models (in the second case, even assuming discrete reproduction) and based on simulations. Finally, all the dynamical phenomena that have been subsequently observed in size-structured consumer-resource systems were first identified in these two papers.

Simplifications of the model presented by Persson, Leonardsson, et al. (1998)—by assuming a completely continuous system with the consumer population consisting of only two stages (juveniles and adults), a simpler resource dynamics, and a simpler adult energy budget (only reproduction allocation)—and expressing the stage-specific competitive ability with a single parameter (q) allowed for a stability analysis of the equilibrium and a closer examination of the ingredients necessary for the different dynamics to occur (de Roos and Persson 2003). This concerned the importance of time delays per se versus food-dependent juvenile development for the presence of juvenile- and adult-driven cycles, respectively (see chapter 9). The simplification of model complexity preserved all of the four basic dynamics (juvenile-driven cycles of two types, adult-driven cycles, and stable equilibrium) that had previously been observed in the more complex model (for a similar analysis concerning cannibalistic systems, see Claessen and de Roos 2003). Finally, a more complete mathematical analysis for the conditions necessary for different forms of cycles to occur at all was provided by Diekmann, Gyllenberg, et al. (2010). A biologically very relevant result from this latter analysis was that juveniles and adults do not have to compete for a common resource in order to generate cohort cycles. This result corresponds to what we found in chapter 10 in our analysis of the effects of varying resource productivities for juveniles and adults when they were feeding on two separate resources (see also the conditions necessary for overcompensation to occur in chapter 3).

The above illustrates how our insights about cohort cycles have progressed over time and how the use of models of different complexities has been pivotal for this progress. Simplifications of model complexity have indeed, as Murdoch, Briggs, and Nisbet (2003) suggested, been key to unraveling the mechanisms behind cohort cycles. In particular, the simplification to describe size/stage-specific competitive ability by only one parameter, q, has undoubtedly allowed a more general analysis of the effects of size-structured interactions on both community structure and population dynamics. Most important, it has paved the way for the development of the, in our view, fundamental concept of ontogenetic asymmetry.

Finally, it is important to make clear that the progress in our understanding of the dynamics of consumer-resource and cannibalistic dynamics is a result of a strong interaction between modeling and analyses of empirical data. These empirical studies include in particular the extensive studies on *Daphnia* dynamics (McCauley, Nisbet, et al. 1999; McCauley, Nelson, and Nisbet 2008; Murdoch and McCauley 1985) and fish populations (Hamrin and Persson 1986; Persson, Byström, and Wahlström 2000; Townsend and Perrow 1989). In these analyses, relatively detailed individual-level information on, for example, individual growth trajectories, which only the fully structured models provide, has been central. The latter pertains especially to the development of insights about cannibalistic dynamics (Claessen, de Roos, and Persson 2000; Claessen, van Oss, et al. 2002; Persson, Claessen, et al. 2004).

INDIVIDUAL-BASED MODELS AND MODEL TESTING

In the above we have discussed how it is possible to move between more complex and more simple models within the same individual-based framework by means of making simpler assumptions about individual energetics from discrete/continuous dynamics with discrete reproduction to continuous reproduction to stage-structured models with simple energy channeling rules, ultimately ending up with the Yodzis and Innes (1992) single-stage model. We have extensively used the Yodzis and Innes model to represent the unstructured individual-based model, because this model can straightforwardly and consistently be derived from the size-structured model as a limiting case.

In contemporary ecology, individual-based approaches have also been termed mechanistic approaches to ecology, defined as the aim to build a population and community ecological theory based on individual-level processes (Schoener 1986). A prominent example of such approaches for unstructured systems is Tilman's (1982) consumer-resource theory. Mechanistic approaches (sensu Schoener 1986) have been contrasted with, for example, the

Lotka-Volterra competition models that lack a clear individual basis. Nonetheless, the competition parameters and carrying capacities in the Lotka-Volterra competition models can still be derived from Tilman's model, even though they are derived at the population level. They can even be interpreted in individual-level terms (the competition factor corresponds to the ratio of per capita consumption rates and the carrying capacity to the ratio of resource renewal rate and per capita consumption rate).

Nonmechanistic, population-level formulations are also commonly used in stage-structured models (cf. de Roos, Persson, and Thieme 2003; Ricker 1954). For example, de Roos, Persson, and Thieme (2003) assumed in a three-stage model that fecundities of adults and maturation rates of juveniles and subadults depended on their respective densities without explicitly accounting for a resource. In contrast to the unstructured Lotka-Volterra competition model, it is, however, difficult for this structured case to make a stringent interpretation of the population-level derived parameters in individual terms.

In chapter 2, we considered four reasons to prefer mechanistic model formulations over phenomenological ones, particularly for structured populations: (1) interdependence of model formulation and the dynamical patterns that the model seeks to explain, (2) model predictions may differ between mechanistic and phenomenological models, (3) realistic model formulations are easier to derive for mechanistic models, and (4) testability against empirical data. The last reason relates in particular to the fact that it is difficult or even impossible to predict a priori the population dynamics from individual-level assumptions about food intake and allocation to growth and reproduction as functions of food density, and even harder to predict the population feedback on these individual-level processes. The two examples of the roach population in Alderfen Broad and the yellow perch population in Crystal Lake that we discussed in chapter 10 are two illustrations of how model predictions of individual growth trajectories can be derived following the route: individual level formulation → prediction of population dynamics → prediction of individual growth rates as a result of population feedbacks, to be subsequently tested against empirical data.

Pioneering studies that tested predictions of physiologically structured models (PSPMs) were largely focused on quantitative testing of the *individual-level* models in the absence of population feedbacks (Gurney, McCauley, et al. 1990; McCauley, Murdoch, et al. 1990), and only afterward also included predictions of *population dynamics* (Murdoch, McCauley, et al. 1992). De Roos and Person (2001) extended the testing procedure to point out the ability of PSPMs to quantitatively predict *individual-level* processes that result from *population feedbacks* (see chapters 10–11).

Furthermore, more detailed quantitative comparisons between model predictions and data have been used in within-season studies to elucidate the

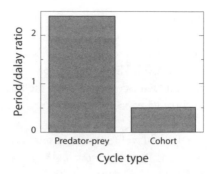

FIGURE 12.7. Experimental estimates of the period/delay ratio in predator-prey cycles and cohort cycles using data on individual growth trajectories to estimate generation time. Based on data from McCauley, Nelson, and Nisbet (2008).

importance of different mechanisms affecting the dynamics of cannibalistic populations (Persson, Claessen, et al. 2004). Our investigations of the dynamics of cannibalistic populations in chapter 11 also elucidate the capacity of PSPMs to discriminate between qualitatively different types of population regulation based on the form, amplitude, and frequency of individual growth trajectories as a result of population feedbacks (figure 11.12). Compared with the quantitative comparisons made in chapter 10, this contrast between qualitatively different types of dynamics will, in our view, represent the most useful way of using physiologically structured population models for model testing in the future. The study by McCauley, Nelson, and Nisbet (2008) illustrates this potential further, once again using *Daphnia* dynamics as the empirical example. Previous work on *Daphnia* dynamics showed that predator-prey cycles and cohort cycles can be distinguished on the basis of cycle amplitude and the juvenile/adult cross-correlation (see chapter 9 and McCauley, Nisbet, et al. 1999). McCauley, Nelson, and Nisbet (2008) extended this analysis by including information on individual growth rate as a result of population feedbacks to estimate the period/juvenile delay ratio, which in their model differed between predator-prey cycles and cohort cycles (figure 12.7). The addition of information on individual growth of *Daphnia* individuals thus added a tool by which to discriminate between different types of dynamics.

DEVELOPMENT VERSUS REPRODUCTION CONTROL: ONTOGENETIC ASYMMETRY

The contents of the different chapters in this book have all aimed to shed light on a blind area in contemporary ecological theory: the consequences of ontogenetic growth on population and community dynamics. The overall results that we have obtained can basically be boiled down to the implications of one concept: ontogenetic asymmetry, resulting in differential limitations caused by maturation and reproduction. In chapter 3, we laid the foundation for how different net-production rates of juveniles and adults led to either development control (net biomass turnover of juveniles is less than the net biomass turnover of adults) or reproduction control (net biomass turnover of juveniles is greater than the net biomass turnover of adults), with both cases reflecting a situation with ontogenetic asymmetry. The only situation with ontogenetic symmetry occurs when the net biomass turnover of juveniles and adults is equal (and both 0 at equilibrium). In the subsequent chapters 4 and 5 we explored how the basic principles discussed in chapter 3 resulted in major consequences for community structure, exemplified by the presence of an emergent Allee effect and predator facilitation. In the following chapters, still dealing with community structure, we showed that the introduction of ontogenetic niches in consumer or predator life history could lead to alternative stable states that included an emergent Allee effect but also other phenomena. In all cases, the patterns observed could nevertheless be related to stage- or size-specific performance with development versus reproduction control as the dominating principles behind the results.

The results presented in the chapters dealing with population dynamics (chapters 9–11) further showed that the basic principles of reproduction and development control carry over to population dynamics, and that cannibalism may tip a reproduction-controlled consumer-resource system into a development-controlled system. It is once again worth pointing out the formal connection between the stage-structured model and the fully size-structured model formulations: the results we have obtained indicate that the two model approaches form two fundamental legs, based on which questions, focused on either population dynamics or community structure, can be addressed. It is also worth reiterating the substantial robustness of the basic outcomes—juvenile- and adult-driven cycles—to specific model assumptions. Finally, our analyses of the dynamics of demand-driven systems in this chapter further extend the domain and potential of the results we have obtained.

The generality and overall robustness of the results stem from the basic principles of development versus reproduction control. In our analyses of different trophic configurations, there are three aspects, besides the effect of resource-specific productivity, that are central in their effect on both community structure and population dynamics: the size scaling of foraging (q or CRD), the birth/maturation body size ratio (z), and the presence and form of ontogenetic niche shifts. Generically, the resulting form of regulation will be a result of both the *individual* state (q, CRD, z) and the condition of the *environment* (resource production of different resources, mortalities). In the case of a simple environment, where all individuals of different stages share the same resource(s) and experience the same mortality, the form of regulation is simply a result of the *i*-state characteristics of different stages in terms of *q* or the form of the CRD. For more complex situations, where individuals of different stages/sizes do not share or only partially share the same resources or experience different mortalities, the form of regulation will result from both *i*-state characteristics and the *environmental condition* and their interaction through the form of ontogenetic niche shifts.

Despite the length of this book, we are far from having provided an analysis that would deserve to be even nearly called a "complete" analysis of the implications of size- and stage-variation on population and community dynamics (if this is at all reachable). In particular, the extension of the theory from a limited number of interacting populations to a many-species, foodweb context is a major challenge for future research (Rudolf and Lafferty 2011). This also concerns an extension to consider ecosystem (consumer-element) and evolutionary dynamics. Furthermore, although we have tried in every chapter to link our modeling results to empirical data, there is at present an imbalance between new insights generated from modeling and empirical investigations (Miller and Rudolf 2011). That said, if we have inspired other scientists to pursue the theoretical and empirical investigations of the fascinating implications of ontogenetic growth on ecological systems, we will feel that we have accomplished our goal.

Technical Appendices

1 BASIC SIZE-STRUCTURED POPULATION MODEL

Assume that the life history state of an individual is fully characterized by its body size. Appropriate measures of this body size may be length or weight, but for the time being we will use a variable s to refer to this individual state variable without making explicit whether this is length or weight. The four basic processes in the individual life history that determine population dynamics are (1) ontogenetic development, (2) reproduction, (3) mortality, and (4) individual impact on the environment. Ontogenetic development is in this case equivalent to growth in body size. We will assume that food density is the only characteristic of the environment in which the individual lives that influences its life history. The impact of the individual on its environment in that case is equivalent to the foraging pressure it exerts on the food. Formulating a basic size-structured population model boils down to deriving functions that describe the rate at which the four life history processes contributing to population dynamics progress. In general, all these rates will depend on both the size of the individual as well as on the food density it encounters.

Growth in body size is modeled by specifying a function $g(R,s)$ representing the growth rate in size for an individual of size s that lives in an environment with food or resource density, which we indicate with R. Given this function the development of the individual follows the following ODE:

$$\frac{ds}{dt} = g(R,s), \qquad s(0) = s_b \qquad\qquad (\text{A.1})$$

This equation describes that the individual's size changes continuously over time and starts at a size at birth, which we have represented with the parameter s_b. Reproduction is modeled by specifying the rate at which an individual produces offspring. This rate, which is equivalent to its fecundity, we refer to with the function $b(R,s)$. As for growth, this rate may also depend on the individual's size as well as on food density. Mortality is modeled with an instantaneous mortality rate $d(R,s)$. The exact interpretation of an instantaneous mortality rate is not straightforward, as it represents a probability per unit time. If mortality were to be constant and hence independent of food density and body size, a mortality rate equal to d would imply that an individual has a probability $\exp(-dt)$ to survive over a time interval of length t. This survival probability takes values between 0 and 1, but the values of d are not restricted to this interval and only have to be positive to make sense biologically. As a last ingredient of the individual life history, we specify the rate $I(R, s)$ with which an individual with body size s feeds on and hence depletes the food density in the environment. Naturally, this feeding rate is also dependent on the actual value of that food density itself.

Given these four elementary functions of individual life history, we are ready to specify the dynamic model for the population of size-structured individuals. Mathematically, the population will be represented by a density function $c(t, s)$. As such, this density function has no direct biological interpretation, unless integrated over a particular range of body sizes. In particular, $c(t, s)ds$ represents the *number* of individuals in a tiny size range ds, while more generally

$$\int_{s_1}^{s_2} c(t, s)\, ds$$

equals the number of individuals with a body size between s_1 and s_2. The dynamics of the density function $c(t, s)$ are described by the system of equations:

$$\frac{\partial c(t,s)}{\partial t} + \frac{\partial g(R,s)c(t,s)}{\partial s} = -d(R,s)c(t,s) \tag{A.2a}$$

$$g(R,s_b)c(t,s_b) = \int_{s_b}^{\infty} b(R,s)c(t,s)\, ds \tag{A.2b}$$

$$c(0,s) = \Phi(s) \tag{A.2c}$$

The partial differential equation (PDE) (A.2a) accounts for the change in the population size distribution over time, expressed by the term $\partial c(t,s)/\partial t$, owing to individual growth in body size, described by the second term $\partial(g(R,s)c(t,s))/\partial s$; and mortality, represented by the last term $-d(R,s)c(t,s)$. The boundary condition (A.2b) accounts for the population increase through reproduction. The left-hand side of this equation represents the *rate* at which newborn individuals with size $s = s_b$ enter the population through reproduction. The right-hand side of this boundary condition represents the total offspring produced by the population as a whole, as it integrates over all sizes s the product of the individual fecundity and the density of individuals, $b(R,s)c(t,s)$. The third and last equation of the system above specifies the initial state of the size-structured population at time $t = 0$. This equation is given here only for the sake of completeness; often when specifying a particular structured population, model details about this initial population state are left unspecified. We will not further discuss here the derivation of the system of equations (A.2) from its underlying principles. For the readers interested in more details about structured population models, we refer to the introduction given in de Roos (1997).

The system of equations (A.2) specifies only how the population of size-structured individuals would change over time and as such does not account for changes in the food density in the environment. Hence, the equations can be used to describe the population dynamics at a constant food density R, which will eventually lead to exponential population growth or decline or, more generally, for any food density $R(t)$, which is given as an explicit function of time. Also, the equations only contain terms that are linear in the population density $c(t, s)$ and hence do not account for any density dependence. Density dependence arises once we account for the changes in food density resulting from population foraging. To specify these food dynamics, we need one additional model ingredient, which describes the dynamics of the food density in the absence of any consumers. This ingredient represents the only system-level model assumption, as all other assumptions pertain to the life history of the size-structured individuals. For the time being, we will model the autonomous dynamics of the food density (i.e., in the absence of consumers) with a function $G(R)$. The dynamics of food density now follows the ordinary differential equation (ODE):

$$\frac{dR}{dt} = G(R) - \int_{s_b}^{\infty} I(R, s)c(t, s)ds \qquad (A.2d)$$

This ODE expresses that food dynamics are the balance between the autonomous dynamics of the food in the absence of consumers and the food consumption by the entire consumer population. As expressed in the second term in the ODE, this total consumption rate equals the accumulated (integrated) intake rate by consumers of different body sizes. In combination with the PDE (A.2a) and its boundary condition (A.2b) this ODE completes the dynamics model of a population of size-structured individuals feeding on a single, shared resource. All model variables, functions, and equations are summarized in box A.1. The model as presented here is a generic formulation. Particular models will differ from each other owing to the specific choices for the different functions describing growth, reproduction, mortality, and feeding of consumer individuals as well as the growth dynamics of resource in the absence of consumers.

BOX A.1

VARIABLES, FUNCTIONS, AND POPULATION-LEVEL EQUATIONS OF THE BASIC
SIZE-STRUCTURED POPULATION MODEL FEEDING ON A SHARED RESOURCE

Dynamic equations	Description
$\dfrac{\partial c(t,s)}{\partial t} + \dfrac{\partial g(R,s)c(t,s)}{\partial s} = -d(R,s)c(t,s)$	Change in size distribution through growth and mortality
$g(R,s_b)c(t,s_b) = \displaystyle\int_{s_b}^{\infty} b(R,s)c(t,s)ds$	Increase in the number of newborn individuals through reproduction
$\dfrac{dR}{dt} = G(R) - \displaystyle\int_{s_b}^{\infty} I(R,s)c(t,s)ds$	Resource biomass dynamics

Function	Description
$G(R)$	Resource growth in absence of consumers
$I(R,s)$	Individual intake rate of resource
$g(R,s)$	Individual growth rate in body size
$b(R,s)$	Individual fecundity
$d(R,s)$	Individual mortality rate

Variables	Description
R	Resource (food) density in environment
s	Individual body size
$c(t,s)$	Density function of individual size distribution

2   DERIVATION OF THE YODZIS AND INNES MODEL

The bioenergetics model formulated by Yodzis and Innes (1992) for the interaction between a consumer and its resource can be derived from the basic size-structured population model (box A.1), provided we make some special assumptions about the model functions representing the individual life history. Most important, assume that resource ingestion by individual consumers is proportional to their body size and can hence be described by a function

$$I(R,s) = \omega_C(R)s \tag{A.3}$$

Furthermore, we have to assume that the net amount of biomass that can be created per unit of time through either reproduction or somatic growth also scales proportionally to body size and can hence be modeled as $v_C(R)s$. Notice that with this latter assumption we assume that all energetic requirements to cover maintenance, costs for growth, and costs for reproduction are already accounted for. If an individual of size s spends a fraction $\kappa(s)$ of this net-biomass-production rate on somatic growth, and hence a fraction $(1 - \kappa(s))$ on reproduction, the growth rate in body size and the fecundity are now given by:

$$g(R,s) = \kappa(s)v_C(R)s \tag{A.4}$$

and

$$b(R,s) = \frac{(1 - \kappa(s))v_C(R)s}{s_b} \tag{A.5}$$

Notice that division by the size at birth s_b in this equation is necessary to convert the biomass production through reproduction into a *number* of new-born offspring. These two equations depend on the function $\kappa(s)$, representing the biomass fraction allocated to somatic growth, but this function will turn out to be irrelevant for the derivation that follows. Finally, we assume that all individuals are suffering from the same background mortality independent of their body size and independent of the resource density, such that

$$d(R,s) = \mu_C \tag{A.6}$$

The total biomass of the population equals an integral over the population density function $c(t,s)$, weighted by the individual body size s of these individuals:

$$C = \int_{s_b}^{\infty} sc(t,s)ds$$

For the total population biomass, an ODE can now be derived by differentiating this expression with respect to time and using the PDE (A.2a) and its boundary condition (A.2b) to simplify it:

$$\frac{dC}{dt} = \int_{s_b}^{\infty} s\frac{\partial c(t,s)}{\partial t}ds$$

$$= -\int_{s_b}^{\infty} s\frac{\partial g(R,s)c(t,s)}{\partial s}ds - \int_{s_b}^{\infty} sd(R,s)c(t,s)ds$$

$$= -sg(R,s)c(t,s)\big|_{s=s_b}^{s\to\infty} + \int_{s_b}^{\infty} g(R,s)c(t,s)ds - \int_{s_b}^{\infty} sd(R,s)c(t,s)ds$$

$$= s_bg(R,s_b)c(t,s_b) + \int_{s_b}^{\infty} g(R,s)c(t,s)ds - \int_{s_b}^{\infty} sd(R,s)c(t,s)ds$$

$$= \int_{s_b}^{\infty} s_bb(R,s)c(t,s)ds + \int_{s_b}^{\infty} g(R,s)c(t,s)ds - \int_{s_b}^{\infty} sd(R,s)c(t,s)ds$$

Going from the second to the third line of this derivation, we have used integration by parts to simplify the first integral in the second line. Going from the third to the fourth line in the derivation, we exploit the fact that no individual lives forever and that hence eventually, when an individual reaches very large sizes ($s \to \infty$), its contribution to the population dynamics becomes negligible. Finally, the last line of the derivation is obtained by substitution of the boundary condition (A.2b). The assumptions about the individual growth rate and fecundity imply that $g(R,s) + s_b b(R,s) = \nu_C(R)s$. Using this identity and the assumption about the individual mortality rate $d(R,s)$ allows us to derive the following ODE:

$$
\begin{aligned}
\frac{dC}{dt} &= \nu_C(R) \int_{s_b}^{\infty} sc(t,s)ds - \mu_C \int_{s_b}^{\infty} sc(t,s)ds \\
&= \nu_C(R)C - \mu_C C
\end{aligned}
\tag{A.7}
$$

Similarly, substitution of expression (A.3) into the general ODE (A.2d) for resource dynamics yields:

$$
\begin{aligned}
\frac{dR}{dt} &= G(R) - \omega_C(R) \int_{s_b}^{\infty} sc(t,s)ds \\
&= G(R) - \omega_C(R)C
\end{aligned}
\tag{A.8}
$$

which is identical to the resource equation for the Yodzis and Innes model given in chapter 3, provided we assume semichemostat resource dynamics in the absence of consumers: $G(R) = \rho(R_{max} - R)$.

The Yodzis and Innes model hence accurately describes the dynamics of a fully size-structured consumer population feeding on a shared resource when the following conditions hold:

1) Consumer ingestion scales linearly with consumer body size.
2) Net biomass production by consumers scales linearly with consumer body size.
3) Production of new biomass through somatic growth and reproduction is equally efficient.
4) Consumer mortality is independent of consumer body size.

3   DERIVATION OF THE STAGE-STRUCTURED BIOMASS MODEL

The stage-structured biomass model, which was introduced in chapter 3 as a stage-structured extension of the Yodzis and Innes model (1992), constitutes a consistent approximation to the size-structured model that was presented in chapter 9 (box 9.1). The approximation is such that all equilibria that are to be found in the stage-structured biomass model have an identical counterpart in the fully size-structured model, and vice versa. Here we provide the mathematical derivation of the stage-structured biomass model from its underlying size-structured population model.

In the size-structured model (see box 9.1), juvenile biomass equals the integral over the juvenile size range of the density function $c(t,s)$ weighted by the individual size s:

$$J(t) = \int_{s_b}^{s_m} sc(t,s)ds$$

Adult biomass is simply the product of the adult density and the adult body size s_m:

$$A(t) = s_m C(t)$$

The ODE describing the dynamics of the resource density can be rewritten in terms of these two biomass quantities as:

$$\frac{dR}{dt} = G(R) - \omega_J(R)J - \omega_A(R)A \tag{A.9}$$

The ODE describing the dynamics of adult biomass $A(t)$ can be derived by multiplying both left- and right-hand side of the ODE describing the dynamics of adult density $C(t)$ (see box 9.1), yielding:

$$\frac{dA}{dt} = s_m g(R,s_m)c(t,s_m) - d_A(R)A$$

$$= s_m g(R,s_m)c(t,s_m) + (v_A(R) - v_A^+(R))A - \mu_A A$$

Following its definition (see box 9.1), we have expressed the adult death rate as $d_A(R) = \mu_A + (v_A^+(R) - v_A(R))$, in which $v_A^+(R)$ represents the adult net production of biomass per unit time $v_A(R)$ but restricted to non-negative values. Notice that the term $v_A^+(R) - v_A(R)$ cancels as long as $v_A(R)$ is positive, that is, as long as $R > H_C/(\sigma_c q M_c/T_C - 1)$, and equals $-v_A(R)$ otherwise.

The ODE for the dynamics of the juvenile biomass can be derived by differentiating the integral expression for $J(t)$ with respect to time and using the PDE and its boundary condition to simplify the result:

$$\frac{dJ}{dt} = \int_{s_b}^{s_m} s\frac{\partial c(t,s)}{\partial t}ds$$

$$= -\int_{s_b}^{s_m} s\frac{\partial g(R,s)c(t,s)}{\partial s}ds - \int_{s_b}^{s_m} sd_J(R)c(t,s)ds$$

$$= -sg(R,s)c(t,s)\Big|_{s=s_b}^{s=s_m} + \int_{s_b}^{s_m} g(R,s)c(t,s)ds - d_J(R) \int_{s_b}^{s_m} sc(t,s)ds$$

$$= s_b g(R,s_b)c(t,s_b) - s_m g(R,s_m)c(t,s_m) + \int_{s_b}^{s_m} g(R,s)c(t,s)ds - d_J(R) \int_{s_b}^{s_m} sc(t,s)ds$$

$$= s_b b(R,s_m)C - s_m g(R,s_m)c(t,s_m) + \int_{s_b}^{s_m} g(R,s)c(t,s)ds - d_J(R) \int_{s_b}^{s_m} sc(t,s)ds$$

Going from the second to the third line in this derivation, we have integrated by parts the integral derived from the growth term in the PDE. The equation can be rewritten in terms of the mass-specific net-production rates $\nu_J(R)$ and $\nu_A(R)$ for juveniles and adults, respectively, using the following expressions for the life history functions: $b(R,s_m) = \nu_A^+(R)s_m/s_b$, $g(R,s) = \nu_J^+(R)s$ and $d_J(R) = \mu_J + (\nu_J^+(R) - \nu_J(R))$, and the definition for the adult biomass $A = s_m C$. Notice that the expression for the juvenile mortality $d_J(R)$, like the corresponding expression for adult mortality, accounts for background mortality under growing conditions and for additional starvation mortality when juvenile net production is negative. Also notice that the use of the rates $\nu_J^+(R)$ and $\nu_A^+(R)$ implies that growth and fecundity are positive as long as juvenile and adult net production, respectively, are positive and equal 0 otherwise. The substitution results in:

$$\frac{dJ}{dt} = \nu_A^+(R)A - s_m g(R,s_m)c(t,s_m) + \nu_J^+(R) \int_{s_b}^{s_m} sc(t,s)ds$$

$$- (\mu_J + (\nu_J^+(R) - \nu_J(R))) \int_{s_b}^{s_m} sc(t,s)ds$$

$$= \nu_A^+(R)A - s_m g(R,s_m)c(t,s_m) + \nu_J^+(R)J - \mu_J J - (\nu_J^+(R) - \nu_J(R))J$$

$$= \nu_A^+(R)A - s_m g(R,s_m)c(t,s_m) + \nu_J(R)J - \mu_J J$$

The first term $\nu_J^+(R)J$ occurring in the second line of this equation represents juvenile biomass increase resulting from somatic growth, while the second one is derived from the expression representing starvation mortality. That these two terms cancel each other reveals one of the underlying assumptions of the size-structured biomass model, namely, that individual mortality is modeled in such a way that the decrease in biomass of a cohort of consumers exactly equals the balance between food assimilation and maintenance rate of that cohort when that balance is negative.

To arrive at a closed set of ODEs for R, J, and A, it is necessary to express the biomass maturation term $s_m g(R,s_m)c(t,s_m)$ in terms of R, J, and A. We approximate this term in a way that ensures the one-to-one correspondence between the equilibria occurring in the stage-structured biomass model and its size-structured counterpart. In the size-structured model, the equilibrium size distribution of juveniles at a particular size s, which we refer to as $\tilde{c}(s)$, is related to its value at s_b, following:

$$g(R,s)\tilde{c}(s) = g(R,s_b)\tilde{c}(s_b)\exp\left(-\int_{s_b}^{s}\frac{\mu_J}{g(R,\xi)}d\xi\right)$$

(e.g., de Roos 1997). Substitution of the expression $g(R,s) = v_J^+(R)s$ for the juvenile growth rate allows evaluation of the integral in this equation, leading to:

$$\tilde{c}(s) = \frac{g(R,s_b)\tilde{c}(s_b)}{v_J(R)s}\exp\left(-\int_{s_b}^{s}\frac{\mu_J}{v_J(R)\xi}d\xi\right) = \frac{g(R,s_b)\tilde{c}(s_b)}{v_J(R)s}\exp\left(-\frac{\mu_J}{v_J(R)}\ln\xi\Big|_{s_b}^{s}\right)$$

$$= \frac{g(R,s_b)\tilde{c}(s_b)}{v_J(R)s}\left(\frac{s}{s_b}\right)^{-\mu_J/v_J(R)}$$

Notice that in equilibrium $v_J^+(R)$ is necessarily positive and equal to $v_J(R)$. In equilibrium the maturation term $s_m g(R,s_m)c(t,s_m)$ can therefore be written as:

$$s_m g(R,s_m)c(t,s_m) = s_m^2 v_J(R)\tilde{c}(s_m) = g(R,s_b)\tilde{c}(s_b)s_b^{\mu_J/v_J(R)}s_m^{1-\mu_J/v_J(R)}$$

Using the equilibrium size distribution, the juvenile biomass at equilibrium can be expressed as:

$$J = \int_{s_b}^{s_m}\xi\tilde{c}(\xi)d\xi = \int_{s_b}^{s_m}\frac{g(R,s_b)\tilde{c}(s_b)}{v_J(R)}\left(\frac{\xi}{s_b}\right)^{-\mu_J/v_J(R)}d\xi = \frac{g(R,s_b)\tilde{c}(s_b)}{v_J(R)}s_b^{\mu_J/v_J(R)}$$

$$\int_{s_b}^{s_m}\xi^{-\mu_J/v_J(R)}d\xi$$

$$= \frac{g(R,s_b)\tilde{c}(s_b)}{v_J(R)-\mu_J}s_b^{\mu_J/v_J(R)}\left(s_m^{1-\mu_J/v_J(R)}-s_b^{1-\mu_J/v_J(R)}\right)$$

This allows us to relate in equilibrium the boundary condition of the PDE to the juvenile biomass J:

$$g(R,s_b)\tilde{c}(s_b) = \frac{J(v_J(R)-\mu_J)}{s_b^{\mu_J/v_J(R)}\left(s_m^{1-\mu_J/v_J(R)}-s_b^{1-\mu_J/v_J(R)}\right)}$$

Substitution of this equality in the expression for the maturation term leads to:

$$s_m g(R,s_m)c(t,s_m) = \frac{J(v_J(R)-\mu_J)}{\left(1-s_b^{1-\mu_J/v_J(R)}s_m^{-1+\mu_J/v_J(R)}\right)}$$

Let z now denote the ratio between size at birth and maturation, $z = s_b/s_m$, and let the maturation function $\gamma(\nu,\mu)$ be defined as:

$$\gamma(\nu,\mu) = \frac{\nu - \mu}{1 - z^{1-\mu/\nu}} \tag{A.10}$$

The ODEs for juvenile and adult biomass dynamics in the stage-structured biomass model can then be expressed in closed form as:

$$\frac{dJ}{dt} = \nu_A^+(R)A - \gamma(\nu_J^+(R),\mu_J)J + \nu_J(R)J - \mu_J J \tag{A.11}$$

$$\frac{dA}{dt} = \gamma(\nu_J^+(R),\mu_J)J + (\nu_A(R) - \nu_A^+(R))A - \mu_A A \tag{A.12}$$

It should be pointed out that the maturation term $\gamma(\nu_J^+(R),\mu_J)J$ is positive as long as $\nu_J^+(R)$ is positive and approaches 0 for $\nu_J^+(R) \downarrow 0$. Hence, it is restricted to non-negative values in the same way as $\nu_J^+(R)$ is. Also, for $\nu_J(R) \to \mu_J$, the maturation function $\gamma(\nu_J^+(R),\mu_J)$ has a regular limit equal to $-\nu_J(R)/\ln(z) = -\mu_J/\ln(z)$.

Along similar lines as shown above, it is possible to derive an approximation for the average size of the juvenile individuals, which we will refer to as $\tilde{s}_J$. In the size-structured model this average juvenile size is defined as:

$$\tilde{s}_J = \frac{\int_{s_b}^{s_m} \xi \tilde{c}(\xi)d\xi}{\int_{s_b}^{s_m} \tilde{c}(\xi)d\xi}$$

The integral in the denominator of this expression represents the total number of juvenile individuals in the population at equilibrium. Substitution of the expression for the equilibrium size distribution into both integrals leads to:

$$\tilde{s}_J = \frac{\int_{s_b}^{s_m} \frac{g(R,s_b)\tilde{c}(s_b)}{\nu_J(R)}\left(\frac{\xi}{s_b}\right)^{-\mu_J/\nu_J(R)}d\xi}{\int_{s_b}^{s_m} \frac{g(R,s_b)\tilde{c}(s_b)}{\nu_J(R)\xi}\left(\frac{\xi}{s_b}\right)^{-\mu_J/\nu_J(R)}d\xi} = \frac{\int_{s_b}^{s_m} \xi^{-\mu_J/\nu_J(R)}d\xi}{\int_{s_b}^{s_m} \xi^{-1-\mu_J/\nu_J(R)}d\xi}$$

$$= \frac{-\mu_J/\nu_J(R)}{1-\mu_J/\nu_J(R)}\frac{s_m^{1-\mu_J/\nu_J(R)} - s_b^{1-\mu_J/\nu_J(R)}}{s_m^{-\mu_J/\nu_J(R)} - s_b^{-\mu_J/\nu_J(R)}}$$

Let z_J now indicate the ratio between the average juvenile size and the maturation or adult size, that is, $z_J = \tilde{s}_J/s_m$. The relationship above can be used to express z_J as a function of the ratio z between size at birth and at maturation, the juvenile net-production rate $\nu_J(R)$, and mortality rate μ_J:

$$z_J = \frac{\mu_J/v_J(R)}{\mu_J/v_J(R)-1} \frac{1-z^{1-\mu_J/v_J(R)}}{1-z^{-\mu_J/v_J(R)}}$$

This ratio has a regular limit for $\mu_J/v_J(R) \to 0$:

$$z_0 = \lim_{\mu_J/v_J(R) \to 0} \frac{\mu_J/v_J(R)}{\mu_J/v_J(R)-1} \frac{1-z^{1-\mu_J/v_J(R)}}{1-z^{-\mu_J/v_J(R)}} = \frac{1-z}{-\ln z}$$

and is a monotonically decreasing function of $\mu_J/v_J(R)$ as shown in figure A.1 for $z = 0.5$, 0.2, and 0.01. The minimum value of the juvenile-adult size z_J of course equals the ratio between the size at birth and at maturation z, which value it approaches when $\mu_J/v_J(R)$ becomes very large:

$$z_\infty = \lim_{\mu_J/v_J(R) \to \infty} \frac{\mu_J/v_J(R)}{\mu_J/v_J(R)-1} \frac{1-z^{1-\mu_J/v_J(R)}}{1-z^{-\mu_J/v_J(R)}} = z$$

Figure A.1 also shows that the expression for the juvenile-adult size ratio z_J yields values between 0 and 1 for negative values of the ratio between juvenile mortality and growth rate $\mu_J/v_J(R)$. However, given the assumption that growth and reproduction stop under nongrowing conditions, it is more realistic to ignore these values. Otherwise, with decreasing values of the growth rate $v_J(R)$ (and hence increasing values of $\mu_J/v_J(R)$) the value of z_J would decrease and approach z and then suddenly take on a much larger value whenever $v_J(R)$ turns negative. Summarizing, the ratio between average juvenile and adult body size is given by the equation:

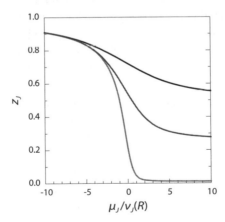

FIGURE A.1. The ratio of average juvenile over adult mass z_J as a function of the ratio between juvenile mortality and growth rate $\mu_J/v_J(R)$ for three different values of the ratio between newborn and adult mass: from top to bottom, $z = s_b/s_m = 0.5$, 0.25, and 0.01.

$$z_J = \begin{cases} \dfrac{\mu_J/\nu_J(R)}{\mu_J/\nu_J(R)-1} \; \dfrac{1-z^{1-\mu_J/\nu_J(R)}}{1-z^{-\mu_J/\nu_J(R)}} & \text{if } \nu_j(R) > 0 \\[3ex] z & \text{otherwise} \end{cases} \qquad (A.13)$$

4 EQUILIBRIUM COMPUTATIONS FOR PHYSIOLOGICALLY STRUCTURED MODELS

Unstructured models of population dynamics are generally formulated in terms of a finite number of variables $x_1, \ldots, x_n$, representing the total number of individuals or total biomass making up each population. If dynamics are described in continuous time, they are determined by a coupled of system of ordinary differential equations of the form:

$$\frac{dx_i}{dt} = h_i(x_1, \ldots, x_n) \qquad i = 1, \ldots, n$$

In this set of ODEs, the functions $h_i(x_1, \ldots, x_n)$ describe how dynamics of every state variable x_i depends on itself and all other state variables. Equilibrium states of the system can be computed by solving for those values $\tilde{x}_i$ that satisfy:

$$h_i(\tilde{x}_1, \ldots, \tilde{x}_n) = 0 \qquad i = 1, \ldots, n$$

Most ecologists prefer analytical solutions for the equilibrium values $\tilde{x}_i$ over solutions that are computed numerically and even question how much insight the latter yield, given that they are only valid for a particular parameter combination. The mathematical study of changes in the qualitative behavior of dynamical systems in terms of ODEs with changes in parameters is known as bifurcation theory (see, for example, Kuznetsov 1995). Bifurcation theory shows that small changes in parameters will in general not lead to a qualitative change in model dynamics except at a limited set of parameter combinations. These threshold values of parameters, at which a qualitative change in model dynamics occurs with a small parameter change, are known as bifurcation points. The qualitative changes in dynamics that can occur at bifurcation points show considerable regularity, as bifurcation theory reveals. Together, these two facts—that most parameter changes do not qualitatively change dynamics and the regularity in changes that can occur—challenge, in our opinion, the reservation about numerically computed equilibrium solutions.

Many of our results are therefore presented in so-called bifurcation graphs, as explained in box 3.5, which reveal the qualitative changes in model dynamics with changes in parameters. For models that are formulated in terms of

ODEs, such as the stage-structured biomass model derived in section 3 above, we used MATCONT, a graphical MATLAB software package (Dhooge, Govaerts, and Yu 2003), to construct these bifurcation graphs. Basically, this software package solves numerically the system of n equations $h_i(\tilde{x}_1, \ldots, \tilde{x}_n) = 0$ for the equilibrium values $\tilde{x}_i$ and subsequently computes the curve of equilibrium points as a function of a particular model parameter, a process also referred to as "curve continuation." In addition, the program incorporates methods to assess whether the computed equilibrium is stable or not. Physiologically structured population models, such as the basic size-structured model shown in box A.1, require more complicated approaches for equilibrium computations. Efficient numerical techniques have only been developed in the past fifteen years (Diekmann, Gyllenberg, and Metz 2003; Kirkilionis, Diekmann, et al. 1997, 2001), including extensions that allow continuation of stability boundaries as a function of two model parameters separating parameter combinations for which a particular equilibrium is stable from those combinations for which the equilibrium is unstable and population cycles occur (de Roos, Diekmann, et al. 2010; Kirkilionis, Diekmann, et al. 1997, 2001). In this section we present some of these methods, which we have used to compute the equilibrium of size-structured population models.

If an explicit expression can be derived for the size-age relationship $s(R, a)$ in equilibrium, possibly depending on resource density, analytical manipulations may yield an equation determining the equilibrium resource density in a general, size-structured population model, such as presented in box A.1 (de Roos 1997). The resulting equation is, however, implicit; does not give much insight into the dependence of the equilibrium resource density on model parameters; and in the end has to be solved numerically. In contrast, the methods we have used to calculate the equilibrium of size-structured models constitute a completely numerical approach, which does not involve analytical manipulations (Diekmann, Gyllenberg, and Metz 2003; Kirkilionis, Diekmann, et al. 1997; see also Kirkilionis, Diekmann, et al. 2001). In a size-structured consumer-resource model such as presented in box A.1, the equilibrium always satisfies the following two conditions: First, the total number of offspring that a consumer individual is expected to produce during its entire life equals 1, as on average every consumer should exactly replace itself in life. Second, the total consumption rate of resource by consumers equals the rate at which the resource is produced. Let $s(\tilde{R}, a)$ denote the relationship between body size s and age a (which we will also refer to as the "size-age relationship") at the equilibrium resource density $\tilde{R}$, and let $H(\tilde{R}, a)$ denote the probability that a consumer individual is still alive at age a. This survival probability may depend on the equilibrium resource density $\tilde{R}$. The first condition can then be expressed formally as:

$$\int_0^\infty b(\tilde{R}, s(\tilde{R}, a)) H(\tilde{R}, a) da = 1 \qquad (A.14)$$

The product $b(\tilde{R}, s(\tilde{R}, a)) H(\tilde{R}, a)$ represents the expected reproduction rate of an individual at age a, taking into account the chance that it may have died before reaching that age. Integrating that expected reproduction over the entire lifetime yields what is known as the expected lifetime reproduction of an individual, which in ecology and epidemiology is commonly indicated with the symbol R_0. Furthermore, if the total population birth rate in equilibrium is indicated with $\tilde{B}$, the second equilibrium condition can be expressed formally as:

$$\tilde{B} \int_0^\infty I(\tilde{R}, s(\tilde{R}, a)) H(\tilde{R}, a) da = G(\tilde{R}) \qquad (A.15)$$

In this equation $\tilde{B}\tilde{H}(R, a)$ represents the density of consumers with age a in equilibrium, and the product $\tilde{B}I(\tilde{R}, s(\tilde{R}, a))\tilde{H}(R, a)$ thus represents their resource intake rate. Integrating that intake rate over all ages yields the intake rate of the total consumer population.

The central idea of the equilibrium computation approach presented here is that the integrals in the conditions (A.14) and (A.15) can be computed from the individual intake rate $I(R, s)$, growth rate $g(R, s)$, reproduction rate $b(R, s)$, and mortality rate $d(R, s)$ by solving numerically a coupled set of ODEs. Given a (necessarily constant) equilibrium resource density $\tilde{R}$, the size-age relationship $s(\tilde{R}, a)$ and the survival probability $H(\tilde{R}, a)$ are the solutions of the system of ODEs:

$$\frac{ds}{da} = g(\tilde{R}, s) \qquad\qquad s(\tilde{R}, 0) = s_b$$

$$\frac{dH}{da} = -d(\tilde{R}, s) H \qquad H(\tilde{R}, 0) = 1$$

Now define two age-dependent functions $\Theta(\tilde{R}, a)$ and $\Psi(\tilde{R}, a)$, representing the expected cumulative reproduction and ingestion, respectively, up to age a. $\Theta(\tilde{R}, a)$ and $\Psi(\tilde{R}, a)$ are defined as:

$$\Theta(\tilde{R}, a) = \int_0^a b(\tilde{R}, s(\tilde{R}, \xi)) H(\tilde{R}, \xi) d\xi$$

$$\Psi(\tilde{R}, a) = \int_0^a I(\tilde{R}, s(\tilde{R}, \xi)) H(\tilde{R}, \xi) d\xi$$

Differentiation of the two sides of these expressions with respect to a shows that $\Theta(\tilde{R}, a)$ and $\Psi(\tilde{R}, a)$ are the solutions of the following system of ODEs:

$$\frac{d\Theta}{da} = b(\tilde{R}, s(\tilde{R}, a)) H(\tilde{R}, a) \qquad \Theta(\tilde{R}, 0) = 0$$

$$\frac{d\Psi}{da} = I(\tilde{R}, s(\tilde{R}, a)) H(\tilde{R}, a) \qquad \Psi(\tilde{R}, 0) = 0$$

Obviously, the two equilibrium conditions (A.14) and (A.15) can be represented in terms of $\Theta(\tilde{R}, a)$ and $\Psi(\tilde{R}, a)$ as:

$$\begin{cases} \Theta(\tilde{R}, \infty) = 1 \\ \tilde{B}\Psi(\tilde{R}, \infty) = G(\tilde{R}) \end{cases} \tag{A.16}$$

This coupled set of equations can be solved by means of standard numerical techniques (i.e., Newton's method) to locate the roots of systems of equations. The only peculiar characteristic of (A.16) is that the left-hand side of these equations is not specified by algebraic expressions, but has to be computed by numerical integration of the following set of ODEs from zero to infinite age:

$$\begin{cases} \dfrac{ds}{da} = g(\tilde{R}, s) & s(\tilde{R}, 0) = s_b \\[2mm] \dfrac{dH}{da} = -d(\tilde{R}, s)H & H(\tilde{R}, 0) = 1 \\[2mm] \dfrac{d\Theta}{da} = b(\tilde{R}, s)H & \Theta(\tilde{R}, 0) = 0 \\[2mm] \dfrac{d\Psi}{da} = I(\tilde{R}, s)H & \Psi(\tilde{R}, 0) = 0 \end{cases} \tag{A.17}$$

Numerical integration of this system of ODEs to an infinite age is of course impossible. In practice, the integration is therefore carried out until the survival probability $H(\tilde{R}, a)$ has reached some minimum value, beyond which further contributions to $\Theta(\tilde{R}, a)$ and $\Psi(\tilde{R}, a)$ can be neglected. In all our computations we have assumed a threshold value of $1.0 \cdot 10^{-9}$ and hence stopped the integration as soon as the survival probability $H(\tilde{R}, a)$ dropped below this limit. We will nonetheless indicate the values that $\Theta(\tilde{R}, a)$ and $\Psi(\tilde{R}, a)$ attain when $H(\tilde{R}, a)$ reaches the threshold value $1.0 \cdot 10^{-9}$ as $\Theta(\tilde{R}, \infty)$ and $\Psi(\tilde{R}, \infty)$, respectively. Alternatively, if the model by assumption already accounts for a maximum life span A_{max} of individual consumers, the integration needs to be continued only until this age A_{max}.

The equilibrium conditions (A.16) show that the first equation does not depend on $\tilde{B}$ and that the value of $\tilde{B}$ can be computed from the second equation as $G(\tilde{R})/\Psi(\tilde{R}, \infty)$ once the equilibrium resource density is known. Therefore, the equilibrium of a size-structured consumer-resource model can be calculated in an iterative manner by first integrating numerically the system of ODEs

(A.17) using an initial estimate of the equilibrium resource density $\tilde{R}^{(0)}$ to obtain $\Theta(\tilde{R}^{(0)}, \infty)$ and $\Psi(\tilde{R}^{(0)}, \infty)$. With the latter values, the conditions (A.16) are evaluated and used to adjust $\tilde{R}^{(0)}$ following Newton's method, yielding a new estimate $\tilde{R}^{(1)}$. With this new estimate, the cycle is repeated until the value of the estimate $\tilde{R}^{(i)}$ converges and the conditions (A.16) are satisfied. Afterward, the equilibrium value of the total reproduction rate $\tilde{B}$ is calculated from the second equation of (A.16).

The computational approach described above for consumer-resource models is easily extended to account for an unstructured predator population foraging on the size-structured consumer (cf. chapter 4). The mortality rate and hence the survival probability of the consumer then depend on the density of predators $\tilde{P}$ and we therefore denote these two functions as $d(\tilde{R}, \tilde{P}, a)$ and $H(\tilde{R}, \tilde{P}, a)$. We also write $\Theta(\tilde{R}, \tilde{P}, a)$ and $\Psi(\tilde{R}, \tilde{P}, a)$, as these quantities will depend on $\tilde{P}$ through their dependence on $H(\tilde{R}, \tilde{P}, a)$. Now, an additional condition has to be satisfied in equilibrium, which stipulates that the encounter rate of predators with suitably sized consumers is sufficient to exactly balance the predator's loss rate through maintenance and mortality with its biomass ingestion through foraging. In other words, if the function $v(s)$ represents the vulnerability of consumers with size s to predation, and the constant C_p represents the food level that predators require to exactly balance their loss rates, the additional equilibrium condition can be expressed as:

$$\tilde{B} \int_0^\infty v(s(\tilde{R}, a)) s(\tilde{R}, a) H(\tilde{R}, \tilde{P}, a) da = C_p$$

In this integral expression, the product $\tilde{B} s(\tilde{R}, a) H(\tilde{R}, \tilde{P}, a)$ represents the biomass density of consumers with body size s. Weighing this biomass density with the vulnerability $v(s)$ and integrating the product over the entire lifetime of consumers yields the density of food available for predators to forage on. This additional equilibrium condition can be dealt with in the same manner as we deal with the conditions (A.14) and (A.15). Hence, define the quantity $\Omega(\tilde{R}, \tilde{P}, a)$ as:

$$\Omega(\tilde{R}, \tilde{P}, a) = \int_0^a v(s(\tilde{R}, \xi)) s(\tilde{R}, \xi) H(\tilde{R}, \tilde{P}, \xi) d\xi$$

Differentiation with respect to a shows that $\Omega(\tilde{R}, \tilde{P}, a)$ is the solution of the ODE:

$$\frac{d\Omega}{da} = v(s(\tilde{R}, a)) s(\tilde{R}, a) H(\tilde{R}, \tilde{P}, a) \qquad \Omega(\tilde{R}, \tilde{P}, 0) = 0 \qquad \text{(A.18)}$$

The system of ODEs (A.17) now has to be extended with the ODE (A.18) and integrated as before from zero to infinite age to obtain $\Theta(\tilde{R}, \tilde{P}, \infty)$, $\Psi(\tilde{R}, \tilde{P},$

∞), and $\Omega(\tilde{R}, \tilde{P}, \infty)$. Using these quantities, the conditions that $\tilde{B}$, $\tilde{R}$, and $\tilde{P}$ have to satisfy in equilibrium can be evaluated:

$$\begin{cases} \Theta(\tilde{R}, \tilde{P}, \infty) = 1 \\ \tilde{B}\Psi(\tilde{R}, \tilde{P}, \infty) = G(\tilde{R}) \\ \tilde{B}\Omega(\tilde{R}, \tilde{P}, \infty) = C_P \end{cases} \qquad (A.19)$$

Because the above equations depend only linearly on $\tilde{B}$, $\tilde{B}$ can be eliminated, reducing the system to only two equations:

$$\begin{cases} \Theta(\tilde{R}, \tilde{P}, \infty) = 1 \\ G(\tilde{R}) \dfrac{\Omega(\tilde{R}, \tilde{P}, \infty)}{\Psi(\tilde{R}, \tilde{P}, \infty)} = C_P \end{cases} \qquad (A.20)$$

Starting from initial estimates of the equilibrium resource density $\tilde{R}^{(0)}$ and the equilibrium predator density $\tilde{P}^{(0)}$, the ODEs (A.17) and (A.18) are integrated numerically to obtain $\Theta(\tilde{R}^{(0)}, \tilde{P}^{(0)}, \infty)$, $\Psi(\tilde{R}^{(0)}, \tilde{P}^{(0)}, \infty)$, and $\Omega(\tilde{R}^{(0)}, \tilde{P}^{(0)}, \infty)$. Subsequently, the conditions (A.20) are evaluated and used to adjust $\tilde{R}^{(0)}$ and $\tilde{P}^{(0)}$, yielding new estimates $\tilde{R}^{(1)}$ and $\tilde{P}^{(1)}$. With these new estimates, the cycle of integration, evaluation, and adjustment is repeated until the values of $\tilde{R}^{(i)}$ and $\tilde{P}^{(i)}$ converge and the conditions (A.20) are satisfied. Afterward, the equilibrium value of the total reproduction rate $\tilde{B}$ is calculated from the second or third equation of (A.19).

The computation of an equilibrium is significantly more complicated in a model with both a size-structured consumer and a size-structured predator population, such as is analyzed in the last part of chapter 4. We then have to integrate a system of ODEs analogous to (A.17) for both consumers and predators to compute their expected lifetime reproduction and ingestion. In addition, however, the predation mortality that consumers are exposed to will generally not depend on a single-valued measure of predator biomass or density, but is likely to depend on the entire predator size distribution $p(t, s)$. More specifically, if the predation rate depends on both prey and predator body size, and this dependence *cannot* be written as a product of two functions, one depending only on prey body size, and the other only on predator body size, then changes in the predator size distribution will translate into disproportionate changes in predation mortality experienced by consumers of different body sizes.

To integrate the system of ODEs for the consumer, we then need next to the equilibrium resource density $\tilde{R}$ an entire, size-dependent function $\mu_P(s)$ as input, which represents the predation mortality that a consumer of size s is exposed to and which varies with changes in the predator size distribution $p(t, s)$. To integrate the system of ODEs for the predator, we furthermore need an entire size-dependent function $C_P(s)$ as input, representing the density of prey available for a predator of body size s to forage on. Instead of a problem with a finite number of input variables, we now face a problem with infinite-dimensional input, that is, two size-dependent functions.

The approach to compute the equilibrium of a physiologically structured model in these cases with an infinite-dimensional input is explained in detail in Claessen and de Roos (2003). Here we only present a summary. As before, the computations are carried out in an iterative manner, in which the input variables are adjusted until convergence and the equilibrium conditions are satisfied. The input to each iteration comprises the equilibrium resource density $\tilde{R}$; the total population birth rate of consumer and predator, $\tilde{B}_C$ and $\tilde{B}_P$, respectively; and a discretized representation of the *relative* predation mortality experienced by consumers at size s. More specifically, this discretization consists of relative mortality rates $\tilde{\mu}_i$ at a range of equally spaced (consumer) body sizes s_i ($i=1,\ldots,N_C$, $s_1=s_b$; as default we used a spacing of 2 mm) that cover the entire range of body sizes consumers can attain. When multiplied with the predator population birth rate $\tilde{B}_P$, these quantities yield the values of the predation mortality function $\mu_P(s_i)$ at the body sizes s_i. We compute the predation mortality experienced at body sizes in between the nodal values s_i by linear interpolation.

With the input variables we integrate a system of ODEs like the one presented in equation (A.17) to evaluate the expected lifetime reproductive output $\Theta_C(\tilde{R},\tilde{B}_P,\tilde{\mu}_i,\infty)$ and the expected lifetime resource ingestion $\Psi_C(\tilde{R},\tilde{B}_P,\tilde{\mu}_i,\infty)$ of the consumer. These two quantities depend on the input equilibrium resource density $\tilde{R}$, as this influences the size-age relation, fecundity, and ingestion of individual consumers. In addition, the two quantities depend on $\tilde{B}_P$ and all $\tilde{\mu}_i$ through their relation with the consumer survival probability $H_C(\tilde{R},\tilde{B}_P,\tilde{\mu}_i,a)$, which itself depends on the input variables $\tilde{B}_P$ and $\tilde{\mu}_i$, as they determine the predation mortality experienced by consumers. Together with the system of equations (A.17) for the size-age relation, survival probability, cumulative reproduction, and cumulative ingestion up to age a, we integrate two additional ODEs:

$$\begin{cases} \dfrac{d\Lambda_C}{da} = H_C & \Lambda_C(\tilde{R},\tilde{B}_P,\tilde{\mu}_i,0) = 0 \\[2mm] \dfrac{d\Xi_C}{da} = sH_C & \Xi_C(\tilde{R},\tilde{B}_P,\tilde{\mu}_i,0) = 0 \end{cases} \tag{A.21}$$

When multiplied with the total consumer population birth rate $\tilde{B}_C$, $\Lambda_C(\tilde{R},\tilde{B}_P,\tilde{\mu}_i,a)$ and $\Xi_C(\tilde{R},\tilde{B}_P,\tilde{\mu}_i,a)$ represent the total number and total biomass, respectively, of all consumers up to age a, while $\Xi_C(\tilde{R},\tilde{B}_P,\tilde{\mu}_i,a)/\Lambda_C(\tilde{R},\tilde{B}_P,\tilde{\mu}_i,a)$

represents their average body size. The integration of the system of ODEs (A.17) and the above system (A.21) for the consumer is carried out in a step-wise manner, stopping and restarting the integration every time the size-at-age value $s(\tilde{R}, a)$ exactly reaches one of the nodal size thresholds s_i. We achieved this stopping and restarting the integration whenever one of the variables crosses a threshold by using an ODE integration technique with event localization (DOPRI5 method; see Hairer, Norsett, and Wanner 1993).

Let a_i denote the age at which the size-at-age function $s(\tilde{R}, a)$ exactly reaches $s_i (a_1 = 0)$. At each subsequent nodal value we then compute the number of consumers with a size between s_{i-1} and s_i and the average size of these consumers as:

$$\begin{cases} \bar{n}_{i-1} = \Lambda_C(\tilde{R}, \tilde{B}_P, \tilde{\mu}_i, a_i) - \Lambda_C(\tilde{R}, \tilde{B}_P, \tilde{\mu}_i, a_{i-1}) \\ \bar{s}_{i-1} = \dfrac{\Xi_C(\tilde{R}, \tilde{B}_P, \tilde{\mu}_i, a_i) - \Xi_C(\tilde{R}, \tilde{B}_P, \tilde{\mu}_i, a_{i-1})}{\bar{n}_{i-1}} \end{cases} \quad i = 2, \dots, N_C$$

After integrating the ODEs that describe the life history of the consumer in equilibrium, we have therefore computed the expected lifetime reproduction of consumers, $\Theta_C(\tilde{R}, \tilde{B}_P, \tilde{\mu}_i, \infty)$; their expected lifetime ingestion of resources, $\Psi_C(\tilde{R}, \tilde{B}_P, \tilde{\mu}_i, \infty)$; and the vectors of $\bar{n}_i$ and $\bar{s}_i$ values, representing the relative density of consumers in the different size intervals and their average body size.

Next, we perform an integration of a system of ODEs for the predator life history. In this case, however, we need a function $C_P(y)$ as input, representing the density of prey available for a predator of body size y to forage on (to avoid confusion with consumer body size, we indicate predator body size in the following with y). Let the function $v(s, y)$ denote the vulnerability of prey with body size s to predators with body size y. We now use the vector of consumer densities $\bar{n}_i$ and their average sizes $\bar{s}_i$ to approximate $C_P(y)$:

$$C_P(y) = \tilde{B}_C \int_0^\infty v(s(\tilde{R}, a), y) s(\tilde{R}, a) H_C(\tilde{R}, \tilde{B}_P, \tilde{\mu}_i, a) da \approx \tilde{B}_C \sum_{i=1}^{N_C-1} v(\bar{s}_i, y) \bar{s}_i \bar{n}_i$$

This particular method to approximate weighted integrals over a population size distribution by finite sums of densities and average body sizes in small subintervals is also a key element of the numerical method to integrate physiologically structured population models (de Roos 1988; de Roos, Diekmann, and Metz 1992). The integration of the ODEs, describing the life history of the consumer in equilibrium, has thus delivered the necessary input to carry out the integration of the ODEs describing predator life history, which can be expressed as:

$$\begin{cases} \dfrac{dy}{da} = g_P(C_P(y), y) & y(C_P(\cdot), 0) = y_b \\[2ex] \dfrac{dH_P}{da} = -d_P(C_P(y), y)H_P & H_P(C_P(\cdot), 0) = 1 \\[2ex] \dfrac{d\Theta_P}{da} = b_P(C_P(y), y)H_P & \Theta_P(C_P(\cdot), 0) = 0 \qquad \text{(A.22)} \\[2ex] \dfrac{d\Lambda_P}{da} = H_P & \Lambda_P(C_P(\cdot), 0) = 0 \\[2ex] \dfrac{d\Xi_P}{da} = yH_P & \Xi_P(C_P(\cdot), 0) = 0 \end{cases}$$

The above ODEs contain the somatic growth rate of predators $g_P(C, y)$, their mortality rate $d_P(C, y)$, and their fecundity $b_P(C, y)$ as key ingredients, which all depend on the prey density C available for a predator of body size y to forage on. The ODEs yield as solutions the size-age relation $y(C_P(\cdot), a)$ of the predator, its survival probability $H(C_P(\cdot), a)$ and cumulative reproduction $\Theta_P(C_P(\cdot), a)$ up to age a, and the relative total density and total biomass of predators younger than a, $\Lambda_P(C_P(\cdot), a)$ and $\Xi_P(C_P(\cdot), a)$, respectively (here, "relative" means that multiplication of the latter two quantities with the predator population birth rate $\tilde{B}_P$ yields the absolute values of total density and biomass). All these quantities depend on the food availability that a predator has encountered in its entire life up to size y, which we indicate with the dependence on the function $C_P(\cdot)$. Moreover, this makes the predator life history variables indirectly dependent on all input variables $\tilde{R}$, $\tilde{B}_C$, $\tilde{B}_P$, and $\bar{\mu}_i$ (cf. the expression for $C_P(y)$ given above). For the predator we do not have to compute its expected lifetime ingestion, because the consumer equilibrium is determined by a condition that does not involve this total ingestion by predators, unlike the resource equilibrium density.

The integration of the ODEs (A.22) is carried out in a stepwise manner, as described above for the integration of the consumer ODEs, stopping and restarting every time the size-age relation reaches a threshold value y_i ($i = 1, \ldots, N_P$, $y_1 = y_b$; all y_i together spanning the range of body sizes predators can attain). At each subsequent threshold value y_i, the relative density of predators in the size range y_{i-1} to y_i and their average body size is computed as:

$$\begin{cases} \bar{m}_{i-1} = \Lambda_P(C_P(\cdot), a_i) - \Lambda_P(C_P(\cdot), a_{i-1}) \\[2ex] \bar{y}_{i-1} = \dfrac{\Xi_P(C_P(\cdot), a_i) - \Xi_P(C_P(\cdot), a_{i-1})}{\bar{n}_{i-1}} & i = 2, \ldots, N_P \end{cases}$$

The integration of the ODEs describing the predator life history thus yields the expected lifetime reproduction of a predator in equilibrium $\Lambda_P(C_P(\cdot), \infty)$

and the vectors of $\bar{m}_i$ and $\bar{y}_i$, representing the relative density of predators in the different size intervals and their average body size. Using these latter two vectors, we can approximate the predation mortality that the entire predator population in equilibrium imposes on consumers of body size s_i:

$$\mu_P(s_i) = \tilde{B}_P \int_0^\infty I_P(C_P(y), s_i, y(C_P(\cdot), a)) H_P(C_P(\cdot), a) da \approx \tilde{B}_P \sum_{i=1}^{N_P-1} I_P(C_P(\bar{y}_i), s_i, \bar{y}_i) \bar{m}_i$$

In this equation we have used the function $I_P(C_P(y), s, y)$ to represent the rate, with which a predator with body size y feeds on a consumer with body size s. In cases where the predator follows a nonlinear functional response, this feeding rate may depend on the total amount of food available to the predator to forage on, hence that $C_P(y)$ is included as an argument of $I_P(C_P(y), s, y)$.

The equilibrium of the structured model now has to satisfy the conditions that (1) the expected lifetime reproduction of the consumer and the predator equal 1, and (2) that the total production of resources equals the foraging by consumers. In other words:

$$\begin{cases} \Theta_C(\tilde{R}, \tilde{B}_P, \tilde{\mu}_i, \infty) = 1 \\ \Theta_P(C_P(\cdot), \infty) = 1 \\ \tilde{B}_C \Psi_C(\tilde{R}, \tilde{B}_P, \tilde{\mu}_i, \infty) = G(\tilde{R}) \end{cases} \quad \text{(A.23)}$$

In addition, the relative predation mortality rates $\tilde{\mu}_i$ that serve as input to the integration have to equal the imposed predation mortality rates that are computed after integrating the predator life history ODEs, that is:

$$\sum_{i=1}^{N_P-1} I_P(C_P(\bar{y}_i), s_i, \bar{y}_i) \bar{m}_i = \tilde{\mu}_i \qquad i = 1, \ldots, N_C \quad \text{(A.24)}$$

Because $\tilde{\mu}_i$ are relative measures of predation mortality, still to be multiplied by the equilibrium predator population birth rate $\tilde{B}_P$ to obtain the actual value of predation mortality, $\tilde{B}_P$ does not show up in the left-hand side of these equations. Together the equations (A.23) and (A.24) represent $N_C + 3$ conditions with as many unknown input variables. As outlined above, in every iteration cycle the left-hand sides of (A.23) and (A.24) are computed by integrating the systems of ODEs (A17) and (A.21) describing the life history of the consumer together with the system of ODEs (A.22) for the predator life history using initial estimates of the input variables $\tilde{R}$, $\tilde{B}_C$, $\tilde{B}_P$, and $\tilde{\mu}_i$. Subsequently, the equations (A.23) and (A.24) are evaluated and used to adjust the estimates until the input variables' convergence and all of the conditions (A.23) and (A.24) are satisfied.

5   COMPUTING PARAMETER BOUNDS TO OVERCOMPENSATION IN THE STAGE-STRUCTURED BIOENERGETICS MODEL

The equilibrium of the stage-structured bioenergetics model is determined by the following three equations:

$$0 = G(\tilde{R}) - \omega_J(\tilde{R})\tilde{J} - \omega_A(\tilde{R})\tilde{A}$$
$$0 = \nu_A(\tilde{R})\tilde{A} + \nu_J(\tilde{R})\tilde{J} - \gamma(\nu_J(\tilde{R}),\mu_J)\tilde{J} - \mu_J\tilde{J}$$
$$0 = \gamma(\nu_J(\tilde{R}),\mu_J)\tilde{J} - \mu_A\tilde{A}$$

To simplify the discussion, we introduce here three functions $F_1(\tilde{R},\tilde{J},\tilde{A})$, $F_2(\tilde{R},\tilde{J},\tilde{A})$, and $F_3(\tilde{R},\tilde{J},\tilde{A})$ to denote the right-hand sides of these equations as a function of the equilibrium biomass density of resource $\tilde{R}$, juvenile $\tilde{J}$, and adult consumers $\tilde{A}$. Obviously, the equilibrium densities $\tilde{R}$, $\tilde{J}$, and $\tilde{A}$, as well as the functions F_1, F_2, and F_3, are dependent on parameter values. We can represent this dependence by writing the equilibrium densities as explicit functions of a particular parameter of interest and by also including this parameter among the independent variables of the functions F_1, F_2, and F_3. For ease of presentation, we will restrict ourselves here to the dependence of the equilibrium on the juvenile mortality rate μ_J, but the following discussion is the same for any other parameter in the model.

We thus have equilibrium densities $\tilde{R}(\mu_J)$, $\tilde{J}(\mu_J)$, and $\tilde{A}(\mu_J)$, as explicit functions of the juvenile mortality rate μ_J, which are determined by the three nonlinear functions $F_1(\tilde{R},\tilde{J},\tilde{A},\mu_J)$, $F_2(\tilde{R},\tilde{J},\tilde{A},\mu_J)$, and $F_3(\tilde{R},\tilde{J},\tilde{A},\mu_J)$, following:

$$\begin{pmatrix} F_1(\tilde{R},\tilde{J},\tilde{A},\mu_J) \\ F_2(\tilde{R},\tilde{J},\tilde{A},\mu_J) \\ F_3(\tilde{R},\tilde{J},\tilde{A},\mu_J) \end{pmatrix} = 0 \qquad (A.25)$$

Here we have used vector notation to more succinctly express the equilibrium condition for the stage-structured bioenergetics model. How the equilibrium densities $\tilde{R}(\mu_J)$, $\tilde{J}(\mu_J)$, and $\tilde{A}(\mu_J)$ change with a change in μ_J can be computed by formally differentiating both sides of equation (A.25) with respect to μ_J. Here we do have to take into account that the functions $F_1(\tilde{R},\tilde{J},\tilde{A},\mu_J)$, $F_2(\tilde{R},\tilde{J},\tilde{A},\mu_J)$, and $F_3(\tilde{R},\tilde{J},\tilde{A},\mu_J)$ may depend not only directly but also indirectly on the parameter of interest, because F_1, F_2, and F_3 are functions of the equilibrium densities $\tilde{R}$, $\tilde{J}$, and $\tilde{A}$, which are themselves dependent on the parameter of interest. The formal differentiation results in:

$$
\begin{pmatrix}
\dfrac{\partial F_1}{\partial R} & \dfrac{\partial F_1}{\partial J} & \dfrac{\partial F_1}{\partial A} \\[2ex]
\dfrac{\partial F_2}{\partial R} & \dfrac{\partial F_2}{\partial J} & \dfrac{\partial F_2}{\partial A} \\[2ex]
\dfrac{\partial F_3}{\partial R} & \dfrac{\partial F_3}{\partial J} & \dfrac{\partial F_3}{\partial A}
\end{pmatrix}
\begin{pmatrix}
\dfrac{\partial \tilde{R}}{\partial \mu_J} \\[2ex]
\dfrac{\partial \tilde{J}}{\partial \mu_J} \\[2ex]
\dfrac{\partial \tilde{A}}{\partial \mu_J}
\end{pmatrix}
+
\begin{pmatrix}
\dfrac{\partial F_1}{\partial \mu_J} \\[2ex]
\dfrac{\partial F_2}{\partial \mu_J} \\[2ex]
\dfrac{\partial F_3}{\partial \mu_J}
\end{pmatrix}
= 0
\qquad (A.26)
$$

Note that all differentials occurring in this and all following equations are evaluated at the equilibrium values $\tilde{R}(\mu_J)$, $\tilde{J}(\mu_J)$, and $\tilde{A}(\mu_J)$. Also note that for brevity, we suppress explicitly representing the function arguments of $\tilde{R}$, $\tilde{J}$, and $\tilde{A}$ (i.e., their dependence on μ_J), as well as the arguments of the functions F_1, F_2, and F_3 (i.e., their dependence on $\tilde{R}$, $\tilde{J}$, $\tilde{A}$, and μ_J) whenever possible. Applying the implicit function theorem to the equation above yields the following expression for the changes in $\tilde{R}$, $\tilde{J}$, and $\tilde{A}$ with changes in μ_J:

$$
\begin{pmatrix}
\dfrac{\partial \tilde{R}}{\partial \mu_J} \\[2ex]
\dfrac{\partial \tilde{J}}{\partial \mu_J} \\[2ex]
\dfrac{\partial \tilde{A}}{\partial \mu_J}
\end{pmatrix}
= -
\begin{pmatrix}
\dfrac{\partial F_1}{\partial R} & \dfrac{\partial F_1}{\partial J} & \dfrac{\partial F_1}{\partial A} \\[2ex]
\dfrac{\partial F_2}{\partial R} & \dfrac{\partial F_2}{\partial J} & \dfrac{\partial F_2}{\partial A} \\[2ex]
\dfrac{\partial F_3}{\partial R} & \dfrac{\partial F_3}{\partial J} & \dfrac{\partial F_3}{\partial A}
\end{pmatrix}^{-1}
\begin{pmatrix}
\dfrac{\partial F_1}{\partial \mu_J} \\[2ex]
\dfrac{\partial F_2}{\partial \mu_J} \\[2ex]
\dfrac{\partial F_3}{\partial \mu_J}
\end{pmatrix}
\qquad (A.27)
$$

Geometrically, this equation specifies the direction of the equilibrium curve in the phase space spanned by $\tilde{R}$, $\tilde{J}$, and $\tilde{A}$ as a function of μ_J. In practice we computed the values of $\partial \tilde{R}/\partial \mu_J$, $\partial \tilde{J}/\partial \mu_J$, and $\partial \tilde{A}/\partial \mu_J$ numerically by solving the linear system of equations given in (A.26) with a standard numerical solver. The differentials of the functions $F_1(\tilde{R}, \tilde{J}, \tilde{A}, \mu_J)$, $F_2(\tilde{R}, \tilde{J}, \tilde{A}, \mu_J)$, and $F_3(\tilde{R}, \tilde{J}, \tilde{A}, \mu_J)$ that make up equation (A.26), however, we computed analytically.

To determine the boundary of the parameter region for which biomass overcompensation can occur, we now require that the derivative of stage-specific biomass with respect to the mortality parameter of interest, as computed from equation (A.27), equals 0. More specifically, to compute the parameter region within which juvenile biomass at equilibrium will increase with juvenile mortality, we solve the following system of equations:

$$
\begin{cases}
F_1(\tilde{R}, \tilde{J}, \tilde{A}, \mu_J^0) = 0 \\[2ex]
F_2(\tilde{R}, \tilde{J}, \tilde{A}, \mu_J^0) = 0 \\[2ex]
F_3(\tilde{R}, \tilde{J}, \tilde{A}, \mu_J^0) = 0 \\[2ex]
\left. \dfrac{\partial \tilde{J}}{\partial \mu_J} \right|_{\mu_J = \mu_J^0} = 0
\end{cases}
\qquad (A.28)
$$

Here μ_J^0 refers to the background mortality rate of juvenile consumers. The first three equations of this system determine the equilibrium state at this background mortality $\mu_J = \mu_J^0$, whereas the last equation of the system stipulates that an increase in juvenile mortality from this background level should neither increase nor decrease the equilibrium juvenile biomass. In other words, the tangent of the curve $\tilde{J}(\mu_J)$ at $\mu_J = \mu_J^0$ is exactly horizontal. The system of four equations (A.28) contains the three equilibrium densities $\tilde{R}$, $\tilde{J}$, and $\tilde{A}$ as unknowns. To compute the boundary in parameter space, we fix all but one parameter and solve the system (A.28) for this single free parameter plus the three equilibrium densities $\tilde{R}$, $\tilde{J}$, and $\tilde{A}$. In particular, to construct the regions in the (q, z)-parameter plane with biomass overcompensation that are presented in chapter 3, we fixed all parameters but q, and solved the system of equations (A.28) for the four unknown quantities q, $\tilde{R}$, $\tilde{J}$, and $\tilde{A}$. The curves that delimit the parameter region with juvenile biomass overcompensation are now obtained by repeating this procedure for a range of values of the newborn-adult body size ratio z.

We have found that the procedure described above reliably locates the parameter limits to stage-specific biomass overcompensation, except for overcompensation in adult biomass in response to increasing adult mortality. In the latter case, which occurs when the equilibrium is mostly governed by development control, adult biomass may first decrease when adult mortality is increased from background levels. For higher values of adult mortality, adult biomass may, however, start to increase again and reach higher values than the biomass density at background mortality rates. To compute the limits in parameter space for this type of biomass overcompensation, it is hence not sufficient to just stipulate that the change in equilibrium adult biomass with an increase in adult mortality, $\partial \tilde{A} / \partial \mu_A$, when starting from background mortality levels is 0. Instead, a more complicated calculation is needed. The equilibrium resource, juvenile, and adult biomass at a background mortality level of $\mu_A = \mu_A^0$ we now denote with $\tilde{R}_0$, $\tilde{J}_0$, and $\tilde{A}_0$, respectively, while the corresponding values at a maximum in the curve relating adult biomass to adult mortality we denote with $\tilde{R}_m$, $\tilde{J}_m$, and $\tilde{A}_m$. The adult mortality at which this maximum occurs we denote with μ_A^m. The values of $\tilde{R}_0$, $\tilde{J}_0$, and $\tilde{A}_0$ are determined by the following system of three equations:

$$\begin{cases} F_1(\tilde{R}_0, \tilde{J}_0, \tilde{A}_0, \mu_A^0) = 0 \\ F_2(\tilde{R}_0, \tilde{J}_0, \tilde{A}_0, \mu_A^0) = 0 \\ F_3(\tilde{R}_0, \tilde{J}_0, \tilde{A}_0, \mu_A^0) = 0 \end{cases} \tag{A.29}$$

The values of $\tilde{R}_m$, $\tilde{J}_m$, and $\tilde{A}_m$, as well as μ_A^m, are determined by the following system of four equations:

$$\begin{cases} F_1(\tilde{R}_m, \tilde{J}_m, \tilde{A}_m, \mu_A^m) = 0 \\ F_2(\tilde{R}_m, \tilde{J}_m, \tilde{A}_m, \mu_A^m) = 0 \\ F_3(\tilde{R}_m, \tilde{J}_m, \tilde{A}_m, \mu_A^m) = 0 \\ \left. \dfrac{\partial \tilde{A}}{\partial \mu_A} \right|_{\mu_A = \mu_A^m} = 0 \end{cases} \tag{A.30}$$

The derivative of the equilibrium adult biomass density with respect to adult mortality, $\partial \tilde{A}/\partial \mu_A$, in the last equation is computed from the direction of the equilibrium curve as a function of μ_A, using an approach similar to the one that resulted in equation (A.27). If we now require that $\tilde{A}_m = \tilde{A}_0$, the maximum in the curve relating adult biomass to adult mortality μ_A exactly equals the value $\tilde{A}_0$ at a background mortality of $\mu_A = \mu_A^0$. Moreover, the seven equations in (A.29) and (A.30) then only contain six unknown quantities: $\tilde{R}_0, \tilde{J}_0, \tilde{A}_0, \tilde{R}_m, \tilde{J}_m$, and μ_A^m. As before, we can then fix all parameters but q and solve the systems of equations (A.29) and (A.30) for the seven unknown quantities $q, \tilde{R}_0, \tilde{J}_0, \tilde{A}_0$, $\tilde{R}_m, \tilde{J}_m$, and μ_A^m, while the curve delimiting the parameter region with adult biomass overcompensation in response to increasing adult mortality is obtained by repeating this procedure for a range of values of the newborn-adult body size ratio z.

6   ONTOGENETIC SYMMETRY AND ASYMMETRY IN ENERGETICS

Assume that the dynamics of a consumer population and its resource are described by the basic size-structured PDE as presented in box A.1, but that individuals mature when they reach a body size s_j. In the following we will equate body size s with individual weight or mass, such that the total biomass of juvenile and adult individuals is given by:

$$J = \int_{s_b}^{s_j} sc(t,s)\,ds$$

and

$$A = \int_{s_j}^{\infty} sc(t,s)\,ds,$$

respectively. For the dynamics of the total population biomass, denoted by C, we can derive the following ODE:

$$\frac{d}{dt}\int_{s_b}^{\infty} sc(t,s)\,ds = \int_{s_b}^{\infty} s\frac{\partial c(t,s)}{\partial t}\,ds = -\int_{s_b}^{\infty} s\frac{\partial g(R,s)c(t,s)}{\partial s}\,ds - \int_{s_b}^{\infty} s\,d(R,s)c(t,s)\,ds$$

$$= s_b g(R,s_b)c(t,s_b) + \int_{s_b}^{\infty} g(R,s)c(t,s)\,ds - \int_{s_b}^{\infty} d(R,s)sc(t,s)\,ds$$

$$= \int_{s_b}^{\infty} \big(b(R,s)s_b + g(R,s) - d(R,s)s\big)c(t,s)\,ds$$

$$(A.31)$$

In this derivation, the PDE is used to replace the term $\partial c(t,s)/\partial t$, while the boundary condition of the PDE is used to substitute the term $s_b g(\tilde{R},s_b)c(t,s_b)$. From this ODE it can be deduced that in equilibrium the following condition should hold:

$$\int_{s_b}^{s_j} \big(g(\tilde{R},s) - d(\tilde{R},s)s\big)\tilde{c}(s)\,ds = -\int_{s_j}^{\infty} \big(b(\tilde{R},s)s_b + g(\tilde{R},s) - d(\tilde{R},s)s\big)\tilde{c}(s)\,ds \quad (A.32)$$

Here $\tilde{R}$ and $\tilde{c}(s)$ represent the equilibrium resource density and population size distribution, respectively. The integral term in the left-hand side of the equation represents the total net biomass turnover in the juvenile stage, as it integrates the difference between the growth rate in biomass $g(\tilde{R},s)$ and the biomass loss rate owing to mortality $d(\tilde{R},s)s$ over the entire juvenile size range. Similarly, the integral in the right-hand side integrates the biomass production through reproduction, $b(\tilde{R},s)s_b$, and growth in body size, $g(\tilde{R},s)$, minus the biomass loss rate through mortality, $d(\tilde{R},s)s$, over all adult body sizes and hence represents the net biomass turnover during the entire adult life stage. In other words, the equality expresses that in equilibrium the turnover of biomass during the juvenile stage always has exactly the opposite sign as the turnover during the adult stage. The analysis hence generalizes the results derived for the stage-structured biomass model in chapter 3 to the more general setting of the basic size-structured PDE of box A.1.

In cases where both the left- and right-hand side of equation (A.32) equal 0, the juvenile and adult stage are both zero-net-producers of biomass. For the juvenile stage, this implies that the overall rate of biomass production through somatic growth exactly equals the total loss rate of biomass within this stage, whereas for adults, the total loss rate of biomass through mortality equals the sum of their biomass production through somatic growth and reproduction. We refer to this particular situation as *juvenile-adult symmetry in biomass turnover*. This symmetry at the level of the juvenile and adult stage does not preclude that within either stage net biomass turnover is positive in some ranges of individual body size and compensated for by net biomass loss in other ranges of body size. Symmetry in biomass turnover over the entire ontogeny, that is, at every body size s, will occur if:

$$b(\tilde{R},s)s_b + g(\tilde{R},s) = d(\tilde{R},s)s \qquad \text{for all } s \geq s_b \qquad \text{(A.33)}$$

We refer to situations in which equation (A.33) holds as *ontogenetic symmetry in biomass turnover*, to distinguish it from symmetry between the juvenile and adult stage only.

Considering the dynamics of juvenile biomass, we can derive in a similar way as above the ODE:

$$\frac{dJ}{dt} = \frac{d}{dt} \int_{s_b}^{s_j} sc(t,s)ds = \int_{s_b}^{s_j} s\frac{\partial c(t,s)}{\partial t}ds = -\int_{s_b}^{s_j} s\frac{\partial g(R,s)c(t,s)}{\partial s}ds - \int_{s_b}^{s_j} sd(R,s)c(t,s)ds$$

$$= s_b g(R,s_b)c(t,s_b) - s_j g(R,s_j)c(t,s_j) + \int_{s_b}^{s_j} g(R,s)c(t,s)ds - \int_{s_b}^{s_j} d(R,s)sc(t,s)ds$$

As before, the PDE is used in this derivation to replace the term $\partial c(t,s)/\partial t$. Equating dJ/dt to 0 shows generally that in equilibrium the following equality holds:

$$s_b g(\tilde{R},s_b)\tilde{c}(s_b) - s_j g(\tilde{R},s_j)\tilde{c}(s_j) = -\int_{s_b}^{s_j} \left(g(\tilde{R},s) - d(\tilde{R},s)s \right)\tilde{c}(s)ds \qquad \text{(A.34)}$$

The left-hand side of this equation represents the difference between total population reproduction, $s_b g(\tilde{R},s_b)c(t,s_b)$, and maturation rate, $s_j g(\tilde{R},s_j)c(t,s_j)$, in terms of biomass. Hence, as already shown in chapter 3 for the stage-structured biomass model, we find for the general class of models, described by the basic size-structured PDE of box A.1, that the difference between the population reproduction and maturation rate in biomass is directly tied to the net biomass turnover in the juvenile stage. In particular, if the juvenile stage constitutes a source of net biomass production and the right-hand side of equation (A.34) is negative, the total population biomass maturation rate is larger than the reproduction rate, whereas the reverse holds if the juvenile stage is a sink of net biomass production.

An expression for the consumer size distribution in equilibrium $\tilde{c}(s)$ can be obtained by formally solving the PDE (see box A.1) after setting $\partial c(t,s)/\partial t$ to 0, which results in (de Roos 1997):

$$\tilde{c}(s) = \frac{g(\tilde{R},s_b)\tilde{c}(s_b)}{g(\tilde{R},s)}\exp\left(-\int_{s_b}^{s} \frac{d(\tilde{R},\xi)}{g(\tilde{R},\xi)}d\xi \right) = \frac{\tilde{B}}{g(\tilde{R},s)}\exp\left(-\int_{s_b}^{s} \frac{d(\tilde{R},\xi)}{g(\tilde{R},\xi)}d\xi \right)$$

$$\text{(A.35)}$$

In this expression we have introduced the variable $\tilde{B} = g(\tilde{R}, s_b)\,\tilde{c}(s_b)$ for the total population reproduction rate in equilibrium. Substitution of this expression into the boundary condition of the PDE yields, after some simplification, the following condition determining the resource density in a consumer-resource equilibrium (de Roos 1997):

$$1 = \int_{s_b}^{\infty} \frac{b(\tilde{R},s)}{g(\tilde{R},s)} \exp\left(-\int_{s_b}^{s} \frac{d(\tilde{R},\xi)}{g(\tilde{R},\xi)} d\xi\right) ds \qquad (A.36)$$

From this condition for the equilibrium resource density, it can be derived that a situation in which there is ontogenetic symmetry in biomass turnover for every size s, that is, when equation (A.33) holds, corresponds under biologically plausible assumptions to an equilibrium state. Given that we assumed $b(\tilde{R}, s)$ equal to 0 for $s < s_j$, the equilibrium condition can be rewritten as:

$$1 = \exp\left(-\int_{s_b}^{s_j} \frac{d(\tilde{R},\xi)}{g(\tilde{R},\xi)} d\xi\right) \int_{s_j}^{\infty} \frac{b(\tilde{R},s)}{g(\tilde{R},s)} \exp\left(-\int_{s_j}^{s} \frac{d(\tilde{R},\xi)}{g(\tilde{R},\xi)} d\xi\right) ds$$

Substitution of condition (A.33) for ontogenetic symmetry in biomass turnover leads to an expression that can be integrated by parts:

$$1 = \exp\left(-\int_{s_b}^{s_j} \frac{1}{\xi} d\xi\right) \int_{s_j}^{\infty} \frac{1}{s_b}\left(s \frac{d(\tilde{R},s)}{g(\tilde{R},s)} - 1\right) \exp\left(-\int_{s_j}^{s} \frac{d(\tilde{R},\xi)}{g(\tilde{R},\xi)} d\xi\right) ds$$

$$= \frac{1}{s_j}\left(\int_{s_j}^{\infty} s \frac{d(\tilde{R},s)}{g(\tilde{R},s)} \exp\left(-\int_{s_j}^{s} \frac{d(\tilde{R},\xi)}{g(\tilde{R},\xi)} d\xi\right) ds - \int_{s_j}^{\infty} \exp\left(-\int_{s_j}^{s} \frac{d(\tilde{R},\xi)}{g(\tilde{R},\xi)} d\xi\right) ds\right)$$

$$= \frac{1}{s_j}\left(-s \exp\left(-\int_{s_j}^{s} \frac{d(\tilde{R},\xi)}{g(\tilde{R},\xi)} d\xi\right) \Bigg|_{s_j}^{\infty}\right)$$

$$= 1 - \lim_{s \to \infty} \frac{s}{s_j} \exp\left(-\int_{s_j}^{s} \frac{d(\tilde{R},\xi)}{g(\tilde{R},\xi)} d\xi\right)$$

In any biologically realistic model, the limit in the last line of this derivation has to vanish. The exponential term in this limit represents the probability that an individual after maturation at size s_j will survive up to size s (de Roos 1997), which decreases with body size as long as the quotient $d(\tilde{R}, s)/g(\tilde{R}, s)$ is bounded away from 0. However, the decrease in this probability should be sufficiently fast to annul the increase in body size itself, otherwise individuals with large (infinite) body size have a non-negligible contribution to the total population biomass, despite the fact that the probability that such individuals occur in the population is negligible. To prevent the occurrence of such gigantic "atto-foxes" (Mollison 1991), either individual growth should limit body size to a maximum or mortality has to be sufficiently strong.

For the remainder of this section, we assume that all individuals experience the same background mortality. Mortality is therefore size-independent, and body size can be dropped as argument of the mortality function, which we hence write as $d(\tilde{R})$. Assume that individuals of all body sizes produce new biomass through somatic growth and reproduction at the same mass-specific rate, that is,

$$\frac{b(R,s)s_b + g(R,s)}{s} = h(R) \qquad \text{for all } s \geq s_b \qquad (A.37)$$

Here $h(R)$ represents some arbitrary function that is possibly resource dependent but not size dependent. We will refer to a situation in which condition (A.37) holds as one with *ontogenetic symmetry in biomass production*. Ontogenetic symmetry in biomass *production* is a less restrictive condition than ontogenetic symmetry in biomass *turnover*, as it does not involve assumptions about the mortality rate. Moreover, where ontogenetic symmetry in biomass turnover only occurs in equilibrium, symmetry in biomass production is independent of equilibrium conditions. In the case of ontogenetic symmetry in biomass production and size-independent mortality, the ODE (A.31) for the total population biomass C simplifies to:

$$\frac{dC}{dt} = \left(h(R) - d(R)\right)C \qquad (A.38)$$

This shows that size-structured populations with ontogenetic symmetry in biomass production and size-independent mortality rates correspond to unstructured population models, in which dynamics are independent of the population size distribution and only influenced by the total population biomass density. Unstructured population models implicitly assume ontogenetic symmetry in biomass production and size-independent mortality, as they do not account for differences between individuals in the population.

From the ODE (A.38) it can be seen that the equilibrium resource density satisfies:

$$d(\tilde{R}) = h(\tilde{R}) = \frac{b(\tilde{R},s)s_b + g(\tilde{R},s)}{s}$$

The combination of ontogenetic symmetry in biomass production and a size-independent mortality rate hence results in an equilibrium with ontogenetic symmetry in biomass turnover.

In the case of ontogenetic symmetry in biomass turnover, we can derive an explicit expression for juvenile biomass in equilibrium in terms of the total population birth rate $\tilde{B}$ and the mortality rate $d(\tilde{R})$:

$$
\begin{aligned}
\tilde{J} &= \int_{s_b}^{s_j} s\tilde{c}(s)\,ds = \tilde{B} \int_{s_b}^{s_j} \frac{s}{g(\tilde{R},s)} \exp\left(-\int_{s_b}^{s} \frac{d(\tilde{R})}{g(\tilde{R},\xi)}\,d\xi \right) ds \\
&= \frac{\tilde{B}}{d(\tilde{R})} \int_{s_b}^{s_j} \exp\left(-\int_{s_b}^{s} \frac{1}{\xi}\,d\xi \right) ds \\
&= \frac{\tilde{B}s_b}{d(\tilde{R})} \ln\left(\frac{s_j}{s_b} \right)
\end{aligned}
\tag{A.39}
$$

This derivation also shows that in equilibrium the density function in the juvenile stage is inversely proportional to squared body size s^{-2}, and consequently the biomass-size distribution is inversely proportional to body size itself:

$$
\tilde{c}(s) = \frac{\tilde{B}}{d(\tilde{R})s} \exp\left(-\int_{s_b}^{s} \frac{1}{\xi}\,d\xi \right) = \frac{\tilde{B}s_b}{d(\tilde{R})} \frac{1}{s^2} \quad \Leftrightarrow \quad s\tilde{c}(s) = \frac{\tilde{B}s_b}{d(\tilde{R})} \frac{1}{s} \tag{A.40}
$$

For equilibrium adult biomass, we cannot derive such a general expression, as it depends on the fraction of biomass production that adults invest in somatic growth versus their investment in reproduction. In cases where adults do not invest in somatic growth, equilibrium adult biomass would equal:

$$
\tilde{A} = \frac{\tilde{B}s_b}{d(\tilde{R})} \tag{A.41}
$$

This expression can be derived from equation (A.32), realizing that $\tilde{B}s_b$ equals the rate at which adults produce biomass through reproduction. Because of the ontogenetic symmetry in biomass turnover of juveniles, $\tilde{B}s_b$ also equals the rate at which new biomass enters the adult stage through maturation. This inflow of new biomass leaves the adult stage through mortality only, which equals $d(\tilde{R})\tilde{A}$. The latter argumentation provides an alternative derivation of expression (A.41) and at the same time leads to the conclusion that equilibrium adult biomass would be larger than $\tilde{B}s_b/d(\tilde{R})$ if adults do invest in somatic growth, as this would increase the rate at which biomass enters the adult stage. In other words, in cases of ontogenetic symmetry in biomass turnover, the ratio between juvenile and adult biomass in equilibrium follows:

$$
\frac{\tilde{J}}{\tilde{A}} \leq \ln\left(\frac{s_j}{s_b} \right) = -\ln z \tag{A.42}
$$

In this equation z equals the ratio of the size at birth and at maturation, $z = s_b/s_j$. The conclusion of this analysis is that even in cases of ontogenetic symmetry in biomass turnover, the juvenile-adult biomass ratio at equilibrium depends on the juvenile size range and on whether or not adults invest in somatic growth. If they don't, juveniles will make up the largest part of total population biomass if $z = s_b/s_j < e^{-1}$.

Now assume that the juvenile stage as a whole is a zero-net-producer of biomass and that hence there is juvenile-adult symmetry in biomass turnover. Consider, however, that for juveniles with a body size up to some threshold size s_0, the somatic growth rate is proportional to body size with proportionality constant $h_0(\tilde{R})$, whereas for juveniles with a body size larger than s_0, the somatic growth rate is proportional to body size with proportionality constant $h_1(\tilde{R})$:

$$g(\tilde{R},s) = \begin{cases} h_0(\tilde{R})s & \text{for } s \leq s_0 \\ h_1(\tilde{R})s & \text{for } s > s_0 \end{cases}$$

For the biomass density J_0 at equilibrium in the juvenile size range between s_b and s_0, we can then derive the following expression:

$$\int_{s_b}^{s_0} s\tilde{c}(s)\,ds = \int_{s_b}^{s_0} s\frac{\tilde{B}}{g(\tilde{R},s)}\exp\left(-\int_{s_b}^{s}\frac{d(\tilde{R})}{g(\tilde{R},\xi)}\,d\xi\right)ds$$

$$= \int_{s_b}^{s_0}\frac{\tilde{B}}{h_0(\tilde{R})}\exp\left(-\frac{d(\tilde{R})}{h_0(\tilde{R})}\int_{s_b}^{s}\frac{1}{\xi}\,d\xi\right)ds = \int_{s_b}^{s_0}\frac{\tilde{B}}{h_0(\tilde{R})}\left(\frac{s}{s_b}\right)^{-\frac{d(\tilde{R})}{h_0(\tilde{R})}}ds$$

$$= \frac{\tilde{B}s_b}{h_0(\tilde{R}) - d(\tilde{R})}\left(\left(\frac{s_0}{s_b}\right)^{1-\frac{d(\tilde{R})}{h_0(\tilde{R})}} - 1\right)$$

Defining $y_0 = 1 - d(\tilde{R})/h_0(\tilde{R})$ and $z_0 = s_b/s_0$, this expression can be rewritten as:

$$J_0 = \frac{\tilde{B}s_b}{d(\tilde{R})}\frac{1-y_0}{y_0}\left(z_0^{-y_0} - 1\right) \tag{A.43}$$

For $y_0 \to 0$, which corresponds to the limit that $h_0(\tilde{R})$ approaches $d(\tilde{R})$, the term $(1-y_0)(z_0^{-y_0} - 1)/y_0$ can be shown to have a regular limit equal to $-\ln(z_0)$. In this limit, J_0 thus approaches the value $\tilde{B}s_b\ln(s_0/s_b)/d(\tilde{R})$, which it would have in cases of ontogenetic symmetry in biomass turnover for all individuals within the size range between s_b and s_0 (cf. equation (A.39)). In addition, the derivative of $(1-y_0)(z_0^{-y_0} - 1)/y_0$ with respect to y_0 equals $(1 + z_0^{-y_0}(y_0^2\ln z_0 - y_0\ln z_0 - 1))/y_0^2$, which for $y_0 \to 0$ approaches the value

$\ln z_0(\ln z_0 + 2)/2$. The derivative of $(1 - y_0)(z_0^{-y_0} - 1)/y_0$ at $y_0 = 0$ is therefore negative as long as $z_0 = s_b/s_0 > e^{-2}$ and positive otherwise. This result implies that, compared with the situation of ontogenetic symmetry in biomass turnover, an *increase* in $h_0(\tilde{R})$ (i.e., an *increase* in y_0) will lead to a *decrease* in the equilibrium biomass in the size range between s_b and s_0 if $s_b/s_0 > e^{-2}$ and will lead to an increase in this biomass otherwise. In other words, if the juvenile size range between s_b and s_0 turns into a net source of biomass production, the equilibrium biomass density in this size range will decrease, unless the size interval is so large that the ratio of its maximum and minimum size exceeds e^2.

The changes in J_0 with changes in $h_0(\tilde{R})$ also result in changes in total juvenile biomass. Given the change in growth rate at $s = s_0$, the equilibrium size distribution (A.35) for juvenile sizes between s_0 and s_j is given by:

$$\tilde{c}(s) = \frac{\tilde{B}}{h_1(\tilde{R})s} \exp\left(-\frac{d(\tilde{R})}{h_0(\tilde{R})} \int_{s_b}^{s_0} \frac{1}{\xi} d\xi\right) \exp\left(-\frac{d(\tilde{R})}{h_1(\tilde{R})} \int_{s_0}^{s} \frac{1}{\xi} d\xi\right)$$

$$= \frac{\tilde{B}}{h_1(\tilde{R})s} z_0^{\frac{d(\tilde{R})}{h_0(\tilde{R})}} \left(\frac{s}{s_0}\right)^{-\frac{d(\tilde{R})}{h_1(\tilde{R})}}$$

In this expression we have used as before $z_0 = s_b/s_0$. In equilibrium the biomass density J_1 of juveniles with a size between s_0 and the maturation size s_j then equals:

$$J_1 = \int_{s_0}^{s_j} s\tilde{c}(s)\, ds = \int_{s_0}^{s_j} \frac{\tilde{B}}{h_1(\tilde{R})} z_0^{\frac{d(\tilde{R})}{h_0(\tilde{R})}} \left(\frac{s}{s_0}\right)^{-\frac{d(\tilde{R})}{h_1(\tilde{R})}} ds$$

$$= \frac{\tilde{B}s_0}{h_1(\tilde{R}) - d(\tilde{R})} z_0^{\frac{d(\tilde{R})}{h_0(\tilde{R})}} \left(\left(\frac{s_j}{s_0}\right)^{1 - \frac{d(\tilde{R})}{h_1(\tilde{R})}} - 1\right)$$

Now define $y_1 = 1 - d(\tilde{R})/h_1(\tilde{R})$, analogous to the earlier definition $y_0 = 1 - d(\tilde{R})/h_0(\tilde{R})$, and $z_1 = s_0/s_j$. The expression for J_1 can then be rewritten as:

$$J_1 = \frac{\tilde{B}s_b}{d(\tilde{R})} z_0^{-y_0} \frac{(1 - y_1)}{y_1} (z_1^{-y_1} - 1) \tag{A.44}$$

Under the assumption that the juvenile stage as a whole is a zero-net-producer of biomass, J_0 and J_1 should also satisfy:

$$h_0(\tilde{R})J_0 + h_1(\tilde{R})J_1 = d(\tilde{R})(J_0 + J_1) \quad \Leftrightarrow \quad \frac{y_0}{1 - y_0} J_0 + \frac{y_1}{1 - y_1} J_1 = 0$$

Substitution of the expressions (A.43) and (A.44) into this identity lead to the following relationship between y_0 and y_1:

$$\frac{\tilde{B}s_b}{d(\tilde{R})}\left((z_0^{-y_0}-1)+z_0^{-y_0}(z_1^{-y_1}-1)\right)=0 \quad \Leftrightarrow \quad z_0^{y_0}=z_1^{-y_1} \quad \Leftrightarrow \quad y_1=-\frac{\ln z_0}{\ln z_1}y_0$$

Given this relation between y_0 and y_1, the total juvenile biomass at equilibrium equals:

$$
\begin{aligned}
J_0+J_1 &= \frac{\tilde{B}s_b}{d(\tilde{R})}\frac{1-y_0}{y_0}\left(z_0^{-y_0}-1\right)+\frac{\tilde{B}s_b}{d(\tilde{R})}z_0^{-y_0}\frac{(1-y_1)}{y_1}\left(z_1^{-y_1}-1\right)\\
&= \frac{\tilde{B}s_b}{d(\tilde{R})}\left[\left(\frac{1}{y_0}-\frac{1}{y_1}\right)\!\left(z_0^{-y_0}-1\right)\right]\\
&= \frac{\tilde{B}s_b}{d(\tilde{R})}\left[\frac{\ln(z_0 z_1)}{y_0\ln z_0}\left(z_0^{-y_0}-1\right)\right]\\
&= \frac{\tilde{B}s_b}{d(\tilde{R})}\ln\!\left(\frac{s_j}{s_b}\right)\!\left[\frac{(1-z_0^{-y_0})}{y_0\ln z_0}\right]
\end{aligned}
$$

For $y_0 \to 0$, the term within brackets has a regular limit equal to 1, which corresponds to the limit of ontogenetic symmetry in biomass turnover, when the sum J_0+J_1 equals the equilibrium juvenile biomass density given by equation (A.39). The derivative of the bracketed term with respect to y_0 equals:

$$\frac{y_0 z_0^{-y_0}\ln z_0 + z_0^{-y_0}-1}{y_0^2 \ln z_0}$$

For $y_0 \to 0$, this derivative has a regular limit equal to $-\ln z_0/2$, which is always positive. Therefore, compared with the situation of ontogenetic symmetry in biomass turnover, an *increase* in $h_0(\tilde{R})$ (which corresponds to an *increase* in y_0) will always lead to an *increase* in total juvenile biomass. In cases where there is ontogenetic symmetry in biomass turnover in the juvenile stage as a whole, but juveniles with a body size between s_b and s_0 are net biomass producers, total juvenile biomass will hence be *higher* than expected on the basis of ontogenetic symmetry in biomass turnover. Vice versa, if under these conditions the size range between s_b and s_0 is a net sink of biomass production, total juvenile biomass will be lower compared with its reference value in cases of ontogenetic symmetry in biomass turnover for all juvenile body sizes.

7   MECHANISMS LEADING TO
BIOMASS OVERCOMPENSATION

In the previous section we studied the consequences of ontogenetic symmetry in biomass turnover for equilibrium densities in the basic size-structured

PDE model of box A.1. In this section we focus on the mechanisms in this basic model that might give rise to biomass overcompensation in response to increases in mortality. For this analysis it is easier to switch from a size-dependent formulation to an age-dependent formulation. To this end, assume that the equilibrium relationship between body size and age is given by the function $\tilde{s}(\tilde{R}, a)$, which is the solution of the following ODE (de Roos 1997):

$$\frac{ds}{da} = g(\tilde{R}, s)$$

Now let $\tilde{a}_j(\tilde{R})$ represent the age at which an individual under equilibrium conditions reaches the maturation size, such that $\tilde{s}(\tilde{R}, \tilde{a}_j(\tilde{R})) = s_j$. Furthermore, define the function $\tilde{F}_J(\tilde{R}, a)$ to represent the survival probability in equilibrium of a juvenile consumer up to age a. $\tilde{F}_J(\tilde{R}, a)$ is determined by the per capita mortality rate following (de Roos 1997):

$$\tilde{F}_J(\tilde{R}, a) = \exp\left(-\int\limits_0^a d(\tilde{R}, \tilde{s}(\tilde{R}, \xi))\, d\xi \right)$$

The survival under equilibrium conditions during the adult life stage up to age a, given that an individual has matured at age $\tilde{a}_j(\tilde{R})$, we will indicate with the function $\tilde{F}_A(\tilde{R}, a - \tilde{a}_j(\tilde{R}))$, defined as:

$$\tilde{F}_A(\tilde{R}, a - \tilde{a}_j(\tilde{R})) = \exp\left(-\int\limits_{\tilde{a}_j(\tilde{R})}^{a - \tilde{a}_j(\tilde{R})} d(\tilde{R}, \tilde{s}(\tilde{R}, \xi))\, d\xi \right)$$

We distinguish these two survival functions because we want to study the changes in equilibrium biomass in response to increases in either juvenile or adult mortality, that is, in response to *decreases* in either $\tilde{F}_J(\tilde{R}, a)$ or $\tilde{F}_A(\tilde{R}, a - \tilde{a}_j(\tilde{R}))$. Notice that all these introduced quantities are functions of the equilibrium resource density $\tilde{R}$, as growth and mortality may depend on resource density.

Using these introduced quantities, the condition (A.36), which determines the equilibrium resource density of the basic size-structured PDE model, can be written as:

$$\tilde{F}_J(\tilde{R}, \tilde{a}_j(\tilde{R})) \int\limits_{\tilde{a}_j(\tilde{R})}^{\infty} b(\tilde{R}, \tilde{s}(\tilde{R}, a)) \tilde{F}_A(\tilde{R}, a - \tilde{a}_j(\tilde{R}))\, da = 1, \qquad (A.45)$$

which simply expresses that the expected lifetime production of offspring of an individual consumer should equal 1; or, in other words, that during life every consumer should on average just replace itself.

The population reproduction rate in equilibrium $\tilde{B}$ is determined by:

$$\tilde{B} = \frac{G(\tilde{R})}{\displaystyle\int_0^{\tilde{a}_j(\tilde{R})} I(\tilde{R}, \tilde{s}(\tilde{R}, a)) \tilde{F}_J(\tilde{R}, a) \, da + \tilde{F}_J(\tilde{R}, \tilde{a}_j(\tilde{R})) \int_{\tilde{a}_j(\tilde{R})}^{\infty} I(\tilde{R}, \tilde{s}(\tilde{R}, a)) \tilde{F}_A(\tilde{R}, a - \tilde{a}_j(\tilde{R})) \, da}$$

$$(A.46)$$

This expression for $\tilde{B}$ is derived by equating the right-hand side of the ODE for the resource density in box A.1 to 0. For a more detailed derivation of these expressions, we refer to de Roos (1997) or Diekmann, Gyllenberg, and Metz (2003). In the above expression we have deliberately distinguished between resource ingestion by a consumer individual during its juvenile (the first integral in the denominator) and its adult phase (the second integral in the denominator), as this will be important in the following analysis.

We will mostly focus on the overcompensation in biomass of juvenile consumers with body sizes in a small size range between the size at birth s_b and some upper limit s_u (we will not further discuss the subtleties associated with the choice of s_u but without loss of generality assume that it is not much larger than s_b). This choice to analyze overcompensation in a small-juvenile size range is inspired by the fact that it can be expected to occur most readily. Using the expression (A.35) for the size distribution at equilibrium, biomass in the juvenile size range between s_b and s_u can be expressed as:

$$\int_{s_b}^{s_u} s\tilde{c}(s) \, ds = \int_{s_b}^{s_u} s \frac{\tilde{B}}{g(\tilde{R}, s)} \exp\left(-\int_{s_b}^{s} \frac{d(\tilde{R}, \xi)}{g(\tilde{R}, \xi)} \, d\xi \right) ds \qquad (A.47)$$

Let's consider the possibility of biomass overcompensation in this juvenile size range owing to an increase in adult mortality, which will lead to an increase in equilibrium resource density $\tilde{R}$ and hence potentially to more rapid growth and lower mortality. We can infer from the equation above that resource-dependent growth in body size plays a dual role as it speeds up the growth through the size range. On the one hand, this causes individuals to leave the size range more quickly, which is expressed by the denominator of the term $\tilde{B}/g(\tilde{R}, s)$. This effect of growth in body size acts against the occurrence of biomass overcompensation, and it will be the dominating consequence of an increased growth rate if juvenile mortality is small or negligible, that is, when the exponential term in the right-hand side of equation (A.47) is equal or close to 1.

On the other hand, if mortality shortly after birth is substantial, more rapid growth through this vulnerable size range will cause a larger proportion of the newborn individuals to reach larger sizes, which is expressed by the occurrence of $g(\tilde{R}, s)$ in the denominator of the exponential term in equation (A.47). This effect of resource-dependent growth in body size will possibly facilitate the occurrence of overcompensation. Furthermore, a decrease in the mortality rate $d(\tilde{R}, s)$ in response to an increase in equilibrium resource density will also promote the occurrence of biomass overcompensation. In summary, considering biomass overcompensation in a size range of small-juvenile individuals owing to an increase in adult mortality, we find that resource-dependent mortality will promote the occurrence of such overcompensation, while resource-dependent growth in body size plays a dualistic role and will in particular counteract overcompensation when mortality is low.

In the stage-structured biomass model and the size-structured Kooijman-Metz model, which we analyzed in chapter 3 for the occurrence of biomass overcompensation, background mortality was low and independent of resource density as well as body size. Hence, this biomass overcompensation was not due to the influence of resource-dependent mortality or the positive effect of more rapid growth through vulnerable size ranges. Rather, the overcompensation resulted from the changes in total population reproduction rate $\tilde{B}$ with changes in mortality, which effect is ignored in the considerations of the previous paragraph. To investigate the occurrence of this type of biomass overcompensation in more detail, we will in the following address the question of what model characteristics may lead to an increase in total population reproduction rate $\tilde{B}$ at equilibrium with an increase in mortality.

In addition to the basic model assumptions, we will now assume that resource ingestion is a separable function of resource density and consumer body size and can hence be written as $I(R, s) = H(R)M(s)$. In this function, $H(R)$ represents the consumer functional response and is assumed to be a nondecreasing function of resource density $R (dH/dR \geq 0)$. The function $M(s)$ represents the dependence of the consumer ingestion rate on consumer body size. Furthermore, we will assume that adult fecundity is proportional to the difference between resource intake and some size-dependent, but resource-independent function, $T(s)$, which, among others, accounts for energy requirements to cover maintenance. Hence, adult fecundity follows $b(R, s) = \chi(H(R)M(s) - T(s))$, in which $T(s)$ is expressed in terms of resource units and χ represents a conversion factor between net energy production allocated to reproduction and the number of offspring produced on the basis of it.

Even though these assumptions are to some extent restrictive, they are made in most dynamic energy budget models that have been formulated to date (e.g., Gurney, McCauley, et al. 1990; Kooijman and Metz 1984; Kooijman 2000; Lika and Nisbet 2000; but see Persson, Leonardsson, et al. 1998 for a counterexample). We will also assume that $G(R)$ is a nonincreasing function of resource density R: $dG/dR \leq 0$, as we are interested in the occurrence of biomass over-compensation that is attributable to changes in the energetic efficiency of the consumer population, as opposed to the result of increases in resource productivity. Furthermore, we will assume that the mortality rate that juveniles are exposed to equals a constant μ_J, while adult individuals experience a constant mortality rate μ_A. The survival functions during the juvenile and adult life stages now become independent of resource density and follow $F_J(a) = e^{-\mu_J a}$ and $F_A(a - \tilde{a}_j(\tilde{R})) = e^{-\mu_A(a - \tilde{a}_j(\tilde{R}))}$, respectively. We make these latter assumptions in correspondence with the assumptions about mortality in both the stage-structured biomass model and the size-structured Kooijman-Metz model.

Using these assumptions, the equilibrium condition (A.45) can be used to solve for the value of $H(\tilde{R})$ at equilibrium:

$$H(\tilde{R}) = \frac{\chi^{-1} + F_J(\tilde{a}_j(\tilde{R})) \int\limits_{\tilde{a}_j(\tilde{R})}^{\infty} T(\tilde{s}(\tilde{R},a))F_A(a - \tilde{a}_j(\tilde{R}))\, da}{F_J(\tilde{a}_j(\tilde{R})) \int\limits_{\tilde{a}_j(\tilde{R})}^{\infty} M(\tilde{s}(\tilde{R},a))F_A(a - \tilde{a}_j(\tilde{R}))\, da}$$

This expression can be used to rewrite the expression (A.46) of the total population reproduction rate $\tilde{B}$ at equilibrium as:

$$\tilde{B} = \Psi^{-1} \frac{G(\tilde{R}) \int\limits_{\tilde{a}_j(\tilde{R})}^{\infty} M(\tilde{s}(\tilde{R},a))F_A(a - \tilde{a}_j(\tilde{R}))\, da}{\left(\int\limits_{0}^{\tilde{a}_j(\tilde{R})} M(\tilde{s}(\tilde{R},a))\frac{F_J(a)}{F_J(\tilde{a}_j(\tilde{R}))}\, da + \int\limits_{\tilde{a}_j(\tilde{R})}^{\infty} M(\tilde{s}(\tilde{R},a))F_A(a - \tilde{a}_j(\tilde{R}))\, da \right)}$$

$$(A.48)$$

The factor Ψ in this expression is defined as:

$$\Psi = \chi^{-1} + F_J(\tilde{a}_j(\tilde{R})) \int\limits_{\tilde{a}_j(\tilde{R})}^{\infty} T(\tilde{s}(\tilde{R},a))F_A(a - \tilde{a}_j(\tilde{R}))\, da$$

Consider a scenario, in which $T(s)$ equals 0, implying that adult fecundity is proportional to adult resource ingestion and the factor Ψ is equal to χ^{-1}, and that growth in body size is independent of resource density, such that we can write $\tilde{s}(a)$ instead of $\tilde{s}(\tilde{R},a)$. In essence, the latter assumption implies that the model is age-structured, as the unique relationship between body size and age $\tilde{s}(a)$ can be used to express all life history functions in terms of individual age, as opposed to size. In particular, these assumptions imply that both the size-scaling of ingestion rate, $M(\tilde{s}(\tilde{R},a))$, and the maturation age $\tilde{a}_j(\tilde{R})$ will be independent of resource density and can hence be denoted as $M(\tilde{s}(a))$ and $\tilde{a}_j$, respectively.

The expression for the total population reproduction rate at equilibrium (A.48) is then an increasing, but saturating function of the integral term:

$$\int_{\tilde{a}_j}^{\infty} M(\tilde{s}(a)) F_A(a - \tilde{a}_j) da$$

This integral, and thereby the total population reproduction rate, will be smaller when the adult survival $F_A(a - \tilde{a}_j)$ is decreased as a result of an increase in μ_A. In other words, increases in adult mortality under these conditions will only decrease total population reproduction and will hence not lead to overcompensation in juvenile biomass. Furthermore, given the constant size-age relationship, the first integral term in expression (A.48) can be written as:

$$\int_0^{\tilde{a}_j} M(\tilde{s}(a)) \frac{F_J(a)}{F_J(\tilde{a}_j)} da = \int_0^{\tilde{a}_j} M(\tilde{s}(a)) e^{\mu_J(\tilde{a}_j - a)} da$$

The right-hand side of the above equation makes clear that this integral term will increase with increases in juvenile mortality rate μ_J. As a consequence, increases in μ_J also result in a decrease in total population reproduction rate at equilibrium (see equation (A.48)). Together, these two results lead to the conclusion that overcompensation in juvenile biomass at equilibrium cannot occur in response to increases in either juvenile or adult mortality when adult fecundity is proportional to adult ingestion rate and growth in body size is independent of resource density.

Therefore, for an increase in total population reproduction with mortality and hence biomass overcompensation to occur, it is necessary that either growth in body size depends on resource density or that adult fecundity increases disproportionally with an increase in adult ingestion. The latter may occur when energy allocation to fecundity equals net energy production and maintenance requirements are not negligible. To disentangle how these two mechanisms operate, consider first the scenario that adult fecundity is proportional to adult ingestion with proportionality factor χ, but that growth in body size is resource dependent. The proportionality between adult fecundity and ingestion implies that the expected lifetime ingestion as an adult equals χ^{-1} times the expected lifetime reproduction, which in equilibrium equals 1. From equations (A.45) and (A.46) we can thus infer that the total population reproduction rate in equilibrium is given by:

$$\tilde{B} = \frac{G(\tilde{R})}{\displaystyle\int_0^{\tilde{a}_j(\tilde{R})} I(\tilde{R}, \tilde{s}(\tilde{R}, a)) \tilde{F}_J(\tilde{R}, a) da + \chi^{-1}}$$

Obviously, total reproduction rate $\tilde{B}$ can only increase with mortality if the integral term in its denominator, representing the expected resource ingestion during the entire juvenile period, decreases. Note, however, that the expected amount of food that is eaten as an adult remains the same, as it equals a constant times the expected number of offspring produced. How can the expected resource ingestion during the juvenile period go down, while the ingestion during the adult phase remains constant? For one, an increase in resource density in response to increased mortality will shorten the juvenile period $\tilde{a}_j(\tilde{R})$. As another factor, however, the increase in resource density may lead to a larger increase in adult than juvenile body size and hence to a larger increase in per capita ingestion as an adult. Both factors, the shortening of the juvenile period and a stronger increase in adult body size and ingestion with an increase in resource density, will contribute to a decline in the expected resource ingestion during the juvenile period, despite an unchanged ingestion during the adult phase. For example, if growth in body size follows a von Bertalanffy growth curve with a maximum size dependent on resource density as in the Kooijman-Metz energy budget model (Kooijman and Metz 1984), the change in size in response to increasing resource density is indeed larger at larger body sizes. In the size-structured Kooijman-Metz model, both factors can thus be shown to play a role in reducing juvenile ingestion with an increase in resource density.

Now consider a scenario in which growth in body size is independent of resource density, such that we can write $\tilde{s}(a)$ instead of $\tilde{s}(\tilde{R},a)$, but that the function $T(s)$ is unequal to 0. The assumption of resource-independent growth leads to the same conclusion as before, that the quotient in expression (A.48) will decrease with increases in either μ_J or μ_A, because this term is an increasing, but saturating function of the integral term

$$\int_{\tilde{a}_j}^{\infty} M(\tilde{s}(a)) F_A(a - \tilde{a}_j)\, da$$

and a decreasing function of $\exp(\mu_J(\tilde{a}_j - a))$, which occurs in the denominator of the term.

However, if $T(s)$ is unequal to 0, increases in μ_J as well as μ_A will also decrease the integral term

$$F_J(\tilde{a}_j(\tilde{R})) \int_{\tilde{a}_j(\tilde{R})}^{\infty} T(\tilde{s}(\tilde{R},a)) F_A(a - \tilde{a}_j(\tilde{R}))\, da$$

that represents requirements for energy-demanding processes, including the energy costs for maintenance, which do not lead to an increase in biomass. As a consequence, the conversion factor Ψ^{-1} in expression (A.48) increases.

Whenever this increase in Ψ^{-1} outweighs the decrease in the second term in expression (A.48), total population reproduction in equilibrium will increase with an increase in mortality. This will occur more readily when $T(s)$ is large, implying that at background mortality rates only a small fraction of the energy ingested by consumers is effectively spent on somatic growth or reproduction. For a further discussion of this route to biomass overcompensation, we refer to Claessen, de Vos, and de Roos (2009), who illustrated in an unstructured, bioenergetics model how increases in mortality may lead to an increase in total population reproduction rate in cases where maintenance requirements at background mortality rates are substantial.

In summary, when an increase in mortality relaxes competition among consumers and raises equilibrium resource densities, overcompensation in juvenile size ranges may arise if lifetime adult offspring production increases faster with resource density than cumulative lifetime resource intake. This leads to an increase in the total population reproduction rate, which is the basis of the overcompensation. One way to achieve this is that adult energetics follow a net-production energy channeling scheme and maintenance costs are not negligible. In this case, the increase in expected adult lifetime reproduction in response to an increase in resource density can be disproportionately large, compared with the increase in cumulative food ingestion as an adult. As a consequence, with increasing mortality, the expected lifetime reproduction as an adult can remain pinned down at 1 (ensuring equilibrium) while expected lifetime food ingestion as an adult decreases. This decrease forms the basis of the positive response in total population reproduction rate in equilibrium.

The second mechanism that can lead to overcompensation in juvenile biomass is food-dependent growth in body size, irrespective of whether adult fecundity is proportional to adult ingestion or not. For this type of overcompensation to occur, juveniles and adults must share the same resource, and growth in body size should lead to larger increases in adult than juvenile ingestion with increases in resource density. As a consequence, with increasing mortality, the expected lifetime reproduction as an adult can remain pinned down at 1 (ensuring equilibrium) and the expected lifetime food ingestion as an adult can remain constant, while the expected food ingestion as a juvenile decreases. This latter decrease forms the basis of the positive response in total population birth rate in equilibrium.

Finally, unless growth in individual body size is resource dependent, the overcompensation in biomass that arises from an increase in the total population reproduction rate will necessarily occur in all possible age intervals or size intervals up to a certain threshold. This threshold occurs at that point in the life history where the product of the population reproduction rate and the individual survival does not change with an increase in additional mortality. Beyond this age or size threshold, the decrease in individual survival will outweigh the increase in total population birth rate. Overcompensation in age or size ranges beyond this threshold, in particular overcompensation in adult biomass, as well as the co-occurrence of a decrease in juvenile biomass and an increase in adult biomass, can only occur if growth in body size depends on resource density.

8   DISCRETE-CONTINUOUS CONSUMER-RESOURCE MODELS

The dynamics of the model studied in chapter 10 are governed by a mixture of both discrete- and continuous-time processes. At the individual consumer level, food ingestion, growth in structural mass, and changes in body reserves all operate in continuous time. The model functions determining these processes are the resource consumption rate $\gamma(w, R)$, the net-production rate of energy $E_g(x, y, R)$, and the allocation function $\kappa(x, y)$, which determines the allocation of the net energy produced to structural mass and reserves. Individual mortality is modeled by the mortality rate $\mu(x, y)$, which is the sum of background and starvation mortality. Mortality is thus determined by a *rate function*, which essentially describes a continuous-time process, despite the fact that at the individual level mortality is inherently a discrete-time process. Reproduction at the individual level occurs as a discrete event at the start of the growth season. The number of offspring produced by an adult individual $(x > x_f)$ is given by the fecundity function $F(x, y)$. Table 10.1 summarizes the functions describing individual life history. Finally, production of resources is assumed to take place continuously throughout the growth season following semichemostat dynamics as described by the function $G(R) = r(K - R)$ (see table 10.2 for parameter values).

The model only examines population dynamics during the growth season, which is assumed to last T time units. Reproduction may hence occur at the moments when $t = nT$ with $n = 1, 2, 3, \ldots$. The changes in consumer and resource populations during the nongrowth season are assumed to be negligible. The population-level model is thus a combination of a continuous-time dynamical system, describing growth and survival of the consumers and production and consumption of the resource during summer, and a discrete map describing the pulsed reproduction of consumers in spring. Analytically, the structured population model can be formulated as a system of integral equations (see the appendix in Persson, Leonardsson, et al. 1998), which represents a way of bookkeeping the dynamics of all individuals making up the population.

To study the dynamics of the population model numerically, its equations were recast into the escalator boxcar train (EBT) framework (de Roos 1988, 1997; de Roos, Diekmann, and Metz 1992). The EBT method is specifically designed to handle the numerical integration of the equations that occur in physiologically structured models. This method is also used to study the dynamics of the completely continuous, size-structured population models from chapters 9 and 12. Using the EBT method to study the dynamics of the discrete-continuous consumer-resource model is simplified by the fact that the consumer population is naturally divided into distinct cohorts of individuals as a consequence of the pulsed reproduction process. All individuals within a single cohort are born with the same irreversible and reversible mass. Moreover, all individuals within a cohort are assumed to grow at the same rate, that is, individuals belonging to a given cohort do not diverge in their allocation to reversible and irreversible masses. As a result, each cohort consists of individuals that will remain identical for the duration of their lives.

Assume that right after the reproduction event at $t = (n - 1)T$ the consumer population consists of M cohorts of individuals. We will denote the total number of individuals in the ith cohort ($i = 1, \ldots, M$) at time t with the variable $N_i(t)$ and the structural mass and reserve mass of these individuals by the variables $X_i(t)$ and $Y_i(t)$, respectively. Notice that we have chosen to express these consumer densities as the number of individuals in a total volume of $V = 1.0 \cdot 10^6 \, \text{m}^3$, whereas the resource biomass density is expressed in a different reference volume, as grams per m^3. During the subsequent growth season, the dynamics of the cohort variables N_i, X_i, and Y_i are fully determined by mortality and growth, following the system of ordinary differential equations (ODEs):

$$
\begin{cases}
\dfrac{dN_i}{dt} = -\mu(X_i, Y_i)N_i \\[2mm]
\dfrac{dX_i}{dt} = \begin{cases} \kappa(X_i, Y_i)E_g(X_i, Y_i, R) & \text{if } E_g(X_i, Y_i, R) > 0 \\ 0 & \text{otherwise} \end{cases} \\[4mm]
\dfrac{dY_i}{dt} = \begin{cases} (1 - \kappa(X_i, Y_i))E_g(X_i, Y_i, R) & \text{if } E_g(X_i, Y_i, R) > 0 \\ E_g(X_i, Y_i, R) & \text{otherwise} \end{cases} \quad i = 1, \ldots, M
\end{cases}
\tag{A.49}
$$

In each cohort, the number of individuals decreases owing to mortality, as reflected by the first ODE for N_i. At the cohort level, mortality is hence a continuous-time process, despite its discrete-time character at the individual level. This is a valid approximation as long as the number of individuals within the cohort is sufficiently large, which we assume is invariably true. The ODE for X_i reflects the assumption that growth in structural mass occurs only when net energy production $E_g(X_i, Y_i, R)$ is positive. Under starvation conditions $(E_g(X_i, Y_i, R) < 0)$, growth in structural mass stops, and all energy demands are covered from reserves, as expressed by the second part of the ODEs for X_i and Y_i. Otherwise, the allocation function $\kappa(X_i, Y_i)$ determines the allocation of net energy produced to growth in structural mass and reserve mass as expressed by the first part of the ODEs for X_i and Y_i. Within-season dynamics of the entire consumer population is thus described by the above system of $3M$ ODEs with three ODEs determining the changes in each individual cohort.

The dynamics of the different cohorts are coupled, because all individuals share the same resource. The changes in resource density R throughout the growth season are described by the ODE:

$$\frac{dR}{dt} = G(R) - \frac{\sum_{i=1}^{M} \gamma((1+q_j)X_i, R)N_i}{V} \qquad (A.50)$$

This ODE expresses the fact that the dynamics of the resource equals the balance between the semichemostat production of resource $G(R)$ and the consumption of resource by all cohorts together, which is expressed by the second term in the right-hand side of the ODE above. This term, representing resource consumption of the total population, involves the division by the total volume V to account for the different units of volume in which we have chosen to express resource and consumer densities.

To track the population dynamics, the system of ODEs (A.49) and (A.50) is integrated numerically from $t = (n-1)T$ until the end of the growth season at $t = nT$, which also corresponds to the beginning of the next growth season. At $t = nT$, reproduction occurs, and a new cohort of individuals is added to the consumer population. This addition implies that the number of variables representing the consumer population, and hence also the number of ODEs describing its dynamics, is increased. The total number of newborn individuals equals the summed fecundity $F(X_i, Y_i)$ of all individuals:

$$\sum_{i=1}^{M} F(X_i, Y_i)N_i$$

The gonad mass that is used for this production of offspring is subtracted from the reserves of all adult consumers that reproduce, leaving them with reversible mass $Y_i = q_j X_i$. Furthermore, the indices of all existing cohorts are renumbered to allow for the addition of the newborn cohort. Hence, at $t = nT$, the following transformation takes place:

$$
\begin{cases}
N_{i+1}(nT^+) = N_i(nT^-) \\
X_{i+1}(nT^+) = X_i(nT^-) \\
Y_{i+1}(nT^+) = \begin{cases}
q_J X_i(nT^-) & \text{if } X_i(nT^-) > x_f \text{ and} \\
& Y_i(nT^-) > q_J X_i(nT^-) \\
Y_i(nT^-) & \text{otherwise} \qquad\qquad i = 1, \dots, M
\end{cases}
\end{cases}
\tag{A.51}
$$

In these renumbering equations, the notations $N_i(nT^-)$ and $N_i(nT^+)$ are used to indicate the values of variables before and after the transformation, respectively. The renumbering operation, increasing the index of cohort i to $i+1$, is primarily a bookkeeping operation, which ensures that the cohorts in the population remain in order according to the age of the individuals in them. Only the value of $Y_i(nT)$ may change, when adult individuals have sufficient reserve mass ($Y_i(nT^-) > q_J X_i(nT^-)$) to invest into reproduction and hence end up with the default amount of reserves ($q_J X_i(nT)$), given their structural mass $X_i(nT)$. At the same time that the transformation (A.51) takes place, the newborn individuals are added to the existing population in the form of a new cohort:

$$
\begin{cases}
N_1(nT^+) = \sum_{i=1}^{M} F(X_i(nT^-), Y_i(nT^-)) N_i(nT^-) \\
X_1(nT^+) = w_b/(1+q_J) \\
Y_1(nT^+) = q_J w_b/(1+q_J)
\end{cases}
\tag{A.52}
$$

Overall, the simulation of the discrete-continuous consumer-resource model of chapter 10 thus involves the numerical integration of a (large) system of ordinary differential equations in between the reproduction events occurring at $t = nT$ ($n = 1, 2, 3, \dots$). The dimension of this system of ODEs changes at a reproduction event through the addition of a new cohort. Simultaneously, some of the variables are reset. The dimension of the system is reduced whenever the number of individuals in a given cohort becomes negligible, at which time this cohort and the differential equations describing its dynamics are removed from the model formulation.

In chapter 10, we also discuss results from variants of the model described above that account for two different resources in two separate parts of the entire lake habitat. Population-level formulations for these model variants differ only slightly from the formulation presented above and are hence readily derived from it. For example, in chapter 10 a model variant was analyzed, which involved a benthic prey population R' living on the bottom surface of the littoral zone of the lake. Consumers were assumed to spend a fraction p of their foraging time in this littoral zone, which was assumed to occupy 10 percent of the total lake volume. The expression for the rate $E(w,R,R')$, at which consumers assimilate resource biomass from foraging, is shown in equation (10.13), while equations (10.14) and (10.15) listed the respective foraging rates, $\gamma(w,R,R')$ and $\gamma'(w,R,R')$, of individual consumers on resource R and R', respectively. Apart from the fact that the new expression for the biomass assimilation rate $E(w,R,R')$ makes the net-production rate $E_g(x,y,R,R')$ dependent on both R and R', these assumptions do not structurally change the system of ODEs (A.49). The most important difference is that these systems of ODEs describing growth and mortality of consumers are, in this variant of the model, coupled to *two*, as opposed to one, ordinary differential equations, describing the dynamics of R and R', respectively. The dynamics of R are described by

$$\frac{dR}{dt} = G(R) - \frac{\sum_{i=1}^{M} \gamma((1+q_J)X_i,R,R')N_i}{0.9 \cdot V} \qquad (A.53)$$

This equation differs from the original ODE (A.50), because the difference in individual-level consumption rate $\gamma(w,R,R')$ of resource R translates into a different consumption rate of this resource by the entire consumer population, as expressed by the term $\sum_{i=1}^{M} \gamma((1+q_J)X_i,R,R')N_i$. In addition, this resource consumption rate is exerted in, and hence scaled by, a smaller volume ($0.9 \cdot V$), because it is assumed that this resource occurs in only 90 percent of the entire lake. In a similar vein, the dynamics of the other resource, the benthic prey population living at the bottom surface in the littoral zone, is described by the ODE:

$$\frac{dR'}{dt} = G(R') - \frac{\sum_{i=1}^{M} \gamma'((1+q_J)X_i,R,R')N_i}{S} \qquad (A.54)$$

This ODE also expresses that the consumption rate of this resource by the entire consumer population, $\sum_{i=1}^{M} \gamma'((1+q_J)X_i,R,R')N_i$, has to be scaled by the value of the habitat size, in this case the total area S of the bottom surface in the littoral lake zone (see box 10.2 for default value), to account appropriately for the feedback of all consumers in the habitat on the density of this resource, which only occurs in part of the habitat. The inclusion of the ODE (A.54) to describe the dynamics of the second resource is the main difference in the population-level formulation of this model variant, compared with the basic formulation discussed previously. All equations relating to the renumbering of cohorts and the production of new cohorts, that is, the transformation (A.51) and (A.52), carry over completely unchanged.

The presentation above also illustrates the key parts of the formulation at the population level, in which different model variants in chapter 10 differ from each other. Basically, for each additional resource an ODE must be included into the continuous-time part of the model to describe its dynamics. In addition, for each resource, the consumption of this resource by the total consumer population must be scaled by the appropriate size of the habitat in which the resource is assumed to occur in order to account correctly for the feedback of the entire consumer population on resource density. Accounting for an additional resource will make the individual-level functions, such as the net production, dependent on this resource, but does not lead to structural changes in the equations describing the dynamics of the consumer cohorts. Hence, the systems of equations (A.49), (A.51), and (A.52) remain the same.

The population-level formulation of the discrete-continuous cannibal-resource model discussed in chapter 11 is similar to the consumer-resource model described above. Accounting for cannibalism implies that the mortality rate function $\mu(x, y)$ of victims of cannibalism includes a term representing mortality through cannibalism on top of background and starvation mortality (equations (11.9) and (11.10)). In addition, the resource consumption rate of cannibals will depend on their encounter rate with victims through their functional response. Lastly, the net-energy-production rate of cannibals will be increased, owing to their additional mass intake from preying on victims. Accounting for cannibalism hence clearly leads to changes in the functions describing the individual life history. The population-level formulation of the cannibal-resource model is, however, the same as for the consumer-resource model, consisting of an alternation between the integration of a large system of ordinary differential equations (equations (A.49) and (A.50)) to account for the dynamics during a growth season and a transformation (equations (A.51) and (A.52)) to account for the discrete reproduction process.

9 A DEMAND-DRIVEN ENERGY BUDGET MODEL

In chapter 12, we analyzed the dynamics of a size-structured consumer-resource model based on a demand-driven principle of individual energetics. In this section we develop the individual-level, dynamic energy budget model that forms the core of this demand-driven population model. For a schematic overview of the energy budget of an individual organism, we refer to figure 12.1.

All individuals are characterized by two measures of individual body size: structural mass s and energy reserves E. The body condition of an individual is taken as the ratio of these two measures of size, E/s. Resource ingestion is assumed to follow a Type II functional response as a function of resource density R with attack or searching rate a and handling time τ. Attack rate and handling time are assumed to be proportional and inversely proportional, respectively, to s^q. A feedback between the body condition of an individual organism and its food ingestion rate is implemented by assuming that the attack rate is a sigmoid, decreasing function of the ratio E/s:

$$a = \frac{2M_C/H_C}{1 + \exp(\eta(E/s - \theta))} s^q$$

In this equation, θ represents the target value of the individual body condition. The parameter η determines how rapidly the resource ingestion responds to changes in body condition. At the target body condition of $E/s = \theta$, the attack rate equals its standard value of $M_C/H_C s^q$, an expression that will be explained below. At very low body conditions, the attack rate approaches twice this value, whereas it decreases to 0 at very high body conditions. The steepness of the relationship is determined by η, such that for very high values of η the attack rate will be maximal as soon as the body condition drops below its target value θ, whereas the individual will stop foraging when the body condition rises above θ. The handling time τ is assumed to be described by:

$$\tau = s^{-q}/M_C$$

Our aim is to formulate a model that is close to the Kooijman-Metz energy budget model (Kooijman and Metz 1984; Kooijman 1993, 2000), which constitutes the core of the size-structured consumer-resource model introduced in chapter 3. In the Kooijman-Metz model, attack rate and handling time scale also with s^q and s^{-q}, respectively, except that the value of q is fixed at 2/3. We will adopt a default value of 2/3 for the parameter q, but will also vary q to change the energetic status of individuals in different stages of their lives. More specifically, with a small value of q, small-size individuals have a more positive energy or mass balance and are thus energetically superior, compared with larger individuals. Vice versa, larger individuals are energetically superior with larger values of q. Substituting the expressions for the attack rate and handling time into the standard expression for a Type II functional response, $aR/(1 + a\tau R)$, leads to the following expression for resource ingestion as function of resource density, energy reserves, and structural mass:

$$I(R,E,s) = \frac{M_C R}{H_C(1 + \exp(\eta(E/s - \theta)))/2 + R} s^q \tag{A.55}$$

In this equation, the parameters M_C and H_C represent the scaling constant of maximum ingestion rate with s^q and the half-saturation resource density, respectively, that are reached at the target body condition $E/s = \theta$. At the target body condition, the ingestion rate is hence analogous to the ingestion rate scaling in the consumer-resource model based on Kooijman-Metz energetics, involving the same parameters. Following ingestion, the food is assumed to be assimilated with conversion efficiency σ_C.

Growth in structural mass with age is assumed to be nonplastic and hence to follow the ordinary differential equation:

$$\frac{ds}{da} = g(s) \tag{A.56}$$

independent of resource density R. This ODE is assumed to have an explicit solution $s(a)$, such that the growth rate function $g(s)$ equals the derivative of the size-age relationship with respect to age, $s'(a)$. Given the expression for the growth in structural mass, the energetic requirements for growth equal:

$$C_G(s) = \varepsilon_G g(s) \tag{A.57}$$

In this expression, ε_G is the proportionality constant that represents the energy requirements for the production of a unit mass by means of somatic growth. To follow the Kooijman-Metz energy budget model as closely as possible, the body length of individual consumers will be assumed proportional to the cubic root of structural mass, and growth in individual length will be assumed to follow a von Bertalanffy growth equation with growth rate γ_C. Hence, the cubic root of structural mass is given by the following function of age:

$$s^{1/3}(a) = s_m^{1/3} - \left(s_m^{1/3} - s_b^{1/3}\right)e^{-\gamma_c a},$$

with s_m representing the maximum structural mass that individuals can reach and s_b representing the structural mass of a newborn. The size-age relationship therefore equals:

$$s(a) = \left(s_m^{1/3} - \left(s_m^{1/3} - s_b^{1/3}\right)e^{-\gamma_c a}\right)^3 \tag{A.58}$$

Taking the derivative $s'(a)$ yields, after some manipulation, the following expression for the growth rate function $g(s)$:

$$g(s) = 3\gamma_C\left(s_m^{1/3} s^{2/3} - s\right) \tag{A.59}$$

Maintenance requirements are assumed to be proportional to total body weight, which equals the sum of energy reserves and structural mass, with proportionality constant T_C:

$$C_M(E,s) = T_C(E + s) \tag{A.60}$$

Energy investment in reproduction is assumed to start after reaching the maturation size s_j. Given the constant relationship between age and size, this size threshold is equivalent to an age threshold. The juvenile delay between birth and the onset of reproduction is therefore constant. Since well-fed organisms can be expected to have higher fecundity than poorly fed ones and larger individuals a higher fecundity than smaller ones, we assume both body condition and individual structural mass to influence the amount of energy invested into reproduction. More specifically, fecundity is taken to be proportional to individual body condition and is assumed to scale with s^q. With the latter assumption, we again closely follow the Kooijman-Metz model, in which fecundity is also proportional to s^q with $q = 2/3$. These assumptions result in the following expression for fecundity as a function of structural mass and body condition:

$$b(E,s) = \alpha_C(E/s)s^q$$

All newborn individuals are assumed to be born with the target body condition θ, such that the energetic costs of reproduction are given by:

$$C_R(E,s) = \varepsilon_R(1+\theta)s_b b(E,s)$$

In this expression ε_R is the proportionality constant that represents the energy requirements for the production of unit mass by means of reproduction.

Taking together the energy assimilation through feeding and all energy expenditures for maintenance, growth, and reproduction, it follows that the dynamics of the energy reserves are described by the ODE:

$$\frac{dE}{da} = \varepsilon_E\left(\sigma_C I(R,E,s) - C_M(E,s) - C_G(s) - C_R(E,s)\right) \qquad \text{(A.61)}$$

In this equation, the parameter ε_E is introduced to describe the conversion efficiency of transforming energy reserve mass into net energy and vice versa. Reserve dynamics can be either anabolic (when energy intake through foraging exceeds total energy demands for growth, maintenance, and reproduction) or catabolic (when energy intake is smaller than the total energy demands). For the sake of simplicity, the conversion efficiency between energy and reserves is assumed to be identical for anabolic and catabolic reserves dynamics, although in reality these efficiencies are likely to differ.

To conclude the formulation of the life history model, individual mortality is assumed to be the sum of background mortality rate μ_C and starvation mortality in the following way (see Persson, Leonardsson, et al. 1998):

$$d(E,s) = \mu_C + \mu_s \max\left((\theta_s s/E - 1),0\right)$$

This formulation implies that individuals experience background mortality when their body condition E/s exceeds the threshold value θ_s and experience increasing mortality when body condition decreases below this threshold, such that death through starvation is certain to occur before individuals have depleted their energy reserves entirely.

The consumer population can now be characterized by its age distribution, which will be denoted by $c(t, a)$. The integral of $c(t, a)$ between a particular lower and upper bound a_1 and a_2, respectively, represents the number of individuals with an age between a_1 and a_2. The population will furthermore be characterized by the function $E(t, a)$, which represents the amount of energy reserves of individuals with an age a. At any given moment in time, the relationship between energy reserves and the age of individuals is unique, given that all individuals are born with the same body condition and individuals born at the same time remain identical throughout life. The relationship $E(t, a)$ between energy reserves and age may, however, vary over time as a consequence of changing resource densities. The dynamics of the consumer population and its resource can be described by a set of two partial differential equations for the functions $c(t, a)$ and $E(t, a)$ and an ODE describing the changes in resource density R (de Roos 1997). Box A.2 summarizes these population-dynamic equations, as well as the life-history functions that constitute the ingredients of this consumer-resource model. Given that individual life history is based on a demand-driven, as opposed to a supply-driven principle of individual energetics, we will refer to this size-structured consumer-resource model as the "demand-driven" consumer-resource model. The population model assumes that autonomous growth of the resource in the absence of consumers follows semichemostat dynamics, similar to the dynamics of the size-structured consumer-resource model based on Kooijman-Metz energetics that is presented in box 3.6.

PARAMETERIZATION OF THE DEMAND-DRIVEN MODEL

The parameterization of the demand-driven consumer-resource model follows as closely as possible the default parameter values of the size-structured consumer-resource model based on Kooijman-Metz energetics, as presented in box 3.7. All derivations discussed below are hence carried out for the default value of $q = 2/3$ that underlies the Kooijman-Metz model. The resulting set of default parameters are summarized in box A.3. Notice, however, that in contrast to the Kooijman-Metz model, in the demand-driven model all mass- and biomass-related parameters are expressed in terms of grams as opposed to milligrams for technical reasons.

Parameter values for H_C, γ_C, μ_C, ρ, and R_{max} occur also in box 3.7 and can hence be used directly. The target body condition θ, the food assimilation efficiency σ_C, the conversion efficiency of energy into offspring mass ε_R, and

BOX A.2

Individual-Level Functions and Population Equations
of the Demand-Driven Consumer-Resource Model

Dynamic equations	Description
$\dfrac{\partial c(t,a)}{\partial t} + \dfrac{\partial c(t,a)}{\partial a} = -d(E(t,a),s(a))c(t,a)$	Consumer age-distribution dynamics
$c(t,0) = \displaystyle\int_0^\infty b(E(t,a),s(a))c(t,a)\,da$	Consumer total population reproduction
$\dfrac{\partial E(t,a)}{\partial t} + \dfrac{\partial E(t,a)}{\partial a} = h(R,E(t,a),s(a))$	Dynamics of the reserves-age relation
$E(t,0) = \theta$	Energy reserves at birth
$\dfrac{dR}{dt} = G(R) - \displaystyle\int_0^\infty I(R,E(t,a),s(a))c(t,a)\,da$	Resource biomass dynamics

Function	Expression	Description
$G(R)$	$\rho(R_{max} - R)$	Resource growth in consumer absence
$s(a)$	$(s_m^{1/3} - (s_m^{1/3} - s_b^{1/3})e^{-\gamma_c a})^3$	Size-age relation
$I(R,E,s)$	$\dfrac{M_C R}{H_C(1+\exp(\eta(E/s-\theta)))/2 + R}s^q$	Resource ingestion rate by consumers
$b(E,s)$	$\begin{cases} \alpha_C(E/s)s^q & \text{if } s \geq s_j \\ 0 & \text{otherwise} \end{cases}$	Consumer fecundity
$d(E,s)$	$\mu_C + \mu_s \max((\theta_s s/E - 1),0)$	Consumer mortality rate
$C_M(E,s)$	$T_C(E+s)$	Energy requirements for maintenance
$C_G(s)$	$3\varepsilon_G\gamma_C(s_m^{1/3}s^{2/3} - s)$	Energy requirements for growth in structural mass
$C_R(E,s)$	$\varepsilon_R(1+\theta)s_b b(E,s)$	Energy requirements for reproduction
$h(R,E,s)$	$\varepsilon_E(\sigma_C I(R,E,s) - C_M(E,s) - C_G(s) - C_R(E,s))$	Consumer reserves dynamics

the starvation mortality parameters θ_s and μ_s are not part of the Kooijman-Metz model shown in boxes 3.6 and 3.7, but do occur as part of the consumer-resource model studied in chapter 10, albeit labeled with different symbols. For these parameters, we use the default values presented in box 10.2, such that $\theta = 0.7$, $\sigma_C = 0.6$, $\varepsilon_R = 2.0$, $\theta_s = 0.2$ and $\mu_s = 0.2$. In the consumer-resource

BOX A.3
DEFAULT PARAMETERS FOR THE DEMAND-DRIVEN
CONSUMER-RESOURCE MODEL IN BOX A.2

Parameter	Default value	Unit	Description
Resource			
ρ	0.1	day^{-1}	Resource turnover rate
R_{max}	0.01	g/L	Resource maximum biomass density
Consumer			
s_b	0.002	g	Structural mass at birth
s_j	7	g	Structural mass at maturation
s_m	30	g	Maximum structural mass
q	2/3	—	Scaling exponent of ingestion and fecundity
M_C	0.33	$g \cdot g^{-q} \cdot day^{-1}$	Scaling constant in maximum ingestion rate
H_C	$1.5 \cdot 10^{-5}$	g/L	Ingestion half-saturation resource density
θ	0.7	—	Target body condition
η	20	—	Steepness in condition scaling of intake rate
σ_C	0.6	—	Resource assimilation efficiency
γ_C	0.006	day^{-1}	Growth rate constant in structural mass
α_C	14	$g \cdot g^{-q} \cdot day^{-1}$	Scalar constant in adult fecundity
T_C	0.03	day^{-1}	Maintenance costs per unit total mass
ε_G	1.0	$g \cdot g^{-1}$	Growth costs per unit structural mass
ε_R	2.0	$g \cdot g^{-1}$	Reproduction costs per unit offspring mass
ε_E	1.0	$g \cdot g^{-1}$	Energy-to-reserves conversion efficiency
μ_C	0.01	day^{-1}	Consumer background mortality rate
μ_s	0.2	day^{-1}	Scalar constant in starvation mortality
θ_s	0.2	—	Condition threshold for starvation mortality

model studied in chapter 10, it was moreover assumed that the food assimilation efficiency included all costs related to conversion between energy and mass, such that the conversion between net energy production and body mass and between reserves (referred to as reversible mass) and net energy did not involve further overheads. We adopt the same principle here and hence assume default values of 1.0 for both ε_G and ε_E.

Estimates for the parameters s_b and s_j can be derived from the parameter values for length at birth and maturation, as presented in box 3.7, in combination with the weight-length relation, $w_C(\ell) = \beta_C \ell^3$, that is assumed in the Kooijman-Metz model of chapter 3. Equating individual weight to the sum of energy reserves and structural mass leads to the relationship $s = \beta_C \ell^3/(1 + \theta)$. This identity can be used to calculate values for s_b and s_j from the corresponding values of ℓ_b and ℓ_j in box 3.7. We consider it unrealistic to base the default parameter estimate for s_m on the maximum length ℓ_m that individuals can reach in the Kooijman-Metz model, as the latter occurs only under unlimited food conditions. We hence base our estimate of s_m on an assumed maximum length of 180 mm.

For $q = 2/3$, the scaling constants relating maximum ingestion to $s^{2/3}$ and ℓ^2 in the demand-driven and Kooijman-Metz models, respectively, can be related to each other, given that $\ell^2 = (\beta_C/(1 + \theta))^{-2/3} s^{2/3}$. Assuming a target body condition θ for reproducing adults, the same relationship between length and structural mass can be exploited to relate the scaling constant for adult fecundity in the demand-driven model to the corresponding constant in the Kooijman-Metz model. In the Kooijman-Metz model the energy requirements for maintenance per unit total mass are subsumed in the compound parameters ℓ_m and γ_C. We can hence not estimate a default value for T_C in the demand-driven model from the Kooijman-Metz model of chapter 3. However, Claessen and de Roos (2003) study a size-structured cannibal-resource model, which in the absence of cannibalism is close to the consumer-resource model based on Kooijman-Metz energetics studied in chapter 3. These authors provide a scaling constant of $2.5 \cdot 10^{-7}$ relating maintenance requirements to cubed length in the Kooijman-Metz model, which we use to derive a default value for T_C in the demand-driven model equal to $2.5 \cdot 10^{-7}/\beta_C \approx 0.03$.

The remaining parameter η determines the steepness in the sigmoid function relating the feeding response (the individual attack rate) to the individual body condition. For this parameter, a default value of 20 is assumed. Independent of the value of η the attack rate always equals 50 percent of its maximum for the target body condition of $\theta - 0.7$. The value of $\eta = 20$ implies that the attack rate is only 10 percent of its maximum value when body condition increases to 0.8, while the attack rate reaches 90 percent of its maximum when body condition drops to 0.6.

References

Abrams, P. A. 1988. How should resources be counted? *Theoretical Population Biology* 33: 226–242.

———. 1996. Dynamics and interactions in food webs with adaptive foragers. *In*: G. A. Polis and K. O. Winemiller, eds., *Food webs: Integration of patterns and dynamics*, 113–121. New York: Chapman and Hall.

———. 2009. When does greater mortality increase population size? The long history and diverse mechanisms underlying the hydra effect. *Ecology Letters* 12: 462–474.

Abrams, P. A., and H. Matsuda. 2005. The effect of adaptive change in the prey on the dynamics of an exploited predator population. *Canadian Journal of Fisheries and Aquatic Sciences* 62: 758–766.

Abrams, P. A., B. A. Menge, G. G. Mittelbach, D. A. Spiller, and P. Yodzis. 1996. The role of indirect effects in food webs. *In*: G. A. Polis and K. O. Winemiller, eds., *Food webs: Integration of patterns and dynamics*, 371–395. New York: Chapman and Hall.

Abrams, P. A., and C. Quince. 2005. The impact of mortality on predator population size and stability in systems with stage-structured prey. *Theoretical Population Biology* 68: 253–266.

Aljetlawi, A. A., E. Sparrevik, and K. Leonardsson. 2004. Prey-predator size-dependent functional response: Derivation and rescaling to the real world. *Journal of Animal Ecology* 73: 239–252.

Allee, W. C. 1931. *Animal aggregations, a study in general sociology*. Chicago: University of Chicago Press.

———. 1938. *The social life of animals*. London: William Heinemann.

Allendorf, F. W., P. R. England, G. Luikart, P. A. Ritchie, and N. Ryman. 2008. Genetic effects of harvest on wild animal populations. *Trends in Ecology and Evolution* 23: 327–337.

Amarasekare, P. 2008. Coexistence of intraguild predators and prey in resource-rich environments. *Ecology* 89: 2786–2797.

Andersen, K. H., and J. E. Beyer. 2006. Asymptotic size determines species abundance in the marine size spectrum. *American Naturalist* 168: 54–61.

Arim, M., and P. A. Marquet. 2004. Intraguild predation: A widespread interaction related to species biology. *Ecology Letters* 7: 557–564.

Armstrong, R. A., and R. McGehee. 1980. Competitive exclusion. *American Naturalist* 115: 151–170.

Begon, M., J. L. Harper, and C. R. Townsend. 1996. *Ecology: Individuals, populations and communities*. 3rd ed. Oxford: Blackwell Scientific Publications.

Beisner, B. E., A. R. Ives, and S. R. Carpenter. 2003. The effects of an exotic fish invasion on the prey communities of two lakes. *Journal of Animal Ecology* 72: 331–342.

Bern, L. 1994. Particle selection over a broad size range by crustacean zooplankton. *Freshwater Biology* 32: 105–112.

Bolker, B., M. Holyoak, V. Krivan, L. Rowe, and O. Schmitz. 2003. Connecting theoretical and empirical studies of trait-mediated interactions. *Ecology* 84: 1101–1114.

Borer, E. T. 2002. Intraguild predation in larval parasitoids: Implications for coexistence. *Journal of Animal Ecology* 71: 957–965.

Breck, J. E., and M. J. Gitter. 1983. Effect of fish size on the reactive distance of bluegill (*Lepomis macrochirus*) sunfish. *Canadian Journal of Fisheries and Aquatic Sciences* 40: 162–167.

Briggs, C. J., S. M. Sait, M. Begon, D. J. Thompson, and H.C.J. Godfray. 2000. What causes generation cycles in populations of stored-product moths? *Journal of Animal Ecology* 69: 352–366.

Broekhuizen, N., W.S.C. Gurney, A. Jones, and A. D. Bryant. 1994. Modeling compensatory growth. *Functional Ecology* 8: 770–782.

Brooks, J. L., and S. I. Dodson. 1965. Predation, body size, and competition of plankton. *Science* 150: 28–35.

Brose, U., R. B. Ehnes, B. C. Rall, O. Vucic-Pestic, E. L. Berlow, and S. Scheu. 2008. Foraging theory predicts predator-prey energy fluxes. *Journal of Animal Ecology* 77: 1072–1078.

Brown, J. H., J. F. Gillooly, A. P. Allen, V. M. Savage, and G. B. West. 2004. Toward a metabolic theory of ecology. *Ecology* 85: 1771–1789.

Bulmer, M. G. 1977. Periodical insects. *American Naturalist* 111: 1099–1117.

Butler, M. I., and C. W. Burns. 1991. Prey selectivity of *Piona exigua*, a planktonic water mite. *Oecologia* 86: 210–222.

Byström, P. 2006. Recruitment pulses induce cannibalistic giants in Arctic char. *Journal of Animal Ecology* 75: 434–444.

Byström, P., and J. Andersson. 2005. Size-dependent foraging capacities and intercohort competition in an ontogenetic omnivore (Arctic char). *Oikos* 110: 523–536.

Byström, P., and E. García-Berthou. 1999. Density dependent growth and size specific competitive interactions in young fish. *Oikos* 86: 217–232.

Byström, P., L. Persson, and E. Wahlström. 1998. Competing predators and prey: Juvenile bottlenecks in whole-lake experiments. *Ecology* 79: 2153–2167.

Calder, W. A., III. 1984. *Size, function and life history*. Cambridge, MA: Harvard University Press.

Cameron, T. C., and T. G. Benton. 2004. Stage-structured harvesting and its effects: An empirical investigation using soil mites. *Journal of Animal Ecology* 73: 996–1006.

Campbell, C. E. 1991. Prey selectivities of threespine sticklebacks (*Gasterosteus aculeatus*) and phantom midge larvae (*Chaoborus* spp.) in Newfoundland lakes. *Freshwater Biology* 25: 155–167.

Carscadden, J. E., and K. T. Frank. 2002. Temporal variability in the condition factors of Newfoundland capelin (*Mallotus villosus*) during the past two decades. *ICES Journal of Marine Science* 59: 950–958.

Carscadden, J. E., K. T. Frank, and W. C. Leggett. 2001. Ecosystem changes and the effects on capelin (*Mallotus villosus*), a major forage species. *Canadian Journal of Fisheries and Aquatic Sciences* 58: 73–85.

Case, T. J. 1978. Evolution and adaptive significance of postnatal-growth rates in terrestrial vertebrates. *Quarterly Review of Biology* 53: 243–282.

Casini, M., M. Cardinale, and J. Hjelm. 2006. Inter-annual variation in herring, *Clupea harengus*, and sprat, *Sprattus sprattus*, condition in the central Baltic Sea: What gives the tune? *Oikos* 112: 638–650.

Casini, M., J. Hjelm, J. C. Molinero, J. Lövgren, M. Cardinale, V. Bartolino, A. Belgrano, and G. Kornilovs. 2009. Trophic cascades promote threshold-like shifts in pelagic marine ecosystems. *Proceedings of the National Academy of Sciences* 106: 197–202.

Caswell, H., and A. M. John. 1992. From the individual to the population in demographic models. *In*: D. L. DeAngelis and L. J. Gross, eds., *Individual-based models and approaches in ecology: Populations, communities and ecosystems*, 36–61. New York: Chapman-Hall.

Chase, J. M., P. A. Abrams, J. P. Grover, S. Diehl, P. Chesson, R. D. Holt, S. A. Richards, R. M. Nisbet, and T. J. Case. 2002. The interaction between predation and competition: A review and synthesis. *Ecology Letters* 5: 302–315.

Chesson, P. 1986. Environmental variation and the coexistence of species. *In*: J. Diamond and T. J. Case, eds., *Community ecology*, 240–256. New York: Harper and Row.

———. 2000. Mechanisms of maintenance of species diversity. *Annual Review of Ecology and Systematics* 31: 343–366.

Childress, M. J., and W. F. Herrnkind. 2001. Influence of conspecifics on the ontogenetic habitat shift of juvenile Caribbean spiny lobsters. *Marine and Freshwater Research* 52: 1077–1084.

Choi, J. S., K. T. Frank, W. C. Leggett, and K. Drinkwater. 2004. Transition to an alternate state in a continental shelf ecosystem. *Canadian Journal of Fisheries and Aquatic Sciences* 61: 505–510.

Christensen, B. 1996. Predator foraging capabilities and prey antipredator behaviours: Pre- versus postcapture constraints on size-dependent predator-prey interactions. *Oikos* 76: 368–380.

Claessen, D., and A. M. de Roos. 2003. Bistability in a size-structured population model of cannibalistic fish: A continuation study. *Theoretical Population Biology* 64: 49–65.

Claessen, D., A. M. de Roos, and L. Persson. 2000. Dwarfs and giants: Cannibalism and competition in size-structured populations. *American Naturalist* 155: 219–237.

———. 2004. Population dynamic theory of size-dependent cannibalism. *Proceedings of the Royal Society B: Biological Sciences* 271: 333–340.

Claessen, D., A. de Vos, and A. M. de Roos. 2009. Bioenergetics, overcompensation and the source/sink status of marine reserves. *Canadian Journal of Fisheries and Aquatic Sciences* 66: 1059–1071.

Claessen, D., C. van Oss, A. M. de Roos, and L. Persson. 2002. The impact of size-dependent predation on population dynamics and individual life history. *Ecology* 83: 1660–1675.

Cohen, J. E. 1978. *Food webs and niche space.* Princeton, NJ: Princeton University Press.

Costantino, R. F., R. A. Desharnais, J. M. Cushing, and B. Dennis. 1997. Chaotic dynamics in an insect population. *Science* 275: 389–391.

Courchamp, F., L. Berec, and J. Gascoigne. 2008. *Allee effects in ecology and conservation.* Oxford: Oxford University Press.

Crowley, P. H., and K. R. Hopper. 1994. How to behave around cannibals: A density-dependent dynamic game. *American Naturalist* 143: 117–154.

Cryer, M., G. Peirson, and C. R. Townsend. 1986. Reciprocal interactions between roach, *Rutilus rutilus*, and zooplankton in a small lake: Prey dynamics and fish growth and recruitment. *Limnology and Oceanography* 31: 1022–1038.

Cushing, J. M. 1991. A simple-model of cannibalism. *Mathematical Biosciences* 107: 47–71.

Dahlgren, C. P., and D. B. Eggleston. 2000. Ecological processes underlying ontogenetic habitat shifts in a coral reef fish. *Ecology* 81: 2227–2240.

Danks, H. V. 2002. Modification of adverse conditions by insects. *Oikos* 99: 10–24.

de Roos, A. M. 1988. Numerical methods for structured population models: The escalator boxcar train. *Numerical Methods for Partial Differential Equations* 4: 173–195.

———. 1997. A gentle introduction to physiologically structured population models. *In*: S. Tuljapurkar and H. Caswell, eds., *Structured population models in marine, terrestrial and freshwater systems*, 119–204. New York: Chapman-Hall.

———. 2008. Demographic analysis of continuous-time life-history models. *Ecology Letters* 11: 1–15.

de Roos, A. M., D. S. Boukal, and L. Persson. 2006. Evolutionary regime shifts in age and size at maturation of exploited fish stocks. *Proceedings of the Royal Society B: Biological Sciences* 273: 1873–1880.

de Roos, A. M., O. Diekmann, P. Getto, and M. A. Kirkilionis. 2010. Numerical equilibrium analysis for structured consumer resource models. *Bulletin of Mathematical Biology* 72: 259–297.

de Roos, A. M., O. Diekmann, and J.A.J. Metz. 1992. Studying the dynamics of structured population models: A versatile technique and its application to *Daphnia*. *American Naturalist* 139: 123–147.

de Roos, A. M., N. Galic, and H. Heesterbeek. 2009. How resource competition shapes individual life history for nonplastic growth: Ungulates in seasonal food environments. *Ecology* 90: 945–960.

de Roos, A. M., K. Leonardsson, L. Persson, and G. G. Mittelbach. 2002. Ontogenetic niche shifts and flexible behavior in size-structured populations. *Ecological Monographs* 72: 271–292.

de Roos, A. M., E. McCauley, R. M. Nisbet, W.S.C. Gurney, and W. W. Murdoch. 1997. What individual life histories can (and cannot) tell about population dynamics. *Aquatic Ecology* 31: 37–45.

de Roos, A. M., E. McCauley, and W. G. Wilson. 1991. Mobility versus density-limited predator-prey dynamics on different spatial scales. *Proceedings of the Royal Society B: Biological Sciences* 246: 117–122.

de Roos, A. M., J.A.J. Metz, E. Evers, and A. Leipoldt. 1990. A size-dependent predator-prey interaction: Who pursues whom? *Journal of Mathematical Biology* 28: 609–643.

de Roos, A. M., and L. Persson. 2001. Physiologically structured models: From versatile technique to ecological theory. *Oikos* 94: 51–71.

———. 2002. Size-dependent life-history traits promote catastrophic collapses of top predators. *Proceedings of the National Academy of Sciences* 99: 12907–12912.

———. 2003. Competition in size-structured populations: Mechanisms inducing cohort formation and population cycles. *Theoretical Population Biology* 63: 1–16.

———. 2005. Unstructured population models: Do population-level assumptions yield general theory? *In*: K. Cuddington and B. E. Beisner, eds., *Ecological paradigms lost: Routes of theory change*, 31–62, San Diego: Elsevier Academic Press.

de Roos, A. M., L. Persson, and E. McCauley. 2003. The influence of size-dependent life-history traits on the structure and dynamics of populations and communities. *Ecology Letters* 6: 473–487.

de Roos, A. M., L. Persson, and H. R. Thieme. 2003. Emergent Allee effects in top predators feeding on structured prey populations. *Proceedings of the Royal Society B: Biological Sciences* 270: 611–618.

de Roos, A. M., T. Schellekens, T. van Kooten, K. van de Wolfshaar, D. Claessen, and L. Persson. 2007. Food-dependent growth leads to overcompensation in stage-specific biomass when mortality increases: The influence of maturation versus reproduction regulation. *American Naturalist* 170: E59–E76.

de Roos, A. M., T. Schellekens, T. van Kooten, and L. Persson. 2008a. Stage-specific predator species help each other to persist while competing for a single prey. *Proceedings of the National Academy of Sciences* 105: 13930–13935.

de Roos, A. M., T. Schellekens, T. van Kooten, K. van de Wolfshaar, D. Claessen, and L. Persson. 2008b. Simplifying a physiologically structured population model to a stage-structured biomass model. *Theoretical Population Biology* 73: 47–62.

Dhooge, A., W. Govaerts, and A. K. Yu. 2003. MATCONT: A MATLAB package for numerical bifurcation analysis of ODEs. *ACM Transactions on Mathematical Software (TOMS)* 29: 141–164.

Diehl, S., and M. Feissel. 2000. Effects of enrichment on three-level food chains with omnivory. *American Naturalist* 155: 200–218.

Diekmann, O., M. Gyllenberg, and J.A.J. Metz. 2003. Steady-state analysis of structured population models. *Theoretical Population Biology* 63: 309–338.

Diekmann, O., M. Gyllenberg, J.A.J. Metz, S. Nakaoka, and A. M. de Roos. 2010. *Daphnia* revisited: Local stability and bifurcation theory for physiologically structured population models explained by way of an example. *Journal of Mathematical Biology* 61: 277–318.

Dong, Q. A., and D. L. DeAngelis. 1998. Consequences of cannibalism and competition for food in a smallmouth bass population: An individual-based modeling study. *Transactions of the American Fisheries Society* 127: 174–191.

Duplisea, D. E. 2005. Running the gauntlet: The predation environment of small fish in the northern Gulf of St. Lawrence, Canada. *ICES Journal of Marine Science* 62: 412–416.

Elgar, M. A., and B. J. Crespi, eds. 1992. *Cannibalism: Ecology and evolution among diverse taxa*. New York: Oxford University Press.

Elliott, J. M. 2004. Contrasting dynamics in two subpopulations of a leech metapopulation over 25 year-classes in a small stream. *Journal of Animal Ecology* 73: 272–282.

Ellner, S., and P. Turchin. 1995. Chaos in a noisy world: New methods and evidence from time-series analysis. *American Naturalist* 145: 343–375.

Elwood, R. 1992. Pup-cannibalism in rodents: Causes and consequences. *In*: M. A. Elgar and B. J. Crespi, eds., *Cannibalism: Ecology and evolution among diverse taxa*, 299–322. New York: Oxford University Press.

Fagan, W. F., and G. M. Odell. 1996. Size-dependent cannibalism in praying mantids: Using biomass flux to model size-structured populations. *American Naturalist* 147: 230–268.

Farlow, J. O. 1976. Consideration of trophic dynamics of a late cretaceous large dinosaur community (Oldman formation). *Ecology* 57: 841–857.

Finke, D. L., and R. F. Denno. 2006. Spatial refuge from intraguild predation: Implications for prey suppression and trophic cascades. *Oecologia* 149: 265–275.

Finstad, A. G., O. Ugedal, and O. K. Berg. 2006. Growing large in a low grade environment: Size dependent foraging gain and niche shifts to cannibalism in Arctic char. *Oikos* 112: 73–82.

Fisher, M. E. 1987. An age-structured fish population-model with coupled size and population-density. *Mathematical Biosciences* 86: 15–34.

Fox, L. R. 1975. Cannibalism in natural populations. *Annual Review of Ecology and Systematics* 6: 87–106.

Frank, K. T., B. Petrie, J. S. Choi, and W. C. Leggett. 2005. Trophic cascades in a formerly cod-dominated ecosystem. *Science* 308: 1621–1623.

Fryxell, J. M., and P. Lundberg. 1993. Optimal patch use and metapopulation dynamics. *Evolutionary Ecology* 7: 379–393.

Fulford, R. S., J. A. Rice, T. J. Miller, and F. P. Binkowski. 2006. Elucidating patterns of size-dependent predation on larval yellow perch (*Perca flavescens*) in Lake Michigan: An experimental and modeling approach. *Canadian Journal of Fisheries and Aquatic Sciences* 63: 11–27.

Gause, G. F. 1934. *The struggle for existence*. Baltimore: Williams and Wilkins.

Gilliam, J. F. 1982. Habitat use and competitive bottlenecks in size-structured populations. PhD diss., Michigan State University.

Gillooly, J. F., J. H. Brown, G. B. West, V. M. Savage, and E. L. Charnov. 2001. Effects of size and temperature on metabolic rate. *Science* 293: 2248–2251.

Gleeson, S. K., and D. S. Wilson. 1986. Equilibrium diet: Optimal foraging and prey coexistence. *Oikos* 46: 139–144.

Gotthard, K. 2008. Adaptive growth decisions in butterflies. *Bioscience* 58: 222–230.

Grover, J. P., and R. D. Holt. 1998. Disentangling resource and apparent competition: Realistic models for plant-herbivore communities. *Journal of Theoretical Biology* 191: 353–376.

Guill, C. 2009. Alternative dynamical states in stage-structured consumer populations. *Theoretical Population Biology* 76: 168–178.

Gurney, W.S.C., S. P. Blythe, and R. M. Nisbet. 1980. Nicholson's blowflies revisited. *Nature* 287: 17–21.

Gurney, W.S.C., E. McCauley, R. M. Nisbet, and W. W. Murdoch. 1990. The physiological ecology of *Daphnia*: A dynamic model of growth and reproduction. *Ecology* 71: 716–732.

Gurney, W.S.C., D.A.J. Middleton, R. M. Nisbet, E. McCauley, W. W. Murdoch, and A. M. de Roos. 1996. Individual energetics and the equilibrium demography of structured populations. *Theoretical Population Biology* 49: 344–368.

Gurney, W.S.C., and R. M. Nisbet. 1985. Fluctuation periodicity, generation separation, and the expression of larval competition. *Theoretical Population Biology* 28: 150–180.

Hairer, E., S. P. Norsett, and G. Wanner. 1993. *Solving ordinary differential equations I: Nonstiff problems*. Berlin: Springer-Verlag.

Hairston, N. G., F. E. Smith, and L. B. Slobodkin. 1960. Community structure, population control, and competition. *American Naturalist* 94: 421–425.

Hall, D. J., W. E. Cooper, and E. E. Werner. 1970. An experimental approach to the production dynamics and structure of freshwater animal communities. *Limnology and Oceanography* 15: 839–928.

Hamrin, S. F., and L. Persson. 1986. Asymmetrical competition between age classes as a factor causing population oscillations in an obligate planktivorous fish. *Oikos* 47: 223–232.

Hansen, P. J., P. K. Bjørnsen, and B. W. Hansen. 1997. Zooplankton grazing and growth: Scaling within the 2–2,000 μm body size range. *Limnology and Oceanography* 42: 687–704.

Hanski, I. 1999. *Metapopulation ecology*. Oxford: Oxford University Press.

Hardin, G. 1960. The competitive exclusion principle. *Science* 131: 1292–1297.

Hardy, A. C. 1924. The herring in relation to its animate environment. Part 1. The food and feeding habits of the herring with special reference to the east coast of England. *Ministry of Agriculture and Fisheries, Fisheries Investigations*, ser. 2, 7: 1–45.

Hassell, M. P., H. N. Comins, and R. M May. 1991. Spatial structure and chaos in insect population dynamics. *Nature* 353: 255–258.

Hastings, A., and R. F. Costantino. 1987. Cannibalistic egg-larva interactions in *Tribolium*: An explanation for the oscillations in population numbers. *American Naturalist* 130: 36–52.

Hausfater, H., and S. B. Hrday, eds. 1984. *Infanticide: Comparative and evolutionary perspectives*, New York: Aldine.

Hendriks, A. J., and C. Mulder. 2008. Scaling of offspring number and mass to plant and animal size: Model and meta-analysis. *Oecologia* 155: 705–716.

Hilborn, D., and M. Mangel. 1997. *The ecological detective: Confronting models with data*. Princeton, NJ: Princeton University Press.

Hin, V., T. Schellekens, L. Persson, and A. M. de Roos. 2011. Coexistence of predator and prey in intraguild predation systems with ontogenetic niche shifts. *American Naturalist* 178: 701–714.

Hjelm, J., and L. Persson. 2001. Size-dependent attack rate and handling capacity: Inter-cohort competition in a zooplanktivorous fish. *Oikos* 95: 520–532.

Holling, C. S. 1966. The strategy of building models of complex systems. *In*: K.E.D. Watt, ed., *System analysis in ecology*, 195–214. New York: Academic Press.

Holt, R. D., and G. R. Huxel. 2007. Alternative prey and the dynamics of intraguild predation: Theoretical perspectives. *Ecology* 88: 2706–2712.

Holt, R. D., and J. H. Lawton. 1994. The ecological consequences of shared natural enemies. *Annual Review of Ecology and Systematics* 25: 495–520.

Holt, R. D., and G. A. Polis. 1997. A theoretical framework for intraguild predation. *American Naturalist* 149: 745–764.

Holyoak, M., and S. Sachdev. 1998. Omnivory and the stability of simple food webs. *Oecologia* 117: 413–419.

Huss, M., and K. A. Nilsson. 2011. Experimental evidence for emergent facilitation: Promoting the existence of an invertebrate predator by killing its prey. *Journal of Animal Ecology* 80: 615–621.

Hutchings, J. A., and R. A. Myers. 1994. What can be learned from the collapse of a renewable resource: Atlantic cod, *Gadus morhua*, of Newfoundland and Labrador. *Canadian Journal of Fisheries and Aquatic Sciences* 51: 2126–2146.

ICES (International Council for the Exploration of the Sea). 2006. Report of the workshop on the decline and recovery of cod stocks throughout the North Atlantic, including trophodynamic effects (WKDRCS). St. John's, Canada. ICES CM 2006/OCC:12. Copenhagen: ICES.

———. 2007. Report of the Baltic fisheries assessment working group (WGBFAS). ICES Headquarters. ICES CM 2007/ACFM:15. Copenhagen: ICES.

Janssen, A., M. W. Sabelis, S. Magalhaes, M. Montserrat, and T. van der Hammen. 2007. Habitat structure affects intraguild predation. *Ecology* 88: 2713–2719.

Jennings, S., J. K. Pinnegar, N.V.C. Polunin, and T. W. Boon. 2001. Weak cross-species relationships between body size and trophic level belie powerful size-based trophic structuring in fish communities. *Journal of Animal Ecology* 70: 934–944.

Jones, A. E., R. M. Nisbet, W.S.C. Gurney, and S. P. Blythe. 1988. Period to delay ratio near stability boundaries for systems with delayed feedback. *Journal of Mathematical Analysis and Applications* 135: 354–368.

Juanes, F. 2003. The allometry of cannibalism in piscivorous fishes. *Canadian Journal of Fisheries and Aquatic Sciences* 60: 594–602.

Kendall, B. E., J. Prendergast, and O. N. Bjørnstad. 1998. The macroecology of population dynamics: Taxonomic and biogeographic patterns in population cycles. *Ecology Letters* 1: 160–164.

Keren-Rotem, T., A. Bouskila, and E. Geffen. 2006. Ontogenetic habitat shift and risk of cannibalism in the common chameleon (*Chamaeleo chamaeleon*). *Behavioral Ecology and Sociobiology* 59: 723–731.

Kirkilionis, M. A., O. Diekmann, B. Lisser, M. Nool, A. M. de Roos, and B. P. Sommeijer. 1997. Numerical continuation of equilibria of physiologically structured population models. I. Theory. Amsterdam: Centre for Mathematics and Computer Science, Technical Report R9714 (MAS).

Kirkilionis, M. A., O. Diekmann, B. Lisser, M. Nool, B. Sommeijer, and A. M. de Roos. 2001. Numerical continuation of equilibria of physiologically structured population models. I. Theory. *Mathematical Models and Methods in Applied Sciences* 11: 1101–1127.

Kleiber, M. 1961. *The fire of life*. New York: John Wiley.

Kooijman, S.A.L.M. 1993. *Dynamic energy budgets in biological systems*. Cambridge: Cambridge University Press.

———. 2000. *Dynamic energy and mass budgets in biological systems*. Cambridge: Cambridge University Press.

Kooijman, S.A.L.M., and J.A.J. Metz. 1984. On the dynamics of chemically stressed populations: The deduction of population consequences from effects on individuals. *Ecotoxicology and Environmental Safety* 8: 254–274.

Köster, F. W., and C. Möllmann. 2000. Trophodynamic control by clupeid predators on recruitment success in Baltic cod? *ICES Journal of Marine Science* 57: 310–323.

Krivan, V. 1996. Optimal foraging and predator-prey dynamics. *Theoretical Population Biology* 49: 265–290.

———. 1997. Dynamic ideal free distribution: Effects of optimal patch choice on predator-prey dynamics. *American Naturalist* 149: 164–178.

———. 2000. Optimal intraguild foraging and population stability. *Theoretical Population Biology* 58: 79–94.

Krivan, V., and S. Diehl. 2005. Adaptive omnivory and species coexistence in tri-trophic food webs. *Theoretical Population Biology* 67: 85–99.

Kuznetsov, Y. A. 1995. *Elements of applied bifurcation theory*. Heidelberg: Springer-Verlag.

Lambert, Y., and J. D. Dutil. 1997. Condition and energy reserves of Atlantic cod (*Gadus morhua*) during the collapse of the northern Gulf of St. Lawrence stock. *Canadian Journal of Fisheries and Aquatic Sciences* 54: 2388–2400.

Landahl, H. D., and B. D. Hansen. 1975. 3 stage population model with cannibalism. *Bulletin of Mathematical Biology* 37: 11–17.

Le Cren, E. D. 1992. Exceptionally big individual perch (*Perca fluviatilis*) and their growth. *Journal of Fish Biology* 40: 599–625.

Leibold, M. A. 1996. A graphical model of keystone predators in food webs: Trophic regulation of abundance, incidence, and diversity patterns in communities. *American Naturalist* 147: 784–812.

Leibold, M., and A. J. Tessier. 1991. Contrasting patterns of body size for *Daphnia* species that segregate by habitat. *Oecologia* 86: 342–348.

Letcher, B. H., J. A. Rice, L. B. Crowder, and K. A. Rose. 1996. Variability in survival of larval fish: Disentangling components with a generalized individual-based model. *Canadian Journal of Fisheries and Aquatic Sciences* 53: 787–801.

Levin, S. A. 1970. Community equilibria and stability, and an extension of the competitive exclusion principle. *American Naturalist* 104: 413–423.

Levins, R. 1966. Strategy of model building in population biology. *American Scientist* 54: 421–431.

Lewontin, R. C. 1964. *The genetic basis of evolutionary change*. New York: Columbia University Press.

Lika, K., and R. M. Nisbet. 2000. A dynamic energy budget model based on partitioning of net production. *Journal of Mathematical Biology* 41: 361–386.

Lilja, C. 1983. A comparative-study of postnatal-growth and organ development in some species of birds. *Growth* 47: 317–339.

Lorena, A., G. M. Marques, S.A.L.M. Kooijman, and T. Sousa. 2010. Stylized facts in microalgal growth: Interpretation in a dynamic energy budget context. *Proceedings of the Royal Society B: Biological Sciences* 365: 3509–3521.

Lundberg, S., and L. Persson. 1993. Optimal body-size and resource density. *Journal of Theoretical Biology* 164: 163–180.

Lundvall, D., R. Svanbäck, L. Persson, and P. Byström. 1999. Size-dependent predation in piscivores: Interactions between predator foraging and prey avoidance abilities. *Canadian Journal of Fisheries and Aquatic Science* 56: 1285–1292.

MacArthur, R. H. 1972. *Geographical ecology*. New York: Harper and Row.

MacArthur, R. H., and R. Levins. 1967. The limiting similarity, convergence and divergence of coexisting species. *American Naturalist* 101: 377–385.

Malthus, T. R. 1798. *An essay on the principle of population*. Harmondsworth, UK: Penguin Books.

Martinele, I., and M. D'Agosto. 2008. Predation and cannibalism among ciliate protozoans (Ciliophora: Entodiniomorphida: Ophryoscolecidae) in the rumen of sheep (*Ovis aries*). *Revista Brasileira de Zoologia* 25: 451–455.

Massie, T. M., B. Blasius, G. Weithoff, U. Gaedke, and G. F. Fussmann. 2010. Cycles, phase synchronization, and entrainment in single-species phytoplankton populations. *Proceedings of the National Academy of Sciences* 107: 4236–4241.

May, R. M. 1974. *Stability and complexity in model ecosystems*. Princeton, NJ: Princeton University Press.

————. 1976. Simple mathematical models with very complicated dynamics. *Nature* 261: 459–467.

May, R. M., and R. H. MacArthur. 1972. Niche overlap as a function of environmental variability. *Proceedings of the National Academy of Sciences* 69: 1109–1113.

McCann, K. 1998. Density-dependent coexistence in fish communities. *Ecology* 79: 2957–2967.

McCauley, E. 1993. Internal and external causes of dynamics in a freshwater plant-herbivore system. *American Naturalist* 141: 428–439.

McCauley, E., and W. W. Murdoch. 1987. Cyclic and stable populations: Plankton as paradigm. *American Naturalist* 129: 97–121.

————. 1990. Predator-prey dynamics in rich and poor environments. *Nature* 343: 455–457.

McCauley, E., W. W. Murdoch, R. M. Nisbet, and W.S.C. Gurney. 1990. The physiological ecology of *Daphnia*: Development of a model of growth and reproduction. *Ecology* 71: 703–715.

McCauley, E., W. A. Nelson, and R. M. Nisbet. 2008. Small-amplitude cycles emerge from stage-structured interactions in *Daphnia*-algal systems. *Nature* 455: 1240–1243.

McCauley, E., R. M. Nisbet, A. M. de Roos, W. W. Murdoch, and W.S.C. Gurney. 1996. Structured population models of herbivorous zooplankton. *Ecological Monographs* 66: 479–501.

McCauley, E., R. M. Nisbet, W. W. Murdoch, A. M. de Roos, and W.S.C. Gurney. 1999. Large-amplitude cycles of *Daphnia* and its algal prey in enriched environments. *Nature* 402: 653–656.

McCoy, M. W., M. Barfield, and R. D. Holt. 2009. Predator shadows: Complex life histories as generators of spatially patterned indirect interactions across ecosystems. *Oikos* 118: 87–100.

McCoy, M. W., and J. F. Gillooly. 2008. Predicting natural mortality rates of plants and animals. *Ecology Letters* 11: 710–716.

Meredith, S. N., V. F. Matveev, and P. Mayes. 2003. Spatial and temporal variability in the distribution and diet of the gudgeon (Eleotridae: *Hypseleotris* spp.) in a subtropical Australian reservoir. *Marine and Freshwater Research* 54: 1009–1017.

Metz, J.A.J., A. M. de Roos, and F. van den Bosch. 1988. Population models incorporating physiological structure: A quick survey of the basic concepts and an application to size-structured population dynamics in waterfleas. *In*: B. Ebenman and L. Persson, eds., *Size-structured populations: Ecology and evolution*, 106–126, Heidelberg: Springer-Verlag.

Metz, J.A.J., and O. Diekmann. 1986. *The dynamics of physiologically structured populations*. Heidelberg: Springer-Verlag.

Miller, T.E.X., and V.H.W. Rudolf. 2011. Thinking inside the box: Community-level consequences of stage-structured populations. *Trends in Ecology and Evolution* 26: 457–466.

Mittelbach, G. G. 1981. Foraging efficiency and body size: A study of optimal diet and habitat use by bluegills. *Ecology* 62: 1370–1386.

————. 1983. Optimal foraging and growth in bluegills. *Oecologia* 59: 157–162.

Mittelbach, G. G., and P. L. Chesson. 1987. Predation risk: Indirect effects on fish populations. *In*: W. C. Kerfoot and A. Sih, eds., *Predation: Direct and indirect impacts on aquatic communities*, 537–555. Hanover, NH: University Press of New England.

Mittelbach, G. G., and C. W. Osenberg. 1993. Stage-structured interactions in bluegill: Consequences of adult resource variation. *Ecology* 74: 2381–2394.

Mittelbach, G. G., and L. Persson. 1998. The ontogeny of piscivory and its ecological consequences. *Canadian Journal of Fisheries and Aquatic Science* 55: 1454–1465.

Mollison, D. 1991. Dependence of epidemic and population velocities on basic parameters. *Mathematical Biosciences* 107: 255–287.

Möllmann, C., G. Kornilovs, M. Fetter, and F. W. Köster. 2004. Feeding ecology of central Baltic Sea herring and sprat. *Journal of Fish Biology* 65: 1563–1581.

————. 2005. Climate, zooplankton, and pelagic fish growth in the central Baltic Sea. *ICES Journal of Marine Science* 62: 1270–1280.

Möllmann, C., B. Müller-Karulis, G. Kornilovs, and M. A. St. John. 2008. Effects of climate and overfishing on zooplankton dynamics and ecosystem structure: Regime shifts, trophic cascade, and feedback loops in a simple ecosystem. *ICES Journal of Marine Science* 65: 302–310.

Montserrat, M., S. Magalhães, M. W. Sabelis, A. M. de Roos, and A. Janssen. 2008. Patterns of exclusion in an intraguild predator-prey system depend on initial conditions. *Journal of Animal Ecology* 77: 624–630.

————. 2012. Invasion success in communities with reciprocal intraguild predation depends on the stage structure of the resident population. *Oikos* 121: 67–76.

Murdoch, W. W., C. J. Briggs, and R. M. Nisbet. 2003. *Consumer-resource dynamics*. Princeton, NJ: Princeton University Press.

Murdoch, W. W., C. J. Briggs, and S. Swarbrick. 2005. Host suppression and stability in a parasitoid-host system: Experimental demonstration. *Science* 309: 610–613.

Murdoch, W. W., B. E. Kendall, R. M. Nisbet, C. J. Briggs, E. McCauley, and R. Bolser. 2002. Single-species models for many-species food webs. *Nature* 417: 541–543.

Murdoch, W. W., and E. McCauley. 1985. Three distinct types of dynamic behaviour shown by a single planktonic system. *Nature* 316: 628–630.

Murdoch, W. W., E. McCauley, R. M. Nisbet, W.S.C. Gurney, and A. M. de Roos. 1992. Individual-based models: Combining testability and generality. *In*: D. L. DeAngelis and L. J. Gross, eds., *Individual-based models and approaches in ecology: Populations, communities and ecosystems*, 18–35. New York: Chapman and Hall.

Murdoch, W. W., and R. M. Nisbet. 1996. Frontiers of population ecology. *In*: R. B. Floyd and A. W. Sheppard, eds., *Frontiers of population ecology*, 31–43, Melbourne: CSIRO Press.

Murdoch, W. W., and M. A. Scott. 1984. Stability and extinction of laboratory populations of zooplankton preyed on by the backswimmer *Notonecta*. *Ecology* 65: 1231–1248.

Myers, R. A., J. A. Hutchings, and N. J. Barrowman. 1997. Why do fish stocks collapse? The example of cod in Atlantic Canada. *Ecological Applications* 7: 91–106.

Mylius, S. D., K. Klumpers, A. M. de Roos, and L. Persson. 2001. Impact of intraguild predation and stage structure on simple communities along a productivity gradient. *American Naturalist* 158: 259–276.

Neill, W. E. 1988. Community responses to experimental nutrient perturbations in oligotrophic lakes. *In*: B. Ebenman and L. Persson, eds., *Size-structured populations: Ecology and evolution*, 203–218. Heidelberg: Springer-Verlag.

Nelson, W. A., E. McCauley, and R. M. Nisbet. 2007. Stage-structured cycles generate strong fitness-equalizing mechanisms. *Evolutionary Ecology* 21: 499–515.

Nicholson, A. J. 1957. The self-adjustment of populations to change. *Cold Spring Harbor Symposium on Quantitative Biology* 22: 153–173.

Nielsen, D. L., T. J. Hillman, and F. J. Smith. 1999. Effects of hydrological variation and planktivorous competition on macroinvertebrate community structure in experimental billabongs. *Freshwater Biology* 42: 427–444.

Nielsen, D. L., T. J. Hillman, F. J. Smith, and R. J. Shiel. 2000. The influence of a planktivorous fish on zooplankton assemblages in experimental billabongs. *Hydrobiologia* 434: 1–9.

Nilsson, K. A., L. Persson, and T. van Kooten. 2010. Complete compensation in *Daphnia* fecundity and stage-specific biomass in response to size-independent mortality. *Journal of Animal Ecology* 79: 871–878.

Nisbet, R. M., and W.S.C. Gurney. 1983. The systematic formulation of population models for insects with dynamically varying instar duration. *Theoretical Population Biology* 23: 114–135.

Nisbet, R. M., E. McCauley, W.S.C. Gurney, W. W. Murdoch, and A. M. de Roos. 1996. Simple representations of biomass dynamics in structured populations. *In*: H. G. Othmer, F. R. Adler, M. A. Lewis, and J. Dallon, eds., *Case studies of mathematical modeling in ecology, physiology and cell biology*, 61–79. Upper Saddle River, NJ: Prentice-Hall.

Nisbet, R. M., E. McCauley, W.S.C. Gurney, W. W. Murdoch, and S. N. Wood. 2004. Formulating and testing a partially specified dynamic energy budget model. *Ecology* 85: 3132–3139.

Noakes, D.L.G., and J.G.J. Godin. 1988. Ontogeny of behavior and concurrent development changes in sensory systems in teleost fishes. *In*: W. S. Hoar and D. J. Randall, eds., *Fish physiology*, vol. 11B, 345–395. New York: Academic Press.

Odenbaugh, J. 2005. The "structure" of population ecology: Philosophical reflections on unstructured and structured models. *In*: K. Cuddington and B. E. Beissner, eds., *Ecological paradigms lost: Routes of theory change*, 63–77. Amsterdam: Elsevier Academic Press.

Oksanen, L., S. D. Fretwell, T. Arruda, and P. Niemelä. 1981. Exploitation ecosystems in gradients of primary productivity. *American Naturalist* 118: 240–261.

Okuyama, T. 2008. Invited views in basic and applied ecology: Intraguild predation with spatially structured interactions. *Basic and Applied Ecology* 9; 135–144.

Olson, M. H. 1996. Ontogenetic niche shifts in largemouth bass: Variability and consequences for first-year growth. *Ecology* 77: 179–190.

Olson, M. H., D. M. Green, and L. G. Rudstam. 2001. Changes in yellow perch (*Perca flavescens*) growth associated with the establishment of a walleye (*Stizostedion vitreum*) population in Canadarago Lake, New York (USA). *Ecology of Freshwater Fish* 10: 11–20.

Olson, M. H., G. G. Mittelbach, and C. W. Osenberg. 1995. Competition between predator and prey: Resource-based mechanisms and implications for stage-structured dynamics. *Ecology* 76: 1758–1771.

Orr, B. K., W. W. Murdoch, and J. R. Bence. 1990. Population regulation, convergence, and cannibalism in *Notonecta* (Hemiptera). *Ecology* 71: 68–82.

Osenberg, C. W., G. G. Mittelbach, and P. C. Wainwright. 1992. Two-stage life histories in fish: The interaction between juvenile competition and adult performance. *Ecology* 73: 255–267.

Österblom, H., M. Casini, O. Olsson, and A. Bignert. 2006. Fish, seabirds and trophic cascades in the Baltic Sea. *Marine Ecology—Progress Series* 323: 233–238.

Pascual, M., and H. Caswell. 1997. From the cell cycle to population cycles in phytoplankton-nutrient interactions. *Ecology* 78: 897–912.

Persson, A., and C. Brönmark. 2002a. Foraging capacity and resource synchronization in an ontogenetic diet switcher, pikeperch (*Stizostedion lucioperca*). *Ecology* 83: 3014–3022.

———. 2002b. Foraging capacities and effects of competitive release on ontogenetic diet shift in bream, *Abramis brama*. *Oikos* 97: 271–281.

Persson, L. 1987. The effects of resource availability and distribution on size class interactions in perch, *Perca fluviatilis*. *Oikos* 48: 148–160.

———. 1988. Asymmetries in competitive and predatory interactions in fish populations. *In*: B. Ebenman and L. Persson, eds., *Size-structured populations: Ecology and evolution*, 203–218, Heidelberg: Springer-Verlag.

Persson, L., P. A. Amundsen, A. M. de Roos, A. Klemetsen, R. Knudsen, and R. Primicerio. 2007. Culling prey promotes predator recovery: Alternative states in a whole-lake experiment. *Science* 316: 1743–1746.

Persson, L., A. Bertolo, and A. M. de Roos. 2006. Temporal stability in size distributions and growth rates of three *Esox lucius* L. populations. A result of cannibalism? *Journal of Fish Biology* 69: 461–472.

Persson, L., P. Byström, and E. Wahlström. 2000. Cannibalism and competition in Eurasian perch: Population dynamics of an ontogenetic omnivore. *Ecology* 81: 1058–1071.

Persson, L., P. Byström, E. Wahlström, A. Nijlunsing, and S. Rosema. 2000. Resource limitation during early ontogeny: Constraints induced by growth capacity in larval and juvenile fish. *Oecologia* 122: 459–469.

Persson, L., D. Claessen, A. M. de Roos, P. Byström, S. Sjögren, R. Svanbäck, E. Wahlström, and E. Westman. 2004. Cannibalism in a size-structured population: Energy extraction and control. *Ecological Monographs* 74: 135–157.

Persson, L., and A. M. de Roos. 2003. Adaptive habitat use in size-structured populations: Linking individual behavior to population processes. *Ecology* 84: 1129–1139.

———. 2006. Food-dependent individual growth and population dynamics in fishes. *Journal of Fish Biology* 69: 1–20.

Persson, L., A. M. de Roos, and A. Bertolo. 2004. Predicting shifts in dynamics of cannibalistic field populations using individual-based models. *Proceedings of the Royal Society B: Biological Sciences* 271: 2489–2493.

Persson, L., A. M. de Roos, and P. Byström. 2007. State-dependent invasion windows for prey in size-structured predator-prey systems: Whole lake experiments. *Journal of Animal Ecology* 76: 94–104.

Persson, L., A. M. de Roos, D. Claessen, P. Byström, J. Lövgren, S. Sjögren, R. Svanbäck, E. Wahlström, and E. Westman. 2003. Gigantic cannibals driving a whole-lake trophic cascade. *Proceedings of the National Academy of Sciences* 100: 4035–4039.

Persson, L., and L. A. Greenberg. 1990. Juvenile competitive bottlenecks: The perch (*Perca fluviatilis*)–roach (*Rutilus rutilus*) interaction. *Ecology* 71: 44–56.

Persson, L., K. Leonardsson, A. M. de Roos, M. Gyllenberg, and B. Christensen. 1998. Ontogenetic scaling of foraging rates and the dynamics of a size-structured consumer-resource model. *Theoretical Population Biology* 54: 270–293.

Peters, R. H. 1983. *The ecological implications of body size*. Cambridge: Cambridge University Press.

Pimm, S. L., and J. H. Lawton. 1978. On feeding on more than one trophic level. *Nature* 275: 542–544.

Pimm, S. L., and J. C. Rice. 1987. The dynamics of multispecies, multi-life-stage models of aquatic food webs. *Theoretical Population Biology* 32: 303–325.

Polis, G. A. 1981. The evolution and dynamics of intraspecific predation. *Annual Review of Ecology and Systematics* 12: 225–251.

———. 1988. Exploitation competition and the evolution of interference, cannibalism and intraguild predation in age/size-structured populations. *In*: B. Ebenman and L. Persson, eds., *Size-structured populations: Ecology and evolution*, 185–202. Heidelberg: Springer Verlag.

———. 1991. Complex trophic interactions in deserts: An empirical critique of food-web theory. *American Naturalist* 138: 123–155.

Polis, G. A., C. A. Myers, and R. D. Holt. 1989. The ecology and evolution of intraguild predation: Potential competitors that eat each other. *Annual Review of Ecology and Systematics* 20: 297–330.

Polis, G. A., and D. R. Strong. 1996. Food web complexity and community dynamics. *American Naturalist* 147: 813–846.

Pough, F. H. 1980. Advantages of ectothermy for tetrapods. *American Naturalist* 115: 92–112.

Preisser, E. L., D. I. Bolnick, and M. F. Benard. 2005. Scared to death? The effects of intimidation and consumption in predator-prey interactions. *Ecology* 86: 501–509.

Rand, D. A., and H. B. Wilson. 1995. Using spatio-temporal chaos and intermediate-scale determinism to quantify spatially extended ecosystems. *Proceedings of the Royal Society B: Biological Sciences* 259: 111–117.

Relyea, R. A. 2003. How prey respond to combined predators: A review and an empirical test. *Ecology* 84: 1827–1839.

Reynolds, S. E. 1990. Feeding in caterpillars: Maximizing or optimizing food aquisition? *In*: J. Mellinger, ed., *Animal nutrition and transport processes 1: Nutrition in wild and domestic animals*, 106–118. Basel: Karger.

Rice, J. A., L. B. Crowder, and E. A. Marschall. 1997. Predation on juvenile fishes: Dynamic interactions between size-structured predators and prey. *In*: R. C. Chambers and E. A. Trippel, eds., *Early life history and recruitment in fish populations*, 333–356. London: Chapman and Hall.

Ricker, W. E. 1954. Stock and recruitment. *Journal of the Fisheries Research Board of Canada* 11: 559–623.

Ricklefs, R. E. 1976. Growth rates of birds in humid New World tropics. *Ibis* 118: 179–207.

Rosenzweig, M. L. 1971. Paradox of enrichment: Destabilization of exploitation ecosystems in ecological time. *Science* 171: 385–387.

Rosenzweig, M. L., and R. H. MacArthur. 1963. Graphical representation and stability conditions of predator-prey interactions. *American Naturalist* 97: 209–223.

Roughgarden, J. 1979. *Theory of population genetics and evolutionary ecology: An introduction*. New York: Macmillan.

Rudolf, V.H.W. 2008. The impact of cannibalism in the prey on predator-prey systems. *Ecology* 89: 3116–3127.

Rudolf, V.H.W., and J. Armstrong. 2008. Emergent impacts of cannibalism and size refuges in prey on intraguild predation systems. *Oecologia* 157: 675–686.

Rudolf, V.H.W., and K. D. Lafferty. 2011. Stage structure alters how complexity affects stability of ecological networks. *Ecology Letters* 14: 75–79.

Sabo, J. L., and M. E. Power. 2002. River-watershed exchange: Effects of riverine subsidies on riparian lizards and their terrestrial prey. *Ecology* 83: 1860–1869.

Sait, S. M., M. Begon, and D. J. Thompson. 1994. Long-term population-dynamics of the indian meal moth *Plodia interpunctella* and its granulosis virus. *Journal of Animal Ecology* 63: 861–870.

Sanderson, B. L., T. R. Hrabik, J. J. Magnuson, and D. M. Post. 1999. Cyclic dynamics of a yellow perch (*Perca flavescens*) population in an oligotrophic lake: Evidence for the role of intraspecific interactions. *Canadian Journal of Fisheries and Aquatic Science* 56: 1534–1542.

Schellekens, T., A. M. de Roos, and L. Persson. 2010. Ontogenetic diet shifts result in niche partitioning between two consumer species irrespective of competitive abilities. *American Naturalist* 176: 625–637.

Schoener, T. W. 1969. Models of optimal size for solitary predators. *American Naturalist* 103: 277–313.

———. 1974. Resource partitioning in ecological communities. *Science* 185: 27–39.

————. 1983. Field experiments on interspecific competition. *American Naturalist* 122: 240–285.

————. 1986. Mechanistic approaches to community ecology: A new reductionism. *American Zoologist* 26: 81–106.

Schreiber, S., and V.H.W. Rudolf. 2008. Crossing habitat boundaries: Coupling dynamics of ecosystems through complex life cycles. *Ecology Letters* 11: 576–587.

Schröder, A., K. A. Nilsson, L. Persson, T. van Kooten, and B. Reichstein. 2009. Invasion success depends on invader body size in a size-structured mixed predation-competition community. *Journal of Animal Ecology* 78: 1152–1162.

Schröder, A., L. Persson, and A. M. de Roos. 2009. Culling experiments demonstrate size-class specific biomass increases with mortality. *Proceedings of the National Academy of Sciences* 106: 2671–2676.

Sebens, K. P. 1987. The ecology of indeterminate growth in animals. *Annual Review of Ecology and Systematics* 18: 371–407.

Sheldon, R., A. Prakash, and W. Sutcliffe. 1972. The size distribution of particles in the ocean. *Limnology and Oceanography* 17: 327–340.

Shurin, J. B., and E. W. Seabloom. 2005. The strength of trophic cascades across ecosystems: Predictions from allometry and energetics. *Journal of Animal Ecology* 74: 1029–1038.

Sih, A., G. Englund, and D. Wooster. 1998. Emergent impacts of multiple predators on prey. *Trends in Ecology and Evolution* 13: 350–355.

Slobodkin, L. B., and S. Richman. 1956. The effect of removal of fixed percentages of the newborn on size and variability in populations of *Daphnia pulicaria* (Forbes). *Limnology and Oceanography* 1: 209–237.

Smith, C., and P. Reay. 1991. Cannibalism in teleost fish. *Reviews in Fish Biology and Fisheries* 1: 41–64.

Sousa, T. N., T. Domingos, J. C. Poggiale, and S.A.L.M. Kooijman. 2010. Dynamic energy budget theory restores coherence in biology. *Proceedings of the Royal Society B: Biological Sciences* 365: 3413–3428.

Starck, J. M., and R. E. Ricklefs, eds. 1998. *Avian growth and development: Evolution within the altricial-precocial spectrum.* New York: Oxford University Press.

Stearns, S. C. 1992. *The evolution of life histories.* Oxford: Oxford University Press.

Stephens, P. A., W. J. Sutherland, and R. P. Freckleton. 1999. What is the Allee effect? *Oikos* 87: 185–190.

Sumari, O. 1971. Structure of the perch populations of some ponds in Finland. *Annales Zoologica Fennica* 8: 406–421.

Tilman, D. 1982. *Resource competition and community structure.* Princeton, NJ: Princeton University Press.

Tonn, W. M., and C. A. Paszkowski. 1986. Size-limited predation, winterkill, and the organization of *Umbra-Perca* fish assemblages. *Canadian Journal of Fisheries and Aquatic Sciences* 43. 194–202.

Townsend, C. R., and M. R. Perrow. 1989. Eutrophication may produce population-cycles in roach, *Rutilus rutilus* (L.), by 2 contrasting mechanisms. *Journal of Fish Biology* 34: 161–164.

Townsend, C. R., W. J. Sutherland, and M. R. Perrow. 1990. A modeling investigation of population-cycles in the fish *Rutilus rutilus*. *Journal of Animal Ecology* 59: 469–485.

Tuffrau, M., G. Fryd-Versavel, H. Tuffrau, and J. Genermont. 2000. Description of *Euplotes versatilis* n. sp., a marine tropical ciliate exhibiting an unusually extensive phenotypic plasticity. *European Journal of Protistology* 36: 355–366.

Turchin, P., and S. P. Ellner. 2000. Living on the edge of chaos: Population dynamics of Fennoscandian voles. *Ecology* 81: 3099–3116.

Turchin, P., and I. Hanski. 2001. Contrasting alternative hypotheses about rodent cycles by translating them into parameterized models. *Ecology Letters* 4: 267–276.

van den Bosch, F., A. M. de Roos, and W. Gabriel. 1988. Cannibalism as a lifeboat mechanism. *Journal of Mathematical Biology* 26: 619–633.

van der Meijden, E., and C.A.M. van der Veen-van Wijk. 1997. Tritrophic metapopulation dynamics: A case study of ragwort, the Cinnabar moth, and the parasitoid *Cotesia popularis*. *In*: I. Hanski and M. E. Gilpin, eds., *Metapopulation biology*, 387–405. New York: Academic Press.

van de Wolfshaar, K. E., A. M. de Roos, and L. Persson. 2006. Size-dependent interactions inhibit coexistence in intraguild predation systems with life-history omnivory. *American Naturalist* 168: 62–75.

———. 2008. Population feedback after successful invasion leads to ecological suicide in seasonal environments. *Ecology* 89: 259–268.

van Kooten, T., J. Andersson, P. Byström, L. Persson, and A. M. de Roos. 2010. Size at hatching determines population dynamics and response to harvesting in cannibalistic fish. *Canadian Journal of Fisheries and Aquatic Sciences* 67: 401–416.

van Kooten, T., L. Persson, and A. M. de Roos. 2007. Size-dependent mortality induces life-history changes mediated through population dynamical feedbacks. *American Naturalist* 170: 258–270.

van Leeuwen, A., A. M. de Roos, and L. Persson. 2008. How cod shapes its world. *Journal of Sea Research* 60: 89–104.

Verhulst, P. F. 1838. Notice sur la loi que la population suit dans son accroissement. *Correspondence Mathematiques et Physiques* 10: 113–121.

Volterra, V. 1926. Fluctuations in the abundance of a species considered mathematically. *Nature* 118: 558–560.

———. 1928. Variations and fluctuations of the number of individuals in animal species living together. *Journal du Conseil/Conseil Permanent International pour l'Exploration de la Mer* 3: 3–51.

Vonder Brink, R. H., and M. J. Vanni. 1993. Demographic and life-history response of the cladoceran *Bosmina longirostris* to variation in predator abundance. *Oecologia* 95: 70–80.

Vonesh, J. R., and C. W. Osenberg. 2003. Multi-predator effects across life-history stages: Non-additivity of egg- and larval-stage predation in an African treefrog. *Ecology Letters* 6: 503–508.

Waage, J. K., and M. P. Hassell. 1982. Parasitoids as biological control agents: A fundamental approach. *Parasitology* 84: 241–268.

Waddell, D. R. 1992. Cannibalism in lower eukaryotes. *In*: M. A. Elgar and B. J. Crespi, eds., *Cannibalism: Ecology and evolution among diverse taxa*, 85–101. New York: Oxford University Press.

Walters, C., and J. F. Kitchell. 2001. Cultivation/depensation effects on juvenile survival and recruitment: Implications for the theory of fishing. *Canadian Journal of Fisheries and Aquatic Science* 58: 39–50.

Werner, E. E. 1986. Species interactions in freshwater fish communities. *In*: J. Diamond and T. Case, eds., *Community ecology*, 344–358. New York: Harper and Row.

———. 1988. Size, scaling and the evolution of complex life cycles. *In*: B. Ebenman and L. Persson, eds., *Size-structured populations: Ecology and evolution*, 60–81. Heidelberg: Springer-Verlag.

———. 1992. Individual behavior and higher-order species interactions. *American Naturalist* 140: S5–S32.

Werner, E. E., and J. F. Gilliam. 1984. The ontogenetic niche and species interactions in size-structured populations. *Annual Review of Ecology and Systematics* 15: 393–425.

Wilbur, H. M. 1980. Complex life cycles. *Annual Review of Ecology and Systematics* 1: 67–93.

———. 1988. Interactions between growing predators and growing prey. *In*: B. Ebenman and L. Persson, eds., *Size-structured populations: Ecology and evolution*, 157–172. Heidelberg: Springer-Verlag.

Wilson, D. S. 1975. Adequacy of body size as a niche difference. *American Naturalist* 109: 769–784.

Woodward, G., and A. G. Hildrew. 2002. Body-size determinants of niche overlap and intraguild predation within a complex food web. *Journal of Animal Ecology* 71: 1063–1074.

Wootton, J. T. 1994. The nature and consequences of indirect effects in ecological communities. *Annual Review of Ecology and Systematics* 25: 443–466.

Yodzis, P., and S. Innes. 1992. Body size and consumer resource dynamics. *American Naturalist* 139: 1151–1175.

Yubuki, N., T. Nakayama, and I. Inouye. 2008. A unique life cycle and perennation in a colorless chrysophyte *Spumella* sp. *Journal of Phycology* 44: 164–172.

Zaret, T. M. 1980. *Predation and freshwater communities*. New Haven, CT: Yale University Press.

Zimmerman, M. S. 2006. Predator communities associated with brook stickleback (*Culaea inconstans*) prey: Patterns in body size. *Canadian Journal of Fisheries and Aquatic Sciences* 63: 297–309.

Index

MONOGRAPHS IN POPULATION BIOLOGY

EDITED BY SIMON A. LEVIN AND HENRY S. HORN